Lecture Notes in Artificial Intelligence   5883

Edited by R. Goebel, J. Siekmann, and W. Wahlster

Subseries of Lecture Notes in Computer Science

T0122302

Roberto Serra   Rita Cucchiara (Eds.)

AI*IA 2009:
Emergent Perspectives
in Artificial Intelligence

XIth International Conference of the
Italian Association for Artificial Intelligence
Reggio Emilia, Italy, December 9-12, 2009
Proceedings

 Springer

Series Editors

Randy Goebel, University of Alberta, Edmonton, Canada
Jörg Siekmann, University of Saarland, Saarbrücken, Germany
Wolfgang Wahlster, DFKI and University of Saarland, Saarbrücken, Germany

Volume Editors

Roberto Serra
Università di Modena e Reggio Emilia
Dipartimento di Scienze Sociali, Cognitive e Quantitative
Via Allegri 9, 42121 Reggio Emilia, Italia
E-mail: roberto.serra@unimore.it

Rita Cucchiara
Università di Modena e Reggio Emilia
Dipartimento di Ingegneria dell'Informazione
Via Vignolese 905, 41100 Modena, Italia
E-mail: rita.cucchiara@unimore.it

Library of Congress Control Number: 2009939107

CR Subject Classification (1998): I.2, F.1, F.4, I.2.6, H.2.8, I.5

LNCS Sublibrary: SL 7 – Artificial Intelligence

ISSN   0302-9743
ISBN-10   3-642-10290-5 Springer Berlin Heidelberg New York
ISBN-13   978-3-642-10290-5 Springer Berlin Heidelberg New York

springer.com

© Springer-Verlag Berlin Heidelberg 2009
Printed in Germany

Typesetting: Camera-ready by author, data conversion by Scientific Publishing Services, Chennai, India
Printed on acid-free paper   SPIN: 12786985   06/3180   5 4 3 2 1 0

Preface

This volume contains the scientific papers accepted for publication at the conference of the Italian Association for Artificial Intelligence (AI*IA), held in Reggio Emilia during December 9-12, 2009. This was the 11th conference of the series, whose previous editions have been held every two years since 1989, in Trento, Palermo, Turin, Florence, Rome, Bologna, Bari, Pisa, Milan and Rome again.

The conference's scope is broad, and covers all the aspects of artificial intelligence. Every edition has, however, an individual flavor, and in this case, both by the choice of some invited speakers and of a dedicated workshop, we stressed the growing importance of complex systems methods and ideas in AI.

Eighty-two papers were submitted from 13 different countries located in four continents, and the Program Committee selected the 50 papers which are published in this volume. I wish to thank all the members of the Program Committee and the other reviewers for their excellent job.

The three invited speakers, who provided a fundamental contribution to the success of the conference, were Shay Bushinsky (Haifa), Marco Dorigo (Brussels) and Paul Verschure (Barcelona).

The conference also hosted a number of interesting workshops, which raised considerable interest and participation:

"Bio-logical: Logic-Based Approaches in Bioinformatics," organized by Stefano Ferilli and Donato Malerba.
"Complexity, Evolution and Emergent Intelligence," organized by Marco Villani and Stefano Cagnoni.
"Evalita 2009: Workshop on Evaluation of NLP and Speech Tools for Italian," organized by Bernardo Magnini and Amedeo Cappelli.
"Intelligent Cultural Heritage," organized by Luciana Bordoni.
"Intelligenza Artificiale ed e-Learning," organized by Giovanni Adorni.
"Pattern Recognition and Artificial Intelligence for Human Behavior Analysis," organized by Luca Iocchi, Andrea Prati and Roberto Vezzani.
"RCRA 2009: Experimental Evaluation of Algorithms for Solving Problems with Combinatorial Explosion," organized by Marco Gavanelli and Toni Mancini.

Special thanks are due to the members of the local Organizing Committee, who carried a heavy organizational burden. I am particularly indebted to my Co-chair Rita Cucchiara, who shared the task of defining all the main choices concerning the conference, and to Andrea Prati, who was of great help in organizing the paper-review process. My colleague and friend Marco Villani deserves special gratitude as he continuously helped in all the issues related to the conference organization with care and competence.

I am also grateful to the members of the local organizing team, my PhD students Alex Graudenzi and Chiara Damiani, later joined by Alessia Barbieri and Luca Ansaloni, who were essential in innumerable aspects, including (but certainly not limited to) the preparation and update of the conference website, the preparation of the camera-ready proceedings volume and the handling of the correspondence.

The conference was hosted in the beautiful palace were the Faculty of Communications and Economics is located; I thank the Faculty and its dean for the availability of these spaces and conference rooms, which required re-scheduling of some lessons. The University of Modena and Reggio Emilia and the Department of Social, Cognitive and Quantitative Sciences are also gratefully acknowledged for their patronage. I also thank Anna Kramer and the staff of Springer's *Lecture Notes in Artificial Intelligence* for their timely and effective editorial work.

A conference like this is also a chance for letting the discipline of AI become better known to laymen and to the general public; several satellite initiatives of this kind were launched, and I thank Shay Bushinsky, Marco Gori, Daniele Nardi and Gianni Zanarini for their support, and the municipality of Reggio Emilia for financial support.

Finally, I am glad to acknowledge the essential role of AI*IA and of its Board of Directors, and in particular of its president Marco Schaerf, who was very active in all the phases of the preparation of this event. I also received invaluable support and advice from my colleagues, the other ex-presidents of the association: Luigia Carlucci Aiello, Oliviero Stock, Pietro Torasso, Marco Gori.

September 2009   Reggio Emilia
                                                                 Roberto Serra

Organization

AI*IA 2009 was organized by the Italian Association for Artificial Intelligence (AI*IA) in cooperation with the department of Social, Cognitive and Quantitative Sciences and the department of Information Engineering of the Universty of Modena and Reggio Emilia

Organizing Committee

Roberto Serra (Modena and Reggio Emilia)	Conference Chair
Rita Cucchiara (Modena and Reggio Emilia)	Co-chair
Franco Zambonelli (Modena and Reggio Emilia)	Local Events Chair
Marco Villani (Modena and Reggio Emilia)	Workshop Coordinator
Andrea Prati (Modena and Reggio Emilia)	Refereeing Coordinator
Marco Mamei (Modena and Reggio Emilia)	Local Arrangements Coordinator

Program Committee

Khurshid Ahmad
Luigia Carlucci Aiello
Stefania Bandini
Roberto Basili
Ernesto Burattini
Stefano Cagnoni
Marc Cavazza
Roberto Cordeschi
Rita Cucchiara
Walter Daelemans
Floriana Esposito
Salvatore Gaglio
Marco Gavanelli
Attilio Giordana
Marco Gori
Nicola Guarino
Evelina Lamma
Marco Mamei

Toni Mancini
Paola Mello
Andrea Omicini
Patrizia Paggio
Teresa Pazienza
Andrea Prati
Andrea Roli
Roberto Serra
Marco Shaerf
Kiril Simov
Giovanni Soda
Steffen Staab
Oliviero Stock
Tokunaga Takenobu
Pietro Torasso
Marco Villani
Roman Yangarber
Franco Zambonelli

Local Arrangements

Alex Graudenzi
Chiara Damiani

Luca Ansaloni
Alessia Barbieri
(Modena and Reggio Emilia University, Italy)

Referees

K. Ahmad	F. Esposito	T. Pazienza
L. Carlucci Aiello	S. Gaglio	A. Prati
M. Andretta	M. Gavanelli	A. Roli
B. Apolloni	A. Giordana	R. Serra
S. Bandini	M. Gori	M. Shaerf
R. Basili	N. Guarino	K. Simov
A. Borghese	E. Lamma	G. Soda
M. Borrotti	P. Liberatore	S. Staab
E. Burattini	M. Mamei	O. Stock
R. Campanini	T. Mancini	T. Takenobu
S. Cagnoni	P. Mello	P. Torasso
M. Cavazza	M. Milano	M. Villani
R. Cordeschi	M. Mirolli	R. Yangarber
R. Cucchiara	A. Omicini	F. Zambonelli
W. Daelemans	P. Paggio	

Sponsoring Institutions

AI*IA (Italian Association for Artificial Intelligence) and University of Modena and Reggio Emilia.

Table of Contents

Knowledge Representation and Reasoning

Machine Learning

Evolutionary Computation

Search

Natural Language Processing

Multi-agent Systems

Application

Bayesian Networks: The Parental Synergy and the Prior Convergence Error

Janneke H. Bolt

Department of Information and Computing Sciences, Utrecht University
P.O. Box 80.089, 3508 TB Utrecht, The Netherlands
janneke@cs.uu.nl

Abstract. In a Bayesian network, for any node its conditional probabilities given all possible combinations of values for its parent nodes are specified. In this paper a new notion, the *parental synergy*, is introduced which is computed from these conditional probabilities. This paper then conjectures a general expression for what we called the *prior convergence error*. This error is found in the marginal prior probabilities computed for a node when the parents of this node are assumed to be independent. The prior convergence error, for example, is found in the prior probabilities as computed by the loopy-propagation algorithm; a widely used algorithm for approximate inference. In the expression of the prior convergence error, the parental synergy is an important factor; it determines to what extent the actual dependency between the parent nodes can affect the computed probabilities. This role in the expression of the prior convergence error indicates that the parental synergy is a fundamental feature of a Bayesian network.

1 Introduction

A Bayesian network is a concise representation of a joint probability distribution over a set of stochastic variables, consisting of a directed acyclic graph and a set of conditional probability distributions [1]. The nodes of the graph represent the variables of the distribution. From a Bayesian network, in theory, any probability of the represented distribution can be inferred. Inference, however, is NP-hard in general [2] and may be infeasible for large, densely connected networks. For those networks, approximate algorithms have been designed. A widely used algorithm for approximate inference with a Bayesian network is the loopy-propagation algorithm [1].

Most research on loopy propagation concerns equivalent algorithms used on undirected networks. By studying loopy propagation in Bayesian networks, in previous work, we could distinguish two types of error in the computed probabilities. Cycling errors and convergence errors [3]. We indicated that a convergence error may arise in the probabilities computed for a node with two or more incoming arcs because the loopy propagation-algorithm assumes the parents of a node to be independent, while, in fact, they may be dependent. Thereafter, for binary networks, we derived an expression for the convergence error found in a node with two incoming arcs, given a network in its prior state. This expression is composed of three factors that capture the degree of dependency between the parents of this node, and of a weighting factor that determines to what extent this degree of dependency can contribute to the prior convergence error.

R. Serra and R. Cucchiara (Eds.): AI*IA 2009, LNAI 5883, pp. 1–10, 2009.

This latter factor, which we indicated with w, is composed of the conditional probabilities specified for the node.

In this paper, the notion of parental synergy is introduced. This notion is a generalisation of the factor w and can be computed for each node, irrespective of its number of parents and irrespective of the cardinality of the involved nodes. Thereafter, the expression for the prior convergence error is generalised to nodes with an arbitrary number of parents and to network with nodes of arbitrary cardinality. In this generalised expression, the parental synergy fulfils the role of weighting factor. The role of the parental synergy in the expression of the prior convergence error indicates that it captures a fundamental feature of the probability landscape in a Bayesian network.

More details about the research described in this paper can be found in [4].

2   General Preliminaries

A *Bayesian network* is a model of a joint probability distribution Pr over a set of stochastic variables $\mathbf{V}$, consisting of a directed acyclic graph and a set of conditional probability distributions.[1] Each variable $A \in \mathbf{V}$ is represented by a node A in the network's digraph.[2] (Conditional) independency between the variables is captured by the digraph's set of arcs according to the d-separation criterion [1]. The strength of the probabilistic relationships between the variables is captured by the conditional probability distributions $\Pr(A \mid \mathbf{p}(A))$, where $\mathbf{p}(A)$ denotes the instantiations of the parents of A. The joint probability distribution is presented by:

$$\Pr(\mathbf{V}) = \prod_{A \in \mathbf{V}} \Pr(A \mid \mathbf{p}(A)) \,.$$

Figure 1(a) depicts the graph of an example Bayesian network. The network includes a node C with n parents $A^1, \ldots, A^n$, $n \geq 0$. The nodes $A^1, \ldots, A^n$ in turn have a common parent D. For $n = 0$ and $n = 1$, no loop is included in the network. For $n = 2$, the graph consists of a simple loop. For $n > 2$, the graph consists of a compound loop; for $n = 3$, this compound loop will be termed a double loop.

For a Bayesian network with a graph as depicted in Figure 1(a), the marginal probability $\Pr(c_i)$ equals:

$$\Pr(c_i) = \sum_{\mathbf{A}} \Pr(c_i \mid \mathbf{A}) \cdot \Pr(\mathbf{A}) \,.$$

Wrongfully assuming independence of the parents $A^1, \ldots, A^n$ would give the approximation $\widetilde{\Pr}(c_i)$:

$$\widetilde{\Pr}(c_i) = \sum_{\mathbf{A}} \Pr(c_i \mid \mathbf{A}) \cdot \Pr(A^1) \cdot \ldots \cdot \Pr(A^n) \,.$$

---

[1] Variables will be denoted by upper-case letters (A, A^i), and their values by indexed lower-case letters (a_i); sets of variables by bold-face upper-case letters ($\mathbf{A}$) and their instantiations by bold-face lower-case letters ($\mathbf{a}$) The upper-case letter is also used to indicate the whole range of values of a variable or a set of variables. Given binary variables, $A = a_1$ is often written as a and $A = a_2$ is often written as $\bar{a}$.

[2] The terms node and variable will be used interchangeably.

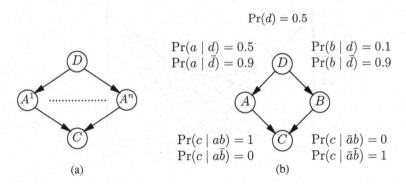

$$\Pr(d) = 0.5$$

$$\Pr(a \mid d) = 0.5 \qquad \Pr(b \mid d) = 0.1$$
$$\Pr(a \mid \bar{d}) = 0.9 \qquad \Pr(b \mid \bar{d}) = 0.9$$

$$\Pr(c \mid ab) = 1 \qquad \Pr(c \mid \bar{a}b) = 0$$
$$\Pr(c \mid a\bar{b}) = 0 \qquad \Pr(c \mid \bar{a}\bar{b}) = 1$$

(a)   (b)

Fig. 1. An example graph of a Bayesian network with a node C with dependent parents $A^1, \ldots, A^n$ (a) and an example Bayesian network with a node C with dependent parents A and B (b)

In the loopy-propagation algorithm [1], a widely used algorithm for approximate inference, indeed the parents of a node are always considered to be independent. With this algorithm, in the network from Figure 1(a) for node C the probabilities $\widetilde{\Pr}(c_i)$ would be yielded.

In [3], we termed the error which arises in the marginal prior probabilities computed for a child node under assumption of independence of its parent nodes, a prior convergence error. Moreover, we analysed the prior convergence error found in a binary network with a graph consisting of a node C with just the parents A and B with the common parent D, as the network depicted in Figure 1(b).

For the prior convergence error $v_i = \Pr(c_i) - \widetilde{\Pr}(c_i)$ in such a network the following expression was found:

$$v_i = l \cdot m \cdot n \cdot w$$

where

$$l = \Pr(d) - \Pr(d)^2$$
$$m = \Pr(a \mid d) - \Pr(a \mid \bar{d})$$
$$n = \Pr(b \mid d) - \Pr(b \mid \bar{d})$$
$$w = \Pr(c_i \mid ab) - \Pr(c_i \mid a\bar{b}) - \Pr(c_i \mid \bar{a}b) + \Pr(c_i \mid \bar{a}\bar{b}) .$$

The factors were illustrated graphically with Figure 2. The line segment in this figure captures the exact probability $\Pr(c)$ as a function of $\Pr(d)$, given the conditional probabilities for the nodes A, B and C from Figure 1(b). $\Pr(d)$ itself is not indicated in the figure, note however, that each particular $\Pr(d)$ has a corresponding $\Pr(a)$ and $\Pr(b)$. The end points of the line segment, for example, are found at $\Pr(d) = 1$ with the corresponding $\Pr(a) = 0.5$ and $\Pr(b) = 0.1$ and at $\Pr(d) = 0$ with the corresponding $\Pr(a) = 0.9$ and $\Pr(b) = 0.9$. The surface captures $\widetilde{\Pr}(c)$ as a function of $\Pr(a)$ and $\Pr(b)$, given the conditional probabilities for node C. The convergence error now equals the distance between the point on the line segment that matches the

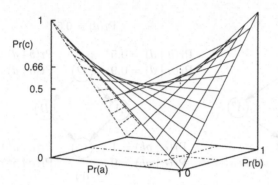

Fig. 2. The line segment capturing $\Pr(c)$ and the surface capturing $\widetilde{\Pr}(c)$, for the example network from Figure 1(b)

probability $\Pr(d)$ from the network and its orthogonal projection on the surface. For $\Pr(d) = 0.5$ the difference between $\Pr(c)$ and $\widetilde{\Pr}(c)$ is indicated by the vertical dotted line segment and equals $0.66 - 0.5 = 0.16$. The factor l reflects the location of the point with the exact probability on the line segment and the factors m and n reflect the location of the line segment. The factor w, to conclude, reflects the curvature of the surface with the approximate probabilities. We argued that the factors l, m and n capture the degree of dependency between the parent nodes A and B and that the factor w acts as a weighting factor, determining to what extent the dependence between A and B can affect the computed probabilities.

The factor w, as used in the expression for the convergence error above, only applies to nodes with two binary parents. In the next section, this factor is extended to a general notion which will be called the *parental synergy*. Subsequently, in Sections 4 and 5, the expression of the prior convergence error is generalised to a node C with an arbitrary number of dependent parents and with parents of arbitrary cardinality. The parental synergy is an important factor of these expressions and, analogous to the factor w, has the function of weighting factor.

3   The Parental Synergy

Before formally defining the parental synergy, the indicator function δ on the joint value assignments $a_{i1}^{1}, \ldots, a_{in}^{n}$ to a set of variables $A^{1}, \ldots, A^{n}$, $n \geq 0$, given a specific assignment $a_{s1}^{1}, \ldots, a_{sn}^{n}$ to these variables is introduced:

$$\delta(a_{i1}^{1}, \ldots, a_{in}^{n} \mid a_{s1}^{1}, \ldots, a_{sn}^{n}) = \begin{cases} 1 & \text{if } \sum_{k=1,\ldots,n} a_{ik}^{k} \neq a_{sk}^{k} \text{ is even} \\ -1 & \text{if } \sum_{k=1,\ldots,n} a_{ik}^{k} \neq a_{sk}^{k} \text{ is odd} \end{cases}$$

where true $\equiv 1$ and false $\equiv 0$. The indicator function compares the joint value assignment $a_{i1}^{1}, \ldots, a_{in}^{n}$ with the joint assignment $a_{s1}^{1}, \ldots, a_{sn}^{n}$, and counts the number of differences: the assignment $a_{i1}^{1}, \ldots, a_{in}^{n}$ is mapped to the value 1 if the number of differences is even and is mapped to -1 if the number of differences is odd. For the binary variables A and B, for example, $\delta(ab \mid ab) = 1$, $\delta(a\bar{b} \mid ab) = -1$, $\delta(\bar{a}b \mid ab) = -1$ and $\delta(\bar{a}\bar{b} \mid ab) = 1$.

Building upon the indicator function δ, the notion of parental synergy is defined as follows:

Definition 1. *Let* $\mathbf{B}$ *be a Bayesian network, representing a joint probability distribution* $\Pr$ *over a set of variables* $\mathbf{V}$. *Let* $\mathbf{A} = \{A^1, \ldots, A^n\} \subseteq \mathbf{V}$, $n \geq 0$, *and let* $C \in \mathbf{V}$ *such that* C *is a child of all variables in the set* $\mathbf{A}$, *that is,* $A^j \rightarrow C$, $j = 1, \ldots, n$. *Let* $\mathbf{a}$ *be a joint value assignment to* $\mathbf{A}$ *and let* c_i *be a value of* C. *Furthermore, let* $\mathbf{X} \subseteq \rho(C) \backslash \mathbf{A}$, *where* $\rho(C)$ *denotes the parents of* C, *and let* $\mathbf{x}$ *be a value assignment to* $\mathbf{X}$. *The* parental synergy *of* $\mathbf{a}$ *with respect to* c_i *given* $\mathbf{X} = \mathbf{x}$, *denoted as* $Y^*_{\mathbf{x}}(\mathbf{a}, c_i)$, *is*

$$Y^*_{\mathbf{x}}(\mathbf{a}, c_i) = \sum_{\mathbf{A}} \delta(\mathbf{A} \mid \mathbf{a}) \cdot \Pr(c_i \mid \mathbf{A}\mathbf{x}) . \qquad \Box$$

Given an an empty value assignment to the nodes $\mathbf{X}$, the parental synergy is denoted by $Y^*(\mathbf{a}, c_i)$.

Example 1. *Consider a node* C *with parents* A *and* B *and with the conditional probabilities for the value* c_i *as listed in Table 1(a). Now* $Y^*(a_2b_2, c_i)$, *for example, is computed from* $\Pr(c_i \mid a_1b_1) - \Pr(c_i \mid a_1b_2) + \Pr(c_i \mid a_1b_3) - \Pr(c_i \mid a_2b_1) + \Pr(c_i \mid a_2b_2) - \Pr(c_i \mid a_2b_3) + \Pr(c_i \mid a_3b_1) - \Pr(c_i \mid a_3b_2) + \Pr(c_i \mid a_3b_3) = 2.0$. *Table 1(b) lists all parental synergies* $Y^*(a_jb_k, c_i)$, $j, k = 1, 2, 3$. $\qquad \Box$

Table 1. The conditional probabilities $\Pr(c_i \mid AB)$ for a node C with the ternary parents A and B (a), and the matching quantitative parental synergies (b)

(a)				(b)			
$\Pr(c_i \mid AB)$	a_1	a_2	a_3	$Y^*(AB, c_i)$	a_1	a_2	a_3
b_1	0.7	0.2	0.3	b_1	2.4	0.4	-0.6
b_2	0.2	1.0	0.8	b_2	-1.2	2.0	-0.2
b_3	0.4	0.1	0.9	b_3	0.8	-0.4	1.4

From the definition of parental synergy, it is readily seen that for a binary parent A^k of C, we have that $Y^*_{\mathbf{x}}(\mathbf{a}a^k, c_i) = -Y^*_{\mathbf{x}}(\mathbf{a}\bar{a}^k, c_i)$ for any value assignments $\mathbf{a}$ and $\mathbf{x}$. For a node C with binary parents only, therefore, for a given $\mathbf{x}$ the parental synergies with respect to some value c_i can only differ in sign. From the definition it further follows that given a binary parent A^k of C, with the values a^k_m and a^k_n, that $Y^*_{\mathbf{x}}(\mathbf{a}a^k_m, c_i) = Y^*_{\mathbf{x}a^k_m}(\mathbf{a}, c_i) - Y^*_{\mathbf{x}a^k_n}(\mathbf{a}, c_i)$.

Example 2. *Consider a node* C *with parents* A *and* B *and with the conditional probabilities for the value* c_i *as listed in Table 2(a); the matching parental synergies are listed in Table 2(b). It is easily verified that* $Y^*(Ab, c_i)$ *equals* $-Y^*(A\bar{b}, c_i)$ *for all possible values of* A. *Furthermore from, for example,* $Y^*(a_1b, c_i) = (0.8 - 0.3 - 0.5) - (0 - 0.4 - 0.9) = 1.3$, $Y^*_b(a_1, c_i) = 0.8 - 0.3 - 0.5 = 0$ *and* $Y^*_{\bar{b}}(a_1, c_i) = 0 - 0.4 - 0.9 = -1.3$, *it is readily verified that* $Y^*(a_1b, c_i) = Y^*_b(a_1, c_i) - Y^*_{\bar{b}}(a_1, c_i)$. $\qquad \Box$

Table 2. The conditional probabilities $\Pr(c_i \mid AB)$ for a node C with the ternary parent A and the binary parent B (a) and the matching quantitative parental synergies (b)

<table>
<tr><td colspan="4">(a)</td><td colspan="4">(b)</td></tr>
<tr><td>$\Pr(c_i \mid AB)$</td><td>a_1</td><td>a_2</td><td>a_3</td><td>$Y^*(AB, c_i)$</td><td>a_1</td><td>a_2</td><td>a_3</td></tr>
<tr><td>b</td><td>0.8</td><td>0.3</td><td>0.5</td><td>b</td><td>1.3</td><td>−0.5</td><td>−1.1</td></tr>
<tr><td>$\bar{b}$</td><td>0.0</td><td>0.4</td><td>0.9</td><td>$\bar{b}$</td><td>−1.3</td><td>0.5</td><td>1.1</td></tr>
</table>

Note, that the parental synergy is related to the concepts of qualitative influence and additive synergy as defined for qualitative probabilistic networks in [5]. Most obviously, in a binary network, given a node C with a single parent A, the sign of the qualitative influence between A and C is computed from $\Pr(c \mid a) - \Pr(c \mid \bar{a})$, which equals $Y_x^*(a, c)$; given a node C with just the parents A and B the sign of the additive synergy of A and B with respect to C is computed from $\Pr(c \mid ab) - \Pr(c \mid a\bar{b}) - \Pr(c \mid \bar{a}b) + \Pr(c \mid \bar{a}\bar{b})$, which equals $Y_x^*(ab, c)$.

4   The Convergence Error Given Binary Nodes

In Section 2 an expression for the prior convergence error found in the marginal prior probabilities of a node C with binary parent nodes A and B with a common binary parent D was given. In Section 4.1 an alternative expression, which is more apt for generalisation, is stated. This expression is generalised to convergence nodes with multiple binary parents in Section 4.2 and is fully generalised in Section 5.

4.1   Two Parent Nodes; an Alternative Expression

The expression for the prior convergence error from Section 2 can also be written as

$$v_i = (s - t) \cdot w$$

where w is as before and

$$s = \sum_D \Pr(a \mid D) \cdot \Pr(b \mid D) \cdot \Pr(D)$$

$$t = \left(\sum_D \Pr(a \mid D) \cdot \Pr(D) \right) \cdot \left(\sum_D \Pr(b \mid D) \cdot \Pr(D) \right).$$

The degree of dependency between the nodes A and B now is captured by $s - t$ instead of by $l \cdot m \cdot n$. Note that the term s equals $\Pr(ab)$ and that the term t equals $\Pr(a) \cdot \Pr(b)$. Note furthermore that $w = \Pr(c_i \mid ab) - \Pr(c_i \mid a\bar{b}) - \Pr(c_i \mid \bar{a}b) + \Pr(c_i \mid \bar{a}\bar{b})$ equals $Y^*(ab, c_i)$.

4.2   Multiple Parent Nodes

Consider the network from Figure 1(a) given that $n = 3$. For three parent nodes, the expression for the prior convergence error v_i is found by subtracting the approximate probability

$$\widetilde{\Pr}(c_i) = \sum_{A^1,A^2,A^3} \Pr(c_i \mid A^1 A^2 A^3) \cdot \Pr(A^1) \cdot \Pr(A^2) \cdot \Pr(A^3)$$

from the exact probability

$$\Pr(c_i) = \sum_{A^1,A^2,A^3,D} \Pr(c_i \mid A^1 A^2 A^3) \cdot \Pr(A^1 \mid D) \cdot \Pr(A^2 \mid D) \cdot \Pr(A^3 \mid D) \cdot \Pr(D).$$

Manipulation of the resulting terms resulted in the following expression

$$v_i = (s_{a^1 a^2 a^3} - t_{a^1 a^2 a^3}) \cdot w + (s_{a^2 a^3} - t_{a^2 a^3}) \cdot w_{\bar{a}^1}(a^2 a^3) +$$
$$(s_{a^1 a^3} - t_{a^1 a^3}) \cdot w_{\bar{a}^2}(a^1 a^3) + (s_{a^1 a^2} - t_{a^1 a^2}) \cdot w_{\bar{a}^3}(a^1 a^2)$$

where

$$s_{a^1 a^2 a^3} = \sum_D \prod_{i=1,2,3} \Pr(a^i \mid D) \cdot \Pr(D)$$

$$t_{a^1 a^2 a^3} = \prod_{i=1,2,3} \sum_D \Pr(a^i \mid D) \cdot \Pr(D)$$

$$w = Y^*(a^1 a^2 a^3, c_i)$$

$$s_{a^m a^n} = \sum_D \Pr(a^m \mid D) \cdot \Pr(a^n \mid D) \cdot \Pr(D)$$

$$t_{a^m a^n} = \left(\sum_D \Pr(a^m \mid D) \cdot \Pr(D) \right) \cdot \left(\sum_D \Pr(a^n \mid D) \cdot \Pr(D) \right)$$

$$w_{\bar{a}^l}(a^m a^n) = Y^*_{\bar{a}^l}(a^m a^n, c_i).$$

The convergence error is composed of the term $(s_{a^1 a^2 a^3} - t_{a^1 a^2 a^3}) \cdot w$ which pertains to the entire double loop, and the three terms $(s_{a^m a^n} - t_{a^m a^n}) \cdot w_{\bar{a}^l}(a^m a^n)$ which pertain to the three simple loops that are included within the double loop. Note that $s_{a^1 a^2 a^3}$ equals $\Pr(a^1 a^2 a^3)$; $t_{a^1 a^2 a^3}$ equals $\Pr(a^1) \cdot \Pr(a^2) \cdot \Pr(a^3)$; $s_{a^m a^n}$ equals $\Pr(a^m a^n)$ and $t_{a^m a^n}$ equals $\Pr(a^m) \cdot \Pr(a^m)$.

Now consider a convergence node C with the binary parents $A^1, \ldots, A^n$ and the common parent D of $A^1, \ldots, A^n$. It is posed as a conjecture that the following expression captures the prior convergence error v_i for the value c_i of C:

$$v_i = \sum_{\mathbf{m}} (s_{\mathbf{am}} - t_{\mathbf{am}}) \cdot w_{\bar{a}^1 \ldots \bar{a}^n \backslash \mathbf{am}}(\mathbf{a^m})$$

where

$$\mathbf{m} \in \mathcal{P}(\{1, \ldots, n\})$$

$$\mathbf{a^m} = a^x \ldots a^y \quad \text{for} \quad \mathbf{m} = \{x, \ldots, y\}$$

$$s_{\mathbf{am}} = \sum_D \prod_{i \in \mathbf{m}} \Pr(a^i \mid D) \cdot \Pr(D)$$

$$t_{\mathbf{am}} = \prod_{i \in \mathbf{m}} \sum_D \Pr(a^i \mid D) \cdot \Pr(D)$$

$$w_{\bar{a}^1 \ldots \bar{a}^n \backslash \mathbf{am}}(\mathbf{a^m}) = Y^*_{\bar{a}^1 \ldots \bar{a}^n \backslash \mathbf{am}}(\mathbf{a^m}, c_i)$$

in which $\bar{a}^1 \ldots \bar{a}^n \backslash \mathbf{a}^{\mathbf{m}}$ denotes the value assignment 'False' to the nodes included in the set $\{A^1, \ldots, A^n\} \backslash \{A^x, \ldots, A^y\}$. This expression is a straightforward generalising of the expression for the prior convergence error given three parent nodes. Note that, analogous to before, the term $s_{\mathbf{a}^{\mathbf{m}}}$ equals $\Pr(a^x \ldots a^y)$ and the term $t_{\mathbf{a}^{\mathbf{m}}}$ equals $\Pr(a^x) \cdot \ldots \cdot \Pr(a^y)$. Further note that the expression includes terms for all possible loops included in the compound loop. The term with $\mathbf{m} = 1, \ldots, n$, pertains to the entire compound loop. With $|\mathbf{m}| = n - 1$, the n compound loops with a single incoming arc of C deleted are considered, and so on.

5   The Convergence Error Given Multiple-Valued Nodes

In the generalisation of the expression of the prior convergence error to multiple-valued nodes, a preliminary observation is that the expressions for this error from Section 4 involve just a single value c_i of the convergence node and therefore are valid for multiple-valued convergence nodes as well. Furthermore is observed that these expressions also provide for a multiple-valued node D. In this section the expression of the convergence error is further generalised to multiple valued parent nodes.

5.1   Two Parent Nodes

Consider a Bayesian network with a graph as the graph of network from Figure 1(b). It is posed as a conjecture that the following expression captures the prior convergence error for C.

$$v_i = \sum_{A,B} (s_{AB} - t_{AB}) \cdot w(AB)/4$$

where

$$s_{AB} = \sum_D \Pr(A \mid D) \cdot \Pr(B \mid D) \cdot \Pr(D)$$

$$t_{AB} = \left(\sum_D \Pr(A \mid D) \cdot \Pr(D) \right) \cdot \left(\sum_D \Pr(A \mid D) \cdot \Pr(D) \right)$$

$$w(AB) = Y^* (AB, c_i) \, .$$

This conjecture was supported by experimental results.

Note that, analogous to the binary case, the term s_{AB} equals $\Pr(AB)$ and the term t_{AB} equals $\Pr(A) \cdot \Pr(B)$. In contrast with the binary case, now all different value combinations of the nodes A and B are considered. The impact of the dependency between a specific combination of values for the nodes A and B on the convergence error is determined by the parental synergy of this combination with respect to the value c_i of node C. Further note that the expression for the convergence error now includes a division by a constant. This constant equals 2^n, where n is the number of loop parents of the convergence node.

5.2 Multiple Parent Nodes

In Section 4.2, the expression for the convergence error was extended to convergence nodes with an arbitrary number of binary parent nodes and in Section 5.1 the expression was extended to convergence nodes with two multiple valued parent nodes. Now, it is posed as a conjecture, that these expressions combine into the following general expression for the prior convergence error. Given a network with a graph, consisting of a convergence node C with the parents $A^1, \ldots, A^n$ and the common parent D of $A^1, \ldots, A^n$, as the graph depicted in Figure 1(a), the convergence error equals

$$v_i = \sum_{\mathbf{m}} \left[\sum_{\mathbf{A^m}} \left((s_{\mathbf{A^m}} - t_{\mathbf{A^m}}) \cdot \sum_{A^1, \ldots, A^n \backslash \mathbf{A^m}} w_{A^1, \ldots, A^n \backslash \mathbf{A^m}}(\mathbf{A^m}) \right) \right] / 2^n$$

where

$$\mathbf{m} \in \mathcal{P}(\{1, \ldots, n\})$$
$$\mathbf{A^m} = A^x, \ldots, A^y, \ \mathbf{m} = \{x, \ldots, y\}$$
$$s_{\mathbf{A^m}} = \sum_D \prod_{i \in \mathbf{m}} \Pr(A^i \mid D) \cdot \Pr(D)$$
$$t_{\mathbf{A^m}} = \prod_{i \in \mathbf{m}} \sum_D \Pr(A^i \mid D) \cdot \Pr(D)$$
$$w_{A^1 \ldots A^n \backslash \mathbf{A^m}}(\mathbf{A^m}) = Y^*_{A^1 \ldots A^n \backslash \mathbf{A^m}}(\mathbf{A^m}, c_i) \, .$$

Again, this conjecture was supported by experimental results.

Note that, as before, the term $s_{\mathbf{A^m}}$ equals $\Pr(A^x \ldots A^y)$ and the term $t_{\mathbf{A^m}}$ equals $\Pr(A^x) \cdot \ldots \cdot \Pr(A^y)$. As in the binary case given multiple parent nodes, all combinations of parent nodes are considered, now however, for each combination of parent nodes also all combinations of value assignments to these parent nodes have to be taken into account. Again, if the number of elements of $\mathbf{m}$ is smaller than two, that is, if just one parent or zero parents are considered, then the term $s_{\mathbf{A^m}}$ equals the term and $t_{\mathbf{A^m}}$ and thus $\sum_{\mathbf{A^m}} \left((s_{\mathbf{A^m}} - t_{\mathbf{A^m}}) \cdot \sum_{A^1, \ldots, A^n \backslash \mathbf{A^m}} w_{A^1, \ldots, A^n \backslash \mathbf{A^m}}(\mathbf{A^m}) \right)$ equals zero. The parental synergy, again, is the weighting factor that determines the impact of the degree of dependency between the parent nodes for a given value assignment, as reflected by $s_{\mathbf{A^m}} - t_{\mathbf{A^m}}$ on the size of the convergence error.

Finally, the fact that in the formula of the convergence error $s_{\mathbf{A^m}} = \Pr(A^x \ldots A^y)$ and $t_{\mathbf{A^m}} = \Pr(A^x) \cdot \ldots \cdot \Pr(A^y)$ may suggest that, also in case of a compound loop with more than one convergence node, the convergence error of a convergence node can be computed using the appropriate probabilities concerning its parents. For convergence nodes which have convergence nodes as ancester, this is not true, however, since errors that have arisen in the ancestor convergence nodes have to be taken into account as well. Some work on the convergence error in compound loops with more than one convergence node can be found in [4].

6 Discussion

In this paper the notion of parental synergy was introduced. This synergy is computed from the conditional probabilities as specified for a node in a Bayesian network.

Moreover, a general expression for the prior convergence error, was proposed. A prior convergence error arises in the prior marginal probabilities computed for a node when its parents are considered to be independent. This type of error arises in the probabilities computed by the loopy-propagation algorithm; a widely used algorithm for approximate probabilistic inference. The expression of the prior convergence error for a node is composed of the parental synergies of this node and of terms that capture the degree of dependence between its parent nodes. The parental synergy acts as weighting factor determining the impact of the degree of dependency between the parent nodes on the size of the convergence error.

In this paper, the parental synergy just features in the expression of the prior convergence error. Its role as weighting factor in this expression, however, indicates that the parental synergy captures a fundamental characteristic of a Bayesian network. It is conceivable, therefore, that the parental synergy has a wider range of application.

Acknowledgments. The research reported in this paper was supported by the Netherlands Organisation for Scientific Research (NWO). I would like to thank Linda van der Gaag for her useful comments on earlier drafts.

References

1. Pearl, J.: Probabilistic Reasoning in Intelligent Systems: Networks of Plausible Inference. Morgan Kaufmann, Palo Alto (1988)
2. Cooper, G.F.: The Computational Complexity of Probabilistic Inference using Bayesian Belief Networks. Artificial Intelligence 42, 393–405 (1990)
3. Bolt, J.H., van der Gaag, L.C.: The Convergence Error in Loopy Propagation. In: AISTA Conference, in cooperation with the IEEE Computer Society, Luxembourg (2004)
4. Bolt, J.H.: Bayesian Networks: Aspects of Approximate Inference. PhD thesis, Department of Information and Computing Sciences, Utrecht University (2008)
5. Wellman, M.P.: Fundamental Concepts of Qualitative Probabilistic Networks. Artificial Intelligence 44, 257–303 (1990)

Constraints for Representing Transforming Entities in Bio-ontologies

C. Maria Keet

KRDB Research Centre, Free University of Bozen-Bolzano, Italy
keet@inf.unibz.it

Abstract. Things change—develop, mature morph—but not everything in the same way. Representing this knowledge in ontologies faces issues on three fronts: what the category of the participating objects are, which type of relations they involve, and where constraints should be added. More precise distinctions can be made by using OntoClean's properties and a novel status property that is generalised from formal temporal conceptual data modeling. Criteria are identified, formulated in 17 additional constraints, and assessed on applicability for representing transformations more accurately. This enables developers of (bio-)ontologies to represent and relate entities more precisely, such as monocyte & macrophage and healthy & unhealthy organs.

1 Introduction

Much effort has been invested in development of ontologies, and in particular of bio-ontologies such as the Gene Ontology, Foundational Model of Anatomy, Cell Cycle, SNOMED and so forth [1,2]. This started from preliminary categorizations and vocabularies of biological entities that primarily focussed on endurants. With the maturing of (bio-)ontologies, ontology development tools, and ontology languages, this scope is broadening to temporal aspects. For instance, the Relation Ontology (RO) [3] contains the hitherto underused *transformation_of* and *derives_from* relations and several more evolution constraints have been identified for temporal conceptual modelling [4] that may find their way into domain ontologies. To be able to say one thing transforms into another, one has to identify the transforming entity x, the transformed entity y, and that the entity preserves its identity irrespective of the transformation while instantiating distinct classes at distinct points in time. What kind of entities are x and y instances of; phased sortals, roles, or merely different states? How should one deal with the temporality to achieve implementable knowledge bases that can handle representations of, and reasoning over, transforming entities? The aim of this paper is to characterise constraints on changing entities in more detail in a way so that it can aid a domain ontology developer to introduce or improve the representation of transforming entities, using an approach that is sufficiently generic to be extensible also to other types of change and to represent as much as possible about temporal aspects in the, to date, a-temporal commonly used ontology languages. We will commence with an analysis from an ontology development

R. Serra and R. Cucchiara (Eds.): AI*IA 2009, LNAI 5883, pp. 11–20, 2009.

perspective to formulate a set of basic constraints (section 2). In section 3, we extract implicit ontological commitments from advances in temporal knowledge representation, which enables us to define additional constraints on the entities and the process of transformation. The resulting 17 constraints are assessed on their usefulness by reconsidering some recent problems in biomedical ontology development in section 4. We discuss and conclude in section 5.

2   Analysis from an Ontology Development Perspective

2.1   Preliminaries

Transformations in the Relation Ontology (RO). Let us recollect the definition of the *transformation_of* relation in the RO [3]: "C *transformation_of* C_1 = [definition] C and C_1 for all c, t, if Cct, then there is some t_1 such that C_1ct_1, and t_1 earlier t, and there is no t_2 such that Cct_2 and C_1ct_2." In an ontology engineering setting, this definition reveals two issues. First, it is ignorant of the distinction between the cases of unidirectional transformations versus where some instance of C_1 may, after transforming into C, transform back into C_1; e.g., the transformations of erythrocytes (red blood cells) into echinocytes and back again, and of a healthy organ into a non-healthy organ and back again. Second, and more important for developing logic-based ontologies (i.e., as artifact), the RO definition does not say how the entities undergoing transformation are able to change and yet keep their identity, other than through the use of the same variable 'c'. This under-specification can lead to unintended models of the theory; e.g., a given particular remains unchanged, but either C or C_1 changes due to increased understanding or correcting a representation error in the domain ontology. Clearly, this does not meet the *intention* of the authors of the above definition. To exclude such unintended models, we have to make explicit in the definition that the individual changes somehow. Let a instantiate universals C_s and C_t—with s for source and t for target the source is transformed into—at the two different times, then we have to assert that 'enough' properties are shared by $a \in C_s$ and $a \in C_t$ so that they can be identified as the same individual—in shorthand notation: $a_s =_i a_t$—while other properties $\pi_1 \ldots \pi_n$ of a are lost or gained so that after the transformation, a instantiates a different universal. A basis for a generic definition might then be:

Definition 1 (C_t transformation_of C_s). *Let C_t be the target and C_s the source universal, x, y range over instances and $t_0, \ldots, t_n$ range over points in time, then $C_t(x)$ transformation_of $C_s(y)$ iff for all x, there exist y, $t_0, \ldots t_n$, if $C_t(x, t_0)$, then there is some t_1 such that $C_s(y, t_1)$, $t_1 < t_0$, C_s and C_t have the same identity criterion ($C_s =_i C_t$), x and y differ in at least one other property π_i, and there does not exist a t_2 such that $C_t(x, t_2)$ and $C_s(y, t_2)$.*

In the next sections, we elaborate on this definition and elicit additional constraints to model and achieve the intended behaviour of *transformation_of* in ontologies and knowledge bases for as much as possible in the a-temporal setting of common ontology languages and tools for the Semantic Web.

Basic Notions from Ontoclean. The main proposal in ontology to represent the entities involved in transformations is that of *phased sortals*. Postulating that C_t and C_s are phased sortals (or another category), brings forward constraints on the participating entities of the transformation relation that, in turn, may affect constraints on the relation itself, and certainly influences ontology development. To arrive at the point where we can specify unambiguously its how and where, we first summarise a section of [5,6] that form the basis of the OntoClean methodology[1] [9]. The relevant aspects for the current scope are the notion of identity, the meta-property rigidity, and identity criteria. Identity of an instance, and by extension also the universal it instantiates, focuses on the problems of distinguishing an instance of one class from other instances of that class by means of a characteristic property, which is unique for that whole instance, and is a relation that each thing has to itself and to nothing else; consequently, the property is essential. The property *rigidity* is based on this notion of essential property; two of the four modes in [5] are relevant for the current scope:

Definition 2 (+R). *A* rigid *property ϕ is a property that is essential to all its instances, i.e.,* $\forall x \phi(x) \rightarrow \Box \phi(x)$

Definition 3 (~R). *An* anti-rigid *property ϕ is a property that is not essential to* all *its instances, i.e.,* $\forall x \phi(x) \rightarrow \neg\Box\phi(x)$

For instance, the properties being a *Patient* and being a *Caterpillar* are ~R and being a *Person* and being a *Herbivore* are +R that may subsume *Patient* and *Caterpillar*, respectively. Objects can keep their identity through the changes it may undergo during its lifetime, thereby exhibiting different properties at different times; thus, we consider *diachronic* identity (cf. synchronic identity) which is about establishing that two entities are the same at two different points in time. Setting aside *how* we can identify instances—it is philosophically difficult [10], but one can always follow common practice and use either identifiers or a collection of OWL object and data properties—the instances and its corresponding universal have identity criteria (IC), which are both necessary and sufficient for identity [5,6]; properties carrying an IC are called *sortals* and only +R properties are sortal properties. In addition, a rigid property ϕ either carries the necessary IC Γ or it carries the sufficient IC Γ (see also [5] definitions 5 and 6). Now we can introduce two other definitions from [5], which are important for phased sortals.

Definition 4 (+I). *A property that is not rigid carries an IC Γ iff it is subsumed by a rigid property carrying Γ.*

Definition 5 (+O). *A property ϕ supplies an IC Γ iff i) it is rigid; ii) it carries Γ; and iii) Γ is not carried by all the properties subsuming ϕ.*

---

[1] There are some refinements but they do not affect the principal method of categorisation (in Fig.1, below). Moreover, the extension toward OntoClean 2.0 [7] with temporal aspects fit well with the temporal language we build upon in section 3 (as has been demonstrated by [8] for rigidity and part-whole relations).

+O	+I	+R	+D / -D	Type		
-O	+I	+R	+D / -D	Quasi-Type		Sortal
-O	+I	~R	+D	Material role		
-O	+I	~R	-D	Phased sortal		
-O	+I	¬R	+D / -D	Mixin		

Fig. 1. Ontological properties [5] (non-sortals omitted). +R: rigid; ∼R anti-rigid and ¬R semi-rigid, +O own identity supplied, +I identity carried, and +D dependent.

Thus, an +O property brings in its own identity criterion as opposed to just carrying it; conversely, properties that do not carry identity or do not supply identity are marked with -I and -O, respectively. Fig.1 summarises a classification of property kinds based on the R, O, and I meta-properties.

2.2   Characterising the Transforming Entities

Given the classification (Fig.1), this translates into the following set of basic constraints CT1-CT5 for phased sortals. CT2 ensures that the entity that changes local IC when going from one 'phase' to another, still can be identified as the same entity (thanks to property inheritance from C_p). CT3 can be derived from CT1, CT2, and Definition 4.

(CT1) A phased sortal does *not* supply an IC, i.e., -O
(CT2) A phased sortal *must* be subsumed by C_p that has +O
(CT3) A phased sortal carries an IC, i.e., +I
(CT4) A phased sortal is a sortal
(CT5) A phased sortal is anti-rigid, i.e., ∼R

It is now straightforward to demonstrate that *if C_t and C_s of the transformation_of relation are both categorised as phased sortals, then:*

(CT6) C_t and C_s both must be subsumed by C_p

As we shall see later with the examples, CT6 is particularly important. It immediately follows from Definition 5, the classification (Fig.1), and CT6 that C_p must be a type (CT7). Further, phased sortals *together* with the *transformation_of* relation cover the implicit requirement that phased sortals never can occur 'alone', because in order to phase, one needs *at least 2* phased sortals (CT8).

(CT7) C_p must be a type (+O+I+R)
(CT8) Each type that subsumes phased sortals, which are related through the *transformation_of* relation, must subsume *at least two* phased sortals

It is possible that C_t and C_s represent universals for different 'states' of a common subsumer C_p, provided the states are distinct enough. However, how does state differ from phase? Intuitively, phases succeed one another and have no circularity, whereas states (e.g., *sensu* DOLCE [11]) permit going back- and forward between alternating states. Let us take the latter assumption, then *if C_t*

transformation_of C_s and C_t and C_s *are categorised as states, then the following constraints must hold*:

(CT9) C_t and C_s must carry identity (+I)

(CT10) If C_t is a transformation of C_s, then it is possible, but not necessary, that at a later point in time C_t transforms back into C_s

(CT11) C_t and C_s have meta-properties that are either $\sim$R or +R

This reduces C_t and C_s to being either types, quasi-types, material roles, or phased sortals. They cannot be roles, because an instance can have more than one role at the same time (e.g., *Patient* and *Employee*), thereby violating Definition 1. C_s and C_t cannot be types either, because then one cannot ensure a common IC for diachronic identity. They can be both quasi-types (-O+I+R), which is a kind that enables one to group entities based on properties that *do not affect identity of* C_s *and* C_t (because of -O), such as being a Herbivore. We add CT12 in order to prevent C_t and C_s to be also types or roles and we make explicit an assumption that transformation of phased sortals is unidirectional (CT13).

(CT12) If C_t and C_s are categorised as states, they are neither both types nor both roles

(CT13) If C_t is a transformation of C_s, C_t and C_s are phased sortals, then it is *not* possible that at a later point in time C_t is a transformation of C_s, i.e., C_t does not transform back

Thus, based on foundational notions of Ontology, CT1-CT13 offers a more precise catergorisation for the relata of the *transformation_of*, as well as their position in a taxonomy.

3   Constraints from Temporal Conceptual Modelling

The challenge of how to represent changing entities has been investigated also from formal and engineering perspectives, most notably with formal temporal conceptual data modelling. Artale et al [4] have formalised the well-known core elements of temporal databases in $\mathcal{DLR}_{\mathcal{US}}$, a temporal description logic that is an expressive fragment of the first order temporal logic $L^{\{\mathbf{Since},\mathbf{Until}\}}$, and a corresponding $\mathcal{ER}_{VT}$ temporal conceptual data modelling language that extends EER. However, because it is intended for conceptual data modelling, they do not take into account the kind of classes like phased sortal—hence, nor the possible consequences on the logical implications either—but they include *evolution constraints* other than transformation, such as dynamic extension and generation. Therefore, we only consider their notion of status classes here. *Status* is associated to a class to log the evolving status of membership of each object in the class and the relation between the statuses. The four possible statuses are *scheduled*, *active* (for backwards compatibility with an a-temporal class), *suspended*, and *disabled* (based on [12]), with various formalised constraints; e.g., that an object that is member of the *disabled* class can never become member of its *active* class again, existence—where *exists* subsumes *scheduled*, *active*, and *suspended*—persist until *disabled*, and *exists* is disjoint from *disabled* [4]. In this setting, one can assert, e.g., that *Caterpillar* is temporally related to *Butterfly*: when at t_0 object $o \in$ *Caterpillar* (and $o \in$ *Scheduled-Butterfly*) starts transforming into an instance of *Butterfly*, then we have at the next time ($\oplus$ in $\mathcal{DLR_{US}}$) transformation at t_1 (with $t_0 < t_1$) that $o \in$ *Disabled-Caterpillar* and $o \in$ *Butterfly*. Thus, status allows us to add additional constraints to C_t and C_s.

Status Property S and Additional Constraints. Given the formal semantics of status classes [4], we introduce the property S and emphasize the core notions of the four status classes and what holds for their respective instances:

Definition 6 (+S). *A property ϕ has status* active *at time t iff $\phi(x)$ holds at time t.*

Definition 7 (-S). *If a property ϕ has status* scheduled *at time t then $\phi(x)$ holds at some time t_0, for $t_0 > t$.*

Definition 8 ($\sim$S). *If a property ϕ has status* suspended *at time t then $\phi(x)$ holds at some time t_0, with $t_0 < t$.*

Definition 9 ($\neg$S). *A property ϕ has status* disabled *at time t iff ϕ holds at some time t_0, with $t_0 < t$, and for all t', such that $t' \geq t$, $\phi(x)$ does not hold.*

This enables us to specify the following constraints for the *transformation_of* relation and its relata, when the instance *cannot* transform back:

(CT14) C_s has +S at the time of transformation and ¬S after transformation
(CT15) C_t has -S at the time of transformation and +S after transformation

If the entity *can* transform back, then CT14 and CT15 have to be replaced with:

(CT14′) C_s has +S at the time of transformation and either ¬S or ∼S after transformation
(CT15′) C_t has either -S or ∼S at the time of transformation and +S after transformation

Thus, theoretical contributions in formal conceptual modelling for temporal databases have the ontological commitment for CT14′ and CT15′, whereas *transformation_of* was ignorant about this aspect. One could argue S is ontologically awkward for it hints toward intentionality, but representing transformations does consider that anyway, and therefore it is preferable to make this explicit. In addition, bio-ontologies are, or should be, represented in some formal ontology language; using S then provides a useful mechanism to represent precisely and explicitly such implicit assumptions about the subject domain, thereby pushing for closer analysis by ontology developers as well as ontological investigation into intentional aspects in a temporal setting. In addition, it offers a way to deal with some underlying ideas about temporal aspects yet representing it in a commonly used a-temporal ontology language such as OWL.

CT14 and CT15 versus CT14′ and CT15′ leads to two distinct sets of constraints on the position of classes in a taxonomy. Let $\forall x(\phi(x) \rightarrow \psi(x))$, henceforth abbreviated as $\phi \rightarrow \psi$, a class' property S indicated with superscript ϕ^+, $\phi^\sim$ and so forth, and 'in the past' denoted with a "$\diamond$", then we can adjust Artale et al's constrains for CT14′ & CT15′ as shown in (1-5). When transformation back is not permitted, then (1), (4), (5), and (6) hold; that is, *suspended* is not permitted, such that (2) is not applicable and $\psi^\sim$ is removed from (3) and therefore replaced by(6).

$$\phi^+ \rightarrow \psi^+ \quad (1) \qquad\qquad \phi^- \rightarrow \neg\psi^- \quad (4)$$

$$\phi^\sim \rightarrow \psi^\sim \vee \psi^+ \quad (2) \qquad\qquad \psi^- \wedge \diamond\phi^+ \rightarrow \phi^- \quad (5)$$

$$\phi^- \rightarrow \psi^- \vee \psi^\sim \vee \psi^+ \quad (3) \qquad\qquad \phi^- \rightarrow \psi^- \vee \psi^+ \quad (6)$$

Combining this with the meta-properties of phased sortals, then the subsumption constraints (1) & (4), CT6, CT14 & CT15 imply C_p^+, because always one of the phased sortals subsumed by C_p is active. Regarding permitting suspension, $\sim$S, then C_p^+ is also implied, because of (1), (2), (4), CT14′ & CT15′.

4 Typical Biomedical Examples Re-examined

In this section we assess if and how the proposed constraints suffice for modelling transformations in biology and biomedicine, using monocytes/macrophages and pathological transformations as examples.

Monocyte/macrophage: alternating states or phased sortals? Multiple cell types in the human body, most notably those involved in hematopoiesis, can be pluri-potent stem cells, progenitor cells that differentiate, mature, and change in other ways, i.e., they seem to transform and are candidates for being categorised as phased sortals. However, looking at the details of the cells and processes, they are neither necessarily transformations nor necessarily phased sortals. For instance, progenitor cells undergo *cell division and differentiation* up to their final types, which is a non-deterministic process up to the point that the cells cease to be progenitor cells. But what about those final types? Here, we analyse the deterministic transformation from monocyte to macrophage, which, curiously, is also called in one compound term "monocyte/macrophage" or "monocyte-macrophage" in recent scientific literature (e.g. [13]) as if we have a cellular version of the Morning/Evening Star. Summarizing the scientific knowledge, Monocyte is part of Blood [2], the cells dock into tissues to transform into macrophages, and are considered part of the ImmuneSystem [14]. In the FMA taxonomy, Monocyte is subsumed by NongranularLeukocyte that is subsumed by Leukocyte, whereas Macrophage is directly subsumed by Leukocyte [2]. Monocytes are considered to be the end stage of the differentiation, and they can change into different types of macrophages, such as microglia (macrophages located in the brain), so that CT14 & CT15 hold. A particular monocyte transforms into a macrophage only when it leaves the blood stream into tissue, changing from "APC, circulating" to "APC, tissue resident", respectively [15] (APC

= Antigen Presenting Cell). A phenotypic difference is that macrophages also stimulates T helper cells [15] whereas both perform phagocytose of foreign organisms. The processes of change involves, among others, physiological response (chemical & DNA) on mechanical pressure [13] and response on the molecule platelet-derived α-chemokine PF4 [16]. Three possible modelling options have been proposed to represent the transformation of monocyte to macrophage, which we reconsider here. *Option 1:* Monocyte and macrophage are phased sortals. The problem with this approach is that there is no known suitable C_p that subsumes both of them only and directly (Leukocyte subsumes a range of other cells), i.e., violating CT6, CT7, CT8. There are multiple similar cases with hemal cells, which, if decided upon nevertheless, require additions of many more *concepts* that do not have a (known) corresponding universal in reality. *Option 2:* Monocyte and macrophage are different states that instances of C_p can have, because of the alternative functional states in the adult stage where each one is best adapted to its own environment: one residing in a fluid with the optimal shape of a sphere, the other in tissue with docking extensions on the cell surface, whilst remaining the same individual cell (cf. multi-cellular life cycles) instantiating one universal throughout the changes. Classifying the two states as distinct classes based on these differences is pursued by the Physiome project [15]. Both are APCs and phagocytes, which are candidates for the $=_i$ in Definition 1. Ascertaining $\pi_1 \ldots \pi_n$, they could be location and shape (attributions -I~R) or the property of stimulation of T helper cells for macrophages, or their differences in activated genes. This approach satisfies CT1, CT3, CT5, CT9, CT10, CT11, CT12, CT14 & CT15. *Option 3:* Take Monocyte and Macrophage as universals, place them somewhere in the taxonomy of cell types and create a new relationship type R to relate them. This does not guarantee diachronic identity of the instances and is therefore inappropriate. Clearly, with the current available information about the mechanisms of transformation from monocyte to macrophage, option 2 offers the more accurate representation.

Pathological transformations. Smith et al [3] include pathological transformations, which complicates both the permissible relata and the *transformation_of* relation itself for two reasons. First, it is ambiguous if the pathological entity may transform back to its healthy form and if this should be under the assumption that nature takes its course or if it also permits medical intervention. Second, it is not the case that for all instances that are transformed into a pathological entity they either all can or all cannot transform back; e.g., 5 out of 100 lung cancer patients survive and keep their lungs. It is true that for all carcinomatous lungs there must have been healthy lungs, but the inverse does not hold for all instances. Moreover, the relata cannot be phased sortals because there is no common subsumer C_p for the healthy and carcinomatous lungs in extant ontologies; e.g., the FMA commits to the "canonical" case, i.e., the assumption is that Lung refers to healthy lungs already. Given these considerations, they lead to the combination of constraints where CT9-CT12 hold together with either CT14 & CT15 for healthy & pathological entities of *non-curable* diseases or CT14' & CT15' for *curable* diseases.

5   Discussion and Conclusions

Regarding the *transformation_of* relation, the basic questions to answer for representing the universals in the subject domain in a domain ontology are: what transforms, how does it transform, and why does it transform? We focussed on the first two questions. Definition 1 in conjunction with constraints CT1-CT15' facilitate the analysis of transforming entities as well as the transformations themselves. For requirements on the participating endurants, the constraints for phased sortals to be subsumed by a type may be too restrictive for representing transforming biological universals most accurately for the main reasons that often either no such C_p is known to exist in reality that readily fits this requirement or it does exist but the criterion that supplies identity is difficult to establish; put differently, states and quasi-types seem to be more adequate for the participating continuants. Nevertheless, phased sortals remain useful kinds for representing stages in organism's life cycle as well as transformations of progenitor cells to differentiated cells.

The property status S helps in understanding and representing more precisely *how* the instances transform, thanks to the constraints it imposes both on status change during transformation and the universals' position in the taxonomy. For the biomedical domain, this is particularly useful for transformations involving pathological entities. More generally, several combinations and exclusions of constraints can be given:

   i. Phased sortals, unidirectional transformation: CT1-CT8, CT13, CT14, CT15;
  ii. States (including quasi-types), unidirectional transformation: CT1-CT9, CT11-CT15;
 iii. States (including quasi-types), transformation back is possible: CT1-CT13, CT14', CT15';
 iv. Pathological transformations, terminal disease: see constraints point ii, permit status change from -S directly into ¬S;
  v. Pathological transformations, reversal possible: see constraints point iii, permit status change from -S directly into ¬S.

One can make a further distinction for point v and define two sub-types of the transformation relation, which is to distinguish between self-healing transformations and those that require medical intervention, i.e., distinguishing between *natural* transformations and *human-mediated* transformations.

To actually use the different constraints and options for representing the *transformation_of* relation, it will be helpful to make the distinctions explicit also in the ontology development software by bringing representation choices of the properties to the foreground, which could be added to OntoClean.

Concluding, to represent changing entities more precisely, we took into account the kind of the participating entities and proposed that more precise distinctions can be made based on some core ideas of the OntoClean approach together with the notion of a status property that was generalised from temporal conceptual data modeling. This resulted in 17 additional constraints, which were assessed on applicability to bio-ontologies by analysing typical examples such as monocyte & macrophage. Currently, we are investigating implications of the interactions between OntoClean's property kinds, the status property, and temporal constraints in $\mathcal{DLR}_{\mathcal{US}}$, and in future works we may consider more complex time modeling, such as with GFO [17].

Acknowledgments. The author would like to thank Alessandro Artale and Barry Smith for useful comments on an earlier draft.

References

1. Gene Ontology Consortium.: The Gene Ontology GO database and informatics resource. Nucleic Acids Research 32(1), D258–D261 (2004)
2. Rosse, C., Mejino Jr., J.L.V.: A reference ontology for biomedical informatics: the foundational model of anatomy. J. of Biomedical Informatics 36(6), 478–500 (2003)
3. Smith, B., et al.: Relations in biomedical ontologies. Genome Biol. 46, R46 (2005)
4. Artale, A., Parent, C., Spaccapietra, S.: Evolving objects in temporal information systems. Annals of Mathematics and Artificial Intelligence 50(1-2), 5–38 (2007)
5. Guarino, N., Welty, C.: A formal ontology of properties. In: Dieng, R., Corby, O. (eds.) EKAW 2000. LNCS (LNAI), vol. 1937, pp. 97–112. Springer, Heidelberg (2000)
6. Guarino, N., Welty, C.: Identity, unity, and individuality: towards a formal toolkit for ontological analysis. In: Proc. of ECAI 2000. IOS Press, Amsterdam (2000)
7. Welty, C., Andersen, W.: Towards ontoclean 2.0: a framework for rigidity. Journal of Applied Ontology 1(1), 107–111 (2005)
8. Artale, A., Guarino, N., Keet, C.M.: Formalising temporal constraints on part-whole relations. In: Proc. of KR 2008, pp. 673–683. AAAI Press, Menlo Park (2008)
9. Guarino, N., Welty, C.: An overview of OntoClean. In: Staab, S., Studer, R. (eds.) Handbook on ontologies, pp. 151–159. Springer, Heidelberg (2004)
10. Noonan, H.: Identity. In: Zalta, E.N., ed.: The Stanford Encyclopedia of Philosophy. Fall 2008 edn. (2008),
    http://plato.stanford.edu/archives/fall2008/entries/identity/
11. Masolo, C., Borgo, S., Gangemi, A., Guarino, N., Oltramari, A.: Ontology library. WonderWeb Deliverable D18 (ver. 1.0, 31-12-2003) (2003)
12. Spaccapietra, S., Parent, C., Zimanyi, E.: Modeling time from a conceptual perspective. In: Proceedings of CIKM 1998 (1998)
13. Yang, J.H., Sakamoto, H., Xu, E.C., Lee, R.: Biomechanical regulation of human monocyte/macrophage molecular function. Am. J. Path. 156, 1797–1804 (2000)
14. Hoffbrand, A.V., Pettit, J.E.: Atlas of Clinical Heamatology, 3rd edn. Elsevier, Amsterdam (2000)
15. Hunter, P., Borg, T.: Integration from proteins to organs: The physiome project. Nature 4(3), 237–243 (2003)
16. Scheuerer, B., et al.: The cxc-chemokine platelet factor 4 promotes monocyte survival and induces monocyte differentiation into macrophages. Blood 95(4), 1158–1166 (2000)
17. Herre, H., Heller, B.: Ontology of time and situoids in medical conceptual modeling. In: Miksch, S., Hunter, J., Keravnou, E.T. (eds.) AIME 2005. LNCS (LNAI), vol. 3581, pp. 266–275. Springer, Heidelberg (2005)

Common-Sense Rule Inference

Ilaria Lombardi and Luca Console

Dipartimento di Informatica - Università di Torino
Corso Svizzera, 185 10149 Torino, Italy
{ilaria.lombardi,luca.console}@di.unito.it

Abstract. In the paper we show how rule-based inference can be made
more flexible by exploiting semantic information associated with the concepts involved in the rules. We introduce flexible forms of common sense
reasoning in which whenever no rule applies to a given situation, the
inference engine can fire rules that apply to *more general* or to *similar*
situations. This can be obtained by defining new forms of match between
rules and the facts in the working memory and new forms of conflict resolution. We claim that in this way we can overcome some of the brittleness
problems that are common in rule-based systems.

1 Introduction

Rule-based systems are widely used to implement knowledge-based systems for
various tasks, problem solving activities and application domains. Several reasons can motivate this choice, e.g., the fact that building knowledge bases is
relatively simple and intuitive, rules can be executed efficiently and many tools
are available and can be integrated with other software components. However,
the intuitive nature of rules hides a number of problems that have to be faced
when building and tuning rule-based problem solvers. A major problem is related to the fact that real knowledge bases are usually incomplete and do not
cover all situations that may be encountered during problem solving. Whatever
the cause of the incompleteness is, the effect is that rule-based systems tend to
suffer of the "brittleness" problem, that is their problem solving ability tends
to degrade abruptly when the portion of the knowledge base which is needed
to solve a problem is incomplete. In most cases they fail, providing no solution,
even if the problem is very close to their domain of expertise.

Various forms of approximate reasoning have been studied and integrated
into rule-based systems. Less frequently, forms of non-monotonic rule inference
have been defined. However, although they contribute significantly to make rule
inference more flexible, they do not solve the problem of brittle degradation
completely. In fact, approximate reasoning can cope in sophisticated ways with
problems such as the incompleteness and uncertainty of data or the uncertainty
of the relation between the premise and the consequent of the rule, but it usually
does not cope with the fact that no rule is applicable to a given problem. The
adoption of non-monotonic formalisms (such as default reasoning), on the other
hand, tends to make the construction of knowledge bases more complex and
inference less intuitive (see the discussion in [13]).

R. Serra and R. Cucchiara (Eds.): AI*IA 2009, LNAI 5883, pp. 21–30, 2009.

What rule-based systems need is the ability to mimic forms of human common sense reasoning, yet keeping the simple and intuitive nature of rules. Whenever there is no specific rule for solving a problem, they could look for rules that solve more general or similar problems; in this way an approximate solution may be computed, which is usually better than no solution at all. Obviously the strategy should impose that such relaxations are performed only when no specific rule is available, i.e., rules that are specific to the problem to be solved should always be preferred.

The aim of this work is to embed such capabilities in traditional rule-based systems. We show that the desired flexibility can be obtained integrating ontologies (taxonomies) into rule-based systems to describe the concepts involved in the rules. The approach is not alternative to other forms of approximate reasoning and can be integrated with any rule-based system and any form of numeric approximate reasoning. In the paper we discuss the approach and present a specific implementation, called "OntoRulE", based on forward chaining inference.

2   A Motivating Example

Let us start by considering a simple example motivating the need for flexible forms of inference. Suppose that S is a rule-based system supporting a car driver and that its knowledge base contains a simple rule such as:

R1) *if*   failure in Electrical System *then* locate and suggest closest workshop

The available data say that the car has a failure in the speed control system.

From the point of view of a traditional rule engine, *R1* cannot be fired as its antecedent does not match data. If the rule base does not contain any rule that is applicable to the specific situation, then it fails to provide an answer. This is an instance of the so called "brittleness" problem: a specific case or a situation was not considered during the design of the knowledge base, possibly because the situation was not foreseen or because there were so many situations that some were abstracted. Whatever the cause of the lack of coverage is, the result is that the system problem solving ability degrades abruptly providing no answer (while the system can provide accurate answers to cases that are covered by the knowledge base). Situations like the one above may occur frequently, especially in complex domains or in cases where it is difficult to enumerate all the situations that could be encountered during problem solving.

One way of explaining why the application of *R1* failed is that the system did not have deep knowledge about the concepts involved in the rules and thus could not reason on the semantics of the rules. One way of getting out of the problem, in fact, is to consider that the speed control system is a special case of electrical system and thus a rule like *R1* could possibly be applied to solve the problem. Maybe the solution that is determined in this way is not the most accurate one but it may approximate it. Conceptually, the idea is that whenever no rule is applicable, the system might look for more general rules and thus try to solve a problem by approximating the current situation with a more general one. Clearly, when the approximation leap is too big the approach may lead to very weak results or in some cases it may even be desirable to block the generalization leap to avoid wrong results. We believe, however, that the approach is potentially very powerful and interesting and these problems can indeed be controlled properly. The solution we propose, therefore, is to:

- Associate with the rule-based system a taxonomy (or an ontology) that defines the atoms in the antecedent and consequent in the rule.
- In the taxonomy one may specify a generalization distance for each ISA arc. The higher the distance is, the more approximate is the solution that is computed. An "infinite" distance can block the generalization leap.
- The notion of match between rules and facts has to be modified to include the notion of generalization leap (the distance between a concept and its generalization). The antecedent of *R1* should match the fact "failure to the speed control system" with a degree of match depending on the distance between speed control and electrical system.
- Define a strategy for conflict resolution which, intuitively, should select the rule that is most specific, i.e., the one that involves a minimal generalization leap. For example, if the knowledge base contains also a rule like

R2) *if*   failure in Speed Control System *then*
          disconnect it and advice the user to contact a workshop

R2 should be preferred to *R1* when the speed control system fails.

Similarly one can exploit also other relations in the taxonomy. For example if we know that both speed control system and brake control system are subclasses of electrical system, one might even apply a rule like

R3) *if*   failure in Brake Control System *then* reduce performance

to the case of the speed control. This is a form of reasoning by analogy: solving a problem for which no rule exists with rules that apply to a similar problem.

The rest of the paper will develop this intuitive ideas.

3   Flexible Rule Inference Using Taxonomies

In this section we formalize in more details the ideas introduced in the previous section. In order to be concrete we focus on a specific, although very generic, rule-based format in which rules have the following format:

Definition 1. *Given a set of taxonomies, a rule has the following format:*
  if <Conjunction of Atoms> then <Conjunction of Atoms>.
Each Atom has the form $p_i(C_j, v_k)$, where:

- *C_j is a reference to a class in one of the taxonomies;*
- *p_i is a property of the class C_j;*
- *v_k is the value of the property p_i or it can be a variable.*

For example, in the rule

$if\ p_1(A,5)\ then\ p_2(M,3)$

the atoms refer to classes in the taxonomies of Fig. 1. If the property p_1 of A has the value 5 then we can infer that the property p_2 of M has the value 3.

The facts in the working memory are atoms and we assume that a score can be associated with each one of them (we shall return later to this point).

In the taxonomies we associate a weight to the ISA arcs (see again Fig. 1) to represent a *distance* between a class and its superclass:

Definition 2. *Given a taxonomy, a weight in the range* $[0,1]$ *is associated with each ISA arc to express the distance (degree of generalization) between a class and its superclass.*

The higher the weight is, the higher the generalization leap is. The idea of having such weights is common to other approaches which exploit taxonomic reasoning and similarity between classes, which also show how such weights can be learned (see, for example, [10], [9]).

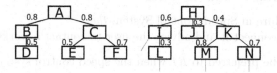

Fig. 1. The taxonomies

3.1   OntoRete

OntoRete derives from the basic Rete [7] algorithm with an "extended" definition of match between rule antecedents and facts, based on taxonomies:

Definition 3. *Given a fact* $p(C_1,v_1)$ *and an atom* $p(C_2,v_2)$ *in the antecedent of a rule,* $p(C_1,v_1)$ *matches* $p(C_2,v_2)$ *if and only if: (i)* C_1 *and* C_2 *represents the same class or* C_1 *is subclass of* C_2 *and (ii)* v_1 *and* v_2 *can be unified.*

The first condition allows performing a flexible match between the antecedents of a rule and the working memory. If C_1 and C_2 refer to the same class, then we have the traditional definition of match; when C_1 is a subclass of C_2 we have an indirect match via the taxonomy. For example, let us consider the rule:

R1) $if\ p_1(A,5)\ then\ p_2(M,3)$

and the fact

$F_1 : p_1(D,5)$

a perfect match is not possible, but an extended one is allowed, since D is a subclass of A (see Fig. 1).

3.2   Measuring Generalization Leap

In order to define preference criteria between rules we need to introduce some metrics to measure how much of generalization is involved in the match between a rule and the facts. In order to define such a degree of generalization, we exploit the distances associated with ISA arcs in the taxonomies. Different formulae could be used; we selected a simple one:

Definition 4. *Given a working memory containing the facts* $F_1, F_2, \ldots F_n$*, the distance between a rule* r *and the working memory is defined as follows:*

$$matchDist(r) = \frac{\sum_{j=1}^{|LHS(r)|} \frac{dist(A_j(r), F_j(r))}{MaxTaxoDist_j}}{|LHS(r)|} \tag{1}$$

where:

- *$LHS(r)$ is the set of atoms in the antecedent of the rule r; $|LHS(r)|$ denotes its cardinality;*
- *$A_j(r)$ is the j-th atom in the antecedent of rule r;*
- *$F_j(r)$ is the fact matching the j-th atom in the antecedent of rule r;*
- *$dist(A_j(r), F_j(r))$ is the distance on the taxonomy between $A_j(r)$ and $F_j(r)$. It is equal to 0 if they coincides. Otherwise it is computed by summing the weights on the arcs in the taxonomy on the path connecting $A_j(r)$ and $F_j(r)$;*
- *$MaxTaxoDist_j$ is the maximum root-leaf distance in the taxonomy $F_j(r)$ and $A_j(r)$ belong to.*

A matching distance of *0* means classical matching. The more generalization is involved, the more the matching distance increases tending to *1*. Referring to previous example, the matching distance for *R1* in case the working memory contains F_1 is:

$$matchDist(R1) = (0.8 + 0.5)/1.5 = 0.867$$

3.3   Conflict Resolution

During the conflict resolution step, in case two or more rules match data, the best one is chosen. Different strategies can be applied, taking into account aspects such as the number of atoms in the antecedents of the rules (rules with many antecedents are usually preferred because they may be more precise), the recentness of the atoms, certainty factors possibly associated with the rules and with the data. In our case there is a further aspect to consider: the matching distance between the rule and the working memory, that is the degree of generalization involved in the rule match. Rules that match the working memory without a generalization step should be preferred to rules that required generalization steps. Furthermore, since we have taxonomies we may prefer rules involving subclasses to rules involving superclasses, i.e., we have another way other than the number of antecedents to evaluate the specificity of a rule.

In order to experiment the strategy, in ONTORULE we implemented a simple heuristic that computes a score *score(r)* for each rule combining:

- the number of antecedents in the rule, as in many rule-based engines;
- the degree of generalization that is associated with the match between the rule and the facts (see Section 3.1);
- the score of the facts matching the atoms in the rule antecedent;
- the specificity of the rule (of its antecedent) with respect to the taxonomies.

Definition 5. *For each rule r we define*

$$score_1(r) = spec(r) - matchDist(r)$$

where $matchDist(r)$ is defined in Section 3.2 (Definition 4) and $spec(r)$ measures the number of atoms in the rule premise. The latter can be defined as the ratio between the cardinality of antecedents in the rule r and the maximum number of antecedents among all the rules in the knowledge base:

$$spec(r) = \frac{|LHS(r)|}{\max_{i=1,n} |LHS(r_i)|} \tag{2}$$

- *$LHS(r)$ and $|LHS(r)|$ are as in Definition 4;*
- *r_i is the i-th rule of the rule set.*

In order to obtain a score for r, $score_1(r)$ should be combined with the scores of the matching facts to obtain the final score:

Definition 6. *The score of a rule r is defined as*

$$score(r) = ((score_1(r) + 1)/2) * antScore(r)$$

where $score_1(r)$ is normalized into the range (0,1] and then combined with ant Score(r), which is the measure of the score of the matching facts. We assume that any formula suitable for conjunctions (e.g., minumum, product) is used for calculating the match of the antecedent of a rule.

In case two rules have the same score we prefer the one that is more specific in the sense that involves facts that are at lower level in the hierarchy. To this purpose we introduce a further score of rules:

Definition 7. *Given a rule r whose antecedents involves the classes $C_1 \ldots C_n$, the specificity of r is defined as follows:*

$$specScore(r) = \sum_{i=1}^{n} d(C_i) \tag{3}$$

where:

- *$d(C_i)$ measures the level of C_i in the taxonomy and is defined as:*
  - *$d(C_i) = 0$ if C_i is a leaf*
  - *$d(C_i) = max(d(C'_1), \ldots d(C'_k)) + 1$ where $C'_1 \ldots C'_k$ are subclasses of C_i*

The final score of a rule r can be obtained as a combination of $score(r)$ and $specScore(r)$. In the prototype we decided to base conflict resolution on $score(r)$, selecting the rule with maximum $score$, using $specScore(r)$ when the score is the same and preferring the rule with minimum *specScore*. Obviously other combinations could be taken into account.

Extending the example of Section 3.1, consider the following set of rules:

R1) *if* $p_1(A,5)$ *then* $p_2(M,3)$
R2) *if* $p_1(E,5)$ *then* $p_2(N,1)$
R3) *if* $p_1(B,5) \wedge p_1(C,5)$ *then* $p_2(N,4)$

and the working memory containing the facts

$$F_1 : p_1(F,5) \; SCORE : 0.6 \qquad F_2 : p_1(E,5) \; SCORE : 0.75$$

The rule *R1* has the lowest $score_1$ (-0.36 both for F_1 and F_2), since it has only one antecedent $p_1(A,5)$ with an high distance to the matching facts. The rule *R2* has an higher $score_1$ (0.5) with respect to *R1* since it has the same number of atoms in the antecedent but a perfect match with the facts. The rule *R3* has the highest $score_1$ (0.66): it has a highest number of antecedents (2) and the distance among the facts and the antecedents is low (see Fig. 1). Combining $score_1$ with the matching facts score, *R2* gets the highest final *score* (0.45), and it is the selected rule.

The score concept is extended from rules to facts to measure how much of generalization was involved in inferring a fact. In a sense such a scoring can be seen as a form of approximate reasoning where we measure whether facts have been inferred using very specific rules or after some generalization step. The higher a score of a fact is, the less generalization is involved in its derivation. A default score 1 is assigned to initial facts (which are assumed to be true or observed). Since the score of a fact inferred with a rule R is computed by combining the score of the facts matching the atoms in the rule premise and the score of R, a high score for an inferred fact means that it has been inferred using a rule without generalization step and whose atoms in the antecedent matched either initial facts or facts which in turn were inferred from initial facts using very specific rules. Notice that in the paper we do not deal with certainty factors associated with facts and rules. In order to deal with numeric approaches to uncertainty, the factors should be also taken into account in the heuristics for rule selection. The form of inference discussed above is intrinsically non-monotonic as it performs a generalization step in the match between rules and facts. The conclusions of the rules are defeasible in the sense that at subsequent steps a more specific rule may make inferences on the same facts. For this reason a dependency recording system is necessary to implement recursive fact retraction. Furthermore, since the fact score computation involves the scores of the facts from which it was inferred, a recursive score updating should be implemented in case a fact score is changed. The dependency recorder implements the principles above and maintains the working memory in a consistent state taking into account the facts that are added and removed and whose score is modified.

3.4 Supporting Reasoning by Analogy

In the sections above we exploited ISA relations in the taxonomy to implement a flexible match between rules and facts based on generalization. The taxonomy could be exploited to implement a second form of flexible match: a match based on reasoning by analogy. The idea is that the notion of flexible match of definition 3 could be extended in such a way that a fact $p(C_1, X)$ can match an atom $p(C_2, X)$ in case C_1 and C_2 are siblings. The metric in Section 3.2 must be modified: one simple approach could be to sum the weights of all the arcs connecting C_1 and C_2 via their common superclass. We believe that rule matching via analogy is weaker and more unreliable than rule matching via generalization. Thus a further heuristic for conflict resolution (see Section 3.3) should be added in order to prefer the former over the latter.

4   Related Work

The paper proposed a novel approach to improve flexibility in rule-based inference, avoiding or at least limiting the brittleness problem that is typical of many knowledge-based systems. The approach extends the inference schemes of traditional rule engines, by adding a form of common sense reasoning. In this sense, the approach merges rule-based and non-monotonic inference. We believe, however, that this is obtained in a very simple and natural way and our approach can be used with any rule knowledge base and integrated in existing rule engines as it only requires a definition of the concepts used in the rules in the form of a taxonomy or ontology. From a theoretical point of view, we introduced a sort of rule inheritance: a rule R defined for a concept A can be applied also to the subconcepts of A, unless the subconcept defines a more specific rule. Up to our knowledge no rule engine implements this principle.

In the paper we discussed only marginally the issue of approximate inference. As we noticed, many formalisms have been proposed in the literature, starting from probabilistic or fuzzy reasoning or on heuristic methods (see [12]). The type of approximate reasoning we introduce is different from the one addressed by these approaches (that, for the sake of clarity, we will reference as "certainty" approaches). We introduce approximation based on the use of generalization. Thus the degree we associate to a fact F measures how much of generalization was involved in the derivation of F. Certainty approaches, on the other hand, measure how much a fact is known to the system (its certainty or probability, due, e.g., to imprecise observations or error in measurement). Similarly, the certainty fact associated with a rule measures how strong is the relation between the antecedent and the consequent. Thus the two approaches are orthogonal and in principle could be integrated; this, however, is not in the scope of this paper and is subject for future investigations.

Recently many approaches focused on the combination of rules and the semantic Web. They faced two different problems: (i) the semantic representation of rules and (ii) the addition of rule inference to semantic web formalisms. The former issue has some similarity with our approach, in the sense that the aim is to provide semantic references for the concepts used in the rule and to represent the rules themselves in semantic formalisms. *Rule markup languages* allow deploying, executing, publishing, and communicating rules on the Web, with the aim of achieving rule interoperability. The most interesting proposals are: RuleML[1], SWRL[2]. Moreover, the Rule Interchange Format (RIF) working group of W3C is currently working on standardizing a rule language[3] grounded in logic programming. The approaches focus on representation issues, mainly with the goal of achieving interoperability of rules, while we defined new inference strategy that can be obtained exploiting the knowledge in these semantic representation. Indeed, we could use any of the formalisms mentioned above to represent our rules.

Adding rule inference to semantic formalisms was proposed as a solution to extend the reasoning capabilities of semantic representation such as ontologies or description logics (DLs). However, many semantic and computational problems have emerged in this combination. Indeed, as shown by the studies in this field [8], [6], [11], decidability and complexity of reasoning are a crucial issue in systems combining DL knowledge bases and rules. With respect to these approaches we took a partially different point of view. We aim at exploiting semantic knowledge to extend the inference capabilities of the rule engine. In a sense, this may be seen as the opposite of the approaches that are aimed at using rules for enriching the inference that can be made in a pure ontology (as described also in [1]).

Clearly the ability of performing some form of generalization in rule application could be obtained in some of the approaches discussed above. For example, one may include in the set of facts all the more general facts that can be inferred using the taxonomy. But in this way it would not be easy to prefer specific inferences over the generic ones like we do. On the other hand, the designers of the knowledge base could be required to provide both specific and generic rules. However, this would make the knowledge acquisition process combinatorial.

Finally, we have to remark that our work inherits some idea from the similarity measures for ontologies and description logics, see for example [10], [5], [2]. These works aim at discovering semantic similarity in a taxonomy by combining corpus statistics with taxonomy structure. What we share with them is the idea of using the values to compute distances between the nodes of a taxonomy. We then exploit such measures in order to improve a taxonomy-based rule engine.

5   Conclusions

Building rule knowledge bases is not an easy task, especially in those cases where the domain is complex and many situations have to be taken into account. Whenever a rule base is incomplete or problems that are at the border of the system expertise have to be solved, there is the risk of a brittle degradation of problem solving ability. We proposed a novel approach to perform flexible and common sense reasoning using rules. We did not aim at building a new rule-based engine; the approach indeed can be applied to any rules knowledge base without requiring any redesign of the knowledge base; it is sufficient to add references to

---

[1] http://www.ruleml.org/
[2] http://www.w3.org/Submission/SWRL/
[3] http://www.w3.org/2005/rules/

possibly external taxonomies. The approach has also the advantage that it keeps separate the rules and the taxonomies supporting independent management of the two knowledge bases.

ONTORULE has been implemented with the aim of testing the ideas discussed in the paper. It has been developed as a Java library, providing essential APIs to integrate it in different applications. The taxonomies representing the domain of interest (in RDFS[4] format) are stored in a Sesame[5] repository and queried using SeRQL ([3]), the RDF query language supported by Sesame.

The approach has been experimented using Context Aware Recommender systems as a test-bench and indeed in this task, where it is difficult to enumerate all detailed situations that may be encountered, the approach proved to be very powerful (see [4] for more details).

References

1. Borgida, A., Brachman, R.J., McGuinness, D.L., Alperin Resnick, L.: Classic: A Structural Data Model for Objects. In: Clifford, J., Lindsay, B.G., Maier, D. (eds.) SIGMOD Conference, pp. 58–67. ACM Press, New York (1989)
2. Borgida, A., Walsh, T., Hirsh, H.: Towards Measuring Similarity in Description Logics. In: Horrocks, I., Sattler, U., Wolter, F. (eds.) Description Logics. CEUR Workshop Proceedings, vol. 147, CEUR-WS.org (2005)
3. Broekstra, J., Kampman, A.: Serql: An RDF Query and Transformation Language. In: Proc. of International Semantic Web Conference (2004)
4. Buriano, L., Marchetti, M., Carmagnola, F., Cena, F., Gena, C., Torre, I.: The Role of Ontologies in Context-Aware Recommender Systems. In: MDM, p. 80. IEEE Computer Society, Los Alamitos (2006)
5. D'amato, C., Staab, S., Fanizzi, N.: On the influence of description logics ontologies on conceptual similarity. In: Gangemi, A., Euzenat, J. (eds.) EKAW 2008. LNCS (LNAI), vol. 5268, pp. 48–63. Springer, Heidelberg (2008)
6. Eiter, T., Lukasiewicz, T., Schindlauer, R., Tompits, H.: Combining Answer Set Programming with Description Logics for the Semantic Web. In: Proc. KR 2004, pp. 141–151. AAAI Press, Menlo Park (2004)
7. Forgy, C.: Rete: A Fast Algorithm for the Many Patterns/Many Objects Match Problem. Artif. Intell. 19(1), 17–37 (1982)
8. Levy, A.Y., Rousset, M.C.: Combining Horn Rules and Description Logics in Carin. Artif. Intell. 104(1-2), 165–209 (1998)
9. Rada, R., Mili, H., Bicknell, E., Blettner, M.: Development and Application of a Metric on Semantic Nets. IEEE Trans SMC 19, 17–30 (1989)
10. Resnik, P.: Using Information Content to Evaluate Semantic Similarity in a Taxonomy. In: XIth IJCAI, pp. 448–453 (1995)
11. Rosati, R.: Dl+log: Tight Integration of Description Logics and Disjunctive Datalog. In: Proc. KR 2006, pp. 68–78. AAAI Press, Menlo Park (2006)
12. Russell, S., Norvig, P.: Artificial Intelligence: a Modern Approach. Prentice-Hall, Englewood Cliffs (2002)
13. Stefik, M.: Introduction to Knowledge Systems. Morgan Kaufmann, San Francisco (1995)

---

[4] http://www.w3.org/TR/rdf-schema/
[5] http://www.openrdf.org/

Hard QBF Encodings Made Easy: Dream or Reality?

Luca Pulina and Armando Tacchella

DIST, Università di Genova, Viale Causa, 13 – 16145 Genova, Italy
Luca.Pulina@unige.it, Armando.Tacchella@unige.it

Abstract. In a recent work we have shown that quantified treewidth is an effective empirical hardness marker for quantified Boolean formulas (QBFs), and that a preprocessor geared towards decreasing quantified treewidth is a potential enabler for the solution of hard QBF encodings.

In this paper we improve on previously introduced preprocessing techniques, and we broaden our experimental analysis to consider other structural parameters and other state-of-the-art preprocessors for QBFs. Our aim is to understand – in light of the parameters that we consider – whether manipulating a formula can make it easier, and under which conditions this is more likely to happen.

1 Introduction

It has been shown that quantified Boolean formulas (QBFs) can provide compact propositional encodings in many automated reasoning tasks, including formal property verification of circuits (see, e.g., [1]) and symbolic planning (see, e.g. [2]). Encoding automated reasoning tasks as QBF decision problems is convenient as long as QBF tools are powerful enough to solve interesting problems, a promise which is largely unfulfilled by current state-of-the-art tools – see, e.g., the results of the latest QBF solvers competition (QBFEVAL'08) [3]. Therefore, knowing exactly what we can expect from current QBF solvers, which formulas they find difficult to solve, and how we can possibly improve their performances, are all important open points in the current QBF research agenda. In [4], as a first step towards reaching a better understanding of hard (and easy) QBF encodings, we studied the properties of quantified treewidth [5] as a marker of empirical hardness. We have shown that approximations of quantified treewidth are the only parameters among several other candidates which succeed consistently in doing so. In [4], we have also considered the related problem of improving the performances of solvers by preprocessing hard encodings to make them easier. Our result has been that QBF solvers may benefit from a preprocessing phase geared towards reducing the treewidth of their input, and that this phase is a potential enabler for the solution of hard QBF encodings.

In this paper, we build on and extend the findings of [4] along several directions. We deepen the study about the relationship between quantified treewidth, prefix structure and empirical hardness of QBFs in prenex conjunctive normal form (prenex CNF). Consequently, we develop a new version of our preprocessor QuBIS (**Qu**antified **B**oolean formula **I**ncomplete **S**olver) in order to take into account prefix structure beyond the linear order imposed by prenexing, at least to some extent. We compare QuBIS and the new and improved QuBIS2.0 with other state-of-the-art preprocessors for QBFs on

R. Serra and R. Cucchiara (Eds.): AI*IA 2009, LNAI 5883, pp. 31–41, 2009.

a selection of representative QBF encodings – the k family of [6] – in order to understand under which conditions hard QBF encodings are made easy by preprocessing, and whether the algorithm of QuBIS2.0 is more effective than QuBIS. Finally, we study the impact of preprocessing with respect to several QBF encodings and considering the overall performances of several state-of-the-art QBF solvers.

In light of the above, we conclude that QuBIS2.0 is always at least as effective as QuBIS and, for specific combinations of solvers/encodings, it can decrease the hardness of the encodings. We also show that in many cases, the improvement brought by QuBIS2.0 and other state-of-the-art preprocessors can be explained by a change in some structural parameters of the formulas after preprocessing. Unfortunately, however effective preprocessing may be, we also show that QuBIS2.0 and other state-of-the-art preprocessors do not improve substantially on our ability to solve difficult encodings, so that making hard encodings easier remains confined to specific cases only.

The paper is structured as follows. In Section 2 we introduce basic definitions. In Section 3 we introduce the notion of empirical hardness, and the relationship between treewidth and useful preprocessing with QuBIS2.0. In Section 4 we evaluate in detail the performances of state-of-the-art QBF solvers and preprocessors on a selection of QBF encodings. In Section 5 we add more QBF encodings to our selection and evaluate the overall performances of QBF solvers and preprocessors.

2   Preliminaries

In this section we consider the definition of QBFs and their satisfiability, and we introduce notation from [5] to define graphs and associated parameters describing the structure of QBFs.

A *variable* is an element of a set P of propositional letters and a *literal* is a variable or the negation thereof. We denote with $|l|$ the variable occurring in the literal l, and with $\bar{l}$ the *complement* of l, i.e., $\neg l$ if l is a variable and $|l|$ otherwise. A literal is *positive* if $|l| = l$ and *negative* otherwise. A *clause* C is an n-ary $(n \geq 0)$ disjunction of literals such that, for any two distinct disjuncts l, l' in C, it is not the case that $|l| = |l'|$. A *propositional formula* is a k-ary $(k \geq 0)$ conjunction of clauses. A *quantified Boolean formula* is an expression of the form $Q_1 z_1 \ldots Q_n z_n \Phi$ where, for each $1 \leq i \leq n$, z_i is a variable, Q_i is either an existential quantifier $Q_i = \exists$ or a universal one $Q_i = \forall$, and Φ is a propositional formula in the variables $\{z_1, \ldots, z_n\}$. The expression $Q_1 z_1 \ldots Q_n z_n$ is the *prefix* and Φ is the *matrix*. A literal l is *existential* if $|l| = z_i$ for some $1 \leq i \leq n$ and $\exists z_i$ belongs to the prefix, and it is *universal* otherwise. A prefix $p = Q_1 z_1 \ldots Q_n z_n$ can be viewed as the concatenation of h *quantifier blocks*, i.e., $p = Q_1 Z_1 \ldots Q_h Z_h$, where the sets Z_i with $1 \leq i \leq h$ are a partition of Z, and consecutive blocks have different quantifiers. To each variable z we can associate a *level* $lvl(z)$ which is the index of the corresponding block, i.e., $lvl(z) = i$ for all the variables $z \in Z_i$. We also say that variable z *comes after* a variable z' in p if $lvl(z) \geq lvl(z')$.

The semantics of a QBF φ can be defined recursively as follows. A QBF clause is *contradictory* exactly when it does not contain existential literals. If the matrix of φ contains a contradictory clause then φ is false. If the matrix of φ has no conjuncts then φ is true. If $\varphi = Qz\psi$ is a QBF and l is a literal, we define φ_l as the QBF obtained from

ψ by removing all the conjuncts in which l occurs and removing $\bar{l}$ from the others. Then we have two cases. If φ is $\exists z\psi$, then φ is true exactly when φ_z or $\varphi_{\neg z}$ are true. If φ is $\forall z\psi$, then φ is true exactly when φ_z and $\varphi_{\neg z}$ are true. The QBF satisfiability problem (QSAT) is to decide whether a given formula is true or false. It is easy to see that if φ is a QBF without universal quantifiers, solving QSAT is the same as solving propositional satisfiability (SAT).

Considering the definitions in [5], to which we refer the reader for further details, given a QBF φ on the set of variables $Z = \{z_1, \ldots, z_n\}$, its *Gaifman graph* has a vertex set equal to Z with an edge (z, z') for every pair of different elements $z, z' \in Z$ that occur together in some clause of φ. A *scheme* for a QBF φ having prefix p is a supergraph (Z, E) of the Gaifman graph of φ along with and ordering $z'_1, \ldots, z'_n$ of the elements of Z such that (i) the ordering $z'_1, \ldots, z'_n$ preserves the order of p, i.e., if $i < j$ then z'_j comes after z'_i in p, and (ii) for any z'_k, its lower numbered neighbors form a clique, that is, for all k, if $i < k$, $j < k$, $(z'_i, z'_k) \in E$ and $(z'_j, z'_k) \in E$, then $(z'_i, z'_j) \in E$. The *width* w_p of a scheme is the maximum, over all vertices z_k, of the size of the set $\{i : i < k, (z_i, z_k) \in E\}$, i.e., the set containing all lower numbered neighbors of z_k. The *treewidth* tw_p of a QBF φ, denoted $tw_p(\varphi)$, is the minimum width over all schemes for φ.

3 Empirical Hardness, Treewidth and QBF Preprocessing

In [4] we defined empirical hardness as follows. Given a set of QBFs Γ, a set of QBF solvers Σ, and an implementation platform Π, *hardness* is a partial function $H_{\Gamma,\Sigma,\Pi}$: $\Gamma \to \{0, 1\}$ such that $H_{\Gamma,\Sigma,\Pi}(\varphi) = 1$ iff no solver in Σ can solve φ on Π – in which case φ is *hard*; and $H_{\Gamma,\Sigma,\Pi}(\varphi) = 0$ iff all solvers in Σ can solve φ on Π – in which case φ is *easy*. Still in [4], we provided empirical evidence that an approximation of tw_p ($\hat{tw}_p$) is a marker of empirical hardness, i.e., the distribution $p(\hat{tw}_p | H = 1)$ is *significantly* different from the distribution $p(\hat{tw}_p | H = 0)$ considering Σ and Γ from the latest QBF solvers competitions (QBFEVAL'06-08). Even considering several other syntactic parameters which are plausible candidates, $\hat{tw}_p$ is the only parameter that succeeds consistently in being a predictor of H, and its significance is not related to some specific solver only.

Here we are concerned with the details of $\hat{tw}_p$ computation since they motivate our improved version of QUBIS. In particular, computing tw is an NP-complete problem [8], and it is not difficult to see that computing tw_p it is also hard. Therefore, in [4] we introduced the tool QUTE that can efficiently compute approximations of tw_p.

In [4] we observed that, since $\hat{tw}_p$ takes into account the prefix structure, $\hat{tw}_p \geq \hat{tw}$. Since we know that $\hat{tw}_p$ is a hardness marker, the gap between $\hat{tw}_p$ and $\hat{tw}$ seems to provide a clear explanation on why QSAT is, in practice, often so much harder than SAT. The key question is if the above state of affairs corresponds or not to some intrinsic characteristics of the QSAT problem. In order to evaluate the impact of the gap between $\hat{tw}_p$ and $\hat{tw}$ in practice, we consider the dataset of hard QBFs from the last QBF solvers competition (QBFEVAL'08).

In Figure 1, for the above dataset, we present the distributions of $\hat{tw}_p$ (QBF – top) and $\hat{tw}$ (SAT – bottom). The distributions in Figure 1 are computed on 563 (out of 790)

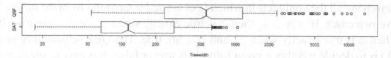

Fig. 1. Treewidth of QBFEVAL'08 hard formulas: $\hat{tw}_p$ (QBF - top) and $\hat{tw}$ (SAT - bottom). We present distributions as box-and-whiskers plot representing the median (bold line), the first and third quartile (bottom and top edges of the box), the minimum and maximum (whiskers at the top and the bottom) of a distribution, on a logarithmic scale. Values laying farther away than the median ± 1.5 times the interquartile range are considered outliers and shown as dots on the plot. In case outliers are detected, the whiskers extend up to the median $+1.5$ (resp. -1.5) times the interquartile range, while the maximum (resp. minimum) value becomes the highest (resp. lowest) outlier. An approximated 95% confidence interval for the difference in the two medians is represented by the notches cut in the boxes: if the notches of two plots do not overlap, this is strong evidence that the two medians differ.

formulas for which QUTE was able to compute an approximation in both cases within 600s and 3GB of RAM[1]. Comparing the distributions, we can see that the median value of $\hat{tw}$ is significantly (one order of magnitude) smaller than the median value of $\hat{tw}_p$.

How much of the difference shown in Figure 1 can be ascribed to a computational gap between SAT and QSAT? To answer this question, we consider a specific setting of our tool QUBIS that enables it to act as a variable-elimination-based QBF solver – and thus also SAT solver – rather than as a preprocessor. As mentioned in [4], QUBIS is based on variable elimination which is also the algorithm whose performances are most sensitive to treewidth values. Therefore, if QUBIS shows a different behaviour when it is taking into account the prefix, i.e., solving a QSAT instance, with respect to when it is not taking into account the prefix, i.e., solving a SAT instance, this means that QUTE is measuring a fundamental difference between the two cases. Allotting a CPU time limit of 3600 seconds and 3GB of memory to QUBIS and running it on the QBFEVAL'08 dataset yields the following results. QUBIS is not able to solve any of the QSAT instances, but it can solve 145 SAT instances. The median value of $\hat{tw}$ on these formulas is 52, while the corresponding median value of $\hat{tw}_p$ is 645. From these results it seems that the difference between $\hat{tw}$ and $\hat{tw}_p$ measured by QUTE corresponds to a definite increase in the difficulty of the instance only to some extent.

The other factor explaining the gap between $\hat{tw}$ and $\hat{tw}_p$ is probably the fact that all the QBFs that we consider are in prenex CNF, meaning that all the bound variables are constrained to a total order. It is known from previous results, see, e.g. [10,11], that the prefix structure can be very often relaxed to a partial order among the bound variables. In our case, knowing the partial order may allow to reduce the above gap in all the cases where it does not correspond to an intrinsic increase in the problem difficulty, because the procedure for tw estimation is not constrained from the prefix order, and thus QUTE would returns a $\hat{tw}_p$ value which is closer to $\hat{tw}$ in QUTE. To take into account this issue in QUBIS2.0, we modify the solution/preprocessing algorithm in order to identify cases in which the total order constraints can be relaxed.

---

[1] All the experiments in this paper, including this one, are carried out on a family of identical Linux workstations comprised of 8 Intel Core 2 Duo 2.13 GHz PCs with 4GB of RAM.

QUBIS2.0 is built on and extends the tool QUBIS described in [4].[2] QUBIS2.0 is based on Q-resolution [13] to perform variable elimination. QUBIS2.0 takes as input a QBF φ and two parameters: (i) an integer deg, which is the maximum degree that a variable can reach considering the Gaifman graph of φ, and (ii) an integer div which is the maximum value of *diversity* [14], corresponding to the number of resolvents generated by variable elimination. In QUBIS2.0, a variable z qualifies for elimination only if (i) z is universal and it comes after all the other variables in φ, or (ii) z is existential and it either (a) comes after all the other variables in φ, or (b) the neighbors of z in the Gaifman graph of φ do not come after z or are in the same level of z. Furthermore, it must be the case that either (c) the degree of z is no larger than deg and the diversity of z is no larger than div, or (d) the diversity of z is zero. Therefore, while QUBIS [4] eliminates existential variables only if they come after all the other variables in φ and they meet the required constraints on degree and diversity, QUBIS2.0 relaxes both conditions. In particular, whenever a variable z sits in a neighborhood wherein all the variables have either the same prefix level or do not come after z, this means that z would be a leaf in the tree representing the prefix as a partial order, and thus it can be safely eliminated. This relaxation goes in the direction of closing the gap between the prenex CNF form and the corresponding quantifier tree as in [10]. On the other hand, allowing elimination of zero-diversity variables, independently from their degree, enables QUBIS2.0 to eliminate variables with very large degree that do not cause a blow-up in the number of generated resolvents. In analogy with QUBIS, QUBIS2.0 is a sound and complete decision procedure for the subclass of QBFs in which variables always qualify for elimination while for all the other formulas QUBIS2.0 behaves like a preprocessor.

4   Preprocessing Can Make QBFs Easier...

In this section we consider the effect of preprocessors from an experimental point of view, with the goal of understanding when and why hard QBFs are made easier by preprocessing them, if at all. To this purpose we consider both QUBIS and QUBIS2.0[3], as well as two other state-of-the-art preprocessors for QBFs, namely PREQUEL [15] and PROVERBOX [16]. In order to evaluate the effect of preprocessing on various kinds of solvers, we also consider several such tools, namely QMRES, QUANTOR, QUBE3.0, QUBE6.5, SKIZZO, and YQUAFFLE (see [9] for references). In choosing the solvers above, the idea was to "cover" the space of mainstream techniques for QSAT: QMRES and QUANTOR are based on variable elimination; QUBE3.0, QUBE6.5, and YQUAF-FLE are based on backtracking search; SKIZZO is based on skolemization.

We start our analysis by considering a set of encodings from modal K satisfiability to corresponding QSAT problems [6]. The motivation behind our choice is that the 378 QBFs in the dataset range from fairly small formulas in terms of, e.g., number of variables, clauses and quantifier blocks, to fairly large ones. In particular, the large number of quantifier blocks helps to highlight the difference between the variable elimination constraints of QUBIS and QUBIS2.0. Moreover, as reported in various QBFEVAL editions, this encoding is challenging for state-of-the-art QBF solvers. Table 1 shows the

---

[2] The latest C++ implementation of QUBIS2.0 is available at [12].

[3] Both preprocessors ran with the setting $deg = 40$ and $div = 500$.

Table 1. Performances of a selection of QBF solvers on k encoding. The table is organized as follows: The first column ("Solver") reports the solver names, and it is followed by five groups of columns. In each group, we consider the performances of QBF solvers on the original formulas (group "Original"), formulas preprocessed using PREQUEL (group "PREQUEL"), formulas preprocessed using PROVERBOX (group "PROVERBOX"), and formulas preprocessed using both QUBIS and QUBIS2.0 (group "QUBIS" and "QUBIS2.0", respectively). For each group, we report the number of formulas solved ("#"), and the total CPU time (in seconds) to solve them ("Time"). In the case of groups related to preprocessors, column "Time" reports the total amount of CPU time spent for preprocessing and solving.

Solver	Original		PREQUEL		PROVERBOX		QUBIS		QUBIS2.0	
	#	Time	#	Time	#	Time	#	Time	#	Time
QMRES	261	1860.91	281	1886.94	275	4659.56	284	1184.59	286	2125.22
QUANTOR	258	883.85	312	1284.21	250	2652.07	293	942.47	293	1031.88
QUBE3.0	114	5049.55	121	6130.04	183	4992.88	254	4030.86	293	3932.22
QUBE6.5	238	6897.46	228	6468.78	262	8873.61	316	1996.71	318	2002.51
SKIZZO	343	6490.71	347	2988.40	335	12375.79	342	6576.04	343	4407.12
YQUAFFLE	142	6362.14	144	5004.48	180	4709.59	266	6887.10	303	6372.66

results of a run of the solvers described above on the encoding k, both on the original QBFs and in the ones obtained after preprocessing.

Looking at Table 1, we can see that QBF solvers improve their performances after preprocessing, in most cases. However there are relevant differences when comparing specific pairs preprocessor/solver. Considering the results of PREQUEL, we can see that it improves the performances of variable-elimination-based solvers, namely QMRES and QUANTOR, but the same is true only to a much smaller extent for other kinds of solvers. Noticeably, in the case of QUBE6.5, 10 encodings are made harder by PREQUEL. Looking now at PROVERBOX, we can see that search-based solvers are the ones which benefit the most. On the other hand, QMRES witnesses only a marginal improvement, while QUANTOR and SKIZZO decrease, albeit not substantially, their performances. From these results, it seems that PREQUEL complements nicely variable-elimination based solvers, while PROVERBOX suits more search-based ones. In terms of number of problems solved, QUBIS and QUBIS2.0 improve the performances of all the solvers but SKIZZO, even if the improvement is more substantial with search-based engines rather than variable-elimination ones. With the only exception of PRE-QUEL with QUANTOR and SKIZZO – which is already the best solver on this dataset, and we are probably witnessing a ceiling effect – QUBIS2.0 turns out to be the best choice overall. Noticeably, the k dataset is characterized by a large number of quantifier blocks, while the number of variables per block is relatively small. SKIZZO, owing to its internal quantifier tree reconstruction procedure, manages to produce relatively small Skolem functions, and its relatively good performances on this dataset indicate that relaxing the total order of the prefix can be the key for efficient reasoning. This is independently confirmed by the different effect that QUBIS and QUBIS2.0 have on the performances of QUBE3.0 and YQUAFFLE – the "purest" search engines. The performances of both solvers are improved by QUBIS and further improved by QUBIS2.0, which means that, for this particular combination of preprocessor/solver, relaxing the constraints of the total order prefix is bound to help substantially. Overall, QUBIS and QUBIS2.0 manage to solve more formulas in the preprocessing phase than those solved

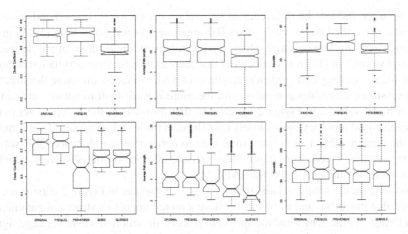

Fig. 2. Distributions of measures computed on Gaifman graphs related to QBFs of the k encoding. The figure is comprised of six box-and-whiskers plots organized in two rows of three columns each, wherein each column corresponds to a structural parameter, i.e., going from left to right, the distributions of μ, π, and $\hat{tw}_p$. The first row reports results about the k-closed dataset, and the second one reports results about the k-open dataset. In each plot, the y axis is in logarithmic scale.

by PREQUEL and PROVERBOX. Indeed, 182 formulas are solved by both preprocessors, and for all of them but seven, it turns out that $\hat{tw}_p \leq 36$. This explains also why both versions of QUBIS improve the performances of search-based solvers more than what happens for variable-elimination-based ones, which already take advantage of small-treewidth encodings.

In order to improve our understanding of the results in Table 1, we analyzed some relevant structural characteristics of the k encodings before and after preprocessing. To this purpose, we considered a subset of 331 formulas such that both the preprocessors and QUTE manage to perform their computation within 600s and 3GB of memory. We further divide this subset into the set "k-closed", which is comprised of 182 QBFs that can be solved by QUBIS, and the set "k-open", which is comprised of the 149 remaining formulas. Considering the Gaifman graphs of each such formulas, we compute $\hat{tw}_p$ as well as the *cluster coefficient* (μ), i.e., a measures of cliqueness in local neighborhoods, and the *average path length* (π), i.e., the path length averaged over all pairs of nodes. We have already discussed the importance of $\hat{tw}_p$ in [4], whereas we consider μ and π – originally defined in [17] to analyze the topology of large networks arising in practice – because they have been used in [18] to establish a relationship between the empirical hardness of SAT encodings and their structural properties. We compute the above measures considering only the results of PREQUEL and PROVERBOX for QBFs in the set k-closed, and we consider also the results of QUBIS and QUBIS2.0 for QBFs in the set k-open.

In Figure 2 we show the results of our analysis on structural parameters. Looking at the figure, we can see that the set k-closed is characterized from a median value of μ and $\hat{tw}_p$ which is smaller than the same value for k-open formulas, whereas the value of π is larger. Indeed, the difference in $\hat{tw}_p$ values between k-closed and k-open formulas is justified on the basis of QUBIS internals. As for μ and π, in [17], a graph characterized by relatively large values of μ and relatively small values of π is defined as a "small world"; in [18] it is conjectured that there exist a positive correlation between small world topology and empirical hardness of SAT instances: the more the structure of the instance resembles a small world, the more it is difficult to solve – at least by search-based solvers. In Figure 2 we can see that formulas in the k-open set match the small world topology more closely than those in the k-closed set. If we consider also the data in Table 1, it is interesting here to notice that search-based QBF solvers find the original k encodings harder than variable-elimination-based solvers, which is in line with the findings of [18] for SAT.

Focusing on the results related to the set k-closed set in Figure 2 (first row), we can see that PREQUEL and PROVERBOX have a mixed impact on structural parameters. With PREQUEL, the center and the spread of μ and π distributions do not change significantly after preprocessing, while $\hat{tw}_p$ is significantly increased on average, albeit the difference in the distributions is not substantial. From a structural viewpoint, PREQUEL is thus leaving the formulas in this set almost unchanged. With PROVERBOX the distribution of $\hat{tw}_p$ does not change significantly, whereas μ and π distribution centers are significantly smaller after preprocessing than before. According to [17] a property of random graphs is to be characterized by a relatively small μ *and* a relatively small π. In this sense, PROVERBOX seems to be making the topology of the encodings closer to a "random world". Still looking at Figure 2, but focusing on the results on k-open formulas (second row) we can immediately see that the distributions of $\hat{tw}_p$ are not significantly different after preprocessing, and this is true of all the preprocessors. This is thus a case where treewidth looses explanatory power in terms of empirical hardness, since the general pattern shown in Table 1 holds also for formulas in the k-open set, i.e., preprocessing does make these encodings easier for most solvers. On the other hand, the distributions of μ and π provide more insight: all the preprocessors, with the notable exception of PREQUEL, decrease significantly both μ and π, on average. Therefore, while the original QBFs structure is close to a small world topology, the structure of the preprocessed formulas is closer to a random world topology. This is essentially the same happening for PROVERBOX on k-closed formulas, and it also explains why PROVERBOX and QUBIS (both versions) seem to be particularly effective with search-based solvers. The performance of PREQUEL, on the other hand, remains largely unexplained. Indeed, see Table 1, with the exception of QUANTOR and sKIZZO, PREQUEL is the least effective among the preprocessors that we tried, but it still manages to improve the performances of almost all the solvers, with the exception of QUBE6.5.

5 ... But It Does Not Improve the State of the Art!

In this section, we try to put the results discussed so far in perspective, by extending our experimentation to other relevant families of QBF encodings, and by considering the performances of the various QBF solvers as a whole. The additional encodings that we consider are a family of instances related to equivalence checking of partial implementations of circuits (adder); a set of encodings from conformant planning

Table 2. Performances of a selection of QBF solvers on encodings preprocessed by using PRE-QUEL, PROVERBOX and QUBIS2.0. The table is split in six horizontal parts, each part is related to a specific encoding, and it is organized as Table 1. We report the number of formulas comprised in each encoding in the value between parentheses.

	Solver	Original #	Original Time	PREQUEL #	PREQUEL Time	PROVERBOX #	PROVERBOX Time	QUBIS2.0 #	QUBIS2.0 Time
	QMRES	19	981.02	20	1191.90	17	1372.94	21	1055.85
	QUANTOR	8	20.67	8	18.53	10	95.92	8	24.47
adder (32)	QUBE3.0	5	2.01	5	1.54	8	142.37	6	2.26
	QUBE6.5	5	1.11	5	1.48	8	58.45	7	4.92
	SKIZZO	15	1366.40	14	713.04	14	913.63	14	826.87
	YQUAFFLE	4	0.87	4	0.83	6	40.64	6	9.73
	QMRES	1	0.31	1	0.32	1	4.27	1	1.73
	QUANTOR	14	874.14	17	1663.75	14	1040.41	14	981.04
cp (24)	QUBE3.0	6	156.61	6	191.72	6	312.83	6	286.43
	QUBE6.5	6	345.75	6	357.44	6	622.95	7	435.55
	SKIZZO	7	141.08	8	798.59	8	962.50	6	221.61
	YQUAFFLE	8	294.92	8	721.98	7	311.79	11	1273.88
	QMRES	7	49.39	6	1068.49	5	84.94	10	1197.09
	QUANTOR	–	–	–	–	–	–	–	–
katz (20)	QUBE3.0	–	–	–	–	–	–	2	174.81
	QUBE6.5	9	714.72	6	54.16	8	937.60	1	561.40
	SKIZZO	–	–	3	1273.70	–	–	2	56.56
	YQUAFFLE	–	–	–	–	–	–	–	–
	QMRES	6	46.13	9	158.58	15	902.76	7	51.08
	QUANTOR	17	1261.29	16	1157.35	21	2771.60	17	1102.04
s (171)	QUBE3.0	1	0.06	1	0.06	15	478.99	1	0.01
	QUBE6.5	56	4645.11	59	5075.17	43	12986.93	58	7143.67
	SKIZZO	19	1714.84	18	955.81	19	2370.71	19	1412.98
	YQUAFFLE	1	0.12	1	0.12	14	465.00	1	0.01
	QMRES	58	2302.88	82	4083.63	60	2710.27	70	2409.68
	QUANTOR	74	635.72	82	867.67	72	730.22	73	486.19
tipdiam (203)	QUBE3.0	71	1107.27	72	1179.69	71	1318.28	93	1115.86
	QUBE6.5	150	3505.25	120	1356.51	135	3099.55	99	961.08
	SKIZZO	133	8352.57	137	5898.20	118	7869.97	100	8872.48
	YQUAFFLE	71	1763.16	71	2882.20	70	2248.16	88	800.22

domains (cp); a family of symbolic reachability encodings for industrially relevant circuits (katz); a family of encodings corresponding to symbolic diameter evaluation of ISCAS89 circuits (s); and a set of symbolic diameter evaluation of various circuits (tipdiam).

In Table 2 we show the results obtained on the encodings mentioned above. Looking at the table, we can see that solvers do not benefit substantially from preprocessing in almost all encodings. There are also a few cases in which the performances of some solvers are deteriorated by preprocessing: this is the case of QUBE6.5 and SKIZZO using PROVERBOX and QUBIS2.0 on the suite tipdiam.

Our next experiment aims to evaluate the impact of preprocessing with respect to the state of the art in QBFs evaluation. In order to do that, we use the state of the art (SOTA) solver, i.e., the ideal solver that always fares the best time among the solvers described in Section 4 on the QBF encodings of Table 1 and Table 2. We compare performances of the SOTA solver both on the original QBFs and on the ones obtained after preprocessing.

Table 3 shows the result of such experiment. Looking at the table, and considering results related to the adder encodings, we can see that preprocessing does not improve the performances of the SOTA solvers in terms of number of solved formulas.

Table 3. Performances of SOTA solver on QBF encodings preprocessed by using PREQUEL, PROVERBOX and QUBIS2.0. The table is organized as follows: The first column reports the SOTAs, and it is followed by six groups of columns, one for each considered encoding. For each group, we report the number of formulas solved ("#"), and the total CPU time (in seconds) to solve them ("Time").

	adder (32)		cp (24)		k (378)		katz (20)		s (171)		tipdiam (203)	
	#	Time	#	Time	#	Time	#	Time	#	Time	#	Time
SOTA	22	1554	15	867	351	1793	10	328	57	4645	169	4425
SOTA-PREQUEL	22	1412	18	1649	350	2478	10	709	60	5027	158	3186
SOTA-PROVERBOX	20	1085	15	913	352	2264	9	356	45	3376	152	2924
SOTA-QUBIS2.0	22	408	16	1252	349	2653	11	1194	59	4264	125	3515

Both SOTA, SOTA-PREQUEL and SOTA-QUBIS2.0 top to 21 solved formulas, while SOTA-PROVERBOX is worse with 19 solved formulas. Considering now cp encoding, we report that SOTA-PROVERBOX solves the same number of formulas of SOTA (14), while SOTA-PREQUEL and SOTA-QUBIS2.0 witness a slight improvement with 17 and 15 formulas solved, respectively. Looking at the results related to the encoding k, we can see the number of formulas solved by SOTA is 351. We report that, in spite of the improvements summarized in Table 1, only SOTA-PROVERBOX improves this performance slightly with 352 formulas solved, while both SOTA-PREQUEL and SOTA-QUBIS2.0 top at 350 and 349 formulas solved, respectively. Regarding the encodings katz and s, we can see that the picture is almost the same of the one related to the encoding cp. Moreover, there is no single preprocessor which is able to improve the SOTA performances for each such encoding. Finally, considering the results on tipdiam encodings, we can see that the performance of *all* the preprocessors SOTAs is worsened with respect to the SOTA solver. Summing up, we can conclude that, in spite of the results reported in Section 4 which show that preprocessing can be effective on specific combinations of solvers and encodings, these improvements are not enough to advance the overall state of the art.

References

1. Mneimneh, M., Sakallah, K.: Computing vertex eccentricity in exponentially large graphs: QBF formulation and solution. In: Giunchiglia, E., Tacchella, A. (eds.) SAT 2003. LNCS, vol. 2919, pp. 411–425. Springer, Heidelberg (2004)
2. Ansotegui, C., Gomes, C.P., Selman, B.: Achille's heel of QBF. In: Proc. of AAAI (2005)
3. Peschiera, C., Pulina, L., Tacchella, A.: QBF comparative evaluation (2008), http://www.qbfeval.org/2008
4. Pulina, L., Tacchella, A.: Treewidth: A useful marker of empirical hardness in quantified boolean logic encodings. In: Cervesato, I., Veith, H., Voronkov, A. (eds.) LPAR 2008. LNCS (LNAI), vol. 5330, pp. 528–542. Springer, Heidelberg (2008)
5. Chen, H., Dalmau, V.: From Pebble Games to Tractability: An Ambidextrous Consistency Algorithm for Quantified Constraint Satisfaction. In: Ong, L. (ed.) CSL 2005. LNCS, vol. 3634, pp. 232–247. Springer, Heidelberg (2005)
6. Pan, G., Vardi, M.Y.: Optimizing a BDD-based modal solver. In: Baader, F. (ed.) CADE 2003. LNCS (LNAI), vol. 2741, pp. 75–89. Springer, Heidelberg (2003)

7. Giunchiglia, E., Narizzano, M., Tacchella, A.: Clause-Term Resolution and Learning in Quantified Boolean Logic Satisfiability. Artificial Intelligence Research 26, 371–416 (2006)
8. Arnborg, S., Corneil, D.G., Proskurowski, A.: Complexity of finding embeddings in a k-tree. SIAM Journal on Algebraic and Discrete Methods, 277–284 (1987)
9. Giunchiglia, E., Narizzano, M., Pulina, L., Tacchella, A.: Quantified Boolean Formulas satisfiability library, QBFLIB (2001), http://www.qbflib.org
10. Benedetti, M.: Quantifier Trees for QBFs. In: Bacchus, F., Walsh, T. (eds.) SAT 2005. LNCS, vol. 3569, pp. 378–385. Springer, Heidelberg (2005)
11. Giunchiglia, E., Narizzano, M., Tacchella, A.: Quantifier Structure in search based procedures for QBFs. IEEE TCAD 26(3) (2007)
12. Pulina, L., Tacchella, A.: MIND-Lab projects and related information (2008), http://www.mind-lab.it/projects
13. Kleine-Büning, H., Karpinski, M., Flögel, A.: Resolution for Quantified Boolean Formulas. Information and Computation 117(1), 12–18 (1995)
14. Rish, I., Dechter, R.: Resolution versus search: Two strategies for sat. Journal of Automated Reasoning 24(1/2), 225–275 (2000)
15. Samulowitz, H., Davies, J., Bacchus, F.: Preprocessing QBF. In: Benhamou, F. (ed.) CP 2006. LNCS, vol. 4204, pp. 514–529. Springer, Heidelberg (2006)
16. Bubeck, U., Büning, H.K.: Bounded Universal Expansion for Preprocessing QBF. In: Marques-Silva, J., Sakallah, K.A. (eds.) SAT 2007. LNCS, vol. 4501, pp. 244–257. Springer, Heidelberg (2007)
17. Watts, D.J., Strogatz, S.H.: Collective dynamics of small-world networks. Nature 393, 440–442 (1998)
18. Walsh, T.: Search in a Small World. In: Proc. of IJCAI (1999)

CONGAS: A COllaborative Ontology Development Framework Based on Named GrAphS

Daniele Bagni, Marco Cappella, Maria Teresa Pazienza, and Armando Stellato

ART Group, Dept. of Computer Science, Systems and Production
University of Rome, Tor Vergata
Via del Politecnico 1, 00133 Rome, Italy
{daniele.bagni,marco.cappella}@gmail.com,
{pazienza,stellato}@info.uniroma2.it

Abstract. The process of ontology development involves a range of skills and know-how often requiring team work of different people, each of them with his own way of contributing to the definition and formalization of the domain representation. For this reason, collaborative development is an important feature for ontology editing tools, and should take into account the different characteristics of team participants, provide them with a dedicated working environment allowing to express their ideas and creativity, still protecting integrity of the shared work. In this paper we present CONGAS, a collaborative version of the Knowledge Management and Acquisition platform Semantic Turkey which, exploiting the potentialities brought by recent introduction of context management into RDF triple graphs, offers a collaborative environment where proposals for ontology evolution can emerge and coexist, be evaluated by team users, trusted across different perspectives and eventually converged into the main development stream.

1 Introduction

The process of ontology development requires different skills and know-how, such as a deep understanding of the domain to be represented and proper expertise in domain modeling, which rarely can be found in one single person. Moreover, assessing the knowledge structure of an ontology typically involves different refinement steps, personal rethinking and discussion. For this reason, even the realization of medium-size ontologies often requires the collaboration of several experts, each of them bringing their own knowledge and skills towards the development of a consistent, hopefully stable, and potentially reusable domain representation.

As a natural consequence for the fact, and thanks to the maturity now reached by ontology development tools and to the proliferation of collaborative solutions brought by the advent of Web 2.0, we have seen in the last years an emerging interest in the research community towards the identification of requirements [1,2] and the proposal of innovative solutions [3,4,5] for collaborative development of ontologies.

The requirements and features which have emerged in these works mainly address the integration of tools supporting communication and discussion among users, the resolution of issues related to concurrent editing, and the definition of standard access

R. Serra and R. Cucchiara (Eds.): AI*IA 2009, LNAI 5883, pp. 42–51, 2009.
© Springer-Verlag Berlin Heidelberg 2009

control and contribution modalities. What lacks thereof is the ability for users to go beyond simple discussion or voting about round-the-corner ontology modifications, to follow or even create arbitrary evolution paths for the ontologies they are working on.

In our research work, we have tried to propose a novel approach to collaborative ontology development which would fill the above gap, by accounting the effort and results of the Semantic Web Interest Group on Named Graphs [6], and by exploiting the possibilities offered by their introduction.

In our approach, ontological knowledge is distributed across different *contexts* (identified by diverse named graphs), which identify the branched workspace of team members as well as the main development trunk shared by all of them. Users can thus freely work in their personal context (which is given by the merge of the named graph assigned to them and of the main development trunk), but can also inspect other *context*s, access their content, and *trust* (part of) the statements contained there, thus virtually importing them into their context. This poses unlimited possibilities to the creativity of each single team member, who can bring his work ahead and lately have it discussed through traditional communication gadgets or even implicitly promoted as he finds other users accepting and *trusting* his proposals.

Thanks to the introduction of Named Graphs into a few of the currently available triple store technologies, such as Jena [7] (through the NG4J [8] library extension), and Sesame 2.0 [9], we have also been able to develop and present here CONGAS, a novel system for collaborative editing of Semantic Web ontologies, which has been developed as a parallel collaborative version of the Knowledge Management and Acquisition platform Semantic Turkey [10].

2 Related Works

The first published result on Named Graphs dates back to the work of Carroll et al. [11], though other works have addressed the problem of data provenance and locality in the years before [12,13]. The introduction of Named Graphs has been a necessary step for the Semantic Web, their ability to express meta-information about RDF graphs is an important ingredient for addressing a range of its important requirements, such as Data syndication, Information Access, RDF Signature [14] and Trust [15], expressing propositional attitudes [16], scoping assertions and logic and managing ontology versioning and evolution. All of the above mainly account for one necessity: the ability to track provenance of single graphs merged into compound RDF repositories.

To our knowledge, no collaborative environment for development of knowledge graphs of the RDF family has widely exploited Named Graphs support, to introduce user spaces. There are however other aspects of collaborative ontology development which have been evidenced and widely experimented in several works. We mention a few of them according to their specific contributions:

– *Wiki adoption*: [17] and [18] offer ontology development collaborative environments based on (modified versions of) wiki platforms. Both of them do not address the general target of ontology development, and are respectively oriented towards Enterprise Modeling and acquiring consensus over the definition of reusable Ontology Design Patterns [19]

- *Ontology modularization*: the Hozo ontology editor [3] enables asynchronous development of ontologies that are subdivided into multiple inter-connected modules. A developer checks out and locks a specific module, edits it locally, and then checks it back in. Each module are however still owned by all team members, thus limiting the freedom of action and rapid drafting of new ontology branches
- *Methodology*: The Cicero tool [20] implements the DILIGENT [21] methodology to for collaborative ontology development. The DILIGENT methodology focuses on the process of argumentation, thus supporting the generation of discussions about ontology maturing, both in general as well as for specific resources
- *Models*: in [22] the authors present an ontology supporting the definition of requirements for collaborative ontology development, while in [23] an workflow ontology can be used to describe different kind of workflows for the same task (they also experimented their model in expressing the DILIGENT methodology cited above).
- *Full integration into complete ontology development tools:* in [5], an extension for the popular Knowledge Management and Acquisition tool Protégé [24] is presented, perfectly integrating several contradistinguish features for collaborative ontology development, into the base tool (and thus beneficiating of all of its traditional editing facilities): these include *user management*, enabling *discussions/annotations*, *workflow support* (through the workflow ontology cited above) and *synchronous/asynchronous editing* of available data.

3   True Collaboration through Interwoven Ontology User Spaces

The objective which has been targeted in the design of CONGAS was to develop a completely new stereotype of collaboration, in which users could discuss, reject or approve modifications to existing ontology resources (from very common ontological axioms, such as classification and *is-a* organization, to detailed triple level analysis) as well as (and this is the novelty with respect to existing tools and methodologies) create and propose entirely new ontology branches, which can then be aligned/merged to the core ontology.

In our model, each user is assigned his dedicated *user space*, and can develop new extensions for the edited ontology without the need to propose them step-by-step on the main development trunk, nor the risk of a totally unrestricted editing, resulting in the production of noisy data which could be entropic for other users and thus for the whole collaborative process. By default, each user is able to view in his *space* the main development trunk and an ontology branch associated to him. The user has full read/write privileges over his personal development branch, while he can only propose changes to the main trunk and discuss them through ordinary argumentation/voting mechanisms (forum, polls etc...). Note that this limitation over the main trunk is confined to what is actually translated into deletion of triples (due to the monotonic discipline of RDF [25]), but addition of axioms referring to resources in the main trunk can be handled through the user space; for example, if the user is willing to add a rdfs:subClassOf relationship between two classes (namely: main:A and main:B) belonging to the main trunk, this is not considered a modification, since it involves the sole creation of the triple:

main:A rdfs:subClassOf main:B

which can thus be stored in his personal ontology branch (that is, in his *user space*). The component which is in charge of projecting the set of triples governed by the rdfs:subClassOf predicate into a tree visualization of ontology classes, will take into account this triple and show the tree accordingly, with class main:A arranged under class main:B.

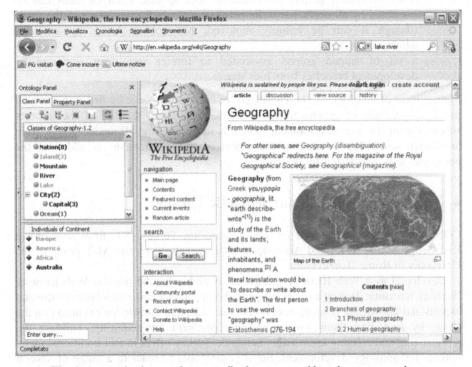

Fig. 1. A user viewing ontology contributions proposed by other team members

What thus happens is that different users could even participate in suggesting changes to the main trunk (which is considered *frozen*) while they can freely contribute to the evolution of the ontology taking their way on extending the set of resources and their related axioms. *User spaces*, though assigned on a per-user basis and granting write-privileges only to their owner, can however be browsed by other users (see Fig. 1, where concepts and instances created by team members, and associated to them by different colors, are browsed by current user): the content of each space can be exported to the spaces of other users who decide to *trust* it and add it to their respective evolution branch. This way, it is easy for new knowledge to be produced by independent users, discussion is supported by forums/polls (which are also enabled for foreign user spaces), while convergence of the result is assured by the trust&import mechanism which allows several users to quickly share (portions of) proposed branches and thus promote them for the next main trunk release. Seen from the perspective of the triple store engine, the RDF repository is an aggregation of:

- several named graphs, representing foreign ontologies (which can only be accessed with *read* privileges by all kind of users) imported by the main trunk
- a core graph, containing data from the frozen main trunk. It is accessible by all users, and it is read-only for all standard users, though granting write permissions to ontology administrators
- a set of named graphs associated to user spaces. Each of them can be inspected by all users, has write permissions only for the owner of the space (though it can be entirely removed, but not modified, by ontology administrators)
- a set of named graphs associated to foreign ontologies imported by development branches from user spaces.

Management of ontologies to be visualized for each user is done by first importing all the first three set of graphs from the above list. Then, the list of owl:imports statements for the user development branch is inspected, and all named graphs from the fourth set which is cited in the object of these statements is added to the ontologies to be visualized for that user.

4   The Hosting Application: Semantic Turkey

CONGAS has been developed on top of Semantic Turkey (ST, from now on), a Knowledge Management and Acquisition System realized by the ART group of the University of Rome, Tor Vergata.

Developed as a Web Browser extension (available for the popular Web Browser Firefox), Semantic Turkey aims at reducing the impedance mismatch between domain experts and knowledge investigators on the one side, and knowledge engineers on the other, by providing them with a unifying platform for acquiring, building up, reorganizing and refining knowledge. Semantic Turkey offers in fact traditional menu options for ontology development, but it is also able to react to a range of several intuitive actions (such as drag'n'dropping objects – text in the standard version – from the web page to the ontology panels) with contextual outcomes (i.e. similar gestures may result in even long sequences of ontology modifications which depend on the kind of object dragged from the Web Page, on the type of resource where it is dragged etc...). ST deeper interaction with Web content is not only limited to the possibility of importing text and other sorts of media from Web Pages; it also features an extension mechanism which covers all of its aspects and technologies: from user interaction, through its UI extension points linked to the same Mozilla Extension mechanism which is at the base of its hosting web browser, to knowledge management and persistence (thorough standard OSGi service extendibility).

In realizing CONGAS, we have first examined the different possibilities which were available for converting the system into a distributed environment. These aspects will be discussed in the next section.

5 Architecture

For economy of space, we will limit ourselves here to describe those architectural changes in Semantic Turkey which have been introduced to realize its collaborative version CONGAS. For a detailed description of Semantic Turkey architecture and knowledge model, the reader may refer to [10], while [26] contains relevant updates related to the extension mechanism.

Semantic Turkey architecture is organized around a three-tier layering, with the presentation layer embodying the true Firefox extension and the other two layers built around java technologies for managing the business logic and data access.

Both the two interlayer interfaces could be, in principle, be separated in a distributed environment (http communication is already adopted between the presentation and middle layer, and data access interface can easily be implemented towards remote RDF repositories) so, when thinking about needed reengineering of this architecture for porting ST to a collaborative framework, we faced the following possibilities:

– Centralizing the sole Persistence layer and realize collaborative client applications with a rewritten business logic to support transaction based communication with the RDF service.
– Keeping the presentation layer for client applications, and move both server and persistency on the centralized collaborative server
– Split the middle layer into two components, one which is bundled with client applications and provides the required business logic for their operations, and the other one which coordinates collaboration between clients and manages all related services

The first one has been discarded, since it would have produced nothing more than a user interface for transaction-based knowledge repositories. The third option would have proven to be the best solution, though one consideration about client technology made us leaning towards the second one: what is reported as the presentation layer in ST architecture, is actually represented by the whole array of technologies supporting browser extendibility. For example, with respect to the current implementation available for Firefox, an extension of the JavaScript language is adopted to support business logic of extensions to be developed; it can thus be used to handle the minimal support required by user interaction, while demanding to the collaborative server most of the necessary computation. Analogous technologies satisfying the minimal computation requirements of user interaction are expected to be found for all classes of browsers, thus not invalidating the architecture on its own.

5.1 Coordination and Synchronization

Coordination between users is important in a collaborative environment, as well as keeping sync between what they see while editing the ontology, and changes to the model which can have been submitted by other users.

A refresh button is present in each client, which has the double function of activating (when depressed) a complete refresh of the graphic view over the ontology and of alerting (by blinking) users when a change has been made by another team member. A log of changes to the whole ontology repository is also available, so that the user can account for these changes in case they generate some conflict with or provide useful input for the work he is doing.

The process of convergence towards a shared evolution for the ontology (i.e. freezing proposed changes in the development *trunk*) is activated through different triggering events: roughly divided as *implicit triggers* (e.g., when a certain percentage of team members has reached consensus over a resource/statement, that is, is trusting, and thus importing, the given resource/statement in its user space) and *explicit* ones (e.g., by explicit approval, through argumentation services such as polls and forums, see next section for details).

Access management divides users according to three main categories:

- *Viewers* who can access the ontology, view resources, and comment or vote on the choices made (this role is usually assigned to domain experts). Those users, then, can "look and speak"
- *Users*: in addition to the rights of viewers, they own a dedicated user space, so that they can "look, speak, and propose".
- *Administrators*. An administrator has all the rights of a user, but it is also able to modify the main development trunk, as well as provide other coordination activity such as moderating forums and accepting proposals. Thus, an administrator can "look, speak, propose and validate ".

Finally, a simple versioning system allows administrators to freeze snapshots of developed ontologies at a certain time and store them with an assigned tag (usually, a version number). A version manager enables then users to retrieve them and inspect their content.

5.2 Services

In building CONGAS, we have tried to integrate several features and tools supporting collaborative development, together with the concept of *user space* and *model trusting* which pervade all of its architectural choices.

A *poll-based mechanism* allows users to express their opinions. They may choose on open arguments for discussion (*open polls*), where both the theme and options for polling are chosen by one of the users, as well as on validity of all statements available in development trunk and branches (*standard polls*). Standard polls are automatically associated by the system to their related statement, and can be easily accessed when inspecting the projection of that statement (a class/property in the tree, an individual in the list of instances of a class, or a valued property in a resource description).

With a similar approach, also a *forum* has been added to the system, enabling both the creation of free thematic discussions about ontology evolution, and of discussions focused on proposed resources and statements – e.g., whenever a user adds a new resource, it is easy to open a discussion on it: the thread is automatically matched by the resource URI and it can lately always be accessed from a context menu option for that resource. Emailing is also supported, by retrieving public mail addresses of registered users.

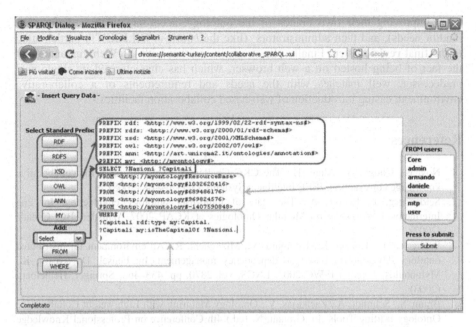

Fig. 2. The SPARQL query panel, with facilities for restricting the domain of the query to specific user contexts

Finally, also query support has been integrated with the user spaces paradigm. A SPARQL interface (Fig. 2) allows users to select which user spaces will be considered (panel on the right) when retrieving tuples from the repository (the main development trunk is put by default in the query template, though it may be excluded by the user by manually changing the query code).

6 Conclusions

Apart from its services and functionalities for supporting collaboration, which are in line with state-of-art tools on collaborative editing (and improve them in some cases, as for the generation of forum threads and polls automatically-linked to proposed ontology resources), the main contribution of CONGAS to collaborative ontology development resides in its coordinated editing of evolution branches proposed by users. The possibilities offered by Named Graphs open up a completely novel scenario in which users may freely (and massively) contribute to the main trunk of development of an ontology, without the risk of over-generating undesired axioms, nor suffering from the impedance brought by a strictly disciplined update procedure. It is this aspect, stressing autonomy and independence of the user, which fills the gap between collaboration methodologies such as the already mentioned DILIGENT, and current implementations of collaboration tools: where these latter (coll. Protégé, or SWOOP [27]) provide support for collaboration and discussion on one single ontology (which may thus implement the **analysis** and **revision** steps of DILIGENT), CONGAS allows for a real **revision** and **local adaptation** phase, with distributed users improving their own branch of the ontology, according to their preferences (and/or needs), and then administrators (like the *board* in DILIGENT) participating to the final **revision** step. Finally, the strong interaction with the Web guaranteed by the fact of being hosted on a web browser, which has characterized its single-user predecessor, well matches with the needs and requirements of a collaborative environment, easing introduction of web-based collaboration facilities.

References

1. Noy, N., Chugh, A., Alani, H.: The CKC Challenge: Exploring tools for collaborative knowledge construction. IEEE Intelligent Systems 23(1), 64–68 (2008)
2. Seidenberg, J., Rector, A.: The state of multi-user ontology engineering. In: 2nd International Workshop on Modular Ontologies at KCAP 2007, Whistler, BC, Canada (2007)
3. Sunagawa, E., Kozaki, K., Kitamura, Y., Mizoguchi, R.: An environment for distributed ontology development based on dependency management. In: Fensel, D., Sycara, K., Mylopoulos, J. (eds.) ISWC 2003. LNCS, vol. 2870, pp. 453–468. Springer, Heidelberg (2003)
4. Braun, S., Schmidt, A., Zacharias, V.: Ontology Maturing with Lightweight Collaborative Ontology Editing Tools. In: Gronau, N. (ed.) 4th Conference on Professional Knowledge Management - Experiences and Visions, Workshop on Productive Knowledge Work (ProKW 2007), GITO, pp. 217–226 (2007)
5. Tudorache, T., Noy, N., Tu, S., Musen, M.: Supporting Collaborative Ontology Development in Protégé. In: Sheth, A.P., Staab, S., Dean, M., Paolucci, M., Maynard, D., Finin, T., Thirunarayan, K. (eds.) ISWC 2008. LNCS, vol. 5318, pp. 17–32. Springer, Heidelberg (2008)
6. Named Graphs / Semantic Web Interest Group. In: World Wide Web Consortium - Web Standards, http://www.w3.org/2004/03/trix/
7. McBride, B.: Jena: Implementing the RDF Model and Syntax Specification. In: Semantic Web Workshop, WWW 2001 (2001)
8. NG4J - Named Graphs API for Jena. In: NG4J - Named Graphs API for Jena, http://www.wiwiss.fu-berlin.de/suhl/bizer/ng4j/
9. Broekstra, J., Kampman, A., van Harmelen, F.: Sesame: A generic architecture for storing and querying RDF and RDF schema. In: Horrocks, I., Hendler, J. (eds.) ISWC 2002. LNCS, vol. 2342, p. 54. Springer, Heidelberg (2002)
10. Griesi, D., Pazienza, M.T., Stellato, A.: Semantic Turkey - a Semantic Bookmarking tool (System Description). In: Franconi, E., Kifer, M., May, W. (eds.) ESWC 2007. LNCS, vol. 4519, pp. 779–788. Springer, Heidelberg (2007)
11. Carroll, J., Bizer, C., Hayes, P., Stickler, P.: Named Graphs, Provenance and Trust. In: WWW 2005: Proceedings of the 14th international conference on World Wide Web, New York, NY, USA, pp. 613–622 (2005)
12. Sintek, M., Decker, S.: Triple - a query, inference, and transformation language for the semantic web. In: Horrocks, I., Hendler, J. (eds.) ISWC 2002. LNCS, vol. 2342, pp. 364–378. Springer, Heidelberg (2002)
13. Guha, R., Fikes, R.: Contexts for the semantic web. In: McIlraith, S.A., Plexousakis, D., van Harmelen, F. (eds.) ISWC 2004. LNCS, vol. 3298, Springer, Heidelberg (2004)
14. Carroll, J.: Signing RDF Graphs. In: Fensel, D., Sycara, K., Mylopoulos, J. (eds.) ISWC 2003. LNCS, vol. 2870, pp. 5–15. Springer, Heidelberg (2003)

15. Bizer, C., Oldakowski, R.: Using context- and content-based trust policies on the semantic web. In: WWW Alt. 2004: Proceedings of the 13th international World Wide Web conference on Alternate track papers & posters, New York, NY, USA, pp. 228–229 (2004)
16. Ibrahim, A.: Agent Communication Languages (ACL) (accessed 2000)
17. Rospocher, M., Ghidini, C., Serafini, L., Kump, B., Pammer, V., Lindstaedt, S.: ollaborative Enterprise Integrated Modelling. In: Semantic Web Applications and Perspectives, FAO UN, Rome, FAO UN, Rome (2008)
18. Daga, E., Presutti, V., Salvati, A.: Ontologydesignpatterns.org and Evaluation WikiFlow. In: Semantic Web Applications and Perspectives, 5th Italian Semantic Web Workshop (SWAP 2008), FAO UN, Rome (2008)
19. Gangemi, A.: Ontology Design Patterns for Semantic Web Content. In: Gil, Y., Motta, E., Benjamins, V., Musen, M. (eds.) Proceedings of the Fourth International Semantic Web Conference, Galway, Ireland (2005)
20. Dellschaft, K., Engelbrecht, H., MonteBarreto, J., Rutenbeck, S., Staab, S.: Cicero: Tracking design rationale in collaborative ontology engineering. In: Proceedings of the ESWC 2008 Demo Session (2008), http://www.uni-koblenz.de/~klaasd/Downloads/papers/Dellschaft2008CTD.pdf
21. Tempich, C., Simperl, E., Luczak, M., Studer, R., Pinto, H.: Argumentation-based ontology engineering. IEEE Intelligent Systems 22(6), 52–59 (2007)
22. Gangemi, A., Lehmann, J., Presutti, V., Nissim, M., Catenacci, C.: C-ODO: an OWL metamodel for collaborative ontology design. In: Workshop on Social and Collaborative Construction of Structured Knowledge, WWW 2007, Banff, Canada (2007)
23. Sebastian, A., Noy, N.F., Tudorache, T., Musen, M.A.: A generic ontology for collaborative ontology-development workflows. In: Gangemi, A., Euzenat, J. (eds.) EKAW 2008. LNCS (LNAI), vol. 5268, pp. 318–328. Springer, Heidelberg (2008)
24. Gennari, J., Musen, M., Fergerson, R., Grosso, W., Crubézy, M., Eriksson, H., Noy, N., Tu, S.: The evolution of Protégé-2000: An environment for knowledge-based systems development. International Journal of Human-Computer Studies, Protege 58(1), 89–123 (2003)
25. Hayes, P.: RDF Semantics. In: World Wide Web Consortium - Web Standards (accessed 2009), http://www.w3.org/TR/rdf-mt/
26. Pazienza, M., Scarpato, N., Stellato, A., Turbati, A.: Din din! The (Semantic) Turkey is served! In: Semantic Web Applications and Perspectives, Rome, Italy (2008)
27. Völkel, M., Enguix, C., Kruk, S., Zhdanova, A., Stevens, R., Sure, Y.: SemVersion - versioning RDF and ontologies. KnowledgeWeb Deliverable D2.3.3.v1, Institute AIFB, University of Karlsruhe (June 2005)
28. Griesi, D., Pazienza, M., Stellato, A.: Gobbleing over the Web with Semantic Turkey. In: Semantic Web Applications and Perspectives, 3rd Italian Semantic Web Workshop (SWAP 2006), Scuola Normale Superiore, Pisa, Italy, December 18-20 (2006)
29. Kalyanpur, A., Parsia, B., Sirin, E., Grau, B., Hendler, J.: Swoop: A web ontology editing browser. Web Semantics: Science, Services and Agents on the World Wide Web 4(2), 144–153 (2004)

Pstable Semantics for Logic Programs with Possibilistic Ordered Disjunction

Roberto Confalonieri, Juan Carlos Nieves, and Javier Vázquez-Salceda*

Universitat Politècnica de Catalunya
Dept. Llenguatges i Sistemes Informàtics
C/ Jordi Girona Salgado 1-3
E - 08034 Barcelona
{confalonieri,jcnieves,jvazquez}@lsi.upc.edu

Abstract. In this paper we define the semantics for capturing possibilistic ordered disjunction programs based on pstable semantics. The pstable semantics, based on paraconsistent logic, allows to treat inconsistency programs. Moreover being closer to possibilistic inference it allows to extend the necessity-values of the clauses to be considered, causing a higher level of comparison at the moment of selecting preferred pstable models of a possibilistic ordered disjunction programs. We compare the possibilistic pstable semantics for ordered disjunction programs with the recently defined possibilistic answer set semantics for the same class of logic programs.

1 Introduction

Logic programs with ordered disjunction (LPODs) are extended logic programs based on answer set semantics with a new *ordered disjunction* logic connector $\times$ which can capture preferences between literals. An ordering between the valid models of a LPOD can then be specified through some comparison criteria, which allow to compare answer sets based on the preferences rules' satisfaction degree [2]. Therefore they are classes of logic programs that fit well in problems such as configuration management, policy monitoring and user preference representation and reasoning [2]. However, in realistic scenarios preference rules can be associated with a degree of relevance (or uncertainty) which can affect the established preference order, preventing the achievement of a single preferred solution, and some of these cases can be handled by possibilistic semantics for ordered disjunction programs [3].

Nevertheless, the syntax of a logic program with ordered disjunction allows the writing of programs such as $b \times a \leftarrow not\ a$, which are considered inconsistent because, under answer set semantics, they do not have any answer sets. Inconsistency in some cases has shown to be a desired feature of logic programs, because sometime the loss of information is worse compared to an inconsistent program not having any model. In the case of ordered disjunction programs the

---

* The authors are named by alphabetic order.

R. Serra and R. Cucchiara (Eds.): AI*IA 2009, LNAI 5883, pp. 52–61, 2009.
© Springer-Verlag Berlin Heidelberg 2009

elimination of an inconsistent rule by the answer set-based reduction defined for this class of programs [2], can either prevent the program to have any solution, or to reach an order between the answer sets. In this paper we propose a *pstable semantics for ordered disjunction programs*, which can handle inconsistent ordered disjunction programs in the sense of no existence of models, and we extend it for capturing *possibilistic ordered disjunction programs* in order to cope with the degree of uncertainty in the reasoning process.

The advantages of our approach are several. First, pstable semantics, being based on paraconsistent logic, is less sensible to inconsistency and allows to treat inconsistent programs [8]. This means that we can provide valid pstable models of an ordered disjunction program whereas the program does not have any answer sets due to the presence of inconsistent rules. Secondly, being closer to possibilistic logic inference, it allows to consider higher degrees of necessity-values associated to the possibilistic atoms in the reasoning process, and sometimes can guarantee a higher level of comparison at the moment of selecting preferred pstable models of a possibilistic ordered disjunction programs. In such cases, we are able to compute a single preferred possibilistic pstable model, whereas the ordered disjunction program based on possibilistic answer set semantics cannot.

The rest of the paper is organized as follows. In the next section we provide some basic definitions. Section 3 describes the pstable semantics for ordered disjunction programs. In Section 4 we propose a possibilistic extension of the pstable semantics and we show the benefits of possibilistic pstable semantics for ordered disjunction. In the last section we outline some conclusions and future work directions. Throughout the paper, we use some simple examples to explain our semantics.

2   Background

This section presents the reader with some basic definitions *w.r.t* extended logic programs, pstable semantics, possibilistic logic, and logic programs with ordered disjunction, which are the basis for our approach.

2.1   Extended Logic Programs

We consider extended logic programs which have two kinds of negation: strong negation $\neg$ and default negation *not*. A signature $\mathcal{L}$ is a finite set of elements that we call atoms, where atoms negated by $\neg$ are called extended atoms. Intuitively, *not a* is true whenever there is no reason to believe a, whereas $\neg a$ requires a proof of the negated atom. In the following we use the concept of atom without paying attention if it is an extended atom or not. A *literal* is either an atom a, called *positive literal*, or the negation of an atom *not a*, called *negative literal*. Given a set of atoms $\{a_1, ..., a_n\}$, we write *not* $\{a_1, ..., a_n\}$ to denote the set of atoms $\{not\ a_1, ..., not\ a_n\}$. An extended *normal* rule, r, is a rule of the form $a \leftarrow b_1, \ldots, b_n, not\ b_{n+1}, \ldots, not\ b_{n+m}$ where a and each of the b_i are atoms for $1 \leq i \leq n + m$. In a slight abuse of notation we will denote such a clause by the formula $a \leftarrow \mathcal{B}^+, not\ \mathcal{B}^-$ where the set $\{b_1, \ldots, b_n\}$ will be denoted by $\mathcal{B}^+$, and the set $\{b_{n+1}, \ldots, b_{n+m}\}$ will be denoted by $\mathcal{B}^-$. A *constraint* is an extended rule of the form: $\leftarrow \mathcal{B}^+$, *not* $\mathcal{B}^-$. We define an *extended logic normal program* P as a finite set of extended normal rules and constraints. If the body of a normal rule is empty, then the clause is known as a *fact* and can be denoted just by a. We write $\mathcal{L}_P$, to denote the set of atoms that appear in the rules of P. We denote by $HEAD(P)$ the set $\{a|a \leftarrow \mathcal{B}^+$, *not* $\mathcal{B}^- \in P\}$. We will manage the strong negation $\neg$ in our logic programs as it is done in Answer Set Programming. Basically, each atom $\neg a$ is replaced by a new atom symbol a' which does not appear in the language of the program and we add the constraint $\leftarrow a, a'$ to the program [1]. For managing the constraints in our logic programs, we will replace each rule of the form $\leftarrow \mathcal{B}^+$ *not* $\mathcal{B}^-$ by the following set of clauses: $v \leftarrow \mathcal{B}^+$, *not* $\mathcal{B}^-$. $w \leftarrow$ *not* y, v. $y \leftarrow$ *not* z, v. $z \leftarrow$ *not* w, v., where v, w, y, z are new atom symbols which do not appear in $\mathcal{L}_P$.[1]

2.2 Pstable Semantics

Pstable semantics is a recently introduced logic programming semantics which is inspired in paraconsistent logic and is defined in terms of a single reduction.

Definition 1 (Pstable reduction [8])
Let P be a normal program and M a set of atoms. The pstable reduction of P is defined as $RED(P, M) := \{a \leftarrow \mathcal{B}^+$, not $(\mathcal{B}^- \cap M)|a \leftarrow \mathcal{B}^+$, not $\mathcal{B}^- \in P\}$.

By considering the reduction $RED(P, M)$, the pstable semantics for normal programs is defined as follows:

Definition 2 (Pstable semantics [8])
Let P be a normal program and M a set of atoms. M is a pstable model of P if $RED(P, M) \Vdash M$.[2]

The pstable semantics allow to treat inconsistent logic programs such as $a \leftarrow$ *not* a. In fact by applying pstable semantics, we can see that $M = \{a\}$ is a pstable model of $a \leftarrow$ *not* a, while under the answer set semantics we would not have obtained any answer sets [5].

2.3 Possibilistic Logic

A *necessity-valued formula* is a pair $(\varphi\ \alpha)$ where φ is a classical logic formula and $\alpha \in (0, 1]$ is a positive number. The pair $(\varphi\ \alpha)$ expresses that the formula φ is *certain* at least to the level α, *i.e.* $N(\varphi) \geq \alpha$, where N is a necessity measure modeling our possibly incomplete state knowledge [4]. α is not a probability (as it is in probability theory) but it induces a certainty (or confidence) scale. This value is determined by the expert providing the knowledge base. A necessity-valued knowledge base is then defined as a finite set (*i.e.* a conjunction) of necessity-valued formulae. Dubois *et al.* [4] introduced a formal system for necessity-valued logic which is based in the following axioms schemata (propositional case):

---

[1] This approach was suggested in [7].
[2] It is written $P \Vdash M$ when $P \vdash_C M$ (logic consequence in classic logic) and M is a classical 2-valued model of P.

(A1) $(\varphi \to (\psi \to \varphi)\ 1)$
(A2) $((\varphi \to (\psi \to \xi)) \to ((\varphi \to \psi) \to (\varphi \to \xi))\ 1)$
(A3) $((\neg\varphi \to \neg\psi) \to ((\neg\varphi \to \psi) \to \varphi)\ 1)$

For the axioms above, the following inference rules are defined:

(GMP) $(\varphi\ \alpha), (\varphi \to \psi\ \beta) \vdash (\psi\ min\{\alpha, \beta\})$
(S) $(\varphi\ \alpha) \vdash (\varphi\ \beta)$ if $\beta \leq \alpha$

We denote by $\vdash_{PL}$ the inference under Possibilistic Logic without paying attention if the necessity-valued formulæ are using either a totally ordered set or a partially ordered set for expressing the levels of uncertainty.

2.4 Logic Programs with Ordered Disjunction

Logic programs with ordered disjunction (LPODs) are extended logic programs which allow the use of an ordered disjunction connector $\times$ in the head of rules to express preferences among its literals [2]. The rule $r = C_1 \times \ldots \times C_n \leftarrow b_1, \ldots, b_m,\ not\ b_{m+1}, \ldots,\ not\ b_{m+k}$ states that if the body is satisfied then some C_i must be in the answer set, if possible C_1, if not then C_2, and so on, and at least one of them must be true. Each of the C_i can be seen as a preference the user is interested into according to a desired order. One interesting characteristic of LPODs is that they provide a mean to represent preferences among answer sets by considering the satisfaction degree [2].

Definition 3. *[2] Let M be an answer set of an ordered disjunction program P. Then M satisfies the rule $r = C_1 \times \ldots \times C_n \leftarrow b_1, \ldots, b_m,\ not\ b_{m+1} \ldots,\ not\ b_{m+k}$*

- *to degree 1 if $b_j \notin M$ for some j $(1 \leq j \leq m)$, or $b_i \in M$ for some i $(m+1 \leq i \leq m+k)$,*
- *to degree j $(1 \leq j \leq n)$ if all $b_l \in M$ $(1 \leq l \leq m)$, no $b_i \in M$ $(m+1 \leq i \leq m+k)$, and $j = min\{r \mid C_r \in M, 1 \leq r \leq n\}$.*

The satisfaction degree of an answer set M w.r.t a rule, denoted by $deg_M(r)$, provides a ranking of the answer sets of a LPOD, and a preference order on the answer sets can be obtained using some proposed combination strategies. In [2], the authors have proposed three criteria for comparing answer sets, respectively *cardinality, inclusion* and *Pareto*. In this paper we keep the three criteria and extend them for comparing possibilistic pstable models.

Example 1. Let P be the following ordered disjunction program.

$r_1 : b \times a \leftarrow not\ a.$ $r_2 : a \leftarrow b.$ $r_3 : b \leftarrow a.$

The program P does not have any answer sets as answer set semantics do not allow to treat inconsistent programs (rule r_1 is inconsistent). In the next section we show how, by defining the pstable semantics for ordered disjunction programs, we can compute pstable models for inconsistent programs.

3   Pstable Semantics for Ordered Disjunction Programs

We specify a pstable semantics for ordered disjunction programs through two reductions based on the pstable semantics of Section 2.2 and on the original reductions defined by Brewka [2].

Definition 4 (Reduction $RED(r, M)_\times$)
Let $r = C_1 \times \ldots \times C_n \leftarrow \mathcal{B}^+$, not $\mathcal{B}^-$ be an ordered disjunction clause and M be a set of atoms. The $\times$ reduct $RED(r, M)_\times$ of r is defined as follows:
$RED(r, M)_\times := \{C_i \leftarrow \mathcal{B}^+, not\ (\mathcal{B}^- \cap M) | C_i \in M \text{ and } M \cap \{C_1, \ldots, C_{i-1}\} = \emptyset\}$.

Definition 5 (Reduction $RED(P, M)_\times$)
Let P be an ordered disjunction program and M a set of atoms. The $\times$ reduct $RED(P, M)_\times$ of P is defined as follows: $RED(P, M)_\times = \bigcup_{r \in N} RED(r, M)_\times$.

Based on the reductions, we can define the pstable semantics extension for ordered disjunction programs.

Definition 6 (Pstable semantics for LPODs)
Let P be an ordered disjunction program P and M a set of atoms. M is pstable model of P if $RED(P, M)_\times \Vdash M$.

To see the benefits of the pstable semantics for ordered disjunction programs let us consider the following example.

Example 2. Let us consider the ordered disjunction program P in Example 1 and the set of atoms $M = \{b, a\}$. We can see that the $RED(P, M)_\times$ is:

$\mathbf{r_1} : b \leftarrow not\ a.$ $\mathbf{r_2} : a \leftarrow b.$ $\mathbf{r_3} : b \leftarrow a.$

Hence, the set $M = \{b, a\}$ is a pstable model of P, as $RED(P, M)_\times \vdash_c M$ and M is a 2-valued model of $RED(P, M)_\times$.

Despite applying pstable semantics, pstable models of ordered disjunction programs are comparable by using the comparison criteria of [2]. Furthermore, the semantics we define is a generalization of the original semantics of Brewka, as formulated in the following theorem.

Theorem 1. *Let P be an ordered disjunction program and M a set of atoms. If M is an answer set of P then M is a pstable model of P.*

4   Pstable Semantics for LPPODs

In this section we recall the syntax of possibilistic ordered disjunction program [3] and extend the pstable semantics for ordered disjunction programs defined in Section 3 with possibilistic logic.

4.1   Syntax of LPPODs

The syntax of a possibilistic ordered disjunction program is based on the syntax of ordered disjunction rules (Section 2.4) and of possibilistic logic (Section 2.3).

A *possibilistic atom* is a pair $p = (a, q) \in \mathcal{A} \times \mathcal{Q}$ where $\mathcal{A}$ is a set of atoms and $(\mathcal{Q}, \leq)$ a finite lattice.[3] The projection $*$ for any possibilistic atom p is defined as: $p^* = a$. Given a set of possibilistic atoms M, the generalization of $*$ over M is defined as: $M^* = \{p^* \mid p \in M\}$. Given $(\mathcal{Q}, \leq)$, a possibilistic ordered disjunction rule r is of the form: $\alpha : C_1 \times \ldots \times C_n \leftarrow \mathcal{B}^+, \; not \; \mathcal{B}^-$, where $\alpha \in \mathcal{Q}$ and $C_1 \times \ldots \times C_n \leftarrow \mathcal{B}^+, \; not \; \mathcal{B}^-$ is an ordered disjunction rule as defined in Section 2.4 with $\mathcal{B}^+ = \{b_1, \ldots, b_m\}$ and $\mathcal{B}^- = \{b_{m+1}, \ldots, b_{m+k}\}$.

The projection $*$ for a possibilistic ordered disjunction rule r, is $r^* = C_1 \times \ldots \times C_n \leftarrow \mathcal{B}^+, \; not \; \mathcal{B}^-$. $n(r) = \alpha$ is a necessity degree representing the certainty level of the information described by r. A possibilistic constraint c is of the form: $\mathcal{TOP}_\mathcal{Q} :\leftarrow \mathcal{B}^+, \; not \; \mathcal{B}^-$, where $\mathcal{TOP}_\mathcal{Q}$ is the top of the lattice $(\mathcal{Q}, \leq)$ and $\leftarrow \mathcal{B}^+, \; not \; \mathcal{B}^-$ is a constraint. Observe that any possibilistic constraint must have the top of the lattice $(\mathcal{Q}, \leq)$. This restriction is motivated by the fact that, like constraints in standard Answer Set Programming, the purpose of the possibilistic constraint is to eliminate possibilistic models. Hence, it is assumed that there is no uncertainty about the information captured by a possibilistic constraint. As in possibilistic ordered disjunction rules, the projection $*$ for a possibilistic constraint c is $c^* = \leftarrow \mathcal{B}^+, \; not \; \mathcal{B}^-$.

A Logic Program with Possibilistic Ordered Disjunction (LPPOD) is a tuple of the form $P := \langle (Q, \leq), N \rangle$ such that N is a finite set of possibilistic ordered disjunction rules and possibilistic constraints. The generalization of $*$ over P is defined as follows: $P^* := \{r^* \mid r \in N\}$. Notice that P^* is an ordered disjunction logic program. Given a possibilistic ordered disjunction program $P := \langle (Q, \leq), N \rangle$, we define the α-cut of P denoted by P_α as $P_\alpha := \{r \mid r \in P, n(r) \geq \alpha\}$.

4.2   Pstable Semantics for LPPODs

We propose a possibilistic logic semantics which is close to the proof theory of possibilistic logic and pstable semantics. As in the pstable semantics definition, our approach is based on a syntactic reduction. We will consider sets of possibilistic atoms as interpretations. Hence, before defining the possibilistic ordered disjunction logic programming semantics, we introduce basic operators between sets of possibilistic atoms and a relation of order between them [6].

Definition 7. *Given $\mathcal{A}$ a finite set of atoms and $(\mathcal{Q}, \leq)$ be a lattice, we consider $\mathcal{PS} = 2^{\mathcal{A} \times \mathcal{Q}}$ as the finite set of all the possibilistic atoms sets induced by $\mathcal{A}$ and $\mathcal{Q}$. Let $A, B \in \mathcal{PS}$, hence we define the operators $\sqcap, \sqcup$ and $\sqsubseteq$ as follows:*

$$A \sqcap B = \{(x, GLB\{q_1, q_2\}) | (x, q_1) \in A \wedge (x, q_2) \in B\}$$
$$A \sqcup B = \{(x, q) | (x, q) \in A \; and \; x \notin B^*\} \cup \{(x, q) | x \notin A^* \; and \; (x, q) \in B\} \cup$$
$$\{(x, LUB\{q_1, q_2\}) | (x, q_1) \in A \; and \; (x, q_2) \in B\}.$$
$$A \sqsubseteq B \Longleftrightarrow A^* \subseteq B^*, \; and \; \forall x, q_1, q_2, (x, q_1) \in A \wedge (x, q_2) \in B \; then \; q_1 \leq q_2.$$

The following reductions are the pstable possibilistic extensions of the reductions in Section 3.

---

[3] In the paper we will consider only finite lattices.

Definition 8 (Reduction $PRED(r, M)_\times$)
Let $r = \alpha : C_1 \times \ldots \times C_n \leftarrow \mathcal{B}^+$, not $\mathcal{B}^-$ be a possibilistic ordered disjunction clause and M be a set of atoms. The $\times$-possibilistic reduct $PRED(r, M)_\times$ of r is defined as follows: $PRED(r, M)_\times := \{\alpha : C_i \leftarrow \mathcal{B}^+, \text{ not } (\mathcal{B}^- \cap M) | C_i \in M \text{ and } M \cap \{C_1, \ldots, C_{i-1}\} = \emptyset\}$.

Definition 9 (Reduction $PRED(P, M)_\times$)
Let $P = \langle(Q, \leq), N\rangle$ be a possibilistic ordered disjunction program and M be a set of atoms. The $\times$-possibilistic reduct $PRED(P, M)_\times$ of P is defined as follows: $PRED(P, M)_\times = \bigcup_{r \in N} PRED(r, M)_\times$.

Example 3. Let P be a possibilistic ordered disjunction program such that $Q = (\{0, 0.1, \ldots, 0.9, 1\}, \leq)$, $\leq$ be the standard relation between rational numbers and $\alpha \in Q$ associated to each rule:

$r_1 = \mathbf{0.3} : a \times b.$ $r_2 = \mathbf{0.8} : b \times a \leftarrow \text{not } a.$

Let $M_1 = \{(a, 0.8)\}$ and $M_2 = \{(b, 0.8)\}$ be sets of possibilistic atoms. Taking M_1 as M we can see that $PRED(P, M)_\times$ is:

$r_1 = \mathbf{0.3} : a.$ $r_2 = \mathbf{0.8} : a \leftarrow \text{not } a.$

By considering the reduction $PRED(P, M)_\times$, we define the possibilistic pstable semantics as follows, which allows to test whether M is a possibilistic pstable model of a possibilistic ordered disjunction program P.

Definition 10 (Possibilistic Pstable Semantics)
Let $P = \langle(Q, \leq), N\rangle$ be a possibilistic ordered disjunction program and M be a set of possibilistic atoms such that M^* is a pstable model of P^*. M is a possibilistic pstable model of P if and only if $PRED(P, M^*)_\times \vdash_{PL} M$ and $\nexists M' \in \mathcal{PS}$ such that $M' \neq M$, $PRED(P, M'^*)_\times \vdash_{PL} M'$ and $M \sqsubseteq M'$.

Example 4. Let us continue with Example 3. We have already reduced the program and now we want to check if M is a possibilistic pstable model of P. First of all it is easy to see that M^* is a pstable model of P^*. Hence, we have to construct a proof in possibilistic logic for the possibilistic atom $(a, 0.8)$.

Let us prove $(a, 0.8)$ from $PRED(P, M^*)_\times$.[4]

Premises from $PRED(P, M^*)_\times$ **From 3 and 1 by S**
1. a **0.3**   4. a **0.8**
2. $\sim a \to a$ **0.8**
From 2 by possibilistic logical equivalence
3. $a \vee a$ **0.8**

Therefore, we can say that $PRED(P, M^*)_\times \vdash_{PL} M$ is true (and similarly it can be proved that $M_2 = \{(b, 0.8)\}$ is a pstable model of P). Notice that it does not exist a possibilistic set M' such that $M' \neq M$, $PRED(P, M^*)_\times \vdash_{PL} M'$ and $M \sqsubseteq M'$, hence we can conclude that M is a possibilistic pstable model of P.

---

[4] When we treat a logic program as a theory, each negative literal *not a* is replaced by $\sim a$, such that $\sim$ is regarded as the negation in classical logic.

From Definition 10, we can observe that there is an important condition $w.r.t$ the definition of a *possibilistic pstable model* of a possibilistic ordered disjunction program.

Proposition 1. *Let* $P = \langle (Q, \leq), N \rangle$ *be a possibilistic ordered disjunction program and* M *be a set of possibilistic atoms. If* M *is a possibilistic pstable model of* P *then* M^* *is a pstable model of* P^*.

Please observe that the pstable semantics for possibilistic ordered disjunction programs is different from the semantics defined in [3] as there are programs where the possibilistic pstable models do not correspond with the possibilistic answer sets. Nevertheless we can identify a relationship between the two semantics.

Proposition 2. *Let* P *be a possibilistic ordered disjunction program and* M *be a set of possibilistic atoms. If* M *is a possibilistic answer set of* P, *then the following conditions hold: (a)* M^* *is a pstable model of* P, *(b) there exists a possibilistic pstable model* M' *of* P *such that* $M \sqsubseteq M'$ *and* $M^* = M'^*$.

The proposition states that, whenever a possibilistic ordered disjunction program P has a possibilistic answer set M, there exists a possibilistic pstable model M' such that the main differences between M and M' are the necessity-values of their elements.

Example 5. Let P be the possibilistic ordered disjunction program of Example 3. We want to show how, under pstable semantics, its possibilistic stable models are $M_1 = \{(a, 0.8)\}$ and $M_2 = \{(b, 0.8)\}$ while, under answer set semantics, its possibilistic answer sets are $M_1 = \{(a, 0.3)\}$ and $M_2 = \{(b, 0.3)\}$. For doing so let us consider the reductions $PRED(P, M)_\times$ (Definition 9) and $P^M_\times$ (Definition 8 in [3]) and apply them using M_1:

$PRED(P, M_1)_\times$: $P^{M_1}_\times$:
$r_1 = \mathbf{0.3} : a.$ $r_1 = \mathbf{0.3} : a.$
$r_2 = \mathbf{0.8} : a \leftarrow not\ a.$

We can see that under answer set semantics we have to cut the inconsistent rule r_2, which triggers a lowering of the necessity degree of the possibilistic answer set, whereas by pstable semantics r_2 is kept and a higher necessity-value is inferred. This result follows from the fact that pstable semantics is closer to possibilistic logic inference than answer set semantics and it can lead to a maximum value choice.

4.3 Preferred Possibilistic Pstable Models

To distinguish between preferred possibilistic pstable models, we take the definition and the notation of satisfaction degree of M $w.r.t$ a rule r as $deg_M(r)$ (see Section 2.4). In the following we define three comparison criteria (as in [2]), adapted to possibilistic pstable models.

Definition 11. *Let* M_1 *and* M_2 *be possibilistic pstable models of a possibilistic ordered disjunction logic program* P. M_1 *is possibilistic cardinality-preferred to* M_2, $(M_1 >_{pc} M_2)$ *iff* $\exists\ i$ *such that* $|\ M_1^{i,\alpha}(P)\ | > |\ M_2^{i,\alpha}(P)\ |$ *and* $\forall j < i$, $|\ M_1^{j,\alpha}(P)\ | = |\ M_2^{j,\alpha}(P)\ |$, *where* $\alpha = min\{n(M_1), n(M_2)\}$.

Definition 12. *Let M_1 and M_2 be possibilistic pstable models of a possibilistic ordered disjunction logic program P. M_1 is possibilistic inclusion-preferred to M_2, $(M_1 >_{pi} M_2)$ iff $\exists\ k$ such that $M_2^{k,\alpha}(P) \subset M_1^{k,\alpha}(P)$ and $\forall\ j < k$, $M_1^{j,\alpha}(P) = M_2^{j,\alpha}(P)$, where $\alpha = min\{n(M_1), n(M_2)\}$.*

Definition 13. *Let M_1 and M_2 be possibilistic pstable models of a possibilistic ordered disjunction logic program P. M_1 is possibilistic pareto-preferred to M_2, $(M_1 >_{pp} M_2)$ iff $\exists\ r \in P$ such that $deg_{M_1}(r) < deg_{M_2}(r)$, and $\nexists\ r' \in P$ such that $deg_{M_1}(r') > deg_{M_2}(r')$, and $n(r) \geq min\{n(M_1), n(M_2)\}$.*

Notice that, applying the pstable semantics, we are able to keep higher degree of necessity-values formulæ. Let us consider the following examples which compare pstable semantics and answer semantics whenever computing preferred pstable models and answer sets respectively. The first example shows how pstable semantics allow to compare more than answer set semantics, while the second example shows how sometimes pstable models are differently preferred than answer sets due to higher cuts produced in the comparison criteria.

Example 6. Let P be the following possibilistic ordered disjunction program:

$$r_1 = \mathbf{0.5} : a \times b \leftarrow not\ c. \quad r_2 = \mathbf{0.5} : b \times a \leftarrow not\ d. \quad r_3 = \mathbf{0.8} : b \times a \leftarrow not\ a.$$

We can first see that $M_1^{pas} = \{(a, 0.5)\}$ and $M_2^{pas} = \{(b, 0.5)\}$ are possibilistic answer sets of P and $M_1^{pstable} = \{(a, 0.8)\}$ and $M_2^{pstable} = \{(b, 0.8)\}$ are possibilistic pstable models of P. We can notice that under answer set semantics no cut is produced in the comparison criteria and M_1^{pas} and M_2^{pas} are not comparable, while considering pstable semantics, the possibilistic necessity-value 0.8 derived from the pstable models triggers a cut of r_1 and r_2. Hence, we can conclude that $M_2^{pstable}$ is preferred to $M_1^{pstable}$ according to all the criteria.

Example 7. Let P be the possibilistic ordered disjunction program of Example 6 with a different necessity-value in rule r_1:

$$r_1 = \mathbf{0.3} : a \times b \leftarrow not\ c. \quad r_2 = \mathbf{0.5} : b \times a \leftarrow not\ d. \quad r_3 = \mathbf{0.8} : b \times a \leftarrow not\ a.$$

As in Example 6 it can be proved that $M_1^{pas} = \{(a, 0.5)\}$ and $M_2^{pas} = \{(b, 0.5)\}$ are possibilistic answer sets of P and $M_1^{pstable} = \{(a, 0.8)\}$ and $M_2^{pstable} = \{(b, 0.8)\}$ are possibilistic pstable models of P. However in this case the necessity-values allow to prefer M_1^{pas} to M_2^{pas} under answer set semantics, while considering pstable semantics the preference relation is inverted and $M_2^{pstable}$ is preferred to $M_1^{pstable}$. This can be justified by the fact that the pstable semantics allow to reach a higher degree of the necessity-values in the possibilistic pstable models and a higher cut is produced.

5   Conclusions

In this paper we have applied pstable semantics to ordered disjunction programs and extended it for capturing possibilistic ordered disjunction programs. The main advantages shown by our approach are essentially that: (a) using pstable semantics (which is based on paraconsistent logic) it is possible to treat inconsistent ordered disjunction rules, thus inconsistent ordered disjunction programs can have a solution, while under answer set semantics they may not have; (b) as pstable semantics is closer to possibilistic logic inference, we are able to reach a higher degree of the necessity-value of possibilistic pstable models as it allows to take the maximum; and finally it allows to compare more models of a program because a higher cut is produced and we have been able to compare possibilistic pstable models whereas possibilistic answer sets are not comparable. Moreover, the semantics used in our approach is a generalization of the ordered disjunction program semantics in [2], as any answer sets of an ordered disjunction program, are pstable models as well (Theorem 1).

The two extensions we defined are computable, and as future work we aim to implement them. In parallel, we are exploring the use of ordered disjunction programs with our semantics in a realistic application to represent and reason about user preferences in the presence of uncertainty to enhance the Web service selection process [3].

Acknowledgements. This work is supported in part by the European Commission ALIVE project (FP7-215890). Javier Vázquez-Salceda is partially supported by the Ramón y Cajal program of the Spanish Ministry of Education and Science. We are grateful to anonymous referees for their useful comments.

References

1. Baral, C.: Knowledge Representation, Reasoning and Declarative Problem Solving. Cambridge University Press, Cambridge (2003)
2. Brewka, G., Niemelä, I., Syrjänen, T.: Logic Programs with Ordered Disjunction. Computational Intelligence 20(2), 333–357 (2004)
3. Confalonieri, R., Nieves, J.C., Vázquez-Salceda, J.: Logic Programs with Possibilistic Ordered Disjunction. Research Report LSI-09-19-R, UPC - LSI (2009)
4. Dubois, D., Lang, J., Prade, H.: Possibilistic Logic. In: Handbook of Logic in Artificial Intelligence and Logic Programming, vol. 3, pp. 439–513. Oxford University Press, Oxford (1994)
5. Gelfond, M., Lifschitz, V.: Classical Negation in Logic Programs and Disjunctive Databases. New Generation Computing 9(3/4), 365–386 (1991)
6. Nicolas, P., Garcia, L., Stéphan, I., Lafèvre, C.: Possibilistic Uncertainty Handling for Answer Set Programming. Annals of Mathematics and Artificial Intelligence 47(1-2), 139–181 (2006)
7. Osorio, M., Arrazola Ramírez, J.R., Carballido, J.L.: Logical Weak Completions of Paraconsistent Logics. Journal of Logic and Computation 18(6), 913–940 (2008)
8. Osorio, M., Pérez, J.A.N., Ramírez, J.R.A., Macías, V.B.: Logics with Common Weak Completions. Journal of Logic and Computation 16(6), 867–890 (2006)

Reasoning about Typicality with Low Complexity Description Logics: The Logic $\mathcal{EL}^{+\perp}\mathbf{T}$

Laura Giordano[1], Valentina Gliozzi[2], Nicola Olivetti[3], and Gian Luca Pozzato[2]

[1] Dip. di Informatica - Università del Piemonte Orientale
laura@mfn.unipmn.it
[2] Dip. di Informatica - Università di Torino
{gliozzi,pozzato}@di.unito.it
[3] LSIS-UMR CNRS 6168 Univ. "P. Cézanne"
nicola.olivetti@univ-cezanne.fr

Abstract. We present an extension of the low complexity Description Logic $\mathcal{EL}^{+\perp}$ for reasoning about prototypical properties and inheritance with exceptions. We add to $\mathcal{EL}^{+\perp}$ a typicality operator $\mathbf{T}$, which is intended to select the "most normal" instances of a concept. In the resulting logic, called $\mathcal{EL}^{+\perp}\mathbf{T}$, the knowledge base may contain subsumption relations of the form "$\mathbf{T}(C)$ is subsumed by P", expressing that typical C-members have the property P. We show that the problem of entailment in $\mathcal{EL}^{+\perp}\mathbf{T}$ is in CO-NP by proving a small model result.

1 Introduction

In Description Logics (DLs), the need of representing prototypical properties and of reasoning about defeasible inheritance of such properties naturally arises. The traditional approach to address this problem is to handle defeasible inheritance by integrating some kind of nonmonotonic reasoning mechanism. This has led to the study of nonmonotonic extensions of DLs [3,4,6,7,8,15]. However, finding a suitable nonmonotonic extension for inheritance with exceptions is far from obvious.

In this work we continue our investigation started in [9], where we have proposed the logic $\mathcal{ALC}+\mathbf{T}$ for defeasible reasoning in the description logic $\mathcal{ALC}$. $\mathcal{ALC}+\mathbf{T}$ is obtained by adding a typicality operator $\mathbf{T}$ to $\mathcal{ALC}$. The intended meaning of the operator $\mathbf{T}$ is that, for any concept C, $\mathbf{T}(C)$ singles out the instances of C that are considered as "typical" or "normal". Thus assertions as "typical football players love football" are represented by $\mathbf{T}(FootballPlayer) \sqsubseteq FootballLover$. The semantics of the typicality operator $\mathbf{T}$ turns out to be strongly related to the semantics of nonmonotonic entailment in KLM logic $\mathbf{P}$ [14].

In our setting, we assume that the TBox element of a KB comprises, in addition to the standard concept inclusions, a set of inclusions of the type $\mathbf{T}(C) \sqsubseteq D$ where D is a concept not mentioning $\mathbf{T}$. For instance, a KB may contain: $\mathbf{T}(Dog) \sqsubseteq Affectionate$; $\mathbf{T}(Dog) \sqsubseteq CarriedByTrain$; $\mathbf{T}(Dog \sqcap PitBull) \sqsubseteq NotCarriedByTrain$; $CarriedByTrain \sqcap NotCarriedByTrain \sqsubseteq \perp$, corresponding

R. Serra and R. Cucchiara (Eds.): AI*IA 2009, LNAI 5883, pp. 62–71, 2009.
© Springer-Verlag Berlin Heidelberg 2009

to the assertions: typically dogs are affectionate, normally dogs can be transported by train, whereas typically a dog belonging to the race of pitbull cannot (since pitbulls are considered as reactive dogs); the fourth inclusion represents the disjointness of the two concepts *CarriedByTrain* and *NotCarriedByTrain*. Notice that, in standard DLs, replacing the second and the third inclusion with *Dog* $\sqsubseteq$ *CarriedByTrain* and *Dog* $\sqcap$ *PitBull* $\sqsubseteq$ *NotCarriedByTrain*, respectively, we would simply get that there are not pitbull dogs, thus the KB would collapse. This collapse is avoided as we do not assume that $\mathbf{T}$ is monotonic, that is to say $C \sqsubseteq D$ does not imply $\mathbf{T}(C) \sqsubseteq \mathbf{T}(D)$.

By the properties of $\mathbf{T}$, some inclusions are entailed by the above KB, as for instance $\mathbf{T}(Dog \sqcap CarriedByTrain) \sqsubseteq$ *Affectionate*. In our setting we can also use the $\mathbf{T}$ operator to state that some domain elements are typical instances of a given concept. For instance, an ABox may contain either $\mathbf{T}(Dog)(fido)$ or $\mathbf{T}(Dog \sqcap PitBull)(fido)$. In the two cases, the expected conclusions are entailed: *CarriedByTrain*(*fido*) and *NotCarriedByTrain*(*fido*), respectively.

In this work, we extend our approach based on the typicality operator to *low complexity* Description Logics, focusing on the logic $\mathcal{EL}^{+^{\perp}}$ of the well known $\mathcal{EL}$ family. The logics of the $\mathcal{EL}$ family allow for conjunction ($\sqcap$) and existential restriction ($\exists R.C$). Despite their relatively low expressivity, a renewed interest has recently emerged for these logics. Indeed, theoretical results have shown that $\mathcal{EL}$ has better algorithmic properties than its counterpart $\mathcal{FL}_0$, which allows for conjunction and value restriction ($\forall R.C$). Also, it has turned out that the logics of the $\mathcal{EL}$ family are relevant for several applications, in particular in the bio-medical domain; for instance, medical terminologies, such as the GALEN Medical Knowledge Base, the Systemized Nomenclature of Medicine, and the Gene Ontology, can be formalized in small extensions of $\mathcal{EL}$.

We present a *small model* result for the logic $\mathcal{EL}^{+^{\perp}} \mathbf{T}$. More precisely, we show that, given an $\mathcal{EL}^{+^{\perp}} \mathbf{T}$ knowledge base, KB, if KB is satisfible, then there is a *small* model satisfying KB, whose size is polynomial in the size of KB. The construction of the model exploits the facts that (1) it is possible to reuse the same domain element (instance of a concept C) to fulfill existential formulas $\exists r.C$ w.r.t. domain elements; (2) we can restrict our attention to a specific class of models which are called multi-linear and include a polynomial number of chains of elements of polynomial length. The construction of the model allows us to conclude that the problem of deciding entailment in $\mathcal{EL}^{+^{\perp}} \mathbf{T}$ is in CO-NP. For reasoning about *irrelevance*, we deal with the inheritance of defeasible properties by introducing default rules. Given the complexity of entailment in $\mathcal{EL}^{+^{\perp}} \mathbf{T}$, credulous reasoning in the resulting default theory is in Σ_2^p. Preliminary results about this work have been presented in [12].

2   The Logic $\mathcal{EL}^{+^{\perp}} \mathbf{T}$

We consider an alphabet of concept names $\mathcal{C}$, of role names $\mathcal{R}$, and of individuals $\mathcal{O}$. The language $\mathcal{L}$ of the logic $\mathcal{EL}^{+^{\perp}} \mathbf{T}$ is defined by distinguishing *concepts* and *extended concepts* as follows: (Concepts) $A \in \mathcal{C}$, $\top$, and $\perp$ are *concepts* of $\mathcal{L}$; if

$C, D \in \mathcal{L}$ and $r \in \mathcal{R}$, then $C \sqcap D$ and $\exists r.C$ are *concepts* of $\mathcal{L}$. (Extended concepts) if C is a concept, then C and $\mathbf{T}(C)$ are extended concepts of $\mathcal{L}$. A knowledge base is a pair (TBox,ABox). TBox contains (i) a finite set of GCIs $C \sqsubseteq D$, where C is an extended concept (either C' or $\mathbf{T}(C')$), and D is a concept, and (ii) a finite set of role inclusions (RIs) $r_1 \circ r_2 \circ \cdots \circ r_n \sqsubseteq r$. ABox contains expressions of the form $C(a)$ and $r(a, b)$ where C is an extended concept, $r \in \mathcal{R}$, and $a, b \in \mathcal{O}$. In order to provide a semantics to the operator $\mathbf{T}$, we extend the definition of a model used in "standard" terminological logic $\mathcal{EL}^{+\perp}$:

Definition 1 (Semantics of T). *A model $\mathcal{M}$ is any structure $\langle \Delta, <, I \rangle$, where Δ is the domain; $<$ is an irreflexive and transitive relation over Δ, and satisfies the following* Smoothness Condition: *for all $S \subseteq \Delta$, for all $a \in S$, either $a \in Min_<(S)$ or $\exists b \in Min_<(S)$ such that $b < a$, where $Min_<(S) = \{a : a \in S$ and $\nexists b \in S$ s.t. $b < a\}$. I is the extension function that maps each extended concept C to $C^I \subseteq \Delta$, and each role r to a $r^I \subseteq \Delta^I \times \Delta^I$. For concepts of $\mathcal{EL}^{+\perp}$, C^I is defined in the usual way. For the $\mathbf{T}$ operator: $(\mathbf{T}(C))^I = Min_<(C^I)$. A model satisfying a KB (TBox,ABox) is defined as usual. Moreover, we assume the unique name assumption.*

Notice that the meaning of $\mathbf{T}$ can be split into two parts: for any a of the domain Δ, $a \in (\mathbf{T}(C))^I$ just in case (i) $a \in C^I$, and (ii) there is no $b \in C^I$ such that $b < a$. In order to isolate the second part of the meaning of $\mathbf{T}$, we introduce a new modality $\square$. The basic idea is simply to interpret the preference relation $<$ as an accessibility relation. By the Smoothness Condition, it turns out that $\square$ has the properties as in Gödel-Löb modal logic of provability G. The interpretation of $\square$ in $\mathcal{M}$ is as follows: $(\square C)^I = \{a \in \Delta \mid$ for every $b \in \Delta$, if $b < a$ then $b \in C^I\}$. We have that a is a typical instance of C ($a \in (\mathbf{T}(C))^I$) iff $a \in C^I$ and, for all $b < a$, $b \notin C^I$, namely we have that $a \in (\mathbf{T}(C))^I$ iff $a \in (C \sqcap \square \neg C)^I$. From now on, we consider $\mathbf{T}(C)$ as an abbreviation for $C \sqcap \square \neg C$. The Smoothness Condition ensures that typical elements of C^I exist whenever $C^I \neq \emptyset$, by preventing infinitely descending chains of elements.

3   Constructing Small Models for $\mathcal{EL}^{+\perp}\mathbf{T}$

We can show that, given a model $\mathcal{M} = \langle \Delta, <, I \rangle$ of a KB, we can build a *small* model of KB whose size is polynomial in the size of the KB. As we will see, this will provide a complexity upper bound for the logic $\mathcal{EL}^{+\perp}\mathbf{T}$.

First of all, we must introduce an appropriate normal form[1] for KBs, in particular for TBoxes. Given a KB=(TBox,ABox), we say that it is normal if:

- all the inclusion relations in TBox have one of the following forms: $C_1 \sqsubseteq D$; $C_1 \sqcap C_2 \sqsubseteq D$; $C_1 \sqsubseteq \exists r.C_2$; $\exists r.C_1 \sqsubseteq D$; $\mathbf{T}(C_1) \sqsubseteq C_2$; $\mathbf{T}(C_1 \sqcap C_2) \sqsubseteq D$; $\mathbf{T}(C_1) \sqsubseteq \exists r.C_2$; $\mathbf{T}(\exists r.C_1) \sqsubseteq D$, where $C_1, C_2 \in \mathcal{C} \cup \{\top\}$ and $D \in \mathcal{C} \cup \{\bot\}$;
- all role inclusions in TBox are of the form $r \sqsubseteq s$ or $r_1 \circ r_2 \sqsubseteq s$.

---

[1] The normal form presented in this paper is not *minimal*. We could further reduce the normal form without reducing the expressivity of the language.

By extending the results presented in [1], we can show that any KB can be turned into a normalized KB' that is a *conservative* extension of KB, that is to say every model satisfying KB' is also a model of KB, whereas every model of KB can be extended to a model of KB' by appropriately choosing the interpretations of the additional concept and role names introduced by the normalization procedure. Furthermore, it can be shown that the size of KB' is linear in the size of KB, and that the normalization procedure can be done in linear time. Without loss of generality, from now on we only refer to normalized KBs. Starting from a normalized KB, we can now prove the following theorem, whose proof lasts until the end of the section.

Theorem 1 (Small model theorem). *Let KB=(TBox,ABox) be an $\mathcal{EL}^{+\perp}\mathbf{T}$ knowledge base. For all models $\mathcal{M} = \langle \Delta, <, I \rangle$ of KB and all $x \in \Delta$, there exists a model $\mathcal{N} = \langle \Delta^{\circ}, <^{\circ}, I^{\circ} \rangle$ of KB such that (i) $x \in \Delta^{\circ}$, (ii) for all $\mathcal{EL}^{+\perp}\mathbf{T}$ concepts C, $x \in C^{I}$ iff $x \in C^{I}$, and (iii) $\mid \Delta^{\circ} \mid$ is polynomial in the size of KB.*

We sketch the proof through the following four steps. In the first step, in order to reduce the size of the model, we cut a portion of it that includes x. In particular, we keep only those domain elements needed to retain the values of formulas in x. Intuitively, we reuse the same domain element to make true existential formulas in different domain elements. Moreover, we need to add new elements to the domain of the constructed model, in order to keep the same evaluation of existential formulas as in the initial model. As it is not guaranteed that the model obtained has a polynomial size in the size of the KB, we have to refine this construction. Our goal is to obtain a multi-linear model (Definition 2 below), that can be further transformed into a model of polynomial size. The second and the third steps are devoted to build a multi-linear model. In the fourth step we show that we can further reduce the size of the model by shortening the length of the linear descending chains to a polynomial size.

(STEP 1). We build a model $\mathcal{M}'$ by means of the following construction. For each atomic concept $C \in \mathcal{C}$ and for each role $r \in \mathcal{R}$ we let $S(C)$ and $R(r)$ be the mappings computed by the algorithm defined in [2] to compute subsumption by means of completion rules, whose meaning is that $D \in S(C)$ implies $C \sqsubseteq D$ and $(C, D) \in R(r)$ implies $C \sqsubseteq \exists r.D$. As usual, for a given individual a in the ABox, we write a^{I} to denote the element of Δ corresponding to the extension of a in $\mathcal{M}$. We make use of three sets of elements: Δ_0 will be part of the domain of the model being constructed, and it contains a portion of the domain Δ of the initial model. All elements introduced in the domain must be processed in order to satisfy the existential formulas. *Unres* is used to keep track of not yet processed elements. Finally, Δ_1 is a set of new elements that will belong to the domain of the constructed model. Each element w_C of Δ_1 is created for a corresponding atomic concept C and is used to satisfy any existential formula $\exists r.C$ throughout the model. In the following, by w_C we mean the element of Δ_1 which is added for the atomic concept C. We provide an algorithmic description of the construction of model $\mathcal{M}'$ from the given model $\mathcal{M}$. Observe that $\mathcal{M}$ can be an infinite model.

1. $\Delta_0 := \{x\} \cup \{a^I \in \Delta \mid a \text{ occurs in the ABox }\}$
2. $Unres := \{x\} \cup \{a^I \in \Delta \mid a \text{ occurs in the ABox }\}$
3. $\Delta_1 := \emptyset$
4. **while** $Unres \neq \emptyset$ **do**
5.   extract one y from $Unres$
6.   **for each** $\exists r.C$ occurring in KB s.t. $y \in (\exists r.C)^I$ **do**
7.   **if** $\nexists w_C \in \Delta_1$ **then**
8.   choose $w \in \Delta$ s.t. $(y, w) \in r^I$ and $w \in C^I$
9. $\Delta_0 := \Delta_0 \cup \{w\}$
10. $Unres := Unres \cup \{w\}$
11.   create a new element w_C associated with C
12. $\Delta_1 := \Delta_1 \cup \{w_C\}$
13.   add $w <' w_C$
14.   add (y, w_C) to $r^{I'}$
15.   **else**
16.   add (y, w_C) to $r^{I'}$
17.   **for each** $y_i \in \Delta$ such that $y_i < y$ **do**
18. $\Delta_0 := \Delta_0 \cup \{y_i\}$
19. $Unres := Unres \cup \{y_i\}$
20. **for each** $w_C, w_D \in \Delta_1$ with $C \neq D$ **do**
21.   **if** $(C, D) \in R(r)$ **then** add (w_C, w_D) to $r^{I'}$

The model $\mathcal{M}' = \langle \Delta', <', I' \rangle$ is defined as follows:

- $\Delta' = \Delta_0 \cup \Delta_1$
- we extend $<'$ computed by the algorithm by adding $u <' v$ if $u < v$, for each $u, v \in \Delta'$;
- the extension function I' is defined as follows:
    - for all atomic concepts $C \in \mathcal{C}$, for all elements in Δ', we define:
        * for each $u \in \Delta_0$, we let $u \in C^{I'}$ if $u \in C^I$;
        * for each $w_D \in \Delta_1$, we let $w_D \in C^{I'}$ if $C \in S(D)$.
    - for all roles r, we extend $r^{I'}$ constructed by the algorithm by means of the following role closure rules:
        * for all inclusions $r \sqsubseteq s \in$ TBox, if $(u, v) \in r^{I'}$ then add (u, v) to $s^{I'}$;
        * for all inclusions $r_1 \circ r_2 \sqsubseteq s \in$ TBox, if $(u, v) \in r_1^{I'}$ and $(v, w) \in r_2^{I'}$ then add (u, w) to $s^{I'}$.
    - I' is extended so that it assigns a^I to each individual a in the ABox.

It is easy to see that relation $<'$ is irreflexive, transitive and satisfies the Smoothness Condition. Moreover, by induction on the structure of C, it can be proved that:

Lemma 1. *Given any $\mathcal{EL}^{+\perp}\mathbf{T}$ concept C occurring in KB, for all $y \in \Delta_0$, $y \in C^I$ iff $y \in C^{I'}$.*

By making use of the above lemma, we can show that:

Lemma 2. $\mathcal{M}'$ *is a model of KB.*

$\mathcal{M}'$ is not guaranteed to have polynomial size in the KB because in line 18 we add an element y_i for each $y_i < y$, then the size of Δ_0 may be arbitrarily large. For this reason, we refine our construction in order to build a multi-linear model, that we will be able to further refine in order to obtain a model of polynomial size. In STEP 2, we replicate some domain elements, namely those belonging to more than one descending chain of $<$.

(STEP 2). We build a model $\mathcal{M}'' = \langle \Delta'', <'', I'' \rangle$ from $\mathcal{M}' = \langle \Delta', <', I' \rangle$. We define $\Delta'_s = \{x\} \cup \{u \in \Delta' \mid \nexists v \in \Delta' \text{ such that } u <' v\}$ to be the set containing x and all the domain elements in Δ' that are not preferred to any other element according to $<'$. Let $k = \mid \Delta'_s \mid$. For each $y_j \in \Delta'_s$ we define $\Delta_{y_j} = \{z \in \Delta' \mid z <' y_j\} \cup \{y_j\}$ the set of all domain elements preferred to y_j. As two sets Δ_{y_j} and Δ_{y_i} with $y_j \neq y_i$, are not guaranteed to be disjoint, we rename the elements of each set Δ_{y_j}, as follows. We consider for each $j = 1, \ldots, k$ a renaming function (i.e. a bijection) f_j whose domain is Δ_{y_j} that makes a copy $\Delta_{f_j(y_j)}$ of Δ_{y_j} which is (i) disjoint from any Δ_{y_l}, and (ii) disjoint from any other $\Delta_{f_l(y_l)}$ with $l \neq j$. Moreover, for $y_j = x$, we let $x = f_j(x)$, that is, x is not renamed in Δ_x.

We define a model $\mathcal{M}'' = \langle \Delta'', <'', I'' \rangle$ as follows: $\Delta'' = \Delta_{f_1(y_1)} \cup \Delta_{f_2(y_2)} \ldots \cup \Delta_{f_k(y_k)}$. The relation $<''$ is defined as follows: $u <'' v$ iff $u, v \in \Delta_{f_j(y_j)}$ so that $u = f_j(z)$ and $v = f_j(w)$, where $z, w \in \Delta_{y_j}$ and $z <' w$. Observe that elements in different components Δ_{y_j} are incomparable w.r.t. $<''$. Finally, given any atomic concept $C \in \mathcal{C}$, for $u \in \Delta_{f_j(y_j)}$ with $u = f_j(w)$, we let $u \in C^{I''}$ iff $w \in C^{I'}$. Moreover, given any role $r \in \mathcal{R}$, for $u \in \Delta_{f_j(y_j)}$ with $u = f_j(w)$ and $v \in \Delta_{f_i(y_i)}$ with $v = f_i(z)$, we let $(u, v) \in r^{I''}$ iff $(w, z) \in r^{I'}$. It can proved that:

Lemma 3. $\mathcal{M}''$ *is a model of KB and, given any concept C of $\mathcal{EL}^{+\perp}\mathbf{T}$, $x \in C^{I''}$ iff $x \in C^{I'}$.*

(STEP 3). First of all, we introduce the notion of *multi-linear* model of a KB. Intuitively, a model $\mathcal{M} = \langle \Delta, <, I \rangle$ is multi-linear if the relation $<$ forms a set of chains of domain elements, that is:

Definition 2 (Multi-linear model). *Given a model $\mathcal{M} = \langle \Delta, <, I \rangle$, we say that it is multi-linear if the following properties hold for every $u, v, z \in \Delta$:*

(i) if $u < z$ and $v < z$ and $u \neq v$, then $u < v$ or $v < u$;
(ii) if $z < u$ and $z < v$ and $u \neq v$, then $u < v$ or $v < u$.

We now define a multi-linear model $\mathcal{M}^* = \langle \Delta'', <^*, I'' \rangle$ as follows: we let $<^*$ be any total order on each $\Delta_{f_j(y_j)}$ which respects $<''$; the elements in different components remain incomparable. More precisely $<^*$ satisfies:

- if $u <'' v$ then $u <^* v$
- for each $u, v \in \Delta_{f_j(y_j)}$, with $u \neq v$, $u <^* v$ or $v <^* u$
- for each $u \in \Delta_{f_i(y_i)}, v \in \Delta_{f_j(y_j)}$, with $i \neq j$, $u \not<^* v$ and $v \not<^* u$

Again, we can prove that:

Lemma 4. $\mathcal{M}^*$ *is a model of KB and, given any concept C of $\mathcal{EL}^{+^\perp}\mathbf{T}$, $x \in C^I$ iff $x \in C^{I''}$.*

(**STEP 4**). We conclude the proof by constructing a model $\mathcal{N} = \langle \Delta^\circ, <^\circ, I^\circ \rangle$ whose domain has polynomial size in the size of KB. Let the size of the initial KB be n. We know that $\mathcal{M}^*$ contains a polynomial number of linear chains of domain elements related by $<^*$, each one starting from a domain elements in Δ_1 (built by the algorithm in STEP 1) or from one domain element in $\{x, a_1^I, \ldots, a_k^I\}$, where $a_1^I, \ldots, a_k^I$ are the domain elements corresponding to the individuals in the ABox. We know that there are $O(n)$ chains, as Δ_1 contains one domain element for each atomic concept in $\mathcal{EL}^{+^\perp}$ and the domain elements $a_1^I, \ldots, a_k^I$ are $O(n)$. However, we have no bound on the length of the chains.

We want to show that the linear chains in the model can be reduced to finite chains of polynomial length in the size of the KB. To this purpose, given $\mathcal{M}^*$, we build a new multi-linear model $\mathcal{N} = \langle \Delta^\circ, <^\circ, I^\circ \rangle$ whose descending chains have polynomial length.

Let us consider a chain $w_0, w_1, w_2, \ldots$ in the multilinear model $\mathcal{M}^*$. Observe that, given two elements w_i and w_j in the chain such that $w_i < w_j$, the set of negated box formulas $\neg\Box\neg C$ of which w_i is an instance is a subset of the set of negated box formulas of which w_j is an instance. We can thus shrink each chain by retaining only those elements w_i and w_j such that $w_i < w_j$ implies that there exists a formula $\neg\Box\neg C$ such that w_j is an instance of $\neg\Box\neg C$, while w_i is not. As there is only a finite polynomial number of such box formulas $\neg\Box\neg C$, each chain will contain only a polynomial number of elements.

The resulting model $\mathcal{N} = \langle \Delta^\circ, <^\circ, I^\circ \rangle$ is defined as follows: Δ° is the set of all the domain elements in Δ^* which have not been removed during the chain transformation process; the relation $<^\circ$ is defined so that, for all $x, y \in \Delta^*$, $x <^\circ y$ if and only if $x <^* y$; the interpretation of atomic concepts in the domain elements is left unchanged. It can be shown that $\mathcal{N}$ is a multi-linear model of the KB and that the valuation in x is the same in $\mathcal{N}$ and in $\mathcal{M}^*$. Since, the number of chains in $\mathcal{N}$ is polynomial in the size of the KB and each chain has polynomial length, the resulting model $\mathcal{N}$ has polynomial size. This concludes the proof of the Small model theorem (Theorem 1).

Given the small model theorem above, we can conclude that, when evaluating the entailment, we can restrict our consideration to small models, namely, to polynomial multi-linear models of the KB. As usual, we write KB $\models \alpha$ to say that a query α holds in all the models of the KB. A query α is either a formula of the form $C(a)$ or a subsumption relation $C \sqsubseteq D$. We write KB $\models_s \alpha$ to say that α holds in all polynomial multi-linear models of the KB.

Theorem 2. *KB $\models \alpha$ if and only if KB $\models_s \alpha$.*

We can prove an upper bound on the complexity of entailment in $\mathcal{EL}^{+^\perp}\mathbf{T}$.

Theorem 3 (Complexity entailment in $\mathcal{EL}^{+^\perp}\mathbf{T}$). *The problem of deciding whether KB $\models \alpha$ is in* CO-NP.

Proof. Let us consider the complementary problem of deciding whether KB $\not\models \alpha$. This problem can be solved by a nondeterministic polynomial time algorithm which guesses a model $\mathcal{N}$ of polynomial size and a domain element x of the model, and then checks in polynomial time that $\mathcal{N}$ is a model of the KB and that x falsifies α.

4   Reasoning about Irrelevance in $\mathcal{EL}^{+\perp}\mathbf{T}$

The logic $\mathcal{EL}^{+\perp}\mathbf{T}$ allows us to capture - through cautious monotonicity [14] - some form of inheritance of typical properties among concepts. For instance, from the KB in the introduction it is possible to conclude that typical dogs that are carried by train are affectionate, i.e., $\mathbf{T}(Dog \sqcap CarriedByTrain) \sqsubseteq Affectionate$. However, there are cases in which cautious monotonicity is not strong enough to derive the intended conclusions. For instance, we would like to conclude that typical red dogs are also affectionate. As the property of being red is not a property neither of all dogs, nor of typical dogs, cautious monotonicity is not applicable to conclude that typical red dogs are affectionate. This is the problem of irrelevance. As the color of a dog is irrelevant with respect to the property of being affectionate, and typical dogs are affectionate, we would like anyhow to conclude that $\mathbf{T}(Dog \sqcap Red) \sqsubseteq Affectionate$. To allow this form of inheritance among concepts, we can introduce default rules like the following:

$$\mathbf{T}(Dog) \sqsubseteq Affectionate \quad : \quad \mathbf{T}(Dog \sqcap Red) \sqsubseteq Affectionate$$
$$\mathbf{T}(Dog \sqcap Red) \sqsubseteq Affectionate$$

If typical dogs are affectionate, and *it is consistent* to assume that typical red dogs are affectionate, then we can conclude that typical red dogs are affectionate.

In general, if $C_1 \sqsubseteq C_2$, we expect that all the defeasible properties of C_2 are inherited by C_1, unless they are inconsistent with other properties of C_1. In the following, we enforce this requirement for all concepts C_2 minimally subsuming C_1, where C_1 and C_2 are any concepts occuring in the normalized KB. Let D be any concept occurring in the KB as the right hand side of an inclusion $T(C) \sqsubseteq D$. We introduce the following default rule:

$$\mathbf{T}(C_2) \sqsubseteq D \quad : \quad \mathbf{T}(C_1) \sqsubseteq D$$
$$\mathbf{T}(C_1) \sqsubseteq D$$

To allow default rules, we extend the KB to a defeasible KB including, in addition to an ABox and a TBox, a defeasible part consisting of a set of default rules. Observe that, as a difference with terminological default theories in [3], here prerequisites, justifications and consequents of default rules are inclusions, rather than concept terms. As defaults are normal, existence of an extension is guaranteed. Multiple extensions are possible due to multiple inheritance. As C_1, C_2 and D are subformulas of the initial KB, a polynomial number of default rules (namely, $O(n^3)$ rules) is needed. Given the complexity of entailment and satisfiability in $\mathcal{EL}^{+\perp}\mathbf{T}$ in the previous section, the verification that an inclusion (for instance, $\mathbf{T}(Dog \sqcap Red) \sqsubseteq Affectionate$) holds in an extension of a defeasible KB (credulous reasoning) is in Σ_2^p, while the verification that an inclusion holds in all the extensions of a defeasible KB (skeptical reasoning) is in Π_2^p.

5   Conclusions and Related Work

We have introduced the description logic $\mathcal{EL}^{+\perp}\mathbf{T}$, obtained by extending $\mathcal{EL}^{+\perp}$ with a tipicality operator $\mathbf{T}$ intended to select the "most normal" instances of a concept. Whereas for $\mathcal{ALC} + \mathbf{T}$ deciding satisfiability (subsumption) is EXP-TIME complete (see [11]), we have shown here that for $\mathcal{EL}^{+\perp}\mathbf{T}$ the complexity is significantly smaller, namely it reduces to NP for satisfiability (and CO-NP for subsumption). This result is obtained by a "small" model property (of a particular kind: multi-linear) that fails for the whole $\mathcal{ALC} + \mathbf{T}$ as well as for $\mathcal{ALC}$. We believe that this bound is also a lower bound, but we have not proved it so far. Although validity/satisfiability for KLM logic $\mathbf{P}$ is known to be (co)NP hard, in $\mathcal{EL}^{+\perp}\mathbf{T}$, we can only directly encode nonmonotonic assertions $A \sim B$ where A is a conjunction of atoms and B is either an atom or $\perp$. As far as we know, the complexity of this fragment of $\mathbf{P}$ is unknown. Thus a lower bound for $\mathcal{EL}^{+\perp}\mathbf{T}$ cannot be obtained from known results about KLM logic $\mathbf{P}$.

The logic $\mathcal{EL}^{+\perp}\mathbf{T}$ in itself is not sufficient for prototypical reasoning and inheritance with exceptions, in particular we need a stronger (nonmonotonic) mechanism to cope with the problem known as *irrelevance*. Concerning the example of the introduction, we would like to conclude that typical red dogs are affectionate, since the color of a dog is irrelevant with respect to the property of being affectionate. However, as the property of being red is not a property neither of all dogs, nor of typical dogs, in $\mathcal{EL}^{+\perp}\mathbf{T}$ we are not able to conclude $\mathbf{T}(Dog \sqcap Red) \sqsubseteq Affectionate$. One possibility is to consider a stronger (nonmonotonic) entailment relation $\mathcal{EL}^{+\perp}\mathbf{T}_{min}$ determined by restricting the entailment of $\mathcal{EL}^{+\perp}\mathbf{T}$ to "minimal models", as defined in [10] for $\mathcal{ALC} + \mathbf{T}$. Intuitively, minimal models are those that maximise "typical instances" of a concept. As shown in [10], for $\mathcal{ALC} + \mathbf{T}_{min}$, minimal entailment can be decided in CO-NEXP^{NP}. We believe that for $\mathcal{EL}^{+\perp}\mathbf{T}_{min}$ we can obtain a smaller complexity upper bound on the base of the results presented here.

Several approaches to handle prototypical reasoning and inheritance with exceptions in DL have been proposed in the literature, all of them are based on the integration of DLs with some nonmonotonic reasoning mechanism: either default logic (see [3,15,4]), or autoepistemic logic (see [7] and [13] for some recents developmentes) containing two epistemic operators, or finally circumscription (see [6] and [5]). As already observed, the use of normal defaults proposed in the previous section is rather different from [3]: we use defaults for handling irrelevant data and not to ascribe prototypical properties to individuals. Circumscription can model inheritance with exeception by means of a set of abnormability predicates. In [5] circumscription is applied to low complexity DLs. Concerning $\mathcal{EL}^{\perp}$ the authors have shown that all reasoning tasks remain EXPTIME-hard, whereas they fall in the second level of the polynomial hierarchy for a restricted type of KB called *left local*. In future work we will deal with the precise relation between our $\mathbf{T}$-DLs with the other nonmonotonic extensions of DLs mentioned above, notably with circumscription. In this setting, a natural question is to compare $\mathcal{EL}^{+\perp}\mathbf{T}_{min}$ with circumscribed $\mathcal{EL}^{\perp}$ and see whether we get the same complexity bounds or not.

Acknowledgements. The work has been partially supported by Regione Piemonte, Project ICT4Law - *ICT Converging on Law: Next Generation Services for Citizens, Enterprises, Public Administration and Policymakers"*.

References

1. Baader, F., Brandt, S., Lutz, C.: Pushing the $\mathcal{EL}$ envelope. In: LTCS-Report LTCS-05-01, Inst. for Theoretical Compute Science, TU Dresden (2005), http://lat.inf.tudresden.de/research/reports.html
2. Baader, F., Brandt, S., Lutz, C.: Pushing the $\mathcal{EL}$ envelope. In: Proc. of IJCAI 2005, Professional Book Center, pp. 364–369 (2005)
3. Baader, F., Hollunder, B.: Embedding defaults into terminological knowledge representation formalisms. J. Autom. Reasoning 14(1), 149–180 (1995)
4. Baader, F., Hollunder, B.: Priorities on defaults with prerequisites, and their application in treating specificity in terminological default logic. J. of Automated Reasoning (JAR) 15(1), 41–68 (1995)
5. Bonatti, P., Faella, M., Sauro, L.: Defeasible inclusions in low-complexity dls: Preliminary notes. In: Proc. of IJCAI 2009 (2009)
6. Bonatti, P.A., Lutz, C., Wolter, F.: Description logics with circumscription. In: Proc. of KR, pp. 400–410 (2006)
7. Donini, F.M., Nardi, D., Rosati, R.: Description logics of minimal knowledge and negation as failure. ACM Trans. Comput. Log. 3(2), 177–225 (2002)
8. Eiter, T., Lukasiewicz, T., Schindlauer, R., Tompits, H.: Combining answer set programming with description logics for the semantic web. In: KR 2004, pp. 141–151 (2004)
9. Giordano, L., Gliozzi, V., Olivetti, N., Pozzato, G.L.: Preferential Description Logics. In: Dershowitz, N., Voronkov, A. (eds.) LPAR 2007. LNCS (LNAI), vol. 4790, pp. 257–272. Springer, Heidelberg (2007)
10. Giordano, L., Gliozzi, V., Olivetti, N., Pozzato, G.L.: Reasoning About Typicality in Preferential Description Logics. In: Hölldobler, S., Lutz, C., Wansing, H. (eds.) JELIA 2008. LNCS (LNAI), vol. 5293, pp. 192–205. Springer, Heidelberg (2008)
11. Giordano, L., Gliozzi, V., Olivetti, N., Pozzato, G.L.: On Extending Description Logics for Reasoning About Typicality: a First Step. Technical Report 116/09, Dip. di Informatica, Univ. di Torino (2009)
12. Giordano, L., Gliozzi, V., Olivetti, N., Pozzato, G.L.: Prototypical reasoning with low complexity Description Logics: Preliminary results. In: Erdem, E., Lin, F., Schaub, T. (eds.) LPNMR 2009. LNCS (LNAI), vol. 5753, pp. 430–436. Springer, Heidelberg (2009)
13. Ke, P., Sattler, U.: Next Steps for Description Logics of Minimal Knowledge and Negation as Failure. In: DL 2008 (2008)
14. Kraus, S., Lehmann, D., Magidor, M.: Nonmonotonic reasoning, preferential models and cumulative logics. Artificial Intelligence 44(1-2), 167–207 (1990)
15. Straccia, U.: Default inheritance reasoning in hybrid kl-one-style logics. In: Proc. of IJCAI, pp. 676–681 (1993)

A Prioritized "And" Aggregation Operator for Multidimensional Relevance Assessment

Célia da Costa Pereira[1], Mauro Dragoni[1], and Gabriella Pasi[2]

[1] Università degli Studi di Milano, DTI
{celia.pereira,mauro.dragoni}@unimi.it
[2] Università degli Studi di Milano Bicocca, DISCO
pasi@disco.unimib.it

Abstract. In this paper a new model is proposed for aggregating multiple criteria evaluations for relevance assessment based on a refinement of the "min" ("and") operator.

The peculiarity of such an operator which also distinguishes it from the traditional "min" aggregation operator is that the extent to which the least satisfied criterion plays a role in determining the overall satisfaction degree depends both on its satisfaction degree and on its importance for the user. If it is not important at all, its satisfaction degree is not considered, while if it is the most important criterion for the user, only its satisfaction degree is considered (like with the traditional "min" operator). The usefulness and effectiveness of such a model are demonstrated by means of a case study on personalized Information Retrieval with multicriteria relevance.

Some preliminary experimental results are also reported.

1 Introduction

The problem of information overload on the Web leads to a demand for effective systems able to locate and retrieve information relevant to a user's individual interests. Systems for content-based access to huge information repositories usually produce a ranked list of documents in response to a user's query; the ranking is based on the assessment of relevance (or probability of relevance) to the user's interests expressed by the query.

On the other hand, as witnessed by both the modeling of the concept of *page popularity* in search engines, and by the recent attention paid by some authors to more clearly defining the notion of *document relevance*, we may assert that relevance can be modeled as a *multidimensional* property of documents [8] Based on this interpretation, the computation of an overall relevance score to be associated with each retrieved document is based on the aggregation of the scores representing the satisfaction of several criteria. The aggregated value may strongly vary depending on the adopted scheme; this important problem is often underestimated.

The traditional non-compensatory "min" operator is suited to situations in which we desire that an alternative satisfies "all" the considered criteria. In

R. Serra and R. Cucchiara (Eds.): AI*IA 2009, LNAI 5883, pp. 72–81, 2009.

such case, the requirement that all the conditions be satisfied is represented by the minimal value among the satisfaction degrees of all criteria. The result is then essentially dominated by the satisfaction degree of just one criterion, that is, the criterion with the worst score. The remaining criteria are only used to discriminate documents with similar scores, which means that a large part of scores is ignored or plays a minor role in the aggregation process.

In this paper, we propose a new model for aggregating multiple criteria evaluations for relevance assessment based on a refinement of the "and" operator. This model can be used for improving document ranking when: (i) we dispose of several relevance criteria for judging if a document might be interesting to the user or not; every requirement (criterion) is essential and no requirement can be dropped without defeating the purpose of the user interests; (iii) we dispose of a user-defined *preference order* on these requirements.

More precisely, our model is to be used in the case of a *rational pessimistic user*, that is, a *pessimistic user who considers the value of the least satisfied criterion as the overall relevance value merely when such criterion is the most important*. Otherwise, the values of the other criteria are also considered, depending on their importance for the user.

An information retrieval context is considered, where relevance is modeled as a multidimensional property of documents. The usefulness and effectiveness of such a model are demonstrated by means of a case study on personalized Information Retrieval (IR). The following criteria are considered to estimate document relevance: *aboutness, coverage, appropriateness*, and *reliability*. Notice that this approach is particularly effective if the user formulates a query inherent to her/his interests.

As in this work we do not consider search engines as a case study but personalized (traditional) IRSs, we do not consider web page popularity among the modeled relevance criteria.

It must be stressed, however, that the proposed aggregation model is general, and it may be applied to managing any set of relevance evaluation criteria.

The paper is organized as follows. Section 2 presents some related works in which the relevance is defined as a multidimensional concept. Section 3 presents the new aggregation method suggested. Section 4 presents some preliminary experiments carried out, which show the suitability of the proposed aggregation and, finally, Section 5 concludes the paper.

2   Related Work

In the first traditional approaches to Information Retrieval (IR), relevance was modeled as "topicality", and its numeric assessment was based on the matching function related to the adopted IR model (*boolean model, vector space model, statistical model* or *fuzzy model*). However, relevance is, in its very nature, the result of several components or dimensions.

Cooper [2] can be considered as one of the first researchers who had intuitions on the multidimensional nature of the concept of relevance. He defined relevance as *topical relevance with utility*. Another work which discusses the advantages of defining relevance as a multidimensional concept is that by Barry [1]. She underlined the need of considering other factors beyond topicality when computing document relevance. Her research is based upon the following assumption: "Each individual does not possess a unique set of criteria by which information is judged. Motivated users evaluating information within the context of a current information need situation will base their evaluations on factors beyond the topical appropriateness of documents."

Mizzaro, who has written an interesting article on the history of relevance [5], proposed a relevance model in which relevance is represented as a four-dimensional relationship between an information resource (surrogate, document, and information) and a representation of the user's problem (query, request, real information need and perceived information need). A further judgment is made according to the: topic, task, or context, at a particular point in time.

The dimensions pointed out by Mizzaro are in line with the five manifestations of relevance suggested by Saracevic [7]: *system or algorithmic relevance, topical or subject relevance, cognitive relevance or pertinence, situational relevance or utility* and *motivational or effective relevance*.

Based on a cognitive approach, Xu and Chen [9] carried out a study in which they focused on the criteria users employ in making relevance judgment beyond topicality. The result consisted in a five-factor model of relevance composed by topicality, novelty, reliability, understandability, and scope.

The work recently proposed by Farah and Vanderpooten [4] is aimed at dealing with *imprecision* underlying criteria design resulting from the fact that there are many formulations of the same criterion. They propose a multicriteria framework using an aggregation mechanism based on decision rules identifying positive and negative reasons for judging whether a document should get a better ranking than another.

More recently, Taylor and colleagues [8] presented a statistical study which extends the research on the relationship between multidimensional user relevance assignments and the stage in the process of completing a task. The obtained results suggest that users consistently identify relevance criteria beyond topical relevance.

The concept of *dimension* we use is somehow different from that used by Mizzaro and Saracevic. They defined several kinds of relevance and call them *dimensions of relevance* while we define relevance as a *concept of concepts*, i.e., as a point in a n-dimensional space composed by n criteria. The relevance score of a document is then the result of a particular combination of those n space components.

3   Prioritized Multicriteria Aggregation

In this section a new approach to the problem of aggregating distinct relevance assessments is defined. We consider this problem as a multicriteria decision problem in which the criteria are the various relevance dimensions and the documents are the list of possible alternatives.

We formally define a prioritized "min" aggregation operator inspired by the one proposed in [10], which considers the existence of a prioritization relationship over criteria. The proposed prioritization is modeled by making the weights associated with a criterion dependent upon the satisfaction of the higher-priority criteria.

3.1   Problem Representation

The presented multicriteria decision making approach has the following components:

- the set C of the n considered criteria: $C = \{C_1, \ldots, C_n\}$, with C_i being the function evaluating the ith criterion;
- the collection of documents D;
- a query q (in our case, a vector of query terms);
- a vector c of user interests;
- the $C_j(d)$ satisfaction scores (of document d with respect to criterion j);
- an aggregation function F to calculate for each document $d \in D$ a score $F(C_1(d), \ldots, C_n(d))$.

The weight associated with each criterion $C_i \in C$, with $i \neq 1$, is document and user-dependent. It depends on the preference order of C_i for the user, and also on both the weight associated to criterion C_{i-1}, and the satisfaction degree of the document with respect to C_{i-1}[1]. Formally, if we consider document d, each criterion C_i has an importance $\lambda_i \in [0, 1]$.

Remark 1. *Different users can have a different preference order over the criteria and, therefore, it is possible to obtain different importance weights for the same document for different users.*

For the sake of simplicity, in the following, we will make the assumption that the criteria are already sorted in descending order according to preference, i.e., that $C_i \succ C_j$ if $i < j$. This assumption implies that, for each document d, if $C_i \succ C_j$, we have $\lambda_i > \lambda_j$.

3.2   The Prioritized "min" Operator

In this section a prioritized "min" (or "and") operator is presented. This operator allows to compute the overall satisfaction degree for a user whose overall satisfaction degree is strongly dependent on the degree of the least satisfied criterion. The peculiarity of such an operator, which also distinguishes it from the traditional "min" operator, is that the extent to which the least satisfied criterion is considered depends on its importance for the user. If it is not important at all, its satisfaction degree should not be considered, while if it is the most important criterion for the user, only its satisfaction degree is considered. This

---

[1] If there are more than one criteria with the same priority order, the average weight and the average satisfaction degree are considered.

way, if we consider a document d, for which the least satisfied criterion C_k is also the least important one, the overall satisfaction degree will be greater than $C_k(d)$; it will not be C_k as it would be the case with the traditional "min" operator — the less important is the criterion, the lower its chances to represent the overall satisfaction degree.

We suppose that:

- for each document d, the weight of the most important criterion C_1 is set to 1, i.e., by definition we have: $\forall d \; \lambda_1 = 1$;
- the weights of the other criteria C_i, $i \in [2, n]$, are calculated as follows:

$$\lambda_i = \lambda_{i-1} \cdot C_{i-1}(d), \tag{1}$$

where $C_{i-1}(d)$ is the degree of satisfaction of criterion C_{i-1} by document d, and λ_{i-1} is the importance weight of criterion C_{i-1}.

The aggregation operator F is defined as follows. $F : [0,1]^n \to [0,1]$ is such that, for all document d,

$$F(C_1(d), \dots, C_n(d)) = \min_{i=1,n} (\{C_i(d)\}^{\lambda_i}). \tag{2}$$

Let us consider an example with the following data: $\forall i \in \{1, \dots, 4\} \; C_i(d_1) = 0.6$; $C_1(d_2) = 0.59$; $C_2(d_2) = 0.9$; $C_3(d_2) = C_4(d_2) = 1$; with $C_2 \succ C_1 \succ C_3 \succ C_4$. By applying Equation 1, we have, for document d_1:

- $\lambda_1 = 1$;
- $\lambda_2 = \lambda_1 \cdot C_1(d_1) = 0.6$;
- $\lambda_3 = \lambda_2 \cdot C_2(d_1) = 0.36$;
- $\lambda_4 = \lambda_3 \cdot C_3(d_1) = 0.216$;

and $F(C_1(d_1), \dots, C_4(d_1)) = 0.6$. Concerning document d_2, we have:

- $\lambda_1 = 1$;
- $\lambda_2 = \lambda_1 \cdot C_1(d_2) = 0.9$;
- $\lambda_3 = \lambda_2 \cdot C_2(d_2) = 0.531$;
- $\lambda_4 = \lambda_3 \cdot C_3(d_2) = 0.531$;

and $F(C_1(d_2), \dots, C_4(d_2)) = 0.62$ which is greater than 0.6 as one would intuitively expect.

Properties of the Prioritized Aggregation Operator. Here, we present some mathematical properties of the proposed operator.

Continuity. The proposed aggregation is a polynomial, as it can be written as

$$F(C_1(d), \dots, C_n(d)) = min(C_1(d), \min_{i=2,n} (\{C_i(d)\}^{\lambda_i})),$$

which is continuous, because, $\forall x, y \in \mathbb{R}$, $min(x,y) = \frac{x+y}{2} - \frac{|x-y|}{2}$. Therefore, the property of continuity holds for F.

Boundary Conditions. The proposed "min" operator satisfies:

- if $\forall i\, C_i(d) = 0$, then $\min_{i=1}^{n} \lambda_i \cdot C_i(d) = 0$;
- if $\forall i\, C_i(d) = 1$, then $\min_{i=1}^{n} \lambda_i \cdot C_i(d) = 1$;
- $\forall i \in \{1, \ldots, n\}$, we have $0 \leq \min_{i=1}^{n} \lambda_i \cdot C_i(d) \leq 1$.

Monotonicity. The proposed "min" operator is not *monotonous*; indeed, if the satisfaction degree of one of the criteria increases, the overall final aggregation value can decrease.

For example, if we have $C_1(d) = 0.7, C_2(d) = 0.1, C_3(d) = 0.3$, from Equation 1 we obtain that $\min(0.7^1, 0.1^{0.7}, 0.3^{0.07}) = 0.1995$. If we replace $C_1(d)$ by a greater value $C_1'(d) = 0.9$, we obtain that $\min(0.9^1, 0.1^{0.9}, 0.3^{0.09}) = 0.125$, which is less than the previous value.

Absorbing Element. The proposed operator has an *absorbing element*, $C_i(d) = 0$. Indeed, if there is a criterion C_i such that $C_i(d) = 0$, we have

$$F(C_1(d), \ldots, C_i(d), \ldots, C_n(d)) = 0.$$

Another consequence of the *absorbing* property is that if a document d slightly satisfies the most important criterion C_1, and satisfies, to some extent, all the other criteria, the score obtained for all the possible aggregations in which C_1 is the most important criterion goes to $C_1(d)$, independently of the satisfaction degrees of the other criteria. This is due to Definitions 1 and 2 and to the fact that $\lim_{k \to 0} x^k = 1$.

Neutral Element. The proposed operator has a *neutral element*, $C_i(d) = 1$. Indeed, if there is a criterion C_i such that $C_i(d) = 1$, we have

$$F(\ldots, C_i(d), \ldots) = F(\ldots, C_{i-1}(d), C_{i+1}(d) \ldots).$$

Indempotence. The proposed operator is *indempotent*. Indeed, if all the criteria have the same satisfaction degree, $x \in [0, 1]$, we have

$$F(C_1(d), \ldots, C_i(d), \ldots C_n(d)) = F(x, \ldots, x)) = x.$$

This holds , if $k \in [0, 1]$, we have $x \leq x^k \leq 1$ for all $x \in [0, 1]$.

4 Experiments

In this paper, in order to illustrate and evaluate the proposed prioritized relevance model, we focus on personalized IR, i.e., an information retrieval approach based on the use of a user model, formally represented by a user profile. In this context, we consider four main relevance criteria: aboutness, coverage, appropriateness, reliability.

The role of *aboutness (topicality)* in information retrieval has been largely studied — it is the most studied among the existing dimensions of relevance and it is also considered as being the crucial one. *Coverage, reliability* and *appropriateness* have been studied more recently and we have decided to use these criteria for our experiments. A detailed definition and explanation of these criteria can be found in [3].

4.1   Assumptions

The preference order of the criteria depends on the user.

Users who just care about the presence of their interests in the document, and who do not care about the other terms in the document — *coverage seekers* — will give more importance to the coverage criterion than to the others. Users who care about both their interests and the entire document content — *appropriate-ness seekers* — will give more importance to the appropriateness criterion than to the others. On the contrary, a user who formulates a query which has no intersection with her/his interests or users who do not have a defined list of interests — *interest neutral* — will not give any importance to the coverage and appropriateness criteria. Users of this kind are just looking for a satisfactory answer to their current concern, as expressed by their query. Finally, users who are cautious about the trustworthiness of the origin of the retrieved documents – *cautious* – will give more importance to the reliability criterion than to the others.

The "kind" of user depends on the preference order (s)he gives to the four criteria. For example, a user with the following preference order *aboutness* $\succ$ *coverage* $\succ$ *reliability* $\succ$ *appropriateness* is called $ACRA_p$ user.

4.2   Experiments on the 'min" Prioritized Operator

In this section, the impact of the proposed prioritized aggregation operator is evaluated in the personalized IR setting. The relevance criteria and their aggregation discussed in the previous sections have been implemented on top of the well-known Apache Lucene open-source API[2]. The Reuteurs RCV1 Collection (over 800,000 documents) has been used, and we applied the Sanderson [6] approach to generate queries, one for each topic code, for a total of 126 queries.

To evaluate the impact of the proposed aggregation scheme, we have considered two kinds of situations: (S_1) in which the users formulate queries in line with their interests and (S_2) in which the users formulate queries only on new interests, that is not in line with their previous interests. The baseline used to compare our results is the rank obtained with the *average* aggregation operator. The aim of the experiments carried out is to show that, with respect to the baseline, the proposed operator improves the ranking of the documents which are related to the user interests in situation S_1. We evaluated the resulting ranking for all possible permutations of the four considered criteria.

In the experiments, we have considered two combinations of user interest degrees: (i) moderate level interests, i.e., all interests between 0.4 and 0.6; and (ii) high level interests, i.e., all interests at 1.

We have carried out experiments for all kinds of users and for all possible aggregations. Due to lack of space, we will just show the graphs concerning one experiment among the aggregations with the same most important criterion, which are (for situation S_1): ACA_pR in Figure 1, $CRAA_p$ in Figure 2, and A_pACR in Figure 4. The results for situation S_2 being very similar, we will just illustrate the results for an ACA_pR user, Figure 3.

---

[2] See URL http://lucene.apache.org/

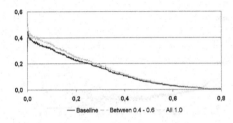

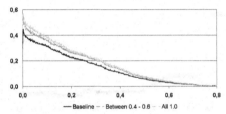

Fig. 1. Precision/Recall graph for ACA_pR users in situation S1

Fig. 2. Precision/Recall graph for $CRAA_p$ users in situation S1

We can see in Figure 1 that for an almost *interest neutral* user like a user who gives more importance to the terms in her/his query than to her/his interests, an ACA_pR user for example, the precision/recall curves for moderate level interests and for high level interests coincide, and are above the baseline curve anyway. The curve for a completely neutral user (without any interest) would correspond to the aboutness.

We can see in Figure 2, that also for a $CRAA_p$ *coverage seeker user* the curves corresponding to the user's moderate interests and high interests are above the baseline, but, unlike for the ACA_pR user, there is a clear distinction between them in the sense that the curve of high interest is clearly above that of moderate interests. This fact shows that when the document selection for a *coverage seeker user* is based on moderate user interests, there are more irrelevant retrieved documents than when the selection is based in a high user interests.

Also interesting is the results obtained for an *appropriateness seeker user*, for example for an A_pACR user, Figure 4. To better understand that curve, we have to recall that *appropriateness* measures the similarity between the document and the user's interest. The consequence is that the amount of documents classified as interesting by the user is in general less than it would be for a *coverage seeker user* for whom it is sufficient that her/his interest are included in a document to deem such a document interesting.

The results obtained in Situation S_2 for an ACA_pR user is illustrate in Figure 3. We can see that the curve lies below the baseline curve. This is due to the absorbing-element property of the proposed operator. Indeed, if a user

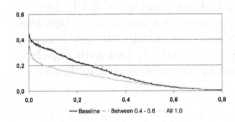

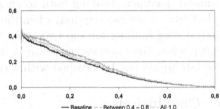

Fig. 3. Precision/Recall graph for ACA_pR users in situation S2

Fig. 4. Precision/Recall graph for A_pACR users in situation S1

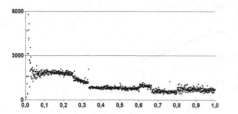 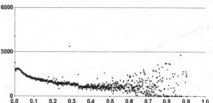

Fig. 5. Gap values between the $CRAA_p$ and the $ARCA_p$ aggregation with medium-value interests

Fig. 6. Gap values between the $CRAA_p$ and the $ARCA_p$ with high-value interests

interest-dependent criterion is not satisfied at all by a document — this is highly possible in Situation S_2 — the document, even with high aboutness or reliability scores but with a null score for the coverage criterion, for example, will not, erroneously, be considered interesting for the user.

It is also interesting to analyze the gap between the rankings obtained for the different types of user. Here, we show the result of such analysis concerning an $ARCA_p$ user and a $CRAA_p$ user for both the case of a medium level interests and high level interests. The curve concerning the case of medium level interests is shown in Figure 5 and that concerning the case of high level interests is shown in Figure 6. The possible satisfaction degrees for the coverage criterion are represented along the X-axis, while the gap between the two typologies of user is represented along the Y-axis.

We have to recall that the *coverage* criterion measures how strongly the user interests are included in a retrieved document. To be completely included in a document, all user interest degrees must be smaller than the index term weights of the document.

In Figure 5, we can see that, in case of medium level interests, the gap is slightly bigger when the coverage measure is low (that is, when the index term weights of the document are even smaller than the user interest level (0.1)). We can then deduce that, in case of low coverage values, the documents proposed to an $ARCA_p$ user slightly "cover" the interests of a $CRAA_p$ user. Instead, we can see that, when the coverage values are high, the gap is smaller. This means that, in case of high coverage values and medium level interests, the system proposes a similar document ranking both to an $ARCA_p$ user and to a $CRAA_p$ user.

In Figure 6, we can see that, when the coverage is low and in case of high level interest, the ranking for the two types of users are similar. The gap is clearly bigger when the coverage measure is high — a $CRAA_p$ user with high level interests would not be satisfied with a ranking proposed to an $ARCA_p$ user.

5   Conclusions

In this paper, a new model for aggregating multiple criteria evaluations for relevance assessment is proposed. Such a model is based on a refinement of the "min" ("and") operator and is suited to improving document ranking when (i) we dispose of several relevance criteria for judging if a document might be interesting for the user or not; (ii) every requirement (criterion) is essential and no requirement can be dropped without defeating the purpose of the user interests; and (iii) we dispose of a user *preference order* on these requirements.

Preliminary experimental results show that the proposed operator improves the ranking of the documents which are related to the user interests, when the user formulates an interest-related query.

References

1. Barry, C.L.: User-defined relevance criteria: an exploratory study. J. Am. Soc. Inf. Sci. 45(3), 149–159 (1994)
2. Cooper, W.S.: On selecting a measure of retrieval effectiveness. Journal of the American Society for Information Science 24(2), 87–100 (1973)
3. da Costa Pereira, C., Dragoni, M., Pasi, G.: Multidimensional relevance: A new aggregation criterion. In: Proceedings of the 31st ECIR 2009, pp. 264–275 (2009)
4. Farah, M., Vanderpooten, D.: A multiple criteria approach for information retrieval. In: Crestani, F., Ferragina, P., Sanderson, M. (eds.) SPIRE 2006. LNCS, vol. 4209, pp. 242–254. Springer, Heidelberg (2006)
5. Mizzaro, S.: Relevance: the whole history. J. Am. Soc. Inf. Sci. 48(9), 810–832 (1997)
6. Sanderson, M.: The reuters collection. In: Proceedings of the 16th BCS IRSG Colloquium (1994)
7. Saracevic, T.: The stratified model of information retrieval interaction: Extension and applications. Journal of American Society for Information Science 34, 313–327 (1997)
8. Taylor, A.R., Cool, C., Belkin, N.J., Amadio, W.J.: Relationships between categories of relevance criteria and stage in task completion. Inf. Process. Manage. 43(4), 1071–1084 (2007)
9. Xu, Y.C., Chen, Z.: Relevance judgment: What do information users consider beyond topicality? J. Am. Soc. Inf. Sci. Technol. 57(7), 961–973 (2006)
10. Yager, R.R.: Prioritized aggregation operators. Int. J. Approx. Reasoning 48(1), 263–274 (2008)

Deriving Information from Sampling and Diving

Michele Lombardi[1], Michela Milano[1], Andrea Roli[2], and Alessandro Zanarini[3]

[1] DEIS, *Alma Mater Studiorum* Università di Bologna
V.le Risorgimento 2, 40136, Bologna, Italy
[2] DEIS - Campus of Cesena, *Alma Mater Studiorum* Università di Bologna
Via Venezia 52, 47023, Cesena, Italy
[3] Département de génie informatique, École Polytechnique de Montréal
C.P. 6079, succ. Centre-ville, Montreal, Canada

Abstract. We investigate the impact of sampling and diving in the solution of constraint satisfaction problems. A sample is a complete assignment of variables to values taken from their domain according to a a given distribution. Diving consists in repeatedly performing depth first search attempts with random variable and value selection, constraint propagation enabled and backtracking disabled; each attempt is called a dive and, unless a feasible solution is found, it is a partial assignment of variables (whereas a sample is a –possibly infeasible– complete assignment). While the probability of finding a feasible solution via sampling or diving is negligible if the problem is difficult enough, samples and dives are very fast to generate and, intuitively, even when they are infeasible, they give some statistic information on search space structure. The aim of this paper is to understand to what extent it is possible to help the CSP solving process with information derived from sampling and diving. In particular, we are interested in extracting from samples and dives precise indications on how good/bad are individual variable-value assignments with respect to feasibility. We formally prove that even uniform sampling could provide precise evaluation of the quality of variable-value assignments; as expected, this requires huge sample sizes and is therefore not useful in practice. On the contrary, diving seems to be much better suited for assignment evaluation purposes. Three dive features are identified and evaluated on a collection of Partial Latin Square instances, showing that diving provides information that can be fruitfully exploited. Many promising direction for future research are proposed.

1 Introduction

Exploiting information collected during search is often one of the key components for the successful solution of Constraint Satisfaction Problems. In this work, we address the issue of formally studying to what extent it is possible to extract information from two kinds of sampling techniques. In particular, we are interested in extracting from random samples and dives precise indications on how good/bad are individual variable-value assignments. In particular we focus on information regarding problem feasibility.

This information can be used for designing search heuristics, for partitioning domains in decomposition based search [9], for over-filtering unpromising values from

R. Serra and R. Cucchiara (Eds.): AI*IA 2009, LNAI 5883, pp. 82–91, 2009.

variable domains, or for defining suitable local moves to repair an infeasible assignment. To the best of our knowledge, a formal analysis of this kind has not yet been published in the literature.

In this paper, we will call a *sample* an assignment of variables to values taken from their domains. In case of random samples, variables and values are chosen according to a uniform distribution. A *dive*, on the contrary, is obtained by tree search in which a variable and a value in its domain are randomly chosen and after each assignment propagation filters out provably infeasible values with the current choices. We will limit our discussion to samples and dives collected at the root node (in a pre-processing phase of the search) and to uniform distributions.

We will first define a formal probabilistic model that makes it possible to estimate through sampling the probability that the problem remains feasible after a specific event occurs, such as a variable-value assignment. This approach shows that it is possible to extract valuable information from random sampling in a sound theoretical framework. Nevertheless, in the case of large-size instances, it has the drawback that a good estimation of the variable-value assignment probability can be achieved only at the price of huge sample sizes.

As a second step, we will show how to extract dive features to measure the potential feasibility and infeasibility of variable-value assignments, using the Partial Latin Square as a sample problem [7]. Parameters such as the average dive length, the number of variable-value occurrences and variations thereof, enable us to extract pieces of information about how good/bad are single variable-value assignments. Experimental results show that the features extracted from dives are very informative, having accuracy (w.r.t. to feasibility) comparable to or better than a state-of-the-art heuristic used in Impact Based Search.

2 Related Work

Uniform sampling of solutions has been recently studied by Gogate and Dechter in [3]. Their approach uses a randomized backtracking procedure guided by an heuristic computed through generalized belief propagation. In order to avoid the rejection problem (ending up to an infeasible solution) they systematically search for a feasible solution using a standard backtracking procedure. In a following work [4] they added a second resampling phase that gives approximation guarantees that the distribution of sampled solutions converges to the uniform distribution.

Gomes et al. in [5] also proposed an algorithm for sampling solutions of SAT problems. They iteratively add streamlining constraints to the SAT formula to reduce the set of solutions (possibly only one) and they return one solution chosen uniformly at random. Our approach mainly differs from the two described above in the fact that we sample assignments that can be either (very unlikely) feasible or infeasible whereas they sample only feasible solutions (thus with an exponential computational effort).

Random dives (in literature often referred to as probes) have been used by Grimes and Wallace in [8] to initialize parameters of a heuristic. One of the two methods proposed employs random dives to collect information on the constraints that led to failures. The information is then used to initialize the weights of the constraints in weighted degree heuristics. Nevertheless, during the diving phase, they do not collect any explicit information regarding variable assignments and dive lengths.

In [12] Ruml proposed to use adaptive probes to iteratively learn the search strategy to employ to explore a search tree. In constraint satisfaction problems, he measures the quality of a dive in terms of the number of instantiated variables (dive length). The algorithm then adaptively adjusts the heuristic used for the following dives.

In [2] random dives are used to initialize an elite set of partial solutions where the number of assigned variables is taken as a quality measure. An elite solution is then randomly chosen to provide heuristic guidance. Note that the heuristic is based solely on a single elite (partial) solution whereas our approach tries to infer general information from the whole set of partial solutions. Refalo in [11] proposed an heuristic (used in Impact Based Search) that estimates the effort of the search as the proportional reduction of the cartesian product of the variable domains after a variable instantiation. This estimation referred to as impact is initialized at the beginning of the search through a dichotomic search where each variable domain is iteratively split in two subdomains.

Note that in none of the methods described above systematic statistic measures are extracted from the random dives; indeed, our approach can be thought as being orthogonal to the described ones, as the emphasis is on evaluating the feasibility related information extracted from samples and dives rather than on providing a top quality heuristic. As a consequence, the method can be easily integrated in some of the mentioned frameworks.

3   Uniform Sampling

Definition 1. *Assume we are given a $CSP = (X, D, C)$ where $X = \{X_1, \ldots, X_n\}$ is the set of variables, $D = \{D_1, \ldots, D_n\}$ the set of domains and $C = \{c_1, \ldots c_k\}$ the set of constraints. A sample s is an element of the cartesian product of the domains of all problem variables, $s \in \prod_{i=1}^{n} D_i$. Constraints are not taken into account in the definition of s.*

The idea is that by sampling the search space randomly, either we find a solution to the problem, and we have done, or we collect only infeasible samples. In the second case, we can try to measure the feasibility or infeasibility degree of single variable-value assignments. Suppose to model the sample generation problem as a stochastic variable with sample space $\Omega = \prod_i D_i$. In this context the set F of feasible solutions and the set of infeasible solutions I are events, such that $F \cup I = \Omega$, $F \cap I = \emptyset$ and $P(F)$ (resp. $P(I)$) is the probability of a random sample to be feasible (resp. infeasible).

Now, let $A_0, A_1, \ldots, A_m$ be a set of events we want to evaluate; one possible event can be a variable-value assignment, another can be the removal of a value from a domain. We want to estimate the probability the problem is still feasible after the event occurs. For example, let us consider the problem of choosing a value v from the domain of a variable X_i, as it would be done by a value selection heuristic; we want the problem to remain feasible after the assignment. More formally, given a threshold value $\theta \in [0, 1)$, we want $P(F \mid X_i = v) > \theta$; that is, we want the probability to randomly pick a feasible solution containing $X_i = v$ to be greater than the specified threshold. As in this work we are mostly concerned with problem feasibility, an assignment $X_i = v$ is considered better than $X_j = u$ if it has a greater value for the probability $P(F \mid X_i = v)$. So, we need a way to estimate the conditional probability $P(F \mid X_i = v)$: this can be done by using probability theory results [10]. From the Bayes theorem we know that:

$$P(F \mid X_i = v) = \frac{P(F)P(X_i = v \mid F)}{P(X_i = v)}$$

where $P(X_i = v \mid F)$ is the probability that $X_i = v$ in a random *feasible* solution. Probability $P(X_i = v)$, instead, is defined on the entire sample space and therefore $P(X_i = v) = 1/|D_i|$, where $|D_i|$ is the cardinality of the domain of X_i. Therefore, in order to identify good assignments we only need a way to compute $P(F)P(X_i = v \mid F)$. For doing this, we consider another result in probability theory which states that, given a set of events $B_0, B_1, \ldots$ such that $B_i \cap B_j = \emptyset$ for each $i \neq j$ and $\bigcup_i B_i = \Omega$, and an arbitrary event A:

$$P(A) = \sum_j P(B_j)P(A \mid B_j)$$

The event A in our case is the assignment $X_i = v$, while events B_j are F and I, that have empty intersection:

$$P(X_i = v) = P(F)P(X_i = v \mid F) + P(I)P(X_i = v \mid I)$$

hence:

$$P(F)P(X_i = v \mid F) = P(X_i = v) - P(I)P(X_i = v \mid I)$$

Note that the probability $P(X_i = v \mid I)$ can be estimated by generating a number of infeasible solutions and counting the occurrences of the $X_i = v$ assignment. Indeed, since in principle we randomly sample the whole space $F \cup I$, also I is randomly sampled. Thus, by counting the frequency of variable-value assignments restricted to the infeasible samples, by definition of conditional probability, we do estimate $P(X_i = v \mid I)$. Therefore:

$$
\begin{aligned}
P(F \mid X_i = v) &= \frac{P(F)P(X_i = v \mid F)}{P(X_i = v)} \\
&= \frac{P(X_i = v) - P(I)P(X_i = v \mid I)}{P(X_i = v)} \\
&= 1 - P(I)\frac{1}{P(X_i = v)}P(X_i = v \mid I)
\end{aligned}
$$

Finally, since $P(X_i = v) = 1/|D_i|$:

$$P(F \mid X_i = v) = 1 - P(I) \, |D_i| \, P(X_i = v \mid I) \tag{1}$$

where $P(I)$ is an unknown, problem dependent, constant factor which has no contribution when used to compare assignments. Note that domain size does influence the evaluation of the assignments. Formula (1) provides us with a way to evaluate assignments of single values to single variables; the result can be generalized to take into account a generic event A and thus target other search strategies as well.

Unfortunately, formula (1) is useless in most practical cases: this is due to the sample size needed to get good probability approximations, especially when there are few feasible solutions. The sample size can be computed with the following formula, used to bound the absolute error on the fraction of a population (for example our solutions) satisfying a specific condition (for example $X_i = v$):

$$n = z_\alpha \frac{p(1-p)}{\delta^2}$$

where p is a rough estimate of the probability that a random chosen solution satisfies the condition, δ is the absolute error we allow and the z_α parameter depends on the confidence level we want to achieve. For example in a Partial Latin Square instance of order 25, with around 2000 feasible solutions and average domain size of 4, in order to get a confidence interval of 10% with a 95% confidence level we need the following sample size:

$$n = 4 \times \frac{P(X_i = v|I)P(X_i \neq v|I)}{(0.1 \times 2000/25^{25})^2} \simeq 4 \times \frac{0.25 \times 0.75}{(0.1 \times 2000/25^{25})^2} \simeq 1.48 \times 10^{65}$$

where 4 is the value of the z_α parameter for a 95% confidence level, $P(X_i = v|I)$ is estimated to be roughly 0.25 since the average domain size is 4, we want an absolute error of around 10% (0.1) and the order of magnitude of the probability of a feasible solution is $(2000/25^{25})$.

Although not often applicable in practice, formula (1) is based on a valid probabilistic model which can be used to draw some conclusions: first of all, the existence of an actual link between the occurrence of an assignment $X_i = v$ in *infeasible* solutions and in *feasible* solutions; then the influence of the domain size in the choice of the best variable.

4 Diving

Dives are different from samples as we take into account constraints in their construction. In the literature, dives are often referred to as probes. Intuitively a dive collects variable assignments (derived both from branching decisions and from propagation) up to a failure. We remove the choice that causes failure, so that the dive is a feasible partial assignment. First, we have to clarify when we consider that a constraint c_j is consistent after some variables are assigned. Suppose we have a set of variables $X_1, \ldots, X_n$ with domains $D_1, \ldots, D_n$. A partial assignment on a subset of m variables $\gamma = (X_{i1} = v_{i1}, \ldots, X_{ik} = v_{im})$ is *consistent* if all the constraints mentioning assigned variables $X_{i1}, \ldots, X_{im}$ are consistent. A constraint c_j is consistent if, after propagating the effects of the assignment, the not yet instantiated variables appearing in the constraints have a non empty domain. We refer to the constraint c_j subject to the assignment γ to as $c_j|_\gamma$. Clearly, some assigned variables may not appear in the constraint. In this case the assignment has no effect.

More formally:

Definition 2. *Given a $CSP = (X, D, C)$ where $X = \{X_1, \ldots, X_n\}$ is the set of variables, $D = \{D_1, \ldots, D_n\}$ the set of domains and $C = \{c_1, \ldots c_k\}$ the set of constraints. A dive γ is an assignment of (some) variables to values $(X_{i1} = v_{i1}, \ldots, X_{im} = v_{im})$ where $m \leq n$ such that for each problem constraint c_j, $c_j|_\gamma$ is consistent*[1].

In our experiments, we consider a special case of dives built as follows. We visit the search tree by choosing randomly at each node a variable and choosing randomly a value in its domain. Propagation is activated, so as to remove from variable domains provably inconsistent values. Possibly some variables become instantiated due to propagation. As soon as a failure occurs, we retract the last assignment and we maintain the consistent set of assignments up to that point. Assignments in a dive are derived both by branching choices and by propagation.

We want to infer from dives some parameters that measure how good/bad are single variable value assignments. In particular, as the focus is on the ability of a search heuristic to keep the problem feasible, for a given variable X_i, an assignment $X_i = v$ is considered "better" than an assignment $X_i = u$ if it is part of more feasible solutions.

All our considerations move from two basic conjectures: first if a variable-value pair $X_i = v$ is part of many feasible solutions it is likely to occur in *longer* partial assignments; second if a variable-value pair $X_i = v$ is part of many feasible solutions it is likely to occur in *many* partial assignments. After preliminary experiments we chose to focus on two features based on the above conjectures: the average dive depth (ADD) and the number of occurrences of a given assignment (OCC). Given a $CSP = (X, D, C)$ and a set of random dives Γ, the *average dive depth* of an assignment $X_i = v_i$ is:

$$ADD(X_i, v_i) = \frac{1}{|\Gamma_{X_i=v_i}|} \sum_{\gamma \in \Gamma_{X_i=v_i}} |\text{dom}(\gamma)|$$

where $\Gamma_{X_i=v_i}$ is the set of dives containing the assignment $X_i = v_i$ and $\text{dom}(\gamma)$ is the set of variables assigned in dive γ. In practice ADD is simply the sample average of the depth of all dives containing a given variable-value pair. With the same notation, the *number of occurrences* as a dive feature is defined as:

$$OCC(X_i, v_i) = |\Gamma_{X_i=v_i}|$$

5   Experimental Results

We performed experiments on 60 Partial Latin Square Instances of order 25 in the phase transition region [7] (41% of holes for instances of this size), with the primary aim to evaluate the extent of the correlation between dive features and the actual quality of the assignments (density in feasible solutions). All the instances were randomly generated following [6]. Instances in the phase transition (i.e. when the number pre-filled cells is between the under-constrained and the over-constrained region) tend to be the hardest.

---

[1] Any form of local consistency can be considered, although it is likely to have an impact both on the quality of the information extracted and on the performance of the dive generation.

Table 1. Average results for the dive features

	dives num	not_max_max	not_max_good	not_min_min	not_min_bad	none_good_good	none_bad_bad	good_bad	bad_good
ADD	2000	.530 (.027)	.403 (.026)	.410 (.024)	.165 (.019)	.163 (.019)	.065 (.012)	.258 (.009)	.100 (.006)
	4000	.495 (.023)	.365 (.023)	.391 (.022)	.153 (.017)	.149 (.015)	.059 (.010)	.244 (.008)	.094 (.006)
	8000	.472 (.020)	.342 (.019)	.374 (.020)	.142 (.015)	.142 (.013)	.056 (.008)	.233 (.007)	.091 (.005)
OCC	2000	.398 (.012)	.275 (.011)	.338 (.019)	.117 (.013)	.165 (.010)	.036 (.005)	.125 (.004)	.123 (.004)
	4000	.396 (.010)	.273 (.010)	.330 (.017)	.112 (.011)	.164 (.008)	.034 (.005)	.122 (.004)	.123 (.003)
	8000	.395 (.009)	.272 (.008)	.324 (.015)	.109 (.010)	.164 (.007)	.034 (.004)	.119 (.003)	.123 (.003)
IMP	—	.464	.335	.344	.147	.214	.053	.167	.132
RND	—	.749	.649	.588	.351	.390	.177	.500	.500

For each instance: (A) the actual occurrence frequency within the feasible solutions was computed for each variable-value pair by complete enumeration; (B) next we collected a number of dives and we used them to extract the ADD and OCC features; (C) then the variable/value rankings obtained by using ADD and OCC as scores were compared to those and those obtained by looking at the actual occurrence densities. (D) Finally, the correlation with the actual occurrence density was evaluated by defining a set of error rate indicators and measuring their value at the root node. The process was repeated several times, with variable number of collected dives (namely 2000, 4000, 8000). All "Aldiff" constraints were tuned to achieve Generalized Arc Consistency.

Table 1 reports the results of a first evaluation; in particular, for each instance and number of dives, dives were collected (and indicators were extracted) 30 times; the mean and standard deviation for each indicator was computed over those 30 iterations, and the resulting values were averaged over the 60 instances. In the table, the outcomes of this process are compared with the results (experimentally measured) of the value selection heuristic used in Impact Based Search [11] (row IMP), and with those of a random value selection heuristic (row RND, computed theoretically). Similarly to what is done in the evaluation of automatic classifiers, *all indicators are error probabilities*, hence the lower the better; the first value in each cell is the mean of the indicator and the standard deviation follows between round brackets.

The *not_max_max* indicator reports the fraction of (unbound) variables where the value with the best score does not match the value with the highest density value in the feasible solutions; equivalently, this measures the probability that the best value according to the heuristic is not actually the more frequently occurring in feasible solutions. More formally:

$$not\_max\_max = \frac{|\{X_i \mid \rho(X_i, v^{max}) \neq \max_j(\rho(X_i, v_j))\}|}{|X|}$$

where v^{max} is the value with highest dive-driven score for variable X_i, and $\rho(X_i, v_j)$ is the density in feasible solutions of the assignment $X_i = v_j$; finally $|X|$ is the number of problem variables. The *not_max_good* indicator reports the fraction of variables where the value with highest score does not correspond to any assignment with good density, i.e., higher than $1/|D_i|$ where $|D_i|$ is the size of the domain of variable X_i. Formally:

$$not\_max\_good = \frac{|\{X_i \mid \rho(X_i, v^{max}) < \frac{1}{|D_i|}\}|}{|X|}$$

Both the indicators can be useful, for example, if we plan to use dives to design a value selection heuristic. The not_min_min and not_min_bad indicators are the counterparts of not_max_max and not_max_good; they tell how often the assignment with lowest score does not have the lowest (resp. low) density. These last indicators can be handful, for instance, if we plan to use the dive to filter out the least promising values.

Indicators $none\_good\_good$ and $none\_bad\_bad$ report the fraction of variables where no value with a good (resp. bad) actual density was classified as good (resp. bad) according to the indicator. If the provided information is to be used as a value selection heuristic, this corresponds to a most critical mistake. The $none\_good\_good$ is formally defined as:

$$none\_good\_good = \frac{|\{X_i \mid \forall v_j \ : \ score(X_i, v_j) < avgscore(X_i) \vee \rho(X_i, v_j) < \frac{1}{|D_i|}\}|}{|X|}$$

where (X_i, v_j) represents the assignment $X_i = v_j$, $score(X_i, v_j)$ is it score according to one of the dive features, $avgscore(X_i)$ is the average score for variable X_i; the definition of $none\_bad\_bad$ is analogous. Finally, $good\_bad$ is the fraction of false positives, i.e. the relative number of assignments classified as good according to their score, while they are bad according to the actual density; similarly, bad_good is the fraction of false negatives. Formally:

$$good\_bad = \frac{|\{(X_i, v_j) \mid score(X_i, v_j) \geq avgscore(X_i) \wedge \rho(X_i, v_j) < \frac{1}{|D_i|}\}|}{\sum_i |D_i|}$$

where $\sum_i |D_i|$ is the sum of all domain sizes (i.e. the number of all problem assignments); the definition of bad_good is analogous. Both those indicators estimate the accuracy we obtain in classifying multiple good and bad assignments, as it happens for example in decomposition-based search.

By looking at the table, first note that the selected indicators *do* provide information about the feasibility of variable-value pairs, as both ADD and OCC dominate RND.

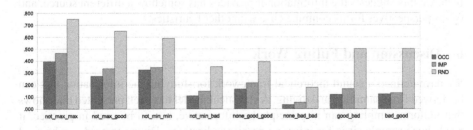

Fig. 1. Quality of the information provided by OCC, IMP, RND

Table 2. Average results for the dive features

	not_max_max	not_max_good	not_min_min	not_min_bad	none_good_good	none_bad_bad	good_bad	bad_good
ADD/OCC	.241 (.016)	.138 (.014)	.203 (.016)	.047 (.009)	.050 (.009)	.013 (.004)	.072 (.004)	.051 (.004)
ADD/IMP	.241 (.016)	.134 (.014)	.163 (.014)	.035 (.008)	.041 (.008)	.010 (.004)	.072 (.004)	.040 (.004)
OCC/IMP	.317 (.008)	.200 (.007)	.210 (.011)	.060 (.007)	.109 (.005)	.022 (.003)	.077 (.003)	.084 (.002)

Moreover, as depicted in figure 1, OCC outperforms impacts w.r.t all selected indicators; this raises a strong interest in using the feature to guide a value selection heuristic (perhaps coupled with impact base variable selection). Conversely, the ADD feature provides lower quality feasibility information compared to impacts, but has a lower chance of missing all good values for a variable (see the $none\_good\_good$ indicator) and of yielding false negatives.

While the OCC features is almost no sensible at all to the number of collected dives, the performance of ADD increases as the number of dives grows. The number of dives has also the well known effect of reducing the variance, although variance is no issue, as all reported standard deviations are indeed very low; this is a very interesting remark, as it suggests we could save time by performing much fewer dives, without drastically loosing information quality.

Finally, a naturally arising question is whether the considered features provide different *quantity* of information of the same *quality*, or whether they are somehow complementary. In the second case one has the opportunity to exploit such complementarity to increase the detection accuracy. On the purpose to answer this question we performed a second set of experiments, evaluating pairs of dive features. The number of collected dives was fixed here to 4000, and the performance indicators were computed by counting the number of detection errors made by both features; in practice we evaluate the intersection of the variable/value pairs were a error occurs for the two features.

The results of this second evaluation are reported in table 2; here one can see how the indicator values for the OCC-IMP pair are close to those of OCC alone, pointing out that OCC and IMP tend to make the same mistakes, with OCC simply failing less often. On the contrary, ADD exhibits a much weaker correlation with the other features; this suggests the information it provides has somehow a different source, and opens perspectives for its combined use with other heuristics.

6   Discussion and Future Work

We have defined a sound theoretical framework for studying the information that can be derived from sampling and diving. A statistical model has been presented that shows that uniform sampling can in principle be informative, but, to be used in practice, it needs huge sample sizes for inferring precise indications. Diving instead can be used for this purpose also in practical cases. We have identified dive features and experimented them on a set of Partial Latin Square instances. Additional experiments have been performed aimed at applying a statistical analysis to measure the confidence level of a comparison between two variable-value assignments. Also, the correlation between dive-driven parameters and assignment density in feasible solutions has been measured. Due to the lack of space these experiments have been dropped and can be found in [13].

A byproduct of this work is the definition of a statistically sound method for evaluating variable-value selection heuristics. A number of open issues remain and are subject of on-going research. We plan to devise new dive features to be compared with current ones. Second, we are investigating the relations between dive derived information and backbones and backdoors; on one hand, to understand which level of accuracy we can reach for such critical variables, and on the other hand to see whether dives can help in their identification.

Other important issues we are considering are the evaluation of the approach on other problems and the estimation of variables by means of dive-driven information, to detect the most critical ones or those for which we have the highest level of confidence. Finally, we plan to devise techniques to make the diving process more efficient via caching and dive re-utilization, or by collecting dives during randomized restart search.

References

1. Katsirelos, G., Bacchus, F.: Generalized NoGoods in CSPs. In: Proc. of AAAI 2005, pp. 390–396 (2005)
2. Beck, J.C.: Multi-Point Constructive Search. In: van Beek, P. (ed.) CP 2005. LNCS, vol. 3709, pp. 737–741. Springer, Heidelberg (2005)
3. Gogate, V., Dechter, R.: A New Algorithm for Sampling CSP Solutions Uniformly at Random. In: Benhamou, F. (ed.) CP 2006. LNCS, vol. 4204, pp. 711–715. Springer, Heidelberg (2006)
4. Gogate, V., Dechter, R.: A Simple Application of Sampling Importance Resampling to Solution Sampling. In: Proc. of CP 2007, pp. 711–715 (2007)
5. Gomes, C., Sabharwal, A., Selman, B.: Near-Uniform Sampling of Combinatorial Spaces Using XOR Constraints. In: Proc. of the 20th Annual Conference on Neural Information Processing Systems, pp. 481–488 (2006)
6. Gomes, C., Shmoys, D.: Completing Quasigroups or Latin Squares: A Structured Graph Coloring Problem. In: Proc. of COLOR 2002, pp. 22–39 (2002)
7. Gomes, C.P., Shmoys, D.B.: Approximations and Randomization to Boost CSP Techniques. Annals OR 130(1-4), 117–141 (2004)
8. Grimes, D., Wallace, R.J.: Learning to Identify Global Bottlenecks in Constraint Satisfaction Search. In: Proc. of FLAIRS-20, pp. 592–598 (2007)
9. van Hoeve, W.J., Milano, M.: Decomposition Based Search. A theoretical and experimental evaluation. LIA Technical Report, University of Bologna (2003)
10. Papoulis, A., Pillai, S.U.: Probability, random variables and stochastic processes. Mac-Graw Hill, Newyork (2001)
11. Refalo, P.: Impact-based search strategies for constraint programming. In: Wallace, M. (ed.) CP 2004. LNCS, vol. 3258, pp. 557–571. Springer, Heidelberg (2004)
12. Ruml, W.: Adaptive Tree Search. Ph.D. thesis, Harvard University (2002)
13. Lombardi, M., Milano, M., Roli, A., Zanarini, A.: Deriving information from sampling and diving. LIA Tech report LIA-006-09,
    http://www-lia.deis.unibo.it/Research/TechReport/lia09006.pdf

Optimal Decision Tree Synthesis for Efficient Neighborhood Computation

Costantino Grana and Daniele Borghesani

Università degli Studi di Modena e Reggio Emilia
Via Vignolese 905/b - 41100 Modena
name.surname@unimore.it

Abstract. This work proposes a general approach to optimize the time required to perform a choice in a decision support system, with particular reference to image processing tasks with neighborhood analysis. The decisions are encoded in a decision table paradigm that allows multiple equivalent procedures to be performed for the same situation. An automatic synthesis of the optimal decision tree is implemented in order to generate the most efficient order in which conditions should be considered to minimize the computational requirements. To test out approach, the connected component labeling scenario is considered. Results will show the speedup introduced using an automatically built decision system able to efficiently analyze and explore the neighborhood.

1 Introduction

Many approaches of artificial intelligence aim at defining an effective decision support systems. Based on some input variables, coming from the environment status, the system must output the optimal decision, i.e. the decision that maximizes a certain utility, or equivalently minimize a certain cost function. In many common cases, like the local pixels neighborhood computation in an image processing algorithm as well as any problem-solving scenarios, the variables can be represented as boolean values. For this reason, the whole set of combinations (in terms of conditions to check and relative actions to take) can be effectively represented as decision tables. In this paper, we are proposing a general approach for this kind of problems based on the compact representation of environmental knowledge through decision tables. In particular, we will deal with decision tables containing possible multiple equivalent procedures to be performed for each action. This is a significant improvement over the classical situation in which for each set of input variables states a single procedure only must be performed. In fact, especially for complex problems, the combinatorial amount of possibilities is unfeasible to be explored exhaustively.

In this paper, we deal with a very specific application representable as a decision system applied to the computer vision, that is the connected component labeling problem. This approach could be naturally extended to every single algorithm requiring an effective way to compute data in the neighborhood of the current pixel under evaluation. Connected component labeling is a fundamental task in several computer vision applications, e.g. for identifying segmented visual objects or image regions. Thus a fast and efficient algorithm, able to minimize its impact on image analysis tasks, is undoubtedly

R. Serra and R. Cucchiara (Eds.): AI*IA 2009, LNAI 5883, pp. 92–101, 2009.

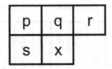

Fig. 1. The pixel mask $\mathcal{M}(x)$ used to compute the label of pixel x, and to evaluate possible equivalences

very advantageous. The research efforts in labeling techniques has a very long story, full of different strategies, improvements and results. Moreover typical applications where labeling is a necessary processing step often have to deal with high resolution images with thousands of labels: complex solutions for document analysis, multimedia retrieval and biomedical image analysis would benefit the speedup of labeling considerably. The proposed technique is able to dramatically reduce the computational cost of most of these algorithms, by means of an intelligent exploration of the neighborhood.

2   Decision Tables and Decision Trees

Briefly, connected components labeling take care of the assignment of an unique integer identifier (label) to every connected component of an image, in order to give the possibility to refer to it in the next processing steps. These algorithm work on a binary image where pixels with value 1 (foreground) represents the objects of interest (the components to be labeled), while pixels with value 0 are background. Typically the connection between foreground components is evaluated by means of *8-connectivity*, that is connectivity with 8-neighbors, while for background regions the *4-connectivity* is used. This usually better matches our usual perception of distinct objects: accordingly to the Gestalt Theory of perception we operate the closure property perceiving objects as a whole even if they are loosely connected as happens in the 8-connectivity case.

These algorithm perform a raster scan over the image. For each visited pixel, a mask of neighboring pixels is considered (usually the one in Fig. 1. During this process, a provisional label is associated to the pixel under analysis, and a structure of class equivalences of label is kept updated. Usually, the most modern and effective algorithm to manage it is the Union-Find algorithm. Once all provisional labels are computed, a second pass over the image is performed in order to solve equivalence and assign to each pixel the representative label of the class to which its label belongs.

The procedure of collecting labels and solving equivalences may be described by a command execution metaphor: the current and neighboring pixels provide a binary command word, interpreting foreground pixels as 1s and background pixels as 0s. A different action must be taken based on the command received. We may identify four different types of actions: *no action* is performed if the current pixel does not belong to the foreground, a *new label* is created when the neighborhood is only composed of background pixels, an *assign* action gives the current pixel the label of a neighbor when no conflict occurs (either only one pixel is foreground or all pixels share the same label), and finally a *merge* action is performed to solve an equivalence between two or more

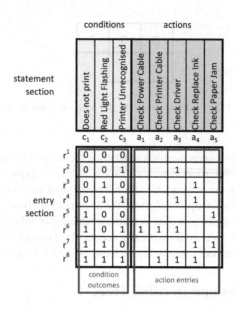

		conditions			actions				
		Does not print (c_1)	Red Light Flashing (c_2)	Printer Unrecognised (c_3)	Check Power Cable (a_1)	Check Printer Cable (a_2)	Check Driver (a_3)	Check Replace Ink (a_4)	Check Paper Jam (a_5)
	r^1	0	0	0					
	r^2	0	0	1			1		
	r^3	0	1	0				1	
entry	r^4	0	1	1			1	1	
section	r^5	1	0	0					1
	r^6	1	0	1	1	1	1		
	r^7	1	1	0				1	1
	r^8	1	1	1		1	1	1	
		condition outcomes			action entries				

Fig. 2. A decision table example, showing a hypothetical troubleshooting checklist for solving printing failures. Note that we use a vertical layout, which is more suitable when dealing with a large number of conditions.

classes and a representative is assigned to the current pixel. The relation between the commands and the corresponding actions may be conveniently described by means of a *decision table*.

A *decision table* is a tabular form that presents a set of conditions and their corresponding actions. A decision table is divided into four quadrants: an example is provided in Fig. 2. The statement section reports a set of conditions which must be tested and a list of actions to perform. Each combination of condition entries (*condition outcomes*) is paired to an *action entry*. In the action entries, a column is marked, for example with a "1", to specify whether the corresponding action is to be performed. If the conditions outcomes may only be true or false, the table is called *limited entry decision table*, and these will be the only tables we will use throughout the manuscript.

More formally, we call $c_1, \ldots, c_L$ the list of conditions. If we call S the *system status* (the lights on a printer, the service quality, current pixel neighborhood, etc...), a condition is a function of S which returns a boolean value. The list of actions is identified by $a_1, \ldots, a_M$, where an action is a procedure or operation which can be executed. Every row in the entry section is called a rule $r^1, \ldots, r^N$, which is a couple of boolean vectors of condition outcomes o_j^i and action entries e_k^i, denoting with i the rules index, with j the conditions index and with k the actions index. A *decision table* may thus be described as

$$\mathcal{DT} = \{r^1, \ldots, r^N\} = \{(\mathbf{o}^1, \mathbf{e}^1), \ldots, (\mathbf{o}^N, \mathbf{e}^N)\} \qquad (1)$$

The straightforward interpretation of a decision table is that the actions a_k corresponding to *true* entries e_k^i should be performed if the outcome $\mathbf{o}^i$ is obtained when testing the conditions. Formally, given the status S, we write

$$\mathbf{c}(S) = \mathbf{o}^i \Leftrightarrow o_j^i = c_j(S), \forall j = 1, \ldots, J, \tag{2}$$

so

$$\text{if } \mathbf{c}(S) = \mathbf{o}^i \text{ then execute } \{a_k | e_k^i = 1\}_{k=1,\ldots,K}. \tag{3}$$

The `execute` operation applied to a set of actions $\{a_k\}$, as in Eq. 3, classically requires the execution of **all** the actions in the set, that is all actions marked with 1s in the action entries vector: we call this behavior an *AND*-decision table. For our problem we define a different meaning for this operation, that is the *OR*-decision table, in which **any** of the actions in the set may be performed in order to satisfy the corresponding condition.

Note that this situation does not imply that the actions are redundant, in the sense that two or more actions are always equivalent. In fact, the result of doing any action in the execution set is the same only when a particular condition is verified.

3 Modeling Raster Scan Labeling with Decision Tables

In order to describe the behavior of a labeling algorithm with a decision table, we need to define the conditions to be checked and the corresponding actions to take. For this

x	p	q	r	s	no action	new label	x=p	x=q	x=r	x=s	x=p+q	x=p+r	x=p+s	x=q+r	x=q+s	x=r+s	x=p+q+r	x=s+p+q	x=p+r+s	x=q+r+s	x=p+q+r+s
0	-	-	-	-	1																
1	0	0	0	0		1															
1	1	0	0	0			1														
1	0	1	0	0				1													
1	0	0	1	0					1												
1	0	0	0	1						1											
1	1	1	0	0							1										
1	1	0	1	0								1									
1	1	0	0	1									1								
1	0	1	1	0										1							
1	0	1	0	1											1						
1	0	0	1	1												1					
1	1	1	1	0													1				
1	1	1	0	1														1			
1	1	0	1	1															1		
1	0	1	1	1																1	
1	1	1	1	1																	1

Fig. 3. The initial decision table providing a different action for every pixel configuration. To produce a more compact visualization, we reduce redundant logic by means of the indifference condition "-", whose values do not affect the decision and always result in the same action.

problem, as we already mentioned, the conditions are given by the fact that the current pixel and the 4 neighboring ones belong to the foreground. The conditions outcomes are given by all possible combinations of 5 boolean variables, leading to a decision table with 32 rules. The actions belong to four classes: *no action*, *new label*, *assign* and *merge*. Fig. 3 shows a basic decision table with these conditions and actions.

The action entries are obtained applying the following considerations:

1. if $x \in B$, then *no action* must be performed;
2. if $x \in F$ and $M(x) \subset B$, then $L(x) \leftarrow$ *new label*;
3. if $x \in F$ and $\exists! y \in M(x) | y \in F$, then $L(x) \leftarrow L(y)$;
4. else $L(x) \leftarrow merge(\{L(y) | y \in M(x) \cap F\})$

Using these considerations or the decision table, the equivalences are solved and a representative (provisional) label is associated to the current pixel x. The process then moves ahead to the next pixel and the next neighborhood accordingly.

Firstly, *merge* operations have a higher computational cost with respect to an *assign*, so we should reduce at the minimum the number of these operations in order to improve the performances of labeling. Similarly a *merge* between two labels is computationally cheaper than a *merge* between three labels. From the knowledge of the provisional labels, we can deduce that a lot of *merge* operations may be useless: for example when $p, q \in F$ and $L(p) \equiv L(q)$, the *merge* operation has no effect. In fact merging an equivalence class with itself obviously returns the same class again. Thus assigning a representative label from the merge outcome or any of $L(p)$ or $L(q)$ has the same result. So in this particular case, since our decision table is an *OR*-decision table we

X	P	Q	R	S	NO ACTION	NEW LABEL	X=P	X=Q	X=R	X=S	X=P+Q	X=P+R	X=P+S	X=Q+R	X=Q+S	X=R+S	X=P+Q+R	X=P+Q+S	X=P+R+S	X=Q+R+S	X=P+Q+R+S
								assign								merge					
0	-	-	-	-	1																
1	0	0	0	0		1															
1	1	0	0	0			1														
1	0	1	0	0				1													
1	0	0	1	0					1												
1	0	0	0	1						1											
1	1	1	0	0			1	1													
1	1	0	1	0							1										
1	1	0	0	1			1			1											
1	0	1	1	0				1	1												
1	0	1	0	1				1		1											
1	0	0	1	1												1					
1	1	1	1	0			1	1	1												
1	1	1	0	1			1	1		1											
1	1	0	1	1								1				1				
1	0	1	1	1				1	1	1											
1	1	1	1	1			1	1	1	1											

Fig. 4. The resulting decision table, after neighborhood optimizations. Bold 1's are selected with the procedure described in section 5.

set to 1 all the action entries of $L(x) \leftarrow L(p)$, of $L(x) \leftarrow L(q)$ and of $L(x) \leftarrow merge(\{L(p), L(q)\})$.

The problem with this reasoning is of course that we would need to add a condition for checking if $L(p) \equiv L(q)$, complicating enormously the decision process, since every condition doubles the number of rules. But, is this condition really necessary? No, in fact we can further notice that if we exploit an algorithm with online equivalences resolution, p and q **cannot have different labels**. Since they are 8-connected, if both of them are foreground, during the analysis of q a label equivalent to $L(p)$ would have been assigned to $L(q)$. This allows us to always remove merge operations between 8-connected pixels, substituting them with assignments of the involved pixels labels.

Extending the same considerations throughout the whole rule set, we obtain an effective "compression" of the table, as shown in Fig. 4. To obtain the table, when an operation could be substituted with a cheaper one, the more costly was removed from the table. Most of the *merge* operations were avoided, obtaining an *OR*-decision table with multiple alternatives between *assign* operations, and only in a single case between *merge* operations. Moreover the reduction leads also to the exclusion of many unnecessary actions (for example, the merge between p and q) without affecting the algorithm outcome.

4   Reducing the Cost of Conditions Testing

Testing the conditions of the decision table has a cost which is related to the number of conditions and the computational cost of testing each one. If we assume that testing each condition has the same cost, which is true in our application, the only parameter we may try to optimize is the number of conditions to be tested.

The definition of decision tables requires all conditions to be tested in order to select the corresponding action to be executed. But it is clear that there are a number of cases in which this is not the best solution. For example in the decision table of Fig. 4, if $x \in B$ we do not need to test all the other conditions, since the outcome will always be *no action*. This consideration suggests that the order with which the conditions are verified impacts on the number of tests required, thus on the total cost of testing.

What we are now looking for is the optimal ordering of conditions tests, which effectively produces a sequence of tests, depending on the outcome of previous tests. This is well represented by a *decision tree*. The sequence requiring the minimum number of tests corresponds to the decision tree with the minimum number of nodes. Our problem now is to transform a decision table in an optimal decision tree. This problem has been deeply studied in the past and we use the Dynamic Programming solution proposed by Schumacher [1].

One of the basic concepts involved in the creation of a simplified tree from a decision table is that if two branches lead to the same action the condition from which they originate may be removed. With a binary notation, if both the condition outcomes 10110 and 11110 require the execution of action 4, we can write that 1–110 requires the execution of action 4, thus removing the need of testing condition 2. The use of a dash implies that both 0 or 1 may be substituted in that condition, representing the concept of *indifference*. The saving given by the removal of a test condition is called *gain* in the algorithm, and we conventionally set it to 1.

The conversion of a decision table (with n conditions) to a decision tree can be interpreted as the partitioning of an n-dimensional hypercube (n-cube in short) where the vertexes correspond to the 2^n possible rules. Including the concept of indifferences, a t-cube corresponds to a set of rules and can be specified as an n-vector of t dashes and $n - t$ 0's and 1's. For example, 01–0– is the 2-cube consisting of the four rules $\{01000, 01001, 01100, 01101\}$. In summary, Schumacher's algorithm proceeds in steps as follows:

- Step 0: all 0-cubes, that is all rules, are associated to a single corresponding action and a starting gain of 0 (this means that if we need to evaluate the complete set of condition, we do not get any computational saving).
- Step t: all t-cubes are enumerated. Every t-cube may be produced by the merge of two $(t - 1)$-cubes in t different ways (for example 01–0– may be produced by the merge of $\{01–00, 01–01\}$ or of $\{0100–, 0110–\}$). For each of these ways of producing the t-cube (denoted as s in the following formulas) we compute the corresponding gain G_s as

$$G_s = G_s^0 + G_s^1 + \delta[A_s^0 - A_s^1] \tag{4}$$

where G_s^0 and G_s^1 are the gains of the two $(t - 1)$-cubes in configuration s, and A_s^0 and A_s^1 are the corresponding actions to be executed. δ is the Kronecker function that provides a unitary gain if the two actions are the same or no gain otherwise, modeling the fact that if the actions are the same we "gain" the opportunity to save a test. The gain assigned to the t-cubes is the maximum of all G_s, which means that we choose to test the condition allowing the maximum saving.

Analogously we have to assign an action to the t-cube. This may be a real action if all rules of the t-cube are associated to the same action, otherwise it is 0, a conventional way of expressing the fact that we need to branch to choose which action to perform. In formulas:

$$A = A_s^0 \cdot \delta[A_s^0 - A_s^1] \tag{5}$$

where s may be chosen arbitrarily, since the result is always the same.

The algorithm continues to execute Step t until $t = n$, which effectively produces a single vector of dashes. The tree may be constructed by recursively tracing back through the merges at each t-cube. A leaf is reached if a t-cube has an action $A \neq 0$.

5   Action Selection in *OR*-Decision Tables

The algorithm described is able to produce an optimal tree from a decision table where every rule leads to *a single action*. Starting from an *AND*-decision table, a single action decision table is straightforward to obtain: for every distinct row of action entries e^i we can define a *complex action* in the form of the set of actions $A_l = \{a_k | e_k^i = 1\}_{k=1,...,K}$. The execution of A_l requires the execution of all actions in A_l. Now we can associate to every conditions outcomes an integer index, which points to the corresponding complex action.

This is not so easy in *OR*-decision tables, since it is useless to perform multiple actions, so we need to select one of the different alternatives provided in e^i. While the execution any of the different actions does not change the result of the algorithm, the optimal tree derived from a decision table implementing these arbitrary choices may be different. So we are now faced with another optimization problem: how do we select the best combination of actions, in order to minimize the final decision tree? To our knowledge, no algorithm exists save exhaustive search, thus we propose an heuristic greedy procedure.

The rationale behind our approach is that the more rules require the execution of the same action, the more likely it will be to find large k-cubes covering that action. For this reason we take a greedy approach in which we count the number of occurrences of each action entry, then we iteratively consider the most common one, and for each rule where this entry is present we remove all others, until no more changes are required. In case two actions have the same number of entries, we arbitrarily chose the one with lower index. The resulting table after applying this process is shown in Fig. 4, with bold faces 1's. The following algorithm formalizes the procedure.

Algorithm 1. Pseudo-code of our heuristic approach to select which action to perform in *OR*-decision tables

```
 1: I = {1, . . . , K}                              ▷ Define actions indexes set
 2: while I ≠ ∅ do
                    N
 3:     k* ← arg max ∑ eₖⁱ                          ▷ Find most frequent action
              k∈I    i=1
 4:     for i = 1, . . . , N do                     ▷ Remove equivalent actions
 5:         if eₖⁱ = 1 then
 6:             eₖⁱ ← 0, ∀k ≠ k*
 7:         end if
 8:     end for
 9:     I = I − k*                                   ▷ This action has been done
10: end while
```

The described approach does not always lead to an optimal selection, but the result is often nearly optimal, based on many different experiments. This is particularly true when the distribution of the actions frequencies is strongly non uniform. For example, from the original *OR*-decision table in Fig. 4, it is possible to derive 3 456 different decision tables, by selecting all permutations of equivalent actions. Using Algorithm 1 only two actions are chosen arbitrarily, leading to 4 possible equivalent decision trees. All of these have the same number of nodes and are optimal (in this case we were able to test all of the 3 456 possibilities).

We provided an algorithmic solution to the optimal neighborhood exploration problem that, with respect to previous approaches, has an important added value: it can be naturally extended to larger problems, without requiring any empirical workaround.

6 Results

The analysis conducted so far evidenced that this approach can be considered general on all the cases in which exists a more effective way to explore a neighborhood, thus when the choice of which condition to verify and the respective order of conditions brings different outcomes in terms of effectiveness (finer result, less elaboration time, etc...). So we selected some connected components algorithms as representative of the most widely used technique proposed in literature measuring their performances. In order to propose a valuable comparison, we mainly exploited a custom dataset composed by the Otsu-binarized versions of 615 images of high resolution (3840x2886) illuminated manuscripts pages, with gothic text, pictures and great floral decorations. This dataset gives us the possibility to test the connected components labeling capabilities (thus the effectiveness of the automatic neighborhood optimization we introduced) with very complex patterns at different sizes, with an average resolution of 10.4 megapixels and up to 80 thousands labels, providing a difficult dataset which heavily stresses the algorithms. The following techniques were selected for the comparison:

- Di Stefano *et al.* [2]. Authors proposed an evolution of the classical Rosenfeld's approach [3], but with the online label resolution algorithm. An array-based structure is used to store the label equivalences, and their solution requires multiple searches over the array at every merge operation.
- Suzuki *et al.* [4]. Authors resumed Haralick's approach [5], including a small equivalence array and providing a linear-time algorithm that in most cases requires 4 passes. The label resolution is performed exploiting array-based data structures, and each foreground pixel takes the minimum class of the neighboring foreground pixels classes. An important addition to this proposal is provided in an appendix in the form of a LUT of all possible neighborhoods, which allows to reduce computational times and costs by avoiding unnecessary Union operations.
- Chang *et al.* [6]. Authors proposed a single pass over the image exploiting contour tracing technique for internal and external contours, with a filling procedure for the internal pixels. This technique proved to be very fast, even because the filling is cache-friendly for images stored in a raster scan order, and the algorithm can also naturally output the connected components contours.
- OpenCV labeling algorithm based on a contour tracing (cvFindContours) followed by a contour filling (cvDrawContours).
- He *et al.* [7]. This is one of the most recent approach for connected components labeling. An efficient data structure is exploited to manage the label resolution by means of three arrays in order to link the sets of equivalent classes without the use of pointers. We applied our optimized tree to this algorithm to test the speed up we could provide by optimally accessing neighbor pixels.

Results of our experiments are shown in Table 1. These tests show that He's approach outperforms all the others, and only the contour tracing technique by Chang can be considered a real competitor. The good results of the He's approach may be further improved by the decision tree automatically generated with our approach. As mentioned before, theoretically the OR-decision table leads to an enormous amount of different decision trees. In simple cases (like this one, where we have to deal with only

Table 1. Time comparisons of different labeling algorithms on a large dataset. The proposed optimization can further speed up the best performing algorithm by reducing the number of conditions which are to be verified.

Algorithm	ms
cvFindContours	1086.67
Suzuki (w mask)	512.04
DiStefano	489.05
Chang	177.25
He	163.55
He (optimized)	145.28

5 variables), the exhaustive search of all the possible combination of equivalent actions is feasible. In our test, we observed that 4 of the 3456 possible combinations lead to a decision tree equivalent to the one obtained with our heuristic.

7   Conclusions

In this paper, we have illustrated a general approach for reducing the time complexity of a broad class of problems based on the compact representation of environmental knowledge through decision tables. In particular, we described *OR*-decision tables, which contain possible multiple equivalent procedures to be performed for each action. Results providing an example application of this technique have been reported.

References

1. Schumacher, H., Sevcik, K.C.: The synthetic approach to decision table conversion. Communications of the ACM 19(6), 343–351 (1976)
2. Di Stefano, L., Bulgarelli, A.: A simple and efficient connected components labeling algorithm. In: International Conference on Image Analysis and Processing, pp. 322–327 (1999)
3. Rosenfeld, A., Pfaltz, J.L.: Sequential operations in digital picture processing. Journal of ACM 13(4), 471–494 (1966)
4. Suzuki, K., Horiba, I., Sugie, N.: Linear-time connected-component labeling based on sequential local operations. Computer Vision and Image Understanding 89, 1–23 (2003)
5. Haralick, R.: Some neighborhood operations. In: Real Time Parallel Computing: Image Analysis, pp. 11–35. Plenum Press, New York (1981)
6. Chang, F., Chen, C.J.: A component-labeling algorithm using contour tracing technique. In: International Conference on Document Analysis and Recognition, pp. 741–745 (2003)
7. He, L., Chao, Y., Suzuki, K.: A linear-time two-scan labeling algorithm. In: International Conference on Image Processing., vol. 5, pp. 241–244 (2007)

Mathematical Symbol Indexing

Simone Marinai, Beatrice Miotti, and Giovanni Soda

Dipartimento di Sistemi e Informatica
University of Florence, Italy
marinai@dsi.unifi.it

Abstract. This paper addresses the indexing and retrieval of mathematical symbols from digitized documents. The proposed approach exploits Shape Contexts (SC) to describe the shape of mathematical symbols. Indexed symbols are represented with a vector space-based method that is grounded on SC clustering. We explore the use of the Self Organizing Map (SOM) to perform the clustering and we compare several approaches to compute the SCs. The retrieval performance are measured on a large collection of mathematical symbols gathered from the widely used INFTY database.

1 Introduction

The growing use of Digital Libraries has made possible a public access to large collections of documents with various scopes and features. Usually, the documents are scanned and stored as images, then they are processed and encoded to make easier the information retrieval process. A common way to process the image is the use of an Optical Character Recognition (OCR) engine to recognize its textual content. The OCR packages perform well on contemporary printed text written in the main languages. However, the recognition of several types of documents, such as early printed books, handwritten text, and graphical documents, is still an open problem. Due to the large number of scientific and technical documents that are nowadays available in digital libraries and due to the wish of being able to recover mathematical formulae from digital libraries, several efforts have been devoted to build systems which are able to recognize the mathematical expressions embedded in printed documents. The recognition of mathematical symbols is particularly difficult for three main reasons: the very large number of symbol classes, the reduced script size for superscripts and subscripts, and the lack of linguistic tools (such as dictionaries) that could help in the recognition.

In [1] and in [2] the limits of commercial OCR tools for the recognition of mathematical expressions are shown. Common OCR engines perform with a great accuracy when processing the textual part of the documents, but fail when dealing with mathematical expressions, images or tables. Both [1] and [2] proposed to use a commercial OCR engine to analyze the textual parts of the documents and then use a character recognition engine specifically developed for mathematical symbols. The main problem related to the conversion of scientific papers from

R. Serra and R. Cucchiara (Eds.): AI*IA 2009, LNAI 5883, pp. 102–111, 2009.

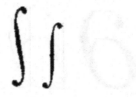

Fig. 1. Two integral symbols with different shapes

printed to electronic form is related to the large variety of mathematical expressions. In a scientific article a formula can appear *embedded*, that is in the same line of text, or *displayed*, that is typed in separate line and usually well separated from the rest of the text [3]. A formula can present different printing levels and therefore a symbol can appear as superscript or as subscript. Usually a formula contains symbols with different sizes and fonts (e.g. from Greek and German alphabet) depending on their meaning.

According to [4], a system for mathematical expression analysis is made up by four steps. The first step pertains to the *Layout Analysis*. The document is segmented and each region is classified as "Text" or "Formula". In [5] is described a bottom-up technique which relies on the connected components extraction. In [6] the top-down projection profile technique is used to deal with academic papers. The second step pertains to the *Symbol Segmentation*. As soon as a formula has been located in the document and recognized as *embedded* or *displayed*, it is necessary to isolate each elementary symbol. The third step concerns the *Symbol Recognition*. The symbols, isolated in the previous step, are classified as belonging to a particular set of characters. To this purpose, symbols are described by means of feature vectors, which can be related to the pixels intensity [7] or to physical peculiarities, such as bounding box, height, width of the symbol [1], [8]. The last step concerns the Structural Analysis. To reconstruct the structure of a mathematical formula, the relation trees among various symbols are built according to the mutual physical positions or to logic considerations [9]. The results are usually verified with context-free grammars [4], [10] or with search algorithms in the space of all the possible spanning-trees [1], [11].

In our work we attempt to use a different approach with respect to the recognition-based one that is adopted in most cases. Rather than extracting the whole mathematical symbol structure we index all the symbols in a page (including the textual parts) and we subsequently identify the mathematical formaulae on the basis of a query by example paradigm.

Document image retrieval techniques have been seldom used to process mathematical expressions. However, several researchers envisage the utility of math search systems that would be able not only to search for text, but also for "fine-grain mathematical data" (e.g. equations and functions) [12]. Most search systems for mathematical documents rely on markup languages designed for mathematical formulae. One example is the MathWebSearch system that harvests the Web for formulae, subsequently indexed using MathML or OpenMath representations [13].

Fig. 2. Process to identify the set P of points in the symbol shape (on the rightmost image)

In this paper, we focus on the symbol indexing and retrieval module of the mathematical equation indexing. In section 2 we describe the indexing and retrieval method that is based on Shape Context. The Infty database is analyzed in Section 3 and used in the experiments described in Section 4. Conclusions and future work are drawn in Section 5.

2 Symbol Indexing

Similarly to words in a text document, an image contains local interest points that concentrate most of its information. The set of symbols on which we investigate is very large, since it includes symbols from Greek, German, Latin alphabets, mathematical symbols and some of them are in different styles, fonts and size. If we consider two images (Fig. 1) of the same symbol which appear rather different, a human observer is able to glean over the differences and basing on the pure shape he can assert these images represent the same symbol.

According to [14] two approaches can be considered to describe an image: feature-based, which involves the use of physical peculiarities of the symbol, such as edge elements and connected components; brightness-based, which uses directly the pixel gray values. Since mathematical symbols in most cases correspond to separated connected components, we use a feature-based approach that is more appropriate to describe the symbol shape. Shape Contexts [14] can be used to describe a shape considering the internal and external symbol contours. Instead of using only the silhouette of the symbol, that gives a limited information because it ignores the internal points, we describe an approach that treats the shape as a set of 2-D points.

In general similar shapes have similar descriptors and therefore different symbols can be compared considering the SCs and then establishing a similarity measure. The symbol image is first processed to find out the internal and external contours that are described as a finite set of points. From this set a subset P of sampled points is extracted as representative of the symbol shape (Figure 2).

The Shape Context for each point p_i in P can be computed by considering the relative position of the other points in P. In particular, the SC for p_i is obtained by computing a coarse histogram h_i whose bins are uniform in log-polar space (Figure 3 (a), (b)). Let m be the cardinality of P and p_j be one of the remaining $m - 1$ points in P. The point p_j is assigned to the appropriate

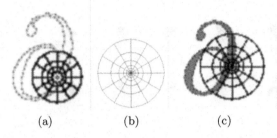

(a)   (b)   (c)

Fig. 3. (a) One example of grid overlapped to the mathematical symbol used to compute the Shape Context. (b) The logarithmic mask. (c) The SC_AllPoints approach.

bin according to the logarithm of the Euclidean distance between p_i and p_j and to the direction of the link between p_i and p_j. The histogram h_i is defined to be the Shape Context of p_i. The m shape context vectors describe the configuration of the whole symbol. Shape contexts are invariant to translations since all the measures are taken with respect to points on the object. SCs can be adapted to be scale invariant as well as rotation invariant [14]. However, in our work we did not considered the rotation invariance to avoid ambiguities in the retrieval process that could bring to confuse, for instance, 6 with 9.

In order to address the peculiarities of mathematical symbol indexing we adapted the SC computation in two ways that are briefly described in the following.

By using large values of the radius it is possible to embrace the whole image and the SC of one point is influenced by points very far from it. On the other hand, with a small radius the points that fall out of the last bins of the mask has to be considered. Two alternatives are possible. In the first case all the external points are included in the last bins. This approach will bring to a SC whose external bins are very dense with values significantly higher than the others. On the opposite, if the external points are not counted the resulting SC will describe a little portion of the symbol shape with the risk to loose information.

To face this problem and to find out the most suitable radius size we performed various experiments which are reported in detail in Section 4. We experimented some values related to the average image size (AWI): a very large radius that is about 4 times larger than AWI, an halfway radius that is a little bit smaller than AWI and finally a small radius which is about the half of AWI. From these experiments it turned out that the second approach should be preferred in most cases.

To increase the robustness against the noise in the symbols, we modified the SC algorithm computing h_i by counting the number of all the symbol points that belong to each bin instead of considering only the points in P (Fig. 3 (c)). In so doing, each SC bin is more populated and more informative. As a matter of fact the symbols belonging to the Infty dataset have a quite small size (on the average 20 x 30 pixel), consequently the number of contour pixels is low and, using the classic algorithm, only a few bins of the corresponding SCs would be filled with some points.

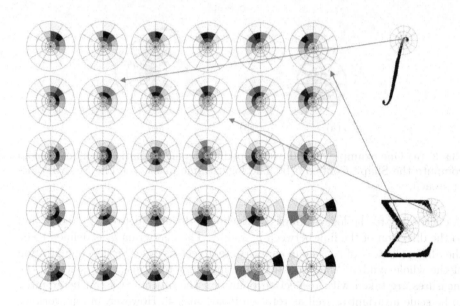

Fig. 4. Visual words obtained by SOM clustering. Each circle is a graphical representation of one cluster centroid. We show also two symbols with a reference to some visual words.

The comparison between the Shape Contexts in a query image and those in each indexed object can provide a very accurate similarity evaluation among objects. However, this process can be time consuming and cannot be considered when dealing with large data-sets. To address the latter problem the shape representation is transformed adopting techniques used in the vector space model of Information Retrieval. The vector quantization is first performed by clustering the vector representations and then identifying each vector by the index of the cluster it belongs to. Going on with the textual analogy, each cluster is considered as a word and each symbol can be represented by the frequencies of each "visual-word" in its description [15].

The peculiarity of the approach described in this paper is the use of Self Organizing Maps (SOM) to perform the vector quantization. One advantage of this clustering method with respect to other clustering methods such as K-means is the topological organization of the cluster centers. As an example Figure 4 depicts a portion of an SOM used to index the mathematic symbols together with two symbols. Each cluster is pictorially depicted by reconstructing a virtual Shape Context that corresponds to the values in the SOM.

3   Infty DataSet

In our experiments we used a dataset built in the INFTY project [16]. This dataset is partitioned in two parts: the InftyCDB-3-A consists of 188,752 symbols (scanned at a resolution of 400 dpi), the InftyCDB-3-B is made up by 70,637 symbols (scanned at a resolution of 600 dpi). Symbols in InftyCDB-3-A have been extracted from 19 articles printed by various publishers, from scanned printouts of fonts used internally on Windows and Macintosh computers, from LaTeX fonts and finally there are some infrequent mathematical fonts taken from mathematical articles. The InftyCDB-3-B dataset consists of images extracted from 20 mathematics articles. Both datasets are delivered with a groundtruth file containing the symbol code of each character (symbols that look identical in print have been assigned the same symbol code, e.g. the summation symbol $\sum$ is not distinguished from the Greek letter Σ). The whole Infty dataset includes 393 different classes.

4 Experiments

In our experiments, the computation of the SC clusters has been performed on a set of $22,923$ symbols, belonging to 53 pages randomly selected from the whole dataset (which consists of 346 pages). From each symbol we extract around 50 SCs so that we used 1,102,049 feature vectors for clustering. In the retrieval, we index both the InftyCDB-3-A and the InftyCDB-3-B datasets with a total of 259,357 symbols. We have performed several experiments to compare the different approaches used to select the size of the SC grid as described in Section 2. To evaluate the retrieval effectiveness we use the *Precision-Recall* curves and a single numerical value: the Precision at 0 % Recall. The Recall is defined as the fraction of the relevant symbols which have been retrieved: $Recall = \frac{|tp|}{|tp+fn|}$. The Precision is defined as the fraction of retrieved symbols which is relevant: $Precision = \frac{|tp|}{|tp+fp|}$, where tp (true positive) are the retrieved symbols which are relevant, fp (false positive) are the retrieved symbols which are not relevant, fn are the relevant symbols which have not been retrieved. We computed the P-R curve for interpolation after estimating the precision when the recall was a multiple of 10.

In the experiments that we carried out we made 392 queries randomly selected from the dataset. To obtain a correct comparison among methods, we always used the same set of 392 queries.

In the first experiment we compare the results that are achieved when varying the radius of the logarithmic mask used to evaluate the SCs. We compare three approaches: in the first (*Range1*) the circular mask has a radius of 90

Table 1. Precision at 0 % Recall for a SOM with 200 centroids, when the radius of the logarithmic mask varies on different ranges

	Range1	Range2	Range 3
SC_Std	42.31	84.97	74.07
SC_AllPoints	86.42	85.54	89.34

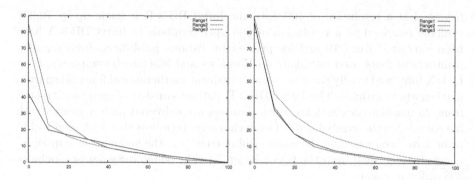

Fig. 5. Precision-Recall curves for SC_Std (left) and SC_AllPoints (right). See also Table 1.

pixels and it embraces the whole symbol. In the second (*Range2*) the mask has a radius of 12 pixels and the points outside the last bin in each direction are collapsed to the last one (the upper bound on the grid is set to infinite). In the last (*Range3*) we consider a radius of 20 pixels and the points outside the last bin are discarted. In this experiment we compare also the standard approach to compute SCs ("SC_Std") and the method that has been described in Section 2 ("SC_AllPoints"). Table 1 reports the results obtained with a clustering based on an SOM containing 200 neurons, when varying the range and the method used to compute the SCs. Figure 5 contains the corresponding P-R curves. From this experiment we can conclude that an intermediate dimension of the logarithmic mask, combined with the SC_AllPoints computation of SCs provides better results.

In the second experiment we used the *Range3* setting and we compare two clustering methods to compute the visual dictionary: SOM and K-means. Table 2 and Figure 6 describe the results of this experiment that basically confirm the findings of the first experiment about the choice of the method used to compute the SC. The SOM clustering also performs better than K-means as we already shown in our preliminary experiments [17].

We further investigate the comparison among the methods used to compute the symbol similarity in the third experiment where we use bins belonging to the *Range3* and the SC_AllPoints approach. In this experiment we compare

Table 2. Value of Precision corresponding at 0 % Recall for the two different methods of SCs evaluation, for 100 centroids

Methods	K-means (%)	SOM (%)
SC_Std	74.69	73.17
SC_AllPoints	79.93	86.81

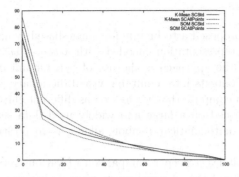

Fig. 6. Precision-Recall curves corresponding to Table 2

Table 3. Value of Precision at 0 % Recall for *SC_AllPoints* on the retrieval of 392 queries, with 100 centroids

K-means	SOM sim	SOM sim4	SOM sim8
79.93	86.81	86.94	87.06

the results achieved with the K-means clustering the SOM-based clustering, and some variants. *sim4* and *sim8* describe the results obtained when considering, in the similarity computation, the topological organization of the SOM (additional details can be found in [17]).

The results of the last experiment have been evaluated for the SC_AllPoints approach with 100 centroids (Table 3). Similar results have been obtained when considering 200 centroids. Even if increasing the number of clusters, the performance of the SOM retrieval approach performs better. Some examples of queries with the corresponding results are shown in Figure 7.

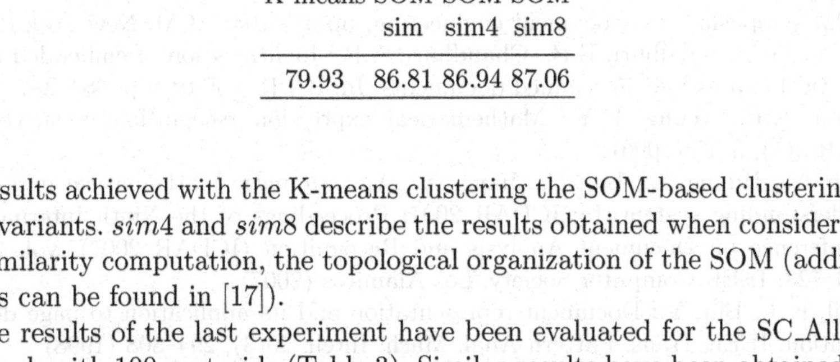

Fig. 7. Some examples of queries with the first 10 retrieved symbols

5   Conclusions

In this paper, we proposed a technique for image-based symbol retrieval based on Shape Context representation encoded with a *bag of visual words* method. The peculiarity of the approach is the use of Self Organizing Maps for clustering the Shape Contexts into a suitable visual dictionary. We also compared various methods to compute the SCs as well as different clustering approaches. Experiments performed on a large and widely used data set containing both alphanumeric and mathematical symbols allows us to positively evaluate the proposed approach.

Future work includes a deeper comparison of various retrieval approaches. We aim also to extend the retrieval mechanism to incorporate the structural information of the formulae in the retrieval algorithm.

References

1. Suzuki, M., Tamari, F., Fukuda, R., Uchida, S., Kanahori, T.: Infty: an integrated ocr system for mathematical documents. In: DocEng 2003: Proceedings of the, ACM symposium on Document engineering, pp. 95–104. ACM, New York (2003)
2. Garain, U., Chaudhuri, B.B., Chaudhuri, A.R.: Identification of embedded mathematical expressions in scanned documents. In: ICPR, vol. 01, pp. 384–387 (2004)
3. Chan, K.F., Yeung, D.Y.: Mathematical expression recognition: a survey. IJDAR 3(1), 3–15 (2000)
4. Guo, Y., Huang, L., Liu, C., Jiang, X.: An automatic mathematical expression understanding system. In: ICDAR 2007: Proceedings of the Ninth International Conference on Document Analysis and Recognition (ICDAR 2007), vol. 2, pp. 719–723. IEEE Computer Society, Los Alamitos (2007)
5. Anil, K.J., Bin, Y.: Document representation and its application to page decomposition. IEEE Trans. Pattern Anal. Mach. Intell. 20(3), 294–308 (1998)
6. Chang, T.Y., Takiguchi, Y., Okada, M.: Physical structure segmentation with projection profile for mathematic formulae and graphics in academic paper images. In: ICDAR 2007: Proceedings of the Ninth International Conference on Document Analysis and Recognition (ICDAR 2007), vol. 2, pp. 1193–1197. IEEE Computer Society Press, Los Alamitos (2007)
7. Takiguchi, Y., Okada, M., Miyake, Y.: A study on character recognition error correction at higher level recognition step for mathematical formulae understanding. In: 18th International Conference on Pattern Recognition, ICPR 2006, vol. 2, pp. 966–969 (2006)
8. Garain, U., Chaudhuri, B.B.: An Approach for Recognition and Interpretation of Mathematical Expressions in Printed Document. PAA, 120–131 (2000)
9. Toyota, S., Uchida, S., Suzuki, M.: Structural analysis of mathematical formulae with verification based on formula description grammar. In: Bunke, H., Spitz, A.L. (eds.) DAS 2006. LNCS, vol. 3872, pp. 153–163. Springer, Heidelberg (2006)
10. Garain, U., Chaudhuri, B.: A syntactic approach for processing mathematical expressions in printed documents. In: ICPR, vol. 04, p. 4523 (2000)
11. Kanahori, T., Suzuki, M.: Refinement of digitized documents through recognition of mathematical formulae. In: DIAL 2006: Proceedings of the Second International Conference on Document Image Analysis for Libraries, Washington, DC, USA, pp. 297–302. IEEE Computer Society Press, Los Alamitos (2006)

12. Youssef, A.M.: Roles of math search in mathematics. In: Borwein, J.M., Farmer, W.M. (eds.) MKM 2006. LNCS (LNAI), vol. 4108, pp. 2–16. Springer, Heidelberg (2006)
13. Kohlhase, M., Sucan, I.: A search engine for mathematical formulae. In: Calmet, J., Ida, T., Wang, D. (eds.) AISC 2006. LNCS (LNAI), vol. 4120, pp. 241–253. Springer, Heidelberg (2006)
14. Belongie, S., Malik, J., Puzicha, J.: Shape matching and object recognition using shape contexts. IEEE Transactions on Pattern Analysis and Machine Intelligence 24(4), 509–522 (2002)
15. Yang, J., Jiang, Y.G., Hauptmann, A.G., Ngo, C.W.: Evaluating bag-of-visual-words representations in scene classification. In: MIR 2007: Proceedings of the international workshop on Workshop on multimedia information retrieval, pp. 197–206. ACM, New York (2007)
16. Suzuki, M., Tamari, F., Fukuda, R., Uchida, S., Kanahori, T.: Infty: an integrated ocr system for mathematical documents. In: DocEng 2003: Proceedings of the, ACM symposium on Document engineering, pp. 95–104. ACM, New York (2003)
17. Marinai, S., Miotti, B., Soda, G.: Mathematical symbol indexing using topologically ordered clusters of shape contexts. In: Int'l Conference on Document Analysis and Recognition, pp. 1041–1045 (2009)

Local Kernel for Brains Classification in Schizophrenia

U. Castellani[1], E. Rossato[1], V. Murino[1], M. Bellani[2],
G. Rambaldelli[2], M. Tansella[2], and P. Brambilla[2,3]

[1] VIPS lab, University of Verona, Italy
[2] Department of Medicine and Public Health, University of Verona, Italy
[3] ICBN Center, University of Udine and Verona, Italy

Abstract. In this paper a novel framework for brain classification is proposed in the context of mental health research. A learning by example method is introduced by combining local measurements with non linear Support Vector Machine. Instead of considering a voxel-by-voxel comparison between patients and controls, we focus on landmark points which are characterized by local region descriptors, namely Scale Invariance Feature Transform (SIFT). Then, matching is obtained by introducing the *local* kernel for which the samples are represented by unordered set of features. Moreover, a new weighting approach is proposed to take into account the discriminative relevance of the detected groups of features. Experiments have been performed including a set of 54 patients with schizophrenia and 54 normal controls on which region of interest (ROI) have been manually traced by experts. Preliminary results on *Dorsolateral PreFrontal Cortex* (DLPFC) region are promising since up to 75% of successful classification rate has been obtained with this technique and the performance has improved up to 85% when the subjects have been stratified by sex.

1 Introduction

Computational neuroanatomy using magnetic resonance imaging (MRI) is a growing research field, which utilizes image analysis methods to quantify morphological characteristics of different brains [1]. Specifically, structural data (i.e., 3DA [2]) are crucial to explore the content of grey and white matter tissue, the volumes of specific structures and the 3D shape morphology of particular brain regions. Therefore, the overall aim is to identify structural brain abnormalities by comparing normal subjects with patients affected by a certain disease. The underline hypothesis consists of the fact that there is a relation between such structural abnormalities and the considered disease. Roughly speaking there are two main categories of methods: (i) methods based on the analysis of Region of Interest (ROI), and (ii) methods based on Voxel-based-Morphometry (VBM)[3]. ROI-based methods are focusing on a limited set of brains subparts which are manually traced by experts. Such regions are in general related to well known functional parts in the brain. Methods based on VBM utilize the whole brain

R. Serra and R. Cucchiara (Eds.): AI*IA 2009, LNAI 5883, pp. 112–121, 2009.

after a normalization procedure which maps the current brain onto a standard reference system, namely the *stereotaxic* space. In this fashion, a voxel-by-voxel correspondence is available among the analyzed subjects. In this case the idea is to exploit also brains subparts which have not a clear meaning a priori known.

In this paper we focus on mental health research by studying subjects affected by schizophrenia. A ROI-based method is proposed by introducing a machine learning [4] approach to classify healthy (i.e., controls) and unhealthy (i.e., patients) subjects. Several works have been proposed recently for human brain classification in the context of schizophrenia research [5,6,7]. Classical methods are based on a comparison of volumetric measurements [3,2] which in general confirmed a statistically significative volume reduction in certain brain structures for schizophrenic patients [2]. Therefore, in order to better identify the structural abnormalities other and more promising approaches are focused on: (i) shape characterization[5], (ii) surface computation [7], and (iii) high dimension pattern classification [6]. In [5] a ROI-based morphometric analysis is introduced by defining a spherical harmonics and a 3D skeleton as shape descriptors. Improvement of such shape-descriptor-based approach with respect to classical volumetric techniques is experimentally shown. In [7] a Support Vector Machine (SVM) has been proposed to classify cortical thickness which has been measured by calculating the Euclidean distance between linked vertices on the inner and outer cortical surfaces. In [6] a new morphological signature has been defined by combining deformation-based morphometry with SVM. In this fashion, multivariate relationships among various anatomical regions have been capture for more effectively characterizing group differences. In this paper we analyze the *Dorso-lateral PreFrontal Cortex* (DLPFC) region since it has already shown its relation with schizophrenia disease [8]. Few and significant landmarks are detected and characterized by local region descriptors. Here, we focus on the Scale Invariant Feature Transform (SIFT) operator [9], which has been proposed in computer vision for object recognition and it has been successfully used also on medical applications for 3D deformable image registration (e.g., in [10]). Therefore, the novelty consists of characterizing brains abnormalities in terms of intra-ROI local pattern not necessarily spatially coherent. The underline hypothesis consist of relaxing the common constraint that morphological anomalies appear at the same voxel location for all the population. Therefore, a new *kernel* of a SVM is designed in order to allow the comparison between a pair of brains represented by an unordered set of features. The proposed method is inspired by the *Bag-of-Words*[11] paradigm which implicitly implements the feature matching within the SVM framework [12]. Such kind of kernels are known as *local kernel* [11,12] in order to emphasize the fact that local information is used to characterize the involved objects. Finally, a weighting function is introduced to define the relevance of the detected groups of features, namely the *visual words*, in discriminating among the two populations (i.e., patients and controls). Moreover, the proposed approach is able to take into account of different morphological abnormalities at the same time, which are possible spread onto the analyzed ROI.

The rest of the paper is organized as following. Section 2 introduces the proposed method by describing the main phases involved in our brain classification pipeline. Section 3 reports the results for the whole dataset and after the stratification by age and sex. Finally, conclusions are drawn in Section 4 and future work is envisage.

2   Proposed Method

The proposed method is based on three main phases: (i) landmarks points detection and description, (ii) *feature* vocabulary construction and relevance computation, (iii) SVM-based brains classification.

2.1   Landmarks Points Detection and Description

Landmarks points are detected robustly by applying the difference of Gaussians (DoG) point detector [9] on each slice, which selects sparse blob-like patches invariant to translation, rotation, and scale. Moreover, since each voxel is observed from three projection planes (i.e., *axial*, *sagittal*, and *coronal*) only those points detected onto at least two projections are kept, in order to increase the robustness of the detected point. Therefore, for each landmark the SIFT descriptor [9] is applied to characterized its local neighborhood. In particular, the SIFT descriptor encodes several histograms of pre-defined image-gradient directions by forming a multidimensional vector (i.e., a $128 - d$ vector). Note that the neighborhood region is automatically defined for each landmark from the detection phase. Figure 1 shows several succeeding slices from the DLPFC region with the extracted landmarks and their region of influence (i.e., the neighborhood). In this fashion we select from each brain the most characteristic patches in terms of strong local pattern variations. Here, the main idea consists of verify whether among those variations there are brain anomalies.

2.2   *Feature* Vocabulary Construction and Relevance Computation

After the landmarks detection each brain is represented by a set of unordered feature vectors. Moreover, such sets generally appear with different cardinality. In order to allow a comparison between a pair of brains a *Bag-of-Words* (BoW) approach is introduced[11]. In the BoW paradigm a *text* (i.e., a document) is represented as an unordered collection of words, disregarding grammar and even word order. The extension of BoW to visual data requires one to build a *visual vocabulary*, i.e., a set of the visual analog of words. Here, as for the 2D image classification [11], the *visual words* are obtained by clustering local point descriptors (i.e., the visual words are the cluster centroids). In practice, the clustering defines a vector quantization of the whole point descriptor space, composed of all the landmarks points extracted from all the brains composing the training set. In order to obtain the clustering, the K-means algorithm is employed [4].

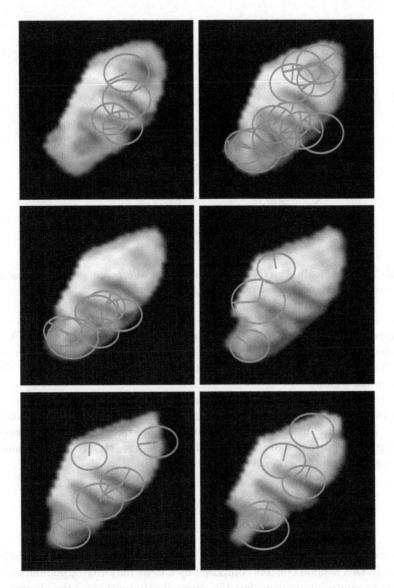

Fig. 1. Six succeeding slices from the Dorso-lateral PreFrontal Cortex ROI: landmarks points are shown with their neighborhood

The number of visual words is defined by fixing the parameter K. In this fashion, each landmark can be easily classified by assigning to it the visual word associated to the closest cluster centroid [11].

Therefore, the set of feature vectors coming from one brain is transformed into a single histogram that counts the frequency of occurrence of each visual word [11]. Such histogram is the BoW representation of the brain.

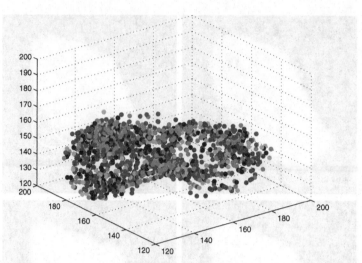

Fig. 2. Clusterized feature vectors: each point is colored according with its *visual word*. Spatial relations are not considered in the clustering.

Figure 2 shows a spatial distribution of the feature points of the DLPFC ROI extracted from the whole dataset. Points are colored according with their *visual word*. Note that spatial relations are not considered in the clustring (as clearly shown in the figure).

As we mentioned before, we are interest in capturing possible relations between feature prototypes and morphological abnormalities due to the analyzed disease. For each cluster of feature we measure its discriminative *relevance* by counting the occurrences of patience and controls respectively. In particular the following weighting function is defined for each *visual word*:

$$w_i(n_p^{c_i}, n_c^{c_i}) = \begin{cases} 1.5 & \text{if } ||n_p^{c_i} - n_c^{c_i}|| \geq \Delta \\ 0.5 & \text{otherwise} \end{cases} \tag{1}$$

where c_i is the ith centroid (i.e., the ith visual word, $i = 1 \ldots K$), $n_p^{c_i}$ and $n_c^{c_i}$ are the percentage of patients and controls in c_i, and Δ is a constant heuristically chosen. In this fashion clusters composed by a clear majority of population (i.e., patients or controls) are considered as more discriminative for the classification purpose.

2.3 SVM-Based Brains Classification

Support Vector Machines [13] are powerful classifiers which have already shown their efficacy in the context of Schizophrenia research [6]. Note that typically a SVM requires a fixed length vector which characterizes *globally* the subject to be classified. Here instead, thanks to the BoW representation a subject (i.e., a brain) is encoded by a set of local features. In particular, the novelty consists of designing a suitable *kernel function* to implicitly implement the feature matching. Note that as mentioned above, such kernels are referred as *local* kernel or *matching* kernels [12].

In order to construct a BoW histogram of a new brain, we compare each of the extracted feature with the visual words w.r.t. the visual vocabulary. In practice, by counting the number of features assigned to each visual word, the BoW representation h^A for brain A is obtained. More in details, given two brains A and B, the kernel function is defined as:

$$k(h^A, h^B) = \sum_{i=1}^{K} w_i \cdot min(h_i^A, h_i^B), \tag{2}$$

where h_i^Θ denotes the count of the ith bin of the histogram h^Θ ($\Theta \in \{A, B\}$), with K bins, and w_i are computed from Eq. (1). Such kernel is called weighted *histogram intersection* function and it is shown to be a valid kernel [12]. Histograms are assumed to be normalized such that $\sum_{i=1}^{n} h_i = 1$.

Note that, as observed in [12], the proposed kernel implicitly encodes the point-to-point matching since corresponding features are likely to belong to the same histogram bin. In other words, when points sharing a bin they are counted as matched, since the size of that bin indicate the farthest distance any two points in it could be form one another. Indeed, the histogram intersection function counts the number of feature matching being intermediated by the visual vocabulary.

3   Results

Dataset is composed of 54 brains of subjects affected by schizophrenia and 54 controls. Subjects were part of the Verona-Udine Brain Imaging and Neuropsychology Program (see [2] for further details). MRI scans were acquired with a

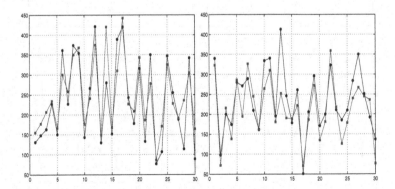

Fig. 3. Histogram of words occurrences for all patients (in red) and all controls (in blue): left and right hemisphere

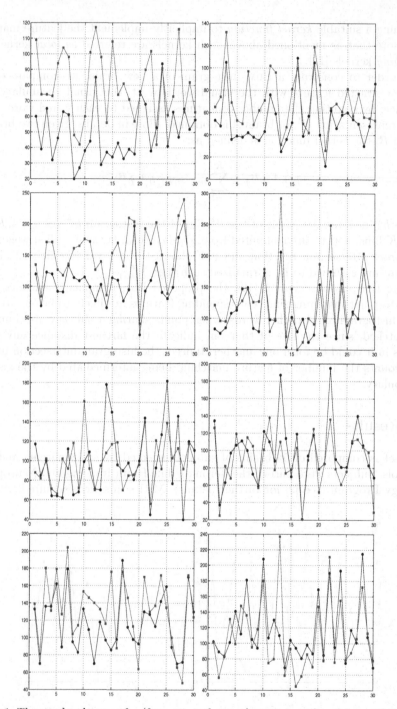

Fig. 4. The words relevance for (from top to bottom) woman, male, senior, and junior. Patients are marked in red while controls in blue. Both the left and right hemispheres are shown.

Table 1. Classification rate. For each experiment the number of involved controls and patients are reported. The scores are reported with weights (w-score) and without weights (score) respectively. Results for both the hemispheres are reported (i.e., l. and r.).

Exp.	n.Cont.	n.Pat.	w-score	score
All l.	54	54	75.00%	62.93%
All r.	54	54	66.38%	59.48%
Wom l.	25	19	84.09%	77.27%
Wom r.	25	19	77.27%	72.73%
Men l.	29	35	60%	44.62%
Men r.	29	35	67.69%	50.77%
Senior l.	23	25	81.25%	73.52%
Senior r.	23	25	70.83%	64.12%
Junior l.	31	29	71.67%	55.27%
Junior r.	31	29	63.33%	51.18%

1.5T Siemens Magnetom Symphony Maestro Class, Syngo MR 2002B. According to standard medical formats each brain-scan is organized on 144 slices each of them is composed by 384×512 pixel. Three projections are available w.r.t the *axial*, *sagittal*, and *coronal* planes respectively.

Standard pre-processing techniques have been applied to the raw data. In particular, brains are manually aligned and the DLPFC region is traced by experts by using *Brains2*[1]. In particular, the ROIs are available for both the left and right hemisphere.

After landmarks extraction[2], feature points are properly clusterized in order to obtain the *visual words*. Here, the Matlab[3] version of the $K-$means algorithm has been used. Note that the K value should be large enough to distinguish relevant changes in structural parts, but not so large as to distinguish irrelevant variations such as noise. Here we fix experimentally $K = 30$. Therefore, the *relevance* of each visual words is computed as described in Section 2.2 by obtaining the weighs w_i. Figure 3 shows the histogram of words occurrences for patients (in red) and controls (in blue). Note that for some visual words a clear majority of population is observed (i.e., visual words $3, 6, 14, 22$ for the left hemisphere and visual words $6, 10, 12, 27$ for the right hemisphere). In order to take into account the intra-class variability the whole dataset has been stratified by sex and age. Figure 3 shows the words relevance for only woman, male, senior (i.e., age$\geq$ 40), and junior respectively.

Table 1 shows the performance of classification. Scores are obtained by leave-one-out cross validation[4]. In general, a drastic improvement is observed when weights have been applied. Satisfactory results are obtained for the classification of the whole dataset. Moreover, performance has been increased when either only women or only senior people have been considered. It is worth nothing

---

[1] http://www.psychiatry.uiowa.edu

[2] We use the SIFT implementation available from http://vision.ucla.edu/vedaldi

[3] http://www.mathworks.com

that, a part of the case of men, better performances are observed from the left hemisphere. This is according with previous studies in the field [14].

4 Conclusions

In this paper the BoW paradigm has been proposed in computational neuroanatomy. A machine learning method has been introduced for brains classification in the context of schizophrenia research. The designed local kernel is able to compare intra-ROI local regions not necessarily with spatial relationship. Local features are encoded by multivariate descriptors which allow a more versatility in capture anatomical variations. Preliminary results are promising since satisfactory scores have been observed from the analysis of the whole dataset and the performance have increased when the subjects have been stratified by sex and age. Finally, the supervised weighed functions has been crucial to augment the score in all the experiments. The proposed results highlight the relevance of the DLPFC region for schizophrenic patients by confirming the presence of structural abnormalities in such regions as already shown on several other works. The proposed approach is able to characterize the structural abnormalities not as generic volumetric variations, but as occurrences of local pattern of pixels randomly located onto the analyzed ROI. In particular, such reliable approach would be instrumental to use innovative processing technique for ameliorating clinical diagnosis in the field of psychiatry. Future work will be addressed to the extension of this analysis to other ROIs. Moreover, the proposed framework can be easily improved by introducing other local descriptors and by adopting a fully 3D approach.

Acknowledgments

We acknowledge financial support from the FET programme within the EU-FP7, under the SIMBAD project (contract 213250). The dataset used in this work is part of a larger database cared by the Research Unit on Brain Imaging and Neuropsychology (RUBIN) at the Department of Medicine and Public Health-Section of Psychiatry and Clinical Psychology of the University of Verona.

References

1. Giuliania, N., Calhon, V., Pearlson, V., Francisd, A., Buchanan, R.: Voxel-based morphometry versus region of interest: a comparison of two methods for analyzing gray matter differences in schizophrenia. Schizophrenia Research 74, 135–147 (2005)
2. Baiano, M., Perlini, C., Rambaldelli, G., Cerini, R., Dusi, N., Bellani, M., Spezzapria, G., Versace, A., Balestieri, M., Mucelli, R.P., Tansella, M., Brambilla, P.: Decreased entorhinal cortex volumes in schizophrenia. Schizophrenia Research 102, 171–180 (2008)

3. Ashburner, J., Friston, K.: Voxel-based morphometry - the methods. Neuroimage 11, 805–821 (2000)
4. Duda, R., Hart, P., Stork, D.: Pattern Classification, 2nd edn. John Wiley and Sons, Chichester (2001)
5. Gerig, G., Styner, M.A., Shenton, M.E., Lieberman, J.A.: Shape versus size: Improved understanding of the morphology of brain structures. In: Niessen, W.J., Viergever, M.A. (eds.) MICCAI 2001. LNCS, vol. 2208, p. 24. Springer, Heidelberg (2001)
6. Fan, Y., Shen, D., Gur, R., Gur, R., Davatzikos, C.: Compare: classification of morphological patterns using adaptive regional elements. IEEE Transaction on Medical Imaging 26(1), 93–105 (2007)
7. Yoon, U., Lee, J., Im, K., Shin, W., Cho, B.H., Kim, I., Kwon, J., Kim, S.: Pattern classification using principal components of cortical thickness and its discriminative pattern in schizophrenia. Neuroimage 34, 1405–1415 (2007)
8. Potkin, S.G., et al.: Working memory and dlpfc inefficiency in schizophrenia: The fbirn study. Schizophrenia Bulletin 35(1), 19–31 (2009)
9. Lowe, D.G.: Distinctive image features from scale-invariant keypoints. International Journal of Computer Vision 60(2), 91–110 (2004)
10. Urschler, M., Bauer, J., Ditt, H., Bischof, H.: SIFT and Shape Context for feature-based nonlinear registration of thoracic CT images. Computer Vision Approach to Medical Image Analysis, 73–84 (2006)
11. Cruska, G., Dance, C.R., Fan, L., Willamowski, J., Bray, C.: Visual categorization with bags of keypoints. In: ECCV Workshop on Statistical Learning in Computer Vision, pp. 1–22 (2004)
12. Grauman, K., Darrell, T.: The pyramid match kernel: Efficient learning with sets of features. Journal of Machine Learning Research 8(2), 725–760 (2007)
13. Burges, C.: A tutorial on support vector machine for pattern recognition. Data Mining and Knowledge Discovery 2, 121–167 (1998)
14. Andreone, N., Tansella, M., Cerini, R., Versace, A., Rambaldelli, G., Perlini, C., Dusi, N., Pelizza, L., Balestrieri, M., Barbui, C., Nosé, M., Gasparini, A., Brambilla, P.: Cortical white-matter microstructure in schizophrenia. diffusion imaging study. Br. J. Psychiatry 191(8), 113–119 (2007)

Improvement of the Classifier Performance of a Pedestrian Detection System by Pixel-Based Data Fusion

Holger Lietz, Jan Thomanek, Basel Fardi, and Gerd Wanielik

Chemnitz University of Technology, Reichenhainer Str. 70
09126 Chemnitz, Germany
holger.lietz@etit.tu-chemnitz.de, jtho@hrz.tu-chemnitz.de

Abstract. This contribution presents an approach how to improve the classifier performance of an existing pedestrian detection system by using pixel-based data fusion of FIR and NIR sensors. The advantage of the proposed method is that the fused images are more suitable for the subsequent feature extraction. Both, the algorithm of the pedestrian detection system and the used pixel-based fusion techniques, are presented. Experimental results show that the detection performance based on a fused image sequence outperforms a detector that is based on just a single sensor.

Keywords: Pixel Fusion, Data Fusion, Classification, Pedestrian Recognition.

1 Introduction

The development of Pedestrian Protection Systems is one of the main focuses in the field of advanced driver assistance systems since the beginning of the 2000s. A state-of-the-art overview is given in [1]. During the last decade, the European Commission funded a series of projects with the objective to reduce the number of collision between vehicles and so called Vulnerable Road Users (VRUs), such as pedestrians and bicyclists. The most challenging issue in the field of active safety systems is the accurate detection of the VRUs. Many different types of sensors, such as different cameras, RADARs and LIDARs, have been investigated in order to find the most suitable sensor for detecting VRUs robustly.

Video sensors are the natural choice for pedestrian detection systems due to the likeness to the human ocular system. Recently, far infrared cameras (FIR) entered the field of pedestrian recognition thanks to the decreasing cost of infrared technology. In many scenarios, FIR sensors are more suited than daylight ones for detecting pedestrians, especially under night or low-illumination conditions. But in some cases, thermal cameras fail in the detection of pedestrians, especially when the weather is hot and sunny, or when pedestrians are not warmer than the background [2].

For this reason, many approaches for a pedestrian detection system used multi-sensor systems, so that the weakness of one sensor can be compensated by an additional sensor. One multi-sensor approach has been realized within the EU project WATCH-OVER [3].

R. Serra and R. Cucchiara (Eds.): AI*IA 2009, LNAI 5883, pp. 122–130, 2009.

2 Pedestrian Detection System

The following sections describe the pedestrian recognition system, which was one part of the WATCH-OVER data fusion system [4].

2.1 System Overview

The WATCH-OVER data fusion system (see figure 1) is based on a near infrared (NIR) camera and two wireless ranging devices. The wireless ranging devices are combined to a virtual sensor that can determine the distance r and the azimuth angle φ to a VRU object. Both types of sensors complement each other in a way that wireless ranging devices deliver precise information about the distance between the VRU and the ego-vehicle, but have low precision concerning its lateral position. The NIR-based pedestrian detection algorithm can accurately determine the position of the VRU in the image, but is not very precise in distance estimation. The measurements of each of the sensors are processed by using a Kalman Filter algorithm, so that a tracker can be realized by means of ego-vehicle movement data. Only one of these measurement vectors is necessary to track a VRU object, the complementary one serves to increase the tracking accuracy. However, for a high tracking accuracy data from both sensor systems are needed.

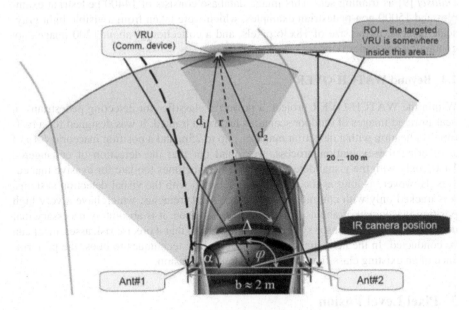

Fig. 1. The WATCH-OVER Data Fusion System

2.2 Pedestrian Classification Algorithm

The visual pedestrian detection system searches for VRUs in different locations in the NIR camera image. The size and position of these regions of interest (ROIs) are dependent from the intrinsic and extrinsic camera parameters. Every ROI is classified by using a trained classifier cascade, which has been introduced by Viola and Jones in [5] and improved by Zhu et al. [6]. This approach was chosen due to its very high detection rate and rapid performance. It consists of a series of Adaboost classifiers [7] that internally consist of pruned decision trees. The cascade output is P if the ROI matches a pedestrian shape and N if not. The cascade algorithm reports a region as pedestrian only if it has passed all its stages, but it can reject a ROI at any stage. This leads to a dramatic performance boost, since the majority of the ROIs are usually non-pedestrians, where most of them are rejected in early cascade stages.

2.3 Features and Training Set

Our approach uses normalized Histograms of Oriented Gradients (HOGs) as input features that were presented by Dalal and Triggs in [8]. The big advantage of these features is their translation and illumination invariance. These features correspond much stronger to the contour of an object than to its color. The feature vector has 2016 dimensions, which is a result of 56 different HOG positions and sizes, and 36 features per HOG block. According to Dalal's approach, each block consists of four histograms; each of those is formed by nine discrete gradient directions.

For the training of the classifier, we used the base version and the additional non-pedestrian images of the *DaimlerChrysler Pedestrian Classification Benchmark Dataset* [9] as training set. This image database consists of 14400 pedestrian examples and 15000 non-pedestrian examples, which were taken from a visible light gray-scale camera, in the size of 18x36 pixels, and a collection of about 1200 images not containing any pedestrians.

2.4 Beyond WATCH-OVER

Within the WATCH-OVER project, a powerful classifier for detecting pedestrians in near infrared images of outdoor scenarios had been trained. It was designed for a real-time application with a detection range r_{max} up to 25m and a position inaccuracy $\Delta\varphi$ of $2°$. During the evaluation process, we found out that the detection of endangered VRUs only with the visual detection system is sometimes too late for evasive maneuvers. However, as long as the VRU is not detected with the visual detection systems, it is tracked only with communication device measurements, which have a very high position inaccuracy of about $\Delta\varphi \approx 30°$. For this reason, it is absolutely necessary that detections from both sensor systems are available so that a precise risk assessment can be conducted. In the next chapter, we will introduce techniques to boost the performance of an existing classifier using pixel-based data fusion.

3 Pixel Level Fusion

3.1 Motivation

Multi-sensor data fusion can be realized on different levels: signal, image, feature, or symbol level. Signal-level fusion refers to the direct combination of several signals. Image-level fusion, also called *pixel-based fusion*, generates a fused image in which each pixel value is computed from a set of pixels in the raw sensor images. Feature-level fusion first employs feature extraction on each of the source images, so that features from every source can be combined and conjointly used for classification. Symbol-level fusion allows combining information from multiple sensors based on a decision rule.

We chose the pixel-based fusion because the applied algorithms are time efficient and can be used within the existing classifier framework (see figure 2). The purpose of pixel-based fusion is to create a single image which contains a more accurate representation of the scene than any of the individual source images. That means the fused image should be more suitable for the subsequent feature extraction [10]. The combination of FIR and NIR images through pixel-based fusion should contain the relevant information from the inputs, and suppress redundant information or data, which are not relevant for the application. Ideally, the fused image shows the contour of pedestrians in all different weather conditions, but it should suppress textures on the pedestrians' clothes, because they are not relevant for the classification.

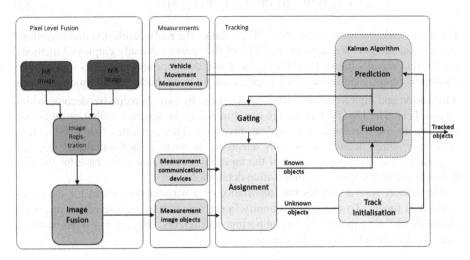

Fig. 2. The Scheme of the Advanced Pedestrian Detection System

3.2 Pixel-Based Fusion Technique

An important pre-processing step for pixel-based fusion is the precise spatial alignment of the far-infrared and near-infrared images, so that the corresponding pixels in both images represent the same location in the real world. Usually, this task refers to as *image registration*. Geometric transformations are frequently used to perform this step. Though, these geometrical transformations are based on 2D point correspondences and map one image onto the other without taking into account the 3D viewing geometry of both cameras. Since objects in the observed scene may appear in different distances, one transformation for each distance plane is necessary for precise image registration. Therefore, we use *image rectification* to transform both camera geometries into a common stereo system with parallel optical axes. Thereby, a perspective transformation projects the plane of the observed scene onto images, which are tilt-free and of the desired scale. Due to the rectification step, the disparities between the images disappear in one direction. To minimize the disparities in the other direction, we mounted both cameras as close as possible.

Since our main goal of the data fusion is to preserve image features from the source images, e.g. the contours of people. Therefore, we used a multi-scale-decomposition based fusion method. The basic idea of all multi-resolution fusion schemes is motivated by the fact that the real-world objects usually consist of structures in different scales and that the extracted features are more clearly depicted in the multi-scale representation than in the spatial domain.

The transform Ψ decomposes the rectified images into the relevant features on different resolutions. These features are combined using some kind of fusion rules ϕ. After that, the fused image I_{fus} is obtained by applying the inverse transform Ψ^{-1}.

$$I_{fus} = \Psi^{-1}\left(\phi\left(\Psi(I_{fir}), \Psi(I_{nir})\right)\right) \tag{1}$$

The *multi-scale transformation block* Ψ computes the multi-resolution decomposition of the input images. We implemented one of the most commonly employed methods, the discrete wavelet transform (DWT). The DWT decomposes the images into a multi-resolution representation (wavelet coefficients y) with both, low frequency coarse information and high frequency detail information, by using a recursive decomposition scheme [11]. The 1D-DWT is used in two dimensions by separately filtering and down sampling in the horizontal and vertical directions. This generates four sub-bands at each scale: three detail images with sensitivity to vertical, horizontal and diagonal frequencies and an approximation of the input image that is used as base for the next decomposition. We used a decomposition depth of 2.

The *activity level* a denotes the importance of a wavelet coefficient for the pixel-based data fusion. To avoid discontinuity in the fused image, the selection of coefficients should also include the neighboring coefficients. We used a window-based energy calculation for the activity level:

$$a(\mathbf{x}) = \sum_{\Delta\mathbf{x}\in W} \left| y(\mathbf{x}+\Delta\mathbf{x})^2 \right| \tag{2}$$

The vector $\mathbf{x}$ denotes the position of the wavelet coefficients within the multi-scale representation of the source images. $\Delta\mathbf{x}$ is the translation vector from a defined 3x3 window W around the corresponding wavelet coefficients.

A *match value* $m(\mathbf{x})$ is used to quantify the similarity between both images. A small match value means that the input images are different on the corresponding position. If $m(\mathbf{x}) = 1$, then the images are identical on the position. It is defined by:

$$m(\mathbf{x}) = \frac{\sum_{\Delta\mathbf{x}\in W} \left| y_{fir}(\mathbf{x}+\Delta\mathbf{x}) \right| \cdot \left| y_{nir}(\mathbf{x}+\Delta\mathbf{x}) \right|}{a_{fir}(\mathbf{x}) + a_{nir}(\mathbf{x})} \tag{3}$$

The applied pixel-based fusion scheme based on Piella's approach [12] is shown in figure 3:

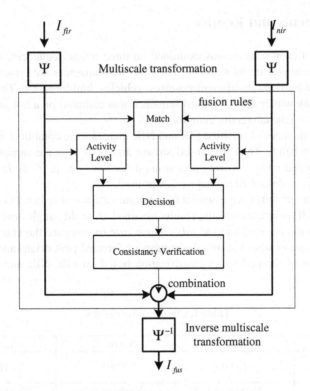

Fig. 3. The Pixel-Based Fusion Scheme

The *decision block* creates a map, which governs the actual combination of the coefficients of the multi-scale decomposition of both images. For the approximation coefficients we used a simple averaging, that means the value of the decision map d is always 0.5. The detail coefficients are combined depending from the activity level and match value. Usually, if the match value is smaller than a threshold, the detail coefficients are combined using the choose-max method: That means the coefficient of the input image, which has the higher activity level, is used. The detail coefficients are averaged if the match value is bigger than the threshold. However, we use a different scheme to combine the detail coefficients which generates smoother images with fewer artifacts.

$$d(\mathbf{x}) = \begin{cases} 1 - m(\mathbf{x}) & \text{if } a_{fir}(\mathbf{x}) > a_{nir}(\mathbf{x}) \\ m(\mathbf{x}) & otherwise \end{cases} \tag{4}$$

After that, the fused image in the wavelet domain is composed as follows

$$y_{fus}(\mathbf{x}) = d(\mathbf{x}) \cdot y_{fir}(\mathbf{x}) + (d(\mathbf{x}) - 1) \cdot y_{nir}(\mathbf{x}) \tag{5}$$

Finally, the fused image is reconstructed by computing the inverse wavelet transformation as described in [11].

4 Experiments and Results

The proposed fusion scheme was examined on three image sequences, two recorded in the winter and one in the summer. The first two sequences are typical city centre scenes with pedestrians in different postures, vehicles, buildings, etc. They were captured on a bleak winter day. The third sequence was captured on a hot and sunny day in the summer on the university campus.

First, the sequences of rectified FIR and NIR images were combined using the proposed fusion scheme. Then, the fused stream and both separate sensor streams, as well, were applied on the existing pedestrian detection system. In all three cases, the same classifier, as described in chapter 2, was used.

Performance measures are essential to determine the possible benefits of the fusion method, as well as to compare the results obtained from the single sensor sequence. We computed the *hit rate* and the *false alarm rate* to compare the classification results. The following table 1 shows the number of detected pedestrians and the number of false alarms of the pedestrian classification based on FIR, NIR and fused image streams:

Table 1. Classification Results

	Scene 1	Scene 2	Scene 3
Detectable pedestrians	1450	2360	1695
	Number of Hits / False Alarms	Number of Hits / False Alarms	Number of Hits / False Alarms
FIR	650 / 21	980 / 18	159 / 6
NIR	442 / 39	309 / 27	196 / 30
FUSED	**828 / 49**	**1017 / 38**	**213 / 9**

The fused image stream provides the best hit rate in each of the three sequences, mainly in the scene 1 where the improvement of the hit rate is almost 25% compared to the FIR scene. Especially in fast moving scenes, where the far-infrared images are often blurred, the fused image obtains the detail information from the near-infrared image (see figure 4/2a-c). Furthermore, if the background is not colder than the pedestrians, e.g. a shop window, a detection based on a single far-infrared camera will also fail (see figure 4/3a-c). Figure 4/4a-c and 4/5a-c show the fading-to-background effect: The pedestrians are almost invisible on the FIR image because of the higher outside temperature in summer.

However, in most cases the pedestrian detection performance in the far-infrared case will outclass the near-infrared one. Particularly under low illumination and low temperature conditions, the fused image will exploit the higher contrast of the far infrared image (see figure 4/1a-c).

Fig. 4. Examples of the Evaluation Scenes: 1–3: winter scenario, 4–5: summer scenario; a) FIR image b) NIR image c) Fused image

5 Conclusion

In this contribution, we have proven that the visual pedestrian classifier, which was used in the WATCHOVER-pedestrian protection system, can not only be applied on NIR images, but also on FIR and fused images. If the outside temperature is low, the classifier performance dramatically increases when the video input comes from a FIR or fused image source. The reason for this phenomenon is probably the stronger pedestrian contour in the FIR image, which results from the higher difference in temperature between the pedestrian and its surroundings. When the weather is hot, the shape of the pedestrian sometimes fades to the background on FIR images, so that the detection rate is lower than in the NIR case. Unless the false-alarm rate is lowest in the FIR case, we think that pixel-fused images are the best choice for a pedestrian protection system, because they seem to be very tolerant against illumination changes and differences concerning the outside temperature.

References

1. Gandhi, T., Trivedi, M.M.: Pedestrian Protection Systems: Issues,Survey, and Challenges. IEEE Transactions on Intelligent Transportations Systems 8(3) (2007)
2. Bertozzi, M., Broggi, A., Felisa, M., Ghidoni, S., Grisleri, P., Vezzoni, G., Gómez, C.H., Del Rose, M.: Multi Stereo- Based Pedestrian Detection by Daylight and Far-Infrared Cameras. In: Augmented Vision Perception in Infrared: Advances in Pattern Recognition, pp. 371–401. Springer, London (2009)
3. Andreone, L., Wanielik, G.: Vulnerable Road Users Thoroughly Addressed in Accident Prevention: TheWATCH-OVER European Project. In: ITS World (2007)
4. Fardi, B., Neubert, U., Giesecke, N., Lietz, H., Wanielik, G.: A Fusion Concept of Video and Communication Data for VRU Recognition. In: Proceedings of the 11th International Conference on Information Fusion (2008)
5. Viola, P., Jones, M.: Rapid Object Detection using a Boosted Cascade of Simple Features. In: Proceedings of the IEEE Computer Society Conference on Computer Vision and Pattern Recognition, vol. 1, p. 511 (2001)
6. Zhu, Q., Avidan, S., Yeh, M., Cheng, K.-T.: Fast human detection using a cascade of Histograms of Oriented Gradients. In: Proceedings of the IEEE Conference on Computer Vision and Pattern Recognition (2006)
7. Freund, Y., Schapire, R.E.: A Decision-Theoretic Generalization of Online Learning and an Application to Boosting. In: Proceedings of the European Conference on Computational Learning Theory, pp. 23–37 (1995)
8. Dalal, N., Triggs, B.: Histograms of Oriented Gradients for Human Detection. In: Proceedings of the IEEE Conference on Computer Vision and Pattern Recognition, vol. II, pp. 886–893 (2005)
9. Munder, S., Gavrila, D.M.: An Experimental Study on Pedestrian Classification. IEEE Transactions on Pattern Analysis and Machine Intelligence 28(11) (2006)
10. Blum, R.S., Xue, Z., Zhang, Z.: An Overview of Image Fusion. In: Multi-Sensor Image Fusion and Its Applications. Taylor & Francis Group, Abington (2006)
11. Mallat, S.G.: A theory for multiresolution signal decomposition: The wavelet representation. IEEE Transactions on Pattern Analysis and Machine Intelligence 11, 674–693 (1989)
12. Piella, G.: A general framework for multiresolution image fusion: From pixel to regions. Information Fusion 4, 259–280 (2003)

Plugging Taxonomic Similarity in First-Order Logic Horn Clauses Comparison

S. Ferilli, M. Biba, N. Di Mauro, T.M.A. Basile, and F. Esposito

Dipartimento di Informatica
Università di Bari
via E. Orabona, 4 - 70125 Bari - Italia
{ferilli,biba,ndm,basile,esposito}@di.uniba.it

Abstract. Horn clause Logic is a powerful representation language exploited in Logic Programming as a computer programming framework and in Inductive Logic Programming as a formalism for expressing examples and learned theories in domains where relations among objects must be expressed to fully capture the relevant information. While the predicates that make up the description language are defined by the knowledge engineer and handled only syntactically by the interpreters, they sometimes express information that can be properly exploited only with reference to a taxonomic background knowledge in order to capture unexpressed and underlying relationships among the concepts described. This is typical when the representation predicates are not purposely engineered but rather derive from the particular words found in a text.

This work proposes the exploitation of a taxonomic background knowledge to better assess the similarity between two First-Order Logic (Horn clause) descriptions, beyond the simple syntactical matching between predicates. To this aim, an existing distance framework is extended by applying the underlying distance measure also to parameters coming from the taxonomic background knowledge. The viability of the solution is demonstrated on sample problems.

1 Introduction

First-Order Logic (*FOL* for short) is a powerful representation language that allows to express relationships among objects, which is often an unnegligible requirement in real-world and complex domains. Logic Programming [11] is a computer programming framework based on a FOL sub-language, which allows to perform reasoning on knowledge expressed in the form of Horn clauses. Inductive Logic Programming (ILP) [13] aims at learning automatically logic programs from known examples of behaviour, and has proven to be a successful Machine Learning approach in domains where relations among objects must be expressed to fully capture the relevant information. Many AI tasks can take advantage from techniques for descriptions comparison: subsumption procedures (to converge more quickly), flexible matching, instance-based classification techniques or clustering, generalization procedures (to focus on the components that are

R. Serra and R. Cucchiara (Eds.): AI*IA 2009, LNAI 5883, pp. 131–140, 2009.

more likely to correspond to each other). In FOL, this is a particularly complex task due to the problem of *indeterminacy* in mapping portions of one formula onto portions of another.

In the traditional approach, predicates that make up the description language are defined by the knowledge engineer that is in charge of setting up the reasoning or learning problem, and are uninterpreted by the systems. The knowledge engineer can also define and provide a background knowledge to be exploited in order to improve performance or effectiveness of the results. However, a particular kind of information that often needs to be expressed in the descriptions is taxonomic information, that can convey implicit relationships among the concepts described. Unfortunately, such a kind of information needs to be interpreted in order to be fully exploited, which requires a proper background knowledge to be set up. Unless the problem domain is very limited, the taxonomic background knowledge to be provided becomes huge: in these cases, the use of existing state-of-the-art taxonomies can be a definite advantage.

This work builds on previous results concerning a framework for similarity assessment beween FOL Horn clauses, where the overall similarity depends on the similarity of the pairs of literals associated by the least general generalization, the similarity of two literals in turn depends on the similarity of their corresponding arguments (i.e., terms), and the similarity between two terms is computed according to the predicates and positions in which they appear. Here, a novel and general approach to the assessment of similarity between concepts in a taxonomy is proposed, and its integration as an extension of the similarity framework for clauses including taxonomic information is described.

The rest of this paper is organized as follows. The next section identifies a sample problem/application in which defining a taxonomic similarity can be of help. Then, Section 3 introduces the basic formula and framework for the overall assessment of similarity between Horn clauses. Section 4 proposes an application of the same formula to compute the taxonomic similarity between two concepts or words, and introduces it in the previous framework. Section 5 shows experiments that suggest the effectiveness of the proposed approach. Lastly, Section 6 concludes the paper and outlines future work directions.

2   Why a Taxonomic Approach: Sample Problems

In this section, one of the many practical problems in which taxonomic information is present and relevant has been selected and discussed, in order to provide the reader with a better understanding of the concepts and methods presented in the paper. The same toy problem will be tackled later using the proposed method to show its behavior and viability. As already pointed out, setting up a general taxonomy is a hard work, for which reason the availability of an already existing resource can be a valuable help in carrying out the task. In this example we will refer to the most famous taxonomy available nowadays, WordNet (WN) [12], that provides both the conceptual and the lexical level.

First of all, let us show a case in which an effective taxonomic similarity assessment can be useful in itself. Consider the following words and concepts:

102330245 *mouse* (animal) : 'any of numerous small rodents typically resembling diminutive rats having pointed snouts and small ears on elongated bodies with slender usually hairless tails'

103793489 *mouse* (device) : 'a hand-operated electronic device that controls the co-ordinates of a cursor on your computer screen as you move it around on a pad; on the bottom of the device is a ball that rolls on the surface of the pad'

103082979 *computer* (device) : 'a machine for performing calculations automatically'

102121620 *cat* (pet) : 'feline mammal usually having thick soft fur and no ability to roar: domestic cats; wildcats'

102127808 *cat* (wild) : 'any of several large cats typically able to roar and living in the wild'

102129604 *tiger* (animal) : 'large feline of forests in most of Asia having a tawny coat with black stripes; endangered'

102084071 *dog* (pet) : 'a member of the genus Canis (probably descended from the common wolf) that has been domesticated by man since prehistoric times; occurs in many breeds'

102374451 *horse* (animal) : 'solid-hoofed herbivorous quadruped domesticated since prehistoric times'

103624767 *horse* (chess) : 'a chessman shaped to resemble the head of a horse; can move two squares horizontally and one vertically (or vice versa)'

One might need to check how close two of such items are, in order to perform further processing such as logic deductions or Natural Language Processing. For instance, one might want to disambiguate a polysemous word (e.g., 'mouse') by comparing its candidate underlying concepts to the other concepts that are present in the same text (e.g., 'cat' and 'dog' rather than 'computer'). Or, one might be interested in ranking a set of candidate concepts by closeness with respect to a given concept (e.g., ranking 'dog (pet)', 'tiger (animal)' and 'cat (wild)' with respect to 'cat (pet)'), etc. etc.

Then, let us show an example in which the taxonomic similarity assessment can support other processes. Specifically, here we consider the problem of similarity assessment between natural language sentences represented by FOL Horn clauses. FOL might be exploited for representing relational features of natural language, such as the syntactic relationships among discourse components. Indeed, although much more computationally demanding than simple bag-of-word approaches traditionally exploited in the literature, techniques that take into account the syntactic structure of sentences are very important to fully capture the information they convey. Reporters know very well that, swapping the subject and the object in a sentence like "The dog bit the man", results in very different interest of the underlying news. Natural language typically requires huge taxonomic information, and the problems of synonimy and polisemy introduce further complexity. For instance, the following sentences:

1. "The boy wants a small dog"
2. "The girl desires a yellow canary"
3. "The hammer hits a small nail"

structurally exhibit the same grammatical pattern, thus no hint is available to assess which is more similar to which. Even worse, at the lexical level, the only common word ('small') appears in sentences 1 and 3, which would suggest they are closer to each other than to sentence 2. However, it becomes clear that the first two are conceptually the most similar to each other as long as one knows and considers that 'boy' and 'girl' are two young persons, 'to want' and 'to desire' are synonyms and 'dog' and 'canary' are two pets.

For demonstration purposes, let us consider a simplified structural description language for natural language sentences:

subj(X,Y) : Y is the subject of sentence X
pred(X,Y) : Y is the predicate of sentence X
dir_obj(X,Y) : Y is the direct object of sentence X
ind_obj(X,Y) : Y is the in direct object of sentence X
noun(X,Y) : Y is a noun appearing in component X of the sentence
verb(X,Y) : Y is a verb appearing in component X of the sentence
adj(X,Y) : Y is an adjective appearing in component X of the sentence
adv(X,Y) : Y is an adverb appearing in component X of the sentence
prep(X,Y) : Y is a preposition appearing in component X of the sentence

Additionally, each noun, verb, adjective or adverb is described by the corresponding concept (or word) in the sentence, plus possible other properties expressed by ordinary unary predicates. For the three sentences reported above one gets:

```
s1 = sentence(s1) :- subj(s1,ss1), pred(s1,ps1), dir_obj(s1,ds1),
     noun(ss1,nss1), boy(nss1), verb(ps1,vps1), want(vps1),
     adj(ds1,ads1), small(ads1), noun(ds1,nds1), dog(nds1).
s2 = sentence(s2) :- subj(s2,ss2), pred(s2,ps2), dir_obj(s2,ds2),
     noun(ss2,nss2), girl(nss2), verb(ps2,vps2), desire(vps2),
     adj(ds2,ads2), yellow(ads2), noun(ds2,nds2), canary(nds2).
s3 = sentence(s3) :- subj(s3,ss3), pred(s3,ps3), dir_obj(s3,ds3),
     noun(ss3,nss3), hammer(nss3), verb(ps3,vps3), hit(vps3),
     adj(ds3,ads3), small(ads3), noun(ds3,nds3), nail(nds3).
```

Syntactically, the generalization between $s2$ and both $s1$ and $s3$ is:

```
sentence(X) :- subj(X,Y), noun(Y,Y1), pred(X,W), verb(W,W1),
               dir_obj(X,Z), adj(Z,Z1), noun(Z,Z2).
```

while the generalization between $s1$ and $s3$ is:

```
sentence(X) :- subj(X,Y), noun(Y,Y1), pred(X,W), verb(W,W1),
               dir_obj(X,Z), adj(Z,Z1), small(Z1), noun(Z,Z2).
```

so that the latter pair, having in the generalization an additional literal with respect to the former pairs, would appear to have a greater similarity value due to just the structural aspects, in spite of the very different content.

3   Similarity Framework

In [6], a framework for computing the similarity between two Datalog Horn clauses has been provided, which is summarized in the following. Let us preliminary recall some basic notions involved in Logic Programming. The *arity* of a predicate is the number of arguments it takes. A *literal* is an *n*-ary predicate, applied to *n* terms, possibly negated. *Horn clauses* are logical formulæ usually represented in Prolog style as l_0 :- $l_1, \ldots, l_n$ where the l_i's are *literals*. It corresponds to an implication $l_1 \wedge \cdots \wedge l_n \Rightarrow l_0$ to be interpreted as "l_0 (called *head* of the clause) is true, provided that l_1 and ... and l_n (called *body* of the clause) are all true". Datalog [3] is, at least syntactically, a restriction of Prolog in which, without loss of generality [15], only variables and constants (i.e., no functions) are allowed as terms. A set of literals is *linked* if and only if each literal in the set has at least one term in common with another literal in the set. We will deal with the case of linked Datalog clauses. In the following, we will call *compatible* two sets or sequences of literals that can be mapped onto each other without yielding inconsistent term associations (i.e., a term in one formula cannot correspond to different terms in the other formula).

Intuitively, the evaluation of similarity between two items i' and i'' might be based both on parameters expressing the amounts of common features, which should concur in a positive way to the similarity evaluation, and of the features of each item that are not owned by the other (defined as the *residual* of the former with respect to the latter), which should concur negatively to the whole similarity value assigned to them [10]:

n , the number of features owned by i' but not by i'' (*residual* of i' wrt i'');
l , the number of features owned both by i' and by i'';
m , the number of features owned by i'' but not by i' (*residual* of i'' wrt i').

A similarity function that expresses the degree of similarity between i' and i'' based on the above parameters, and that has a better behaviour than other formulæ in the literature in cases in which any of the parameters is 0, is [6]:

$$sf(i', i'') = \mathrm{sf}(n, l, m) = 0.5 \frac{l+1}{l+n+2} + 0.5 \frac{l+1}{l+m+2} \qquad (1)$$

It takes values in $]0, 1[$, which resembles the theory of probability and hence can help human interpretation of the resulting value. When $n = m = 0$ it tends to the limit of 1 as long as the number of common features grows. The full-similarity value 1 is never reached, being reserved to two items that are exactly the same ($i' = i''$), which can be checked in advance. Consistently with the intuition that there is no limit to the number of different features owned by the two descriptions, which contribute to make them ever different, it is also always strictly greater than 0, and will tend to such a value as long as the number of non-shared features grows. Moreover, for $n = l = m = 0$ the function evaluates to 0.5, which can be considered intuitively correct for a case of maximum uncertainty. Note that each of the two terms refers specifically to one of the two items under comparison, and hence they could be weighted to reflect their importance.

In FOL representations, usually terms denote objects, unary predicates represent object properties and *n*-ary predicates express relationships between objects; hence, the overall similarity must consider and properly mix all such components. The similarity between two clauses C' and C'' is guided by the similarity between their structural parts, expressed by the n-ary literals in their bodies, and is a function of the number of common and different objects and relationships between them, as provided by their least general generalization $C = l_0 :- l_1, \ldots, l_k$. Specifically, we refer to the θ_{OI} generalization model [5]. The resulting formula is the following:

$$\mathrm{fs}(C', C'') = \mathrm{sf}(k' - k, k, k'' - k) \cdot \mathrm{sf}(o' - o, o, o'' - o) + \mathrm{avg}(\{\mathrm{sf}_s(l_i', l_i'')\}_{i=1,\ldots,k})$$

where k' is the number of literals and o' the number of terms in C', k'' is the number of literals and o'' the number of terms in C'', o is the number of terms in C and $l_i' \in C'$ and $l_i'' \in C''$ are generalized by l_i for $i = 1, \ldots, k$. The similarity of the literals is smoothed by adding the overall similarity in the number of overlapping and different literals and terms.

The similarity between two compatible n-ary literals l' and l'', in turn, depends on the multisets of n-ary predicates corresponding to the literals directly linked to them (a predicate can appear in multiple instantiations among these literals), called *star*, and on the similarity of their arguments:

$$\mathrm{sf}_s(l', l'') = \mathrm{sf}(n_s, l_s, m_s) + \mathrm{avg}\{\mathrm{sf}_o(t', t'')\}_{t'/t'' \in \theta}$$

where θ is the set of term associations that map l' onto l'' and S' and S'' are the stars of l' and l'', respectively:

$$n_s = |S' \setminus S''| \qquad l_s = |S' \cap S''| \qquad m_s = |S'' \setminus S'|$$

Lastly, the similarity between two terms t' and t'' is computed as follows:

$$\mathrm{sf}_o(t', t'') = \mathrm{sf}(n_c, l_c, m_c) + \mathrm{sf}(n_r, l_r, m_r)$$

where the former component takes into account the sets of properties (unary predicates) P' and P'' referred to t' and t'', respectively:

$$n_c = |P' \setminus P''| \qquad l_c = |P' \cap P''| \qquad m_c = |P'' \setminus P'|$$

and the latter component takes into account how many times the two objects play the same or different roles in the n-ary predicates; in this case, since an object might play the same role in many instances of the same relation, the *multisets* R' and R'' of roles played by t' and t'', respectively, are to be considered:

$$n_r = |R' \setminus R''| \qquad l_r = |R' \cap R''| \qquad m_r = |R'' \setminus R'|$$

4   Taxonomic Similarity

A lot of research has been devoted to develop and test similarity measures for concepts in a taxonomy (a survey for WN can be found in [2]). The most exploited relationship is generalization/specialization, relating concepts or classes to their super-/sub-concepts or classes. Various proposals are based on the length of the paths that link the concepts to be compared to their closest common ancestor, according to the intuition that, the closer a common ancestor of two concepts, the more they can be considered as similar to each other.

Since this work aims at extending a general similarity framework by considering taxonomic information, for compatibility and smooth integration we exploited the same function as the base framework[1]. In our case, (1) requires three parameters: one expressing the common information between the two objects to be compared, and the others expressing the information carried by each of the two but not by the other. If the taxonomy is a hierarchy, and hence can be represented as a tree, the path connecting any node (concept) to the root (the most general concept) is unique: given two concepts c' and c'', let us call $< p'_1, \ldots, p'_{n'} >$ the path related to c', and $< p''_1, \ldots, p''_{n''} >$ the path related to c''. Thus, given any two concepts, their closest common ancestor is uniquely identified, as the last element in common in the two paths: suppose this is the k-th element (i.e., $\forall i = 1, \ldots, k : p'_i = p''_i = p_i$). Consequently, three sub-paths are induced: the sub-path in common, going from the root to such a common ancestor ($< p_1, \ldots, p_k >$), and the two trailing sub-paths ($< p'_{k+1}, \ldots, p'_{n'} >$ and $< p''_{k+1}, \ldots, p''_{n''} >$). Now, the former can be interpreted as the common information, and the latter as the residuals, and hence their lengths ($n' - k, k, n'' - k$) can serve as arguments (n, l, m) to apply the similarity formula. This represents a novelty with respect to other approaches in the literature, where only (one or both of) the trailing parts are typically exploited, and is also very intuitive, since the longest the path from the top concept to the common ancestor, the more they have in common, and the higher the returned similarity value.

Actually, in real-world domains the taxonomy is a heterarchy, not just a hierarchy, since by multiple inheritance a concept can specialize many other concepts. This is very relevant as regards the similarity criterion above stated, because the closest common ancestor and the paths linking two nodes are no more unique. Hence, many incomparable common ancestors and paths between concepts can be found, and going to the single common one would very often result in overgeneralization. Our novel solution to this problem is computing the whole set of ancestors of either concept, and then considering as common information (yielding l) the intersection of such sets, and as residuals (yielding n, m) the two symmetric differences. Again this is intuitive, since the number of common ancestors can be considered a good indicator of the shared features between the two concepts, just as the number of different ancestors can provide a reasonable estimation of the different information and features they own.

Dealing with natural language words, instead of explicit concepts, due to the problem of polysemy (a word may correspond to many concepts), their similarity must somehow combine the similarities between each pair of concepts underlying the words. Such a combination can consist, for instance, in the average or maximum similarity among such pairs, or can exploit the domain of discourse. A distance between groups of words (if necessary) can be obtained by couplewise working on the closest (i.e., taxonomically most similar) words in each group. Note that, in case of synonymy or polysemy, assuming consistency of domain among the

---

[1] According to the definition in [2], this yields a similarity measure rather than a full semantic relatedness measure, but we are currently working to extend it by taking into account other relationships as well.

words used in a same context [9], the similarity measure, by couplewise comparing all concepts underlying two words, can also suggest a ranking of which are the most probable senses for each, this way serving as a simple Word Sense Disambiguation [7] procedure, or as a support to a more elaborate one.

Since the taxonomic predicates represent further information about the objects involved in a description, in addition to their properties and roles, term similarity is the component where the corresponding similarity can be introduced in the overall framework. Of course, we assume that there is some way to distinguish taxonomic predicates from ordinary ones, so that they can be handled separately by the procedures. The similarity between two terms becomes:

$$\mathrm{sf}_o(t', t'') = \mathrm{sf}(n_c, l_c, m_c) + \mathrm{sf}(n_r, l_r, m_r) + \mathrm{sf}(n_t, l_t, m_t)$$

where the additional component refers to the number of common and different ancestors of the two concepts associated to the two terms, as specified above. In case the taxonomic information is expressed in the form of words instead of concepts, according to the one-domain-per-discourse assumption, these values can be referred to the closest pair of concepts associated to those words.

5   Application to the Sample Problems

Let us now go back to the sample problems presented in Section 2, and show how they can be tackled by properly setting and exploiting the general similarity framework proposed above. As to the list of concepts/words, Table 1 reports the similarity values corresponding to some more interesting couples. At the level of concepts, the similarity ranking is quite intuitive, in that less related concepts receive a lower value. The closest pairs are 'wild cat'-'tiger' and 'pet cat'-'tiger', followed by 'mouse animal'-'pet cat', then by 'mouse device'-'computer device', by 'pet cat'-'dog pet' and by 'dog pet'-'horse animal', all with similarity values above 0.5. Conversely, all odd pairs, mixing animals and devices or objects (including polysemic words), get very low values, below 0.4.

Table 1. Sample similarity values between WordNet words/concepts

Concept	Concept	Similarity
mouse (animal) [102330245]	computer (device) [103082979]	0.394
mouse (device) [103793489]	computer (device) [103082979]	0.727
mouse (device) [103793489]	cat (pet) [102121620]	0.384
mouse (animal) [102330245]	cat (pet) [102121620]	0.775
cat (domestic) [102121620]	computer (device) [103082979]	0.384
cat (pet) [102121620]	tiger (animal) [102129604]	0.849
cat (wild) [102127808]	tiger (animal) [102129604]	0.910
cat (pet) [102121620]	dog (pet) [102084071]	0.627
dog (pet) [102084071]	horse (domestic) [102374451]	0.542
horse (domestic) [102374451]	horse (chess) [103624767]	0.339
mouse (animal) [102330245]	mouse (device) [103793489]	0.394

As to the natural language sentences, the similarity between words is:

boy-girl = 0.75	boy-hammer = 0.436	girl-hammer = 0.436
want-desire = 0.826	want-hit = 0.361	desire-hit = 0.375
yellow-small = 0.563	small-small = 1	
dog-canary = 0.668	dog-nail = 0.75	canary-nail = 0.387

which allows to overcome the problem of wrong similarity assessment according to syntactic comparisons only[2]. Indeed, the first two sentences neatly get the largest similarity value with respect to the other combinations:

$$fs(s1,s2) = 1.770 \qquad fs(s1,s3) = 1.739 \qquad fs(s2,s3) = 1.683$$

This specific application can be compared to other works that combine in various shapes and for different purposes structural descriptions of sentences, some kind of similarity and WN. [16] concerns Question Answering: sentences are translated into first-order descriptions by directly mapping them on the taxonomy concepts and relations, rather than describing their syntactic structure; the similarity algorithm is original but based on the classical Dice's coheffi-cient and on a proprietary, domain-specific ontology, while WN is exploited to disambiguate word meanings with the user's intervention. Other works concern Textual Entailment. [8] uses WN but not the hyperonimy relation as in our case, focussing on relations that are considered more meaningful for entailment. [4] exploits WN's hyperonymy relation only for finding a direct implication between terms (and similarity and glosses for the rest). [1] does not consider the grammatical structure in word overlap, and exploits classical WN similarities based on synsets (they use the most common sense for each word, while we choose the maximum similarity among all possible couples of senses). [14] exploits exact structural correspondences between the two sentences (while our framework can suggest proper associations even for indeterminate structural parts), applies taxonomic similarities among all possible pairs of words and then chooses the best ones (while we selectively compute similarity between structurally corresponding words only), and uses other WN relations than hyperonimy.

6   Conclusions

Horn clause Logic is a powerful representation language for automated learning and reasoning in domains where relations among objects must be expressed to fully capture the relevant information. While the predicates in the description language are defined by the knowledge engineer and handled only syntactically by the interpreters, they sometimes express information that can be properly

---

[2] Note that all similarities agree with intuition, except the pair dog-nail that gets a higher similarity value than dog-canary, due to the interpretations of 'dog' as 'a hinged catch that fits into a notch of a ratchet to move a wheel forward or prevent it from moving backward' and 'nail' as 'a thin pointed piece of metal that is hammered into materials as a fastener'. Nevertheless, the overall correct similarity ranking between sentences is not affected.

exploited only with reference to a taxonomic background knowledge in order to capture unexpressed and underlying relationships among the concepts described. This paper presented a general strategy for evaluating the similarity between Horn clauses in which taxonomic descriptors are used in support of normal ones, and provided toy experiments to show its effectiveness.

Future work will concern fine-tuning of the taxonomic similarity computation methodology by exploiting other taxonomic relationships, and its application to other problems, such as Word Sense Disambiguation in phrase structure analysis and building refinement operators for incremental ILP systems.

References

[1] Agichtein, E., Askew, W., Liu, Y.: Combining lexical, syntactic, and semantic evidence for textual entailment classification. In: Proc. 1st Text Analysis Conference (TAC) (2008)

[2] Budanitsky, A., Hirst, G.: Semantic distance in wordnet: An experimental, application-oriented evaluation of five measures. In: Proc. Workshop on Word-Net and Other Lexical Resources, 2nd meeting of the North American Chapter of the Association for Computational Linguistics, Pittsburgh (2001)

[3] Ceri, S., Gottlöb, G., Tanca, L.: Logic Programming and Databases. Springer, Heidelberg (1990)

[4] Clark, P., Harrison, P.: Recognizing textual entailment with logical inference. In: Proc. 1st Text Analysis Conference (TAC) (2008)

[5] Esposito, F., Fanizzi, N., Ferilli, S., Semeraro, G.: A generalization model based on oi-implication for ideal theory refinement. Fundamenta Informaticæ 47(1-2), 15–33 (2001)

[6] Ferilli, S., Basile, T.M.A., Biba, M., Di Mauro, N., Esposito, F.: A general similarity framework for horn clause logic. Fundamenta Informaticæ 90(1-2), 43–46 (2009)

[7] Ide, N., Vronis, J.: Word sense disambiguation: The state of the art. Computational Linguistics 24, 1–40 (1998)

[8] Inkpen, D., Kipp, D., Nastase, V.: Machine learning experiments for textual entailment. In: Proc. 2nd PASCAL Recognising Textual Entailment Challenge (RTE-2) (2006)

[9] Krovetz, R.: More than one sense per discourse. In: NEC Princeton NJ Labs., Research Memorandum (1998)

[10] Lin, D.: An information-theoretic definition of similarity. In: Proc. 15th International Conf. on Machine Learning, pp. 296–304. Morgan Kaufmann, San Francisco (1998)

[11] Lloyd, J.W.: Foundations of Logic Programming, 2nd edn. Springer, Berlin (1987)

[12] Miller, G.A.: Wordnet: A lexical database for English. Communications of the ACM 38(11), 39–41 (1995)

[13] Muggleton, S.: Inductive logic programming. New Generation Computing 8(4), 295–318 (1991)

[14] Pennacchiotti, M., Zanzotto, F.M.: Learning shallow semantic rules for textual entailment. In: Proc. International Conference on Recent Advances in Natural Language Processing (RANLP 2007) (2007)

[15] Rouveirol, C.: Extensions of inversion of resolution applied to theory completion. In: Inductive Logic Programming, pp. 64–90. Academic Press, London (1992)

[16] Vargas-Vera, M., Motta, E.: An ontology-driven similarity algorithm. Tech Report kmi-04-16. Knowledge Media Institute (KMi), The Open University, UK

Empirical Assessment of Two Strategies for Optimizing the Viterbi Algorithm

Roberto Esposito and Daniele P. Radicioni

Università di Torino, Dipartimento di Informatica
{esposito,radicion}@di.unito.it

Abstract. The Viterbi algorithm is widely used to evaluate sequential classifiers. Unfortunately, depending on the number of labels involved, its time complexity can still be too high for practical purposes. In this paper, we empirically compare two approaches to the optimization of the Viterbi algorithm: Viterbi Beam Search and `CarpeDiem`. The algorithms are illustrated and tested on datasets representative of a wide range of experimental conditions. Results are reported and the conditions favourable to the characteristics of each approach are discussed.

1 Introduction

Different from the *propositional* setting, where classifiers can optimize the assignment of labels individually, in the setting of *Supervised Sequential Learning* (SSL) objects are assumed to be part of a sequence. SSL classifiers explore the relationships between previous and subsequent objects, in order to improve the classification accuracy. For instance, in the optical character recognition task, the labelling *"learning"* would receive higher score than *"1earn1n9"*, even though the description of the first, sixth and eight characters taken in isolation might suggest otherwise. A SSL classifier would cope with such ambiguities by exploiting the higher sequential correlation of bigrams such as *le* with respect to *1e*.

Since evaluating *all* possible dependencies would result in a complexity exponential in the length of sequences, a Markov assumption of order 1 is usually adopted so that dependencies are assumed to span over adjacent objects only. This allows one using the Viterbi algorithm to find the optimal sequence of labels in $\Theta(TK^2)$ time [1], where T is the length of the sequence and K is the number of possible labels. Unfortunately, even the terrific reduction in complexity achieved by the Viterbi algorithm may be not sufficient in some domains, such as the cases of web-logs related tasks, of music analysis, and of activity monitoring through body sensors [2,3]. In all these cases the number of possible labels is so large that the classification time is prohibitively high [4].

Viterbi Beam Search (VBS) algorithms [5,6,7] use the same optimization approach as the Viterbi algorithm. Given a beam width b, such algorithms perform a breadth-first search, in which only the b most promising solutions are retained at each step. An alternative optimization approach has been proposed that is based on the distinction between information inherent to states (*vertical* information) and information inherent to state transitions (*horizontal* information) [8].

R. Serra and R. Cucchiara (Eds.): AI*IA 2009, LNAI 5883, pp. 141–150, 2009.

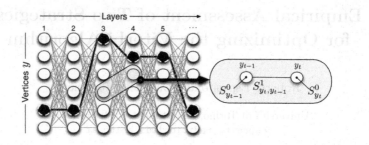

Fig. 1. $S_{y_t}^0$ and $S_{y_t\ _1}^0$ denote per vertex (vertical) weights. $S_{y_t,y_t\ _1}^1$ denotes per edge (horizontal) weights.

The algorithm `CarpeDiem` prunes the search space by exploiting vertical information as much as possible, resorting to the use of horizontal features only in case this is really necessary to the classification purposes. Besides the aforementioned strengths, the given approaches have weaknesses as well: VBS approach renounces to guarantee optimal solutions, whilst `CarpeDiem` relies on having at disposal strong vertical features.

This work focuses on comparing such approaches: to this aim, the mentioned algorithms are tested on a widely varied set of experimental conditions.

2   Preliminaries and Algorithms

The problem of finding the best sequence of labels is often represented as a search for the optimal path in a layered and weighted graph (Figure 1).

Layered graph. A layered graph is a connected graph where vertices are partitioned into a set of "layers" such that: *i*) edges connect only vertices in adjacent layers; *ii*) any vertex in a given layer is connected to all vertices of the successive layer. We indicate the layer to which a vertex belongs with a subscript to the vertex name, so that y_t denotes *a* vertex in layer t. We associate to each vertex y_t a weight $S_{y_t}^0$, and to each edge (y_{t-1}, y_t) a weight $S_{y_t,y_t\ _1}^1$ (Figure 1).

In the following we use the term "vertical" in referring to "per node" properties. For instance, we will use the expressions "vertical weight" of y_t and "vertical information" to refer to $S_{y_t}^0$ and to the information provided by evidence related to vertices, respectively. Similarly, we use the term "horizontal" in referring to "per edge" properties. For instance, we will use the expression "horizontal weight" in referring to the weight associated to a given transition. The distinction between vertical and horizontal information is fundamental in the following: in facts, while in principle the Viterbi algorithm does not distinguish between vertical and horizontal information, one major idea in `CarpeDiem` is to exploit vertical information to avoid considering the horizontal one.

Given a layered and weighted graph with T layers and K vertices per layer, a *path* is a sequence of vertices $y_1, y_2, \ldots, y_t$ $(1 \le t \le T)$. The reward for a path is the sum of the vertical and horizontal weights associated to the path:

$$\text{reward}(y_1, y_2, \ldots, y_t) = \left(\sum_{u=1}^{t-1} S_{y_u}^0 + S_{y_{u+1}, y_u}^1 \right) + S_{y_t}^0 \tag{1}$$

The Viterbi Algorithm. Let us define $\gamma(y_t)$ as the maximal reward associated to any path from any node in layer 1 to y_t, and consider the problem of picking the maximal path from the leftmost layer to the rightmost layer. The reward of the best path to node y_t can be computed recursively as: *i)* the reward of the best path to the predecessor $\pi(y_t)$ on the optimal path to y_t; *ii)* plus the reward for transition $S_{y_t, \pi(y_t)}^1$; *iii)* plus the weight of node y_t. By noticing that $\pi(y_t)$ is the vertex y_{t-1} (in layer $t-1$) that maximizes the quantity $\gamma(y_{t-1}) + S_{y_t, y_{t-1}}^1$, the recursive definition of $\gamma(y_t)$ can be formulated as follows:

$$\gamma(y_t) = \begin{cases} S_{y_t}^0 & \text{if } t = 1 \\ \max_{y_{t-1}} \left(\gamma(y_{t-1}) + S_{y_t, y_{t-1}}^1 + S_{y_t}^0 \right) & \text{otherwise} \end{cases} \tag{2}$$

the Viterbi algorithm proceeds from left to right storing the values of γ into an array $\mathbb{G}$ as soon as such values are computed. Assuming that $\forall y_{t-1} : \mathbb{G}(y_{t-1}) = \gamma(y_{t-1})$, then $\mathbb{G}(y_t)$ is computed as:

$$\mathbb{G}(y_t) = \max_{y_{t-1}} \left(\mathbb{G}(y_{t-1}) + S_{y_t, y_{t-1}}^1 + S_{y_t}^0 \right)$$

The weight of the best possible path (and, with few additions to the definitions, also the best path itself) can be retrieved by searching in $\mathbb{G}$ the maximal entry corresponding to layer T.

Viterbi Beam Search. Many problems can be represented as path-finding problems, and solved by the Viterbi algorithm. In particular, the Viterbi algorithm has been applied to with large states spaces arising in the speech recognition field. This is a typical case where approximate solutions really pay off: suboptimal paths can be tolerated and tight time constraints would prevent exhaustive search. A popular approach in this field is the *Viterbi beam search* (VBS) [5,9,10]: essentially, VBS performs a breadth-first suboptimal search in which only the most promising solutions are retained at each step. The pseudo code for the algorithm is reported in Algorithm 1 (note that it falls back to the Viterbi algorithm if the beam size is set to 100%). Many improvements over this basic strategy have been proposed to refine either the computational performance or the accuracy of the solution, e.g., [6]. In most cases domain-based knowledge –such as language constraints– is used to restrict the search efforts to some relevant regions of the search space [11]. Also, in recent years, several algorithms have been proposed that overcome the difficulties inherent in heuristic ranking strategies by learning ranking functions specifically optimized for the problem at hand [12].

The VBS approach does not come without difficulties. For instance, [7] propose Viterbi beam search to improve the performances of the perceptron algorithm on the particular problem of natural language parsing. Interestingly, the

Algorithm 1. The Viterbi Beam Search.

Data: a beam size b

Result: The best endpoint y_T^* to layer T using beam size b

forall y_1 **do**
$\quad|\quad \mathbb{G}(y_1) \leftarrow S_{y_1}^0$;
end
for $t = 2$ to T **do**
$\quad|\quad$ **for** $y_t \in beam(b)$ **do**
$\quad|\quad\quad|\quad \mathbb{G}(y_t) \leftarrow \max_{y_{t-1}} \left(\mathbb{G}(y_{t-1}) + S_{y_t, y_{t-1}}^1 + S_{y_t}^0\right)$;
$\quad|\quad$ **end**
end
$y_T^* \leftarrow \arg\max_{y_T} \mathbb{G}(y_T)$;

authors note how the sub-optimality of the beam search can badly affect the learning performances. The problem arises when a sub-optimal sequence is used instead of the optimal one to update the weights of the features. In order to alleviate this issue, the authors stop the search – during learning – as soon as the beam does not contain the optimal solution. In such case only the partial sequence, up to when the stopping occurred, is used to update the weights. This prevents from training the perceptron using "bad" predictions, but it still has the drawback of exploiting only partially the training sequences. In such system, then, the sub-optimality of Viterbi beam search has *two* drawbacks: at learning time, it hinders the process of finding better classifiers (or at least it slows the process down); at testing time, it yields sub-optimal classifications.

CarpeDiem. CarpeDiem is a replacement for the Viterbi algorithm that, under mild conditions, allows for dramatic speed-ups and still guarantees optimal results. The main idea underlying CarpeDiem stems from noting that in many application domains not all features really depend on both y_t, y_{t-1}. On the contrary, very often the characteristics of the example are, by themselves, very relevant for predicting y_t. This justifies the distinction between two kinds of features: the *vertical* features that do not require to know the previously predicted label –and work, thus, under a zero-order Markov assumption–, and the *horizontal* features that do need it –thereby working under a first-order Markov assumption. CarpeDiem exploits vertical information as much as possible, resorting to the use of horizontal features only in case this is really necessary to the classification purposes. The algorithm can be best described as a twofold search strategy. The main *forward* search strategy scans the layers in the graph from left to right. For each layer CarpeDiem finds the node with the best possible reward and stops as soon as this node can be determined. In so doing, it possibly leaves a layer without having evaluated the exact reward for reaching all nodes of that layer. The *backward* strategy is called only when it is necessary to evaluate the reward associated to the nodes (i.e., it evaluates $\gamma(y_t)$ for a given y_t) opening the least possible number of previous nodes. Opening a node may imply the need to go back to nodes in previous layer(s) to gather information left unspecified during the forward step.

In the best case, the algorithm scans only the most promising node for each layer, and never calls the backward strategy. In such case, the cost of the algorithm would be $\Theta(K \log(K)T)$, where the factor $K \log(K)$ is due to the time spent for sorting the nodes in each layer. In the worst case (no vertical features), the algorithm has a complexity of $\Theta(TK^2)$, i.e., the algorithm is never asymptotically worse than the Viterbi algorithm (for further details, please refer to [8]).

3   Experimentation

To compare VBS and `CarpeDiem` is not an easy task: VBS performances are affected by the beam size used, while `CarpeDiem` is parameter free. Clearly for very low beam sizes the VBS approach is likely to run very fast (at the risk of outputting suboptimal results), while for very large beam sizes it is likely to output the same result of the Viterbi algorithm at the price of running in almost the same time. On the other side, `CarpeDiem` guarantees optimality of the result, and it may 'silently' degrade to the performances of the Viterbi algorithm in cases where vertical information is not good enough to foster the optimization. In the following, then, we will comment the results in terms of both accuracy and running times. Additionally, it must be recalled that VBS needs a heuristics to guide the search. In the present work, we experiment using a generic heuristic modelled after the one implemented in `CarpeDiem`: it selects labels according to their vertical weight (the higher, the better). Clearly, in domains where better *ad-hoc* heuristics are known the result may vary in a significant way.

In order to systematically assess how VBS and `CarpeDiem` compare when tested on a varied range of situations, it is crucial to take into consideration all parameters affecting the performances of the two algorithms. As regards as `CarpeDiem`, this implies considering datasets with different mixtures of vertical and horizontal information. As regards as VBS this implies considering its performance with different beam size settings.

3.1   Dataset Design

Artificial datasets have been generated by using a Hidden Markov process built using a software tool developed *ad-hoc*. It allows parameterizing the generation of the dataset by specifying, among the others, the amount of vertical and horizontal information to be used. These two parameters have been varied in order to generate two hundred datasets (one hundred learning sets, one hundred test sets). In the following we will use the symbols v and h to denote the amount of vertical and horizontal information specified when building the dataset. Both v and h have been varied from 10% to 100% in steps of 10%. Further parameters required by the dataset generation tool have been set to the values shown in Table 1. The dataset generation tool implements a standard hidden Markov process and is not worth of a detailed description. However, the following three distributions are of interest: *i*) the distribution used for generating the initial state; *ii*) the distribution

Table 1. The parameters used for generating the datasets

Attribute Name	Description	Value
num_attributes	Number of attributes per row	5
num_labels	Number of labels	30
num_values_per_attribute	Number of values for each attribute	10
num_learn_sequences	Number of sequences in the learning set	10
num_test_sequences	Number of sequences in the test set	10
avg_events_per_sequence	Average number of events per sequence	100

governing the transition between labels; *iii*) the distribution that models the dependence of attributes values from current label. Distribution *i*) is uniform over the set of labels. Distribution *ii*) is built as a function of h. The higher h the lesser the uncertainty about the next transition given the current state. Distribution *iii*) is built as a function of v. The higher v the lesser the uncertainty about the output symbols given the current state. The dataset generator sources and a richer description of its working are available for download.[1]

3.2   Experiments Design

Two hundred datasets have been generated: one hundred for learning (we will refer to those datasets with $L(v,h)$, for $v, h \in \{0.1, 0.2, \ldots, 1.0\}$), and one hundred for testing purposes ($T(v,h)$ for $v, h \in \{0.1, 0.2, \ldots, 1.0\}$). Both were generated according to the procedure described above. We used the voted perceptron algorithm to acquire one hundred classifiers $\{H_{v,h}\}$, one per each learning set in $\{L(v,h)\}$ [13]. Each classifier $H_{v,h}$ has been tested over the parallel (same v and h) test set $T(v,h)$ by "plugging" into it the required inference algorithm (Carpe-Diem or VBS). For each execution we recorded the run times and the reward (see Equation 1) of the inferred sequences. In case of VBS we repeated the experiment varying the beam size in $\{0.01, 0.1, 0.2, \ldots, 1.0\}$, for a total of 11 experiments per v, h pair. The resulting number of experiments is $1,200 = 11$ experiments $\times$ 100 datasets for VBS + 1 experiment $\times$ 100 datasets for CarpeDiem.

3.3   Results and Discussion

For space reasons we cannot report about *all* experiments. We selected some experiments representative of interesting cases.[2] We do consider three datasets that lie on the diagonal of the parameter space, namely $T(50,50)$, $T(70,70)$ and $T(90,90)$. In Figure 2 we report in logarithmic scale (with base 10) both the time performances of VBS as the beam size grows and the time performances of CarpeDiem.

---

[1] http://www.di.unito.it/~esposito/software/markov_ds_generator.tar.gz
[2] The interested reader can retrieve the full set of results at:
http://www.di.unito.it/~esposito/Experiments/aixia09.

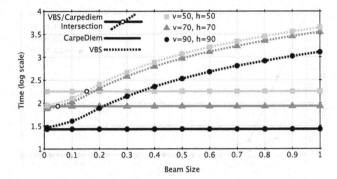

Fig. 2. Time performances of VBS against `CarpeDiem` for parameter settings v = 50, h = 50, v = 70, h = 70, and v = 90, h = 90. Time figures are expressed in logarithmic scale (with base 10).

The results show that VBS performances improve on datasets where the amount of information is higher. This is rather unintuitive since VBS time performances formally depend only on the beam size, the length of the sequence and the number of possible labels. To explain this behaviour, let us note that the voted perceptron sets the weight to zero for features never asserted. In noisy datasets (i.e., ones having low v or low h), chances are that many irrelevant features are asserted (and that their weight is set to non-zero) only because of the noise. On the contrary, in datasets where the noise is low, only relevant features get to have non-zero weights. The reported improvement in running times is not due to the VBS strategy itself, it rather depends on the fact that more features have zero weight as the amount of information grows. In our implementation, in facts, all zero weighted features are not evaluated at all.

Results show that when compared with VBS using medium and large beam sizes, the `CarpeDiem` algorithm runs faster. Moreover, the intersection points (see ○ symbols in Figure 2) shifts to the left as the amount of information available (both vertical and horizontal) increases. This provides evidence of the benefits due to using `CarpeDiem` algorithm when vertical information is very relevant. Indeed, the experiment with parameters v = 90, h = 90 is a case where `Carpe-Diem` runs always faster than VBS, even when the beam size is set to 1% (i.e., a situation where VBS considers only the two most promising labels).[3]

Since VBS could output suboptimal sequences, it is interesting to consider the number of differing labels output by VBS w.r.t. the optimal path returned by the `CarpeDiem` algorithm. Figure 3 shows the ratio between the number of labels for which VBS prediction differs from `CarpeDiem`, and the the total number of predictions. In agreement with the intuition, the experiments show that for small beam sizes higher differences w.r.t. the optimal path are in place. Interestingly,

---

[3] All datasets are built using a set of 30 labels (see Section 3.1). Actually, since restricting to 1% of 30 labels would result in considering less than one label, our implementation defaults the number of considered labels to 2.

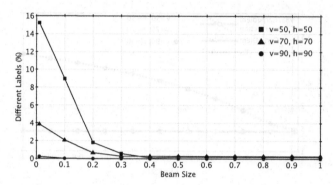

Fig. 3. Ratio of differing labels for the best sequence found by VBS and `CarpeDiem` for increasing beam sizes. Results are provided for datasets with parameters $v = 50, h = 50$, $v = 70, h = 70$, and $v = 90, h = 90$.

even high percentage of differences in the output labels actually corresponds to small differences in the *scores* of the output sequences. For instance, the reported 15% of different labels (the worst case reported in Figure 2) corresponds to a sequence having a score which is less than 1.5% worse than the optimal one. Also, the 15% of different labels cannot be interpreted as 15% of classification error. In fact, no mention is being made here about the "correct" sequence, and it is indeed possible that the suboptimal sequence returned by the VBS approach is more accurate than the one having an optimal score. This is a phenomenon we actually observed in very noisy datasets where the smaller beam size *simplifies* the hypothesis and counterbalances overfitting.

To better understand the effects of suboptimal results provided by VBS on a real dataset as well as to check whether the above findings could apply to real cases, we tested VBS and `CarpeDiem` over the music harmony analysis problem introduced in [8]. Results are reported in Figure 4. The figures are plotted using the same criteria used in previous figures with the notable exception of the values reported on the y axis of Figure 4(a) that reports the accuracy of inferred sequences, rather than the percentage of differences w.r.t. the optimal sequence.

The VBS accuracies increase from about 70% to about 80% as the beam size grows. The fact that it never attains `CarpeDiem` accuracy is due to the fact that more than one optimal sequence can exist. In the reported experiment, the sequence found by VBS (for beam size equal to 100%) is slightly worse than the one found by `CarpeDiem` (the two algorithms find different sequences as a result of different ways of breaking ties). However, to attain competitive accuracies, the VBS algorithm needs to consider 50% of the labels. Figure 4(b) shows that for such beam size, the time performances of VBS are about five times worse than `CarpeDiem`. Moreover, choosing the *correct* beam size is an hard task per se, one that is often solved by costly trial/error iterations. Also, once this parameter has been chosen it is hardly changed afterwards. This may result in suboptimal systems, or the necessity of retuning the classifier, whenever the data in input is subject to non predictable changes over time.

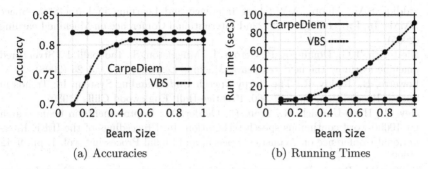

Fig. 4. Comparison of VBS and CarpeDiem algorithms over a natural dataset. The problem considered is performing tonal harmony analysis over a dataset of 4-parts chorales harmonized by J.S. Bach (1685-1750).

4   Conclusions and Future Work

We presented an extensive experimentation to assess the differences between VBS and CarpeDiem algorithms under a variety of different settings. Over 1,200 experiments have been carried out. Overall, the results show that under reasonable conditions CarpeDiem is a good choice: it finds the optimal sequence and is fast. On the other side, VBS ensures a bounded-time solution: a constraint that can be mandatory and that CarpeDiem is unable to satisfy. Also, to choose the best beam size for the problem at hand is sometimes not a trivial problem. One of the lessons learned by the presented experimentation is that under reasonable conditions CarpeDiem runs faster unless the beam is very small. Since in these cases VBS is also expected to produce suboptimal results, is may be suggested to adopt CarpeDiem unless very tight time constraints are in place.

It is apparent that the knowledge of v and h could be exploited to foretell the performances of the CarpeDiem algorithm. How to estimate v and h as well as how to use them to estimate CarpeDiem performances are problems open for future research efforts.

Acknowledgment

This research has been partly supported by the postdoctoral research grant Università di Torino – A.A,200.102 POSTDOC.

References

1. Viterbi, A.J.: Error Bounds for Convolutional Codes and an Asymptotically Optimum Decoding Algorithm. IEEE Transaction on Information Theory 13, 260–269 (1967)
2. Felzenszwalb, P.F., Huttenlocher, D.P., Kleinberg, J.M.: Fast Algorithms for Large-State-Space HMMs with Applications to Web Usage Analysis. In: Advances in Neural Information Processing Systems (2003)

3. Siddiqi, S.M., Moore, A.W.: Fast Inference and Learning in Large-State-Space HMMs. In: Proceedings of the 22nd International Conference on Machine Learning (2005)
4. Dietterich, T.G., Domingos, P., Getoor, L., Muggleton, S., Tadepalli, P.: Structured machine learning: the next ten years. Machine Learning 73(1), 3–23 (2008)
5. Lowerre, B., Reddy, R.: The Harpy Speech Understanding System. In: Trends in Speech Recognition, pp. 340–360. Prentice-Hall, Englewood Cliffs (1980)
6. Ney, H., Haeb-Umbach, R., Tran, B., Oerder, M.: Improvements in beam search for 10000-word continuous speech recognition. In: Proceedings of the IEEE International Conference on Acoustics, Speech, and Signal Processing, vol. 1, pp. 9–12 (1992)
7. Collins, M., Roark, B.: Incremental parsing with the perceptron algorithm. In: Proceedings of the Association for Computational Linguistics, pp. 111–118 (2004)
8. Esposito, R., Radicioni, D.P.: CarpeDiem: an Algorithm for the Fast Evaluation of SSL Classifiers. Proceedings of the 24th Annual International Conference on Machine Learning, IC(ML 2007) (2007)
9. Spohrer, J.C., Brown, P.F., Hochschild, P.H., Baker, J.K.: Partial traceback in continuous speech recognition. In: Proc. IEEE Int Cong. Cybernetics and Societ, Boston, MA (1980)
10. Bridle, J.S., Brown, M.D., Chamberlain, R.M.: An algorithm for connected word recognition. In: Acoustics, Speech, and Signal Processing, IEEE International Conference on ICASSP 1982, pp. 899–902 (1982)
11. Ney, H., Mergel, D., Noll, A., Paeseler, A.: Data driven search organization for continuous speech recognition. IEEE Transactions on Signal Processing 40, 272–281 (1987)
12. Xu, Y., Fern, A.: On learning linear ranking functions for beam search. In: Ghahramani, Z. (ed.) Proceedings of the 24th International Conference on Machine Learning, pp. 1047–1054 (2007)
13. Collins, M.: Discriminative Training Methods for Hidden Markov Models: Theory and Experiments with Perceptron Algorithms. In: Proceedings of the Conference on Empirical Methods in Natural Language Processing (2002)

Approximate Frequent Itemset Discovery from Data Stream

Anna Ciampi, Fabio Fumarola, Annalisa Appice, and Donato Malerba

Dipartimento di Informatica, Università degli Studi di Bari
via Orabona, 4 - 70126 Bari - Italy
{aciampi,ffumarola,appice,malerba}@di.uniba.it

Abstract. Traditional algorithms for frequent itemset discovery are designed for static data. They cannot be straightforwardly applied to data streams which are continuous, unbounded, usually coming at high speed and often with a data distribution which changes with time. The main challenges of frequent pattern mining in data streams are: avoiding multiple scans of the entire dataset, optimizing memory usage and capturing distribution drift. To face these challenges, we propose a novel algorithm, which is based on a sliding window model in order to deal with efficiency issues and to keep up with distribution change. Each window consists of several slides. The generation of itemsets is local to each slide, while the estimation of their approximate support is based on the window. Efficiency in the generation of the itemsets is ensured by the usage of a synopsis structure, called SE-tree. Experiments prove the effectiveness of the proposed algorithm.

Keywords: Frequent Itemset Discovery, Data Stream Mining, Sliding Window Model.

1 Introduction

Data streams are sequences of time-stamped transactions which arrive on-line, at consecutive time points. They are common to many applications, such as network traffic monitoring and sensor data analysis. Therefore, there is a growing need of algorithms that can properly analyze data streams in order to extract interesting and valid patterns. Characteristics of data streams prevent the application of traditional data mining algorithms, which are designed to extract knowledge from static data only. First, data stream applications demand for a real-time response, while the traditional data mining process requires all data are collected before being analyzed. Second, the high data generation rate prevents multiple scans on data, since even simply preserving these transactions for future use can cause storage management problems. Third, when the data stream is open-ended, only approximations of the correct patterns can be computed. Fourth, when data distribution changes along time, it is convenient to mine only recent data in order to take this drift into account.

In this work, we face the new challenges posed by data stream for the task of frequent itemset discovery. We propose a novel algorithm, called SLIM (**SL**iding

R. Serra and R. Cucchiara (Eds.): AI*IA 2009, LNAI 5883, pp. 151–160, 2009.

Window Algorithm for Approximate Itemset Mining), which is a false positive oriented algorithm, i.e., SLIM does not discover any false negative frequent itemset. SLIM is based on the sliding window model [4] according to which a data stream is broken down into slides of transactions arriving in series. A sliding window is composed by a fixed number of consecutive slides. At each time point, only the last slides form the sliding window. Each time a slide flows in, itemsets are generated locally to the slide and enumerated in a synopsis structure called *Set-Enumeration tree* (SE-tree) [12]. In general, the use of SE-trees provides a unifying search-based framework, which guarantees complete, irredundant and prioritized search and supports efficient pruning based on anti-monotonicity property of the support. In SLIM, the local itemsets enumerated by the SE-tree, are also used to approximate the set of frequent itemsets in the current sliding window. This way, multiple scans due to itemset evaluation are limited to slides, thus making the discovery process more efficient both in time and space.

The paper is organized as follows. Section 2 presents some related works. The algorithm SLIM is detailed ins Section 3, while the experimental evaluation of the proposed algorithm on a benchmark data stream is reported in Section 4.

2   Related Work

Frequent itemset discovery has been deeply investigated in the literature on data stream mining. Proposed algorithms differ either in the type of mined itemsets (e.g., maximal or closed) or in the dimension and the type of window model used (e.g., landmark window, dumped window, progressive time window, sliding window [3]). Manku *et al.* [8] proposed an algorithm that, for a given time t, finds a subset of frequent itemsets over the entire data stream up to t (landmark window). The algorithm ensures that the returned subset does not include false negatives, i.e., itemsets that are frequent on the entire data set and infrequent in the landmark window. Conversely, the 1-pass algorithm presented by Yu *et al.* [14] ensures that the returned subset does not include false positives, i.e., itemsets which are frequent in the landmark window and infrequent on the entire data set. Chang *et al.* [1] presented an algorithm that mines recent frequent itemsets where the frequency is defined by an aging function (dumped window approach). Giannella *et al.* [5] proposed an approximate algorithm for frequent itemset discovery with progressive time windows. An in-memory data structure, called FP-stream, is used to store and update historic information about frequent itemsets together with their frequency over time, and an aging function is used to update the entries so that more recent entries are weighted more.

The sliding window model used in this work, presents some advantages over the three models mentioned above. Indeed, stale data are completely discarded, thus saving memory storage and facilitating the detection of the distribution drift. This window model is common to several other algorithms. *Moment* [2] finds closed frequent itemsets over the last N elements, where N is the window size. It uses a synopsis data structure called CET (Closed Enumeration Tree) to monitor the boundary between closed frequent itemsets and the rest of the itemsets. In [10] the data stream is split according to the number of elements, so that all blocks composing a window have the same size. The algorithm returns the frequent itemsets for each window. Golab *et al.* [6] propose an efficient variant based on the computation of the top-k frequent itemsets for each block.

The algorithm proposed by Ren and Li [11] discovers frequent itemsets by splitting the data stream into several slides. A sliding window is composed by a fixed number of the most recent slides. Both a heap structure and a transactions table are used to maintain all the frequent itemsets over the slides of the current sliding window. Lin *et al.* [7] propose a variant based on two table structures which are used to store frequent itemsets in slides and windows respectively. The support of an "unstored" itemset over a slide which can no longer be exactly computed because data have been discarded, is estimated as the minimum support threshold. In AP_{Stream} [13] the same support is estimated on the basis of the support of "similar" itemsets.

As in [7], we investigate the use of sliding windows, but differently from [7,13], we utilize the SE-Tree [12] to store and retrieve discovered itemsets. In addition, we propose an improved estimation of the support for "unstored" itemsets which reduces the number of false positive.

3   The Algorithm

The framework of SLIM is reported in Figure 1. A buffer continuously consumes the stream transactions and pours them slide-by-slide into the SLIM system. After a slide goes through SLIM, it is discarded. SLIM operations consist in generating local itemsets over the slide and estimating an approximation of their support over the sliding window. The minimum support threshold σ, the number p of transactions in a slide, the number w of the slides composing the sliding window and the maximum depth $MaxTreeDepth$ are given before SLIM starts. Details of the local frequent itemset discovery over a slide and the approximate frequent itemeset generation over a sliding window are reported in the next sub-sections.

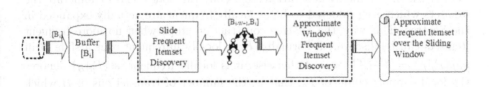

Fig. 1. SLIM framework

3.1   Local Frequent Itemset Discovery over a Slide

The discovery of the itemsets which are frequent over a slide B is performed by exploring level-by-level the lattice of candidate itemsets ordered according to the containment relation $\subseteq$. Formally, given two itemsets P_1 and P_2, $P_2 \subseteq P_1$ denotes

Fig. 2. The SE-tree over the items $A = \{a, b, c\}$ to search the itemsets a, b, c, ab, ac, bc, abc

that P_1 (P_2) is more general (specific) than P_2 (P_1). The containment relation defines a quasi-ordering of the lattice of candidate itemsets which satisfies the reflexivity and transitivity property, but not the anti-symmetric property. This quasi-ordering is searched from the most general itemset (the empty one) by iteratively alternating the candidate generation phase and the candidate evaluation phase [9]. The search is performed in an SE-Tree framework, where itemsets are enumerated as nodes of the SE-tree synopsis structure.

Originally an SE-tree that contains only the root node is constructed. The root is the top node which enumerates the empty itemset. The search starts from the root of the SE-tree and proceeds recursively by exploring a node g of the SE-tree, and generating the child nodes of g in the SE-tree. Nodes are generated according a lexicographic order of the nodes which are potentially added to the SE-tree. This way, distinct itemsets in the search space are enumerated only once. Practically, a node g of a SE-tree is represented as a group comprising both the *head* ($h(g)$), i.e., the itemset P_g enumerated at g, and the *tail* ($t(g)$), i.e., the ordered set consisting of all items which can be potentially appended to g in order to form an itemset enumerated by a sub-node of g. A child g_c of g is formed by taking an item $q \in t(g)$ and appending it to $h(g)$. Therefore, $t(g_c)$ contains all items in $t(g)$ that follows q (see Figure 2). Given this child expansion policy, the SE-tree enumerates all possible itemsets and prevents the generation and evaluation of equivalent candidates, which differ only in the order of items (e.g., "abc" and "bac"). In practice, SLIM exploits the property of anti-monotonicity of $\subseteq$ with respect to support in order to prune unfrequent itemsets in the candidate generation step. Only the nodes which enumerate the itemsets which are evaluated to be as locally frequent are actually expanded in the SE-tree. This way, the SE-tree construction stops when no more frequent itemsets can be discovered. Formally, P_g is locally infrequent over a slide B iff $sup(P_g|B) = \frac{sc(P_g|B)}{|B|} < \sigma$. Otherwise, it is locally frequent. $sc(P_g|B)$ denotes the local support count of P_g, that is, the number of transactions of B which contain P. A final pruning criterion stops the search when a maximum depth (*MaxTreeDepth*) of the SE-tree is reached, where *MaxTreeDepth* is a user-defined parameter.

3.2   Approximate Frequent Itemset Discovery over a Window

Naively, a distinct SE-tree should be constructed for each slide. This would lead to enumerate several times itemesets which are discovered in distinct slides. To reduce memory usage, SLIM maintains a single SE-tree for the entire window. To this aim, each node g of the SE-tree maintains a w sized sliding vector $sv(g)$, which stores one support for each slide in the window. By default, the local support values which are stored in $sv(g)$ are set to unknown. According to the sliding model, when a new slide flows in the buffer, the support vector is shifted on the left in order to remove the expired support. This way, only the last w support values are maintained in the nodes of the SE-tree.

The maintenance of the SE-tree proceeds as follows. When an itemset P_g is discovered over a slide B, we distinguish between two cases, namely, P_g is enumerated in the SE-tree or not. In the latter case, the SE-tree is expanded with the new node g which enumerates P_g, while in the former case the node g already exists in the SE-tree and $sv(g)$ is shifted on the left. In both cases, the value of support count $sc(P_g|B)$ is computed over B and then it is stored in the last position of $sv(n)$. Finally, nodes are pruned when they enumerate itemsets which are unknown on each slide of the window.

Once the SE-tree is updated on the current slide, approximate frequent itemsets are evaluated over the sliding window. The local support counts stored in the SE-tree are used to identify the approximate support of itemsets enumerated by the SE-tree. An itemset P_g is approximately frequent over the window iff the approximate support computed over the window is greater than or equal to σ. The approximate support is estimated on the basis of the local support counts of P_g stored in the SE-tree. More precisely:

$$approximateSup(P_g|W) = \frac{\sum_{B \in W} sc(P_g|B)}{p \times w}. \tag{1}$$

If the support count of P_g is unknown over a slide B, it is estimated by using the known support count of one of its most specific ancestor itemset $Q_{g'}$ ($P_g \subseteq Q_{g'}$ and $sc(Q_{g'}|B) \neq UNKNOWN$) that is enumerated in the SE-tree. Theoretically, the complete set of $2^k - 1$ ancestors should be explored, where k denotes the itemset length. Anyway, this solution may be impractical for high value of k. To improve efficiency, only the ancestors along the path from the node containing P to the root are truly explored. This way, the time complexity of the search is $O(k)$. Since the SE-tree enumerates itemsets, there are several possibilities to estimate P_g. In the case $Q_{g'}$ is infrequent over B, the support count of P_g is overestimated by the support count of $Q_{g'}$ over B, while in the case $Q_{g'}$ is frequent, the support of P_g is correctly determined as zero. This assignment with a zero value is motivated by the monotonicity based policy applied to the generation of candidates enumerated in the SE-tree. According to this policy, a candidate itemset $Q_{g'}$ is anyhow stored in the SE-tree. In the case $Q_{g'}$ is infrequent, the level-wise search stops at $Q_{g'}$.

An example of the the discovery of approximate frequent itemsets over a sliding window is reported in Example 1.

Example 1. Let us consider the data stream in Figure 3, $p = 5$, $w = 2$ and $\sigma = 0.7$. SLIM constructs the SE-tree from the first slide (B_1) of p transactions

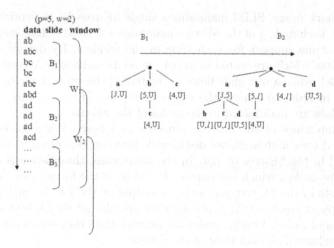

Fig. 3. The SE-Tree constructed from B_1 and then maintained according to B_2. $\sigma = 0.7$.

which income the stream and then it maintains the SE-tree over the second slide (B_2). "ad" is one of the itemsets enumerated in this SE-tree. $sc(ad|_{B_1})$ is unknown, hence it is estimated by means of $sc(a|_{B_1}) = 3$ where "a" is the most specific ancestor of "ad" in the path to the top of SE-tree with known value of support count. Differently $sc(ad|_{B_2}) = 5$. Approximate support of "ad" is estimated as $approximateSup(ad|_{B_1 B_2}) = \frac{3+5}{5\times2} = 0.8$.

4   Experiments

We evaluate SLIM over 150,000 transactions of the KDD99 Network Intrusion Detection Dataset[1]. This dataset is converted into a data stream by taking the data input order as the order of streaming. A transaction of the stream describes an individual TCP connection by means of nine basic features (e.g., duration, service, protocol and so on). Continuous features (e.g., number of bytes transmitted from the source) are pre-discretized. Experiments are run by varying p ($p = 120, 240$ transactions) and w ($w = 2400, 4800, 7200$ transactions). σ is set to 0.6 and $MaxTreeDepth$ is set to 9.

Frequent itemsets discovered by SLIM are compared with itemsets discovered by an implementation that we have done of the algorithm proposed by Lin et al. [7]. Initially, we analyze the total number of false positive itemsets which are discovered over the sliding windows of the entire stream. No false negative itemset is discovered by both SLIM and Lin et al.'s algorithm due to the overestimation of the support. The number of false positive itemsets is reported in Table 1. False positives are those approximate itemsets which are not included in the set of true frequent itemsets we have directly discovered over the entire window.

[1] Available at: http://kdd.ics.uci.edu/databases/kddcup99/kddcup99.html

Table 1. The number of false positive itemsets discovered over the entire stream

Experimental Settings	True itemsets	Number of windows	False positive	
			SLIM	Lin et Al.
$p = 120$ trans $w = 20$	315230		64	192
$p = 120$ trans $w = 40$	307013	1250	0	128
$p = 120$ trans $w = 60$	299481		0	192
$p = 240$ trans $w = 10$	155352		0	0
$p = 240$ trans $w = 20$	153634	625	0	64
$p = 240$ trans $w = 30$	149932		0	64

These results confirm that SLIM discovers a lower number of false positive than Lin et al.'s algorithm. This is due to a better approximation of unknown local support counts (Lin et al. naively approximate unknown support counts with $p \times \sigma$). As expected, the number of false positive itemsets increases by enlarging the window size and/or reducing the number of transactions in p.

Further considerations are suggested by the analysis of the absolute error of the approximated support, averaged over the true positive itemsets. Only the sliding windows where the error is greater than zero are plotted in Figure 4. Error shows that SLIM always outperforms the baseline algorithm. In particular, error decreases to zero when the slide size increases. This confirms the effectiveness of the SLIM estimation of the approximate support that is significantly better than Lin et al.'s algorithm. Additionally, SLIM takes advantages from the ability

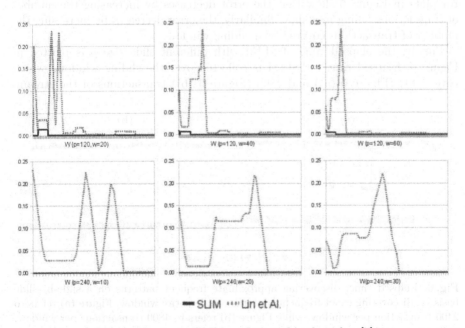

Fig. 4. Error rate: SLIM vs. Lin et al.'s algorithm [7]

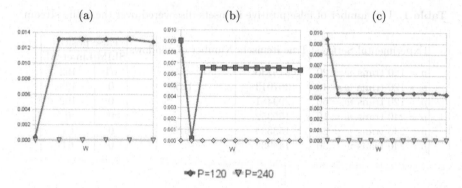

Fig. 5. Error of SLIM over sliding windows which cover the same data: (a) W covers 2400 transactions segmented with $p = 120$ ($w = 20$), $p = 240$ ($w = 10$) (b) W covers 4800 transactions segmented with $p = 120$ ($w = 40$), $p = 240$ ($w = 20$) (c) W covers 7200 transactions segmented with $p = 120$($w = 60$), $p = 240$($w = 30$)

of approximating unknown support counts with their exact zero value when the ancestor of the itemset is frequent over the slide. This motivates the fact that the error of SLIM is often zero over the sliding windows.

A different perspective of the results is offered by the comparison of the itemsets discovered by both SLIM and Lin et al. over the sliding windows which cover the same portion of data stream, but are generated with different slide size. Error plots in Figure 5 show that the error decreases by increasing the number of transactions falling in a slide. Similarly, the error decreases by increasing the number of transactions covered by a sliding window.

Finally, the elapsed time of SLIM with different slide size p is plotted in Figure 6. Elapsed time for SLIM is collected over the sliding windows which cover 2400 (Figure 6 (a)) and 4800 (Figure 6 (b)) transactions of the stream.

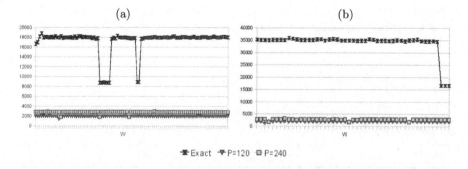

Fig. 6. Elapsed time: discovering approximate frequent patterns on a slide-by-slide basis vs. discovering exact frequent patterns on the entire window. Figure (a) refers to 2400 transaction per window, while Figure (b) refers to 4800 transactions per window. Only the first 100 windows are plotted.

The discovery of approximate frequent patterns on a slide-by-slide basis is more efficient than the discovery of exact frequent patterns on the entire window. As expected, elapsed time decreases by reducing the slide period.

5 Conclusions

We presented SLIM, an algorithm for the approximate frequent itemset discovery in sliding windows of data streams. SLIM is based on the idea that a data stream is segmented into slides of consecutive transactions. A fixed number of consecutive slides form a sliding window. The algorithm generates itemsets locally to each slide and estimates the support on the entire sliding window. Experiments on the KDD99 Network Intrusion Detection Dataset are reported. For this benchmark dataset, SLIM proves to be both accurate and efficient. As future work, we plan to evaluate the proposed algorithm on additional benchmark data streams.

Acknowledgments

This work is supported by both the Strategic Project PS121 "Telecommunication Facilities and Wireless Sensor Networks in Emergency Management" funded by Apulia Region, and the project DDTA "Apulian Textile-Clothing Digital District" funded by the National Department for Innovation and Technology.

References

1. Chang, J.H., Lee, W.S.: Finding recent frequent itemsets adaptively over online data streams. In: KDD 2003, pp. 487–492. ACM Press, New York (2003)
2. Chi, Y., Wang, H., Yu, P.S., Muntz, R.R.: Moment: Maintaining closed frequent itemsets over a stream sliding window. In: Perner, P. (ed.) ICDM 2004. LNCS (LNAI), vol. 3275, pp. 59–66. Springer, Heidelberg (2004)
3. Gaber, M.M., Zaslavsky, A., Krishnaswamy, S.: Mining data streams: a review. SIGMOD Rec 34(2), 18–26 (2005)
4. Ganti, V., Gehrke, J., Ramakrishnan, R.: Mining data streams under block evolution. SIGKDD Explorations 3(2), 1–10 (2002)
5. Giannella, C., Han, J., Pei, J., Yan, X., Yu, P.: Mining frequent patterns in data streams at multiple time granularities, technical report, computer science department, indiana university (2002)
6. Golab, L., Dehaan, D., Demaine, E.D., Lopez-Ortiz, A., Munro, J.I.: Identifying frequent items in sliding windows over on-line packet streams. In: Proceedings of the Internet Measurement Conference, pp. 173–178. ACM Press, New York (2003)
7. Lin, C., Chiu, D., Wu, Y.: Mining frequent itemsets from data streams with a time-sensitive sliding window. In: SDM 2005 (2005)
8. Manku, G.S., Motwani, R.: Approximate frequency counts over data streams. In: VLDB 2002, pp. 346–357 (2002)
9. Mannila, H., Toivonen, H.: Levelwise search and borders of theories in knowledge discovery. Data Mining and Knowledge Discovery 1(3), 241–258 (1997)

10. Mozafari, B., Thakkar, H., Zaniolo, C.: Verifying and mining frequent patterns from large windows over data streams. In: DE 2008, pp. 179–188 (2008)
11. Ren, J., Li, K.: Find recent frequent items with sliding windows in data streams. In: IIH-MSP 2007, pp. 625–628. IEEE Computer Society Press, Los Alamitos (2007)
12. Rymon, R.: An se-tree based characterization of the induction problem. In: ICML 1993, pp. 268–275. Morgan Kaufmann, San Francisco (1993)
13. Silvestri, C., Orlando, S.: Approximate mining of frequent patterns on streams. Intell. Data Anal. 11(1), 49–73 (2007)
14. Yu, J.X., Chong, Z., Lu, H., Zhou, A.: False positive or false negative: mining frequent itemsets from high speed transactional data streams. In: VLDB 2004, VLDB Endowment, pp. 204–215 (2004)

Kernel-Based Learning for Domain-Specific Relation Extraction

Roberto Basili[1], Cristina Giannone[2], Chiara Del Vescovo[2],
Alessandro Moschitti[3], and Paolo Naggar[2]

[1] University of Roma, Tor Vergata, Rome, Italy
basili@info.uniroma2.it
[2] CM Sistemi s.p.a., Rome, Italy
{cristina.giannone,chiara.delvescovo,paolo.naggar}@gruppocm.it
[3] University of Trento, Trento, Italy
moschitti@disi.unitn.it

Abstract. In a specific process of business intelligence, i.e. investigation on organized crime, empirical language processing technologies can play a crucial role. The analysis of transcriptions on investigative activities, such as police interrogatories, for the recognition and storage of complex relations among people and locations is a very difficult and time consuming task, ultimately based on pools of experts. We discuss here an inductive relation extraction platform that opens the way to much cheaper and consistent workflows. The presented empirical investigation shows that accurate results, comparable to the expert teams, can be achieved, and parametrization allows to fine tune the system behavior for fitting domain-specific requirements.

1 Analysis of Investigative Texts

The semi-automated extraction of information from data of a textual nature has become of interest in recent years for different theoretical and applicative contexts, relevant both in the investigative and decisional stages of judicial processes. Starting from the results of the research project "*ASTREA*, Information and Communication for Justice" [1], the *REVEAL* (Relation Extraction for inVEstigating criminAL enterprises) project has been carried out for setting up a relation extraction system and putting it on trial with unstructured judicial documents (such as questioning or confession reports). Currently the population process is executed by teams of analysts that, by reading each document, annotate all quotations about the facts and the involved subjects according to a conceptual schema. The amount of daily texts produced affects the scalability and timeliness of the overall process. This asks for the automation of the recognition of the specific relations whereas machine learning methods are highly beneficial. First, statistical learning methods represent the state-of-the-art in several Information Extraction and Semantic Parsing tasks as systematic benchmarking in the NLP area have shown in international challenges ([2]). Second, the adoption

R. Serra and R. Cucchiara (Eds.): AI*IA 2009, LNAI 5883, pp. 161–171, 2009.
© Springer-Verlag Berlin Heidelberg 2009

Table 1. The set of targeted Relationship classes

Description	Abbreviated Form
A physical person knows another physical person	PP KNOWS PP
A physical person belongs to a criminal enterprise	PP BELONGS TO CE
A physical person photographically identifies a physical person	PP IDENTIFIES PP
A criminal enterprise includes a criminal enterprise	CE INCLUDES CE
A physical person hangs out at a place	PP HANGS OUT AT PL
A means of communication is linked to a juridical person	MC IS LINKED TO JP
A means of communication is linked to a physical person	MC IS LINKED TO PP

of inductive methods enables an incremental approach where the interleaving between the automatic learning for tagging and human validation allows to scale up in a much cheaper fashion. Moreover, annotations provide links between individual relational facts and their textual realisations that are very important in the long term perspective.

This paper presents the main results of the *REVEAL* project. The project aims to substitute the manual *analysis phase* employing the REVEAL system for real-time information extraction. Early results confirm the wide applicability of the proposed approach.

1.1 Definition of the Task

Text mining for the crime investigation domain is a complex semantic task where textual, linguistic and domain knowledge are critically involved. Relation extraction is at the cross-road of all these knowledge sources. Typical textual phenomena of interest for crime investigation refer to classes of entities and specific relationships, as those reported in Table 1, have been choose to carried out the experimental activities. The set of documents relevant for the peculiar analysis in this domain are of different types like questioning reports or transcribed confessions, land registry documents or telephone printouts. As result, they exhibit many different phenomena that make them semantically and syntactically highly heterogeneous. A typical case is represented by some target information (e.g. the connection between people expressed by the relation KNOWS), that is not always realized within single sentences, but can span much larger textual units. Moreover, extra-linguistic information plays a role in establishing the correctness of some relations. For example, several criminal enterprises take their name from the place of origin and a systematic ambiguity arises.

In order to support the investigative analysis, the *REVEAL* project focuses on the collection of living examples from real texts. This phase corresponds to the first stage of analysts' work. They are currently required to annotate every fact of interest directly on the target document, that is to collect *quotations*. In machine learning, this corresponds to create training examples for automating the analysis phase, that is keeping track of the link between the target structured information to be extracted and the originating (host) text.

The manual annotation phase has thus been designed with the development of specific guidelines for the annotators. In collecting instances, the annotators have been asked to mark the exact text boundaries of all accepted relation instances. Several conceptual and linguistic problems emerged in this phase as a clear consequence of the text complexity.

Linguistic complexity. The natural language phenomena exhibited by the texts are highly heterogeneous. Most of the linguistic problems are related to the use of specific forms, as dialectal and jargon expressions, that open a variety of ambiguities to the interpretation. This renders the application of a syntactic parser very problematic as for coverage at the level of lexical and grammatical phenomena. A crucial problem is that interpretations are often open to subjectivity. Take, for example, a sentence like *"Ne parlai con Mario e Giorgio"*[1] that was differently annotated by individual analysts. One of them accepted the relation KNOWS between the speaker and both entities Mario and Giorgio, and produced, in this way, three annotations for the three pairs of physical persons (PP): (speaker,Giorgio), (speaker,Mario) and (Giorgio,Mario). This interpretation clearly assumed that a meeting had taken place between the three. On the opposite, a second annotator outlined that no information could be found in the sentence confirming that the speaker met both subjects at the same time. This alternative interpretation results into just two annotations between the speaker and each mentioned PP.

Consistency. Although trained about very specific rules and guidelines, annotators are often seen not to follow them strictly. This is largely due to the combinatorial explosion of some phenomena that is difficult to be pointed out exhaustively. This leaves some free space for the annotators to neglect some cases, thus reducing coverage. An example of such inconsistent behavior is the analysis of an excerpt like *"All'incontro a Roma erano presenti: Andrea, Barbara, Claudio, Daniela, Ettore e Francesca"*.[2] It is obviously true that this sentence suggests binary relations (KNOWS relation) between all pairs of the mentioned physical persons, and between people and the location (i.e. Rome, with 6 HANGS OUT relations between PPs and PLACE). One annotator pointed out for this sentence only the last 6 relations.

2 Automatic Relation Extraction for Investigative Text Analysis

Given a finite set of entity types $\mathcal{O}$ and binary relation types $\mathcal{R}$ over $\mathcal{O}$, the relation extraction task corresponds to a function $f(e_i, T_i, e_j, T_j, t_{ij}) \rightarrow \mathcal{R} \cup \{\perp\}$, where, t_{ij} is the text fragment including the mentions to two entities $e_i, e_j \in \mathcal{O}$,

---

[1] *I told it to Mario and Giorgio.*

[2] *Andrea, Barbara, Claudio, Daniela, Ettore and Francesca attended the meeting in Rome.*

whose types are T_i, T_j respectively. f thus accepts a relation $r_i \in \mathcal{R}$ and the special type $\perp$ represents the rejection of all relations.

The adoption of an empirical view on Relation Extraction from texts has been already studied within the machine learning community, as in [3,4,5], where variants of Support Vector Machines ([6]) are applied. The common idea of these works is that the computation of the function f is seen as a classification task. The targeted entities e_i and e_j are here mapped into a vector x_{ij} of properties expressing different types of features of the text unit t_{ij} (i.e. a potential quotation) in which they appear. A boolean standpoint can be thus taken, where $f(e_i, T_i, e_j, T_j, t_{ij}) = r_k$ only when $H_k(x_{ij}) = true$: in other words, the recognition is embodied by the hypothesis function $H_k(.)$, to be learnt, that accepts or rejects x_{ij} as an instance of the relation r_k. Functions $H_k(.)$ are binary classifiers for relations r_k and can be acquired from collections of annotated examples.

SVM classifiers learn a decision boundary between two data classes that maximizes the minimum distance, or *margin*, from the training points of each class to the boundary. The involved distance and the feature space in which the boundary is set are determined by the choice of the kernel function ([7]).

These methods have the advantage that combinations of kernel functions can be easily integrated into SVM as they are still kernels. Kernel combinations are very useful to mix the knowledge provided by the original features whereas these characterize feature spaces with quite different topological properties, for example acting on different perspectives on the original objects like the lexical and syntagmatic properties of the textual units. Feature mapping compositions are thus useful to derive expressive kernels.

A particular class of kernel function, successfully applied to relation extraction tasks [3,4,5], is the *string* (or *sequence*) *kernel* one ([8]). String kernels compute the similarity between instances according to their common sparse subsequences as observed in the targeted textual units t_{ij} used to represent them. Learning proceeds through the matching of such subsequences as they are exhibited by training examples. During classification, common sequences (with gaps) are efficiently matched according to dynamic programming techniques. A decay factor λ is imposed to lower the contributions to the overall score of those characterized by longer gaps([8]).

Analogously in our work we used a similar structuring. A quotation here is intended as a text window that includes the two target entities. Usually, a structured representation in three segments is adopted. A *Fore-Between* segment (FB) is made by words in the sentence appearing between the n-th position before and the n-th position after the earlier entity in the text. The *Between* segment (B) is made of words that appear between the positions of the two entity mentions. Finally, the *Between-After* segment (BA) includes words appearing between the n-th position before and the n-th position after the latter of the two entities in the text. These annotations (left-to-right FB to BA) are thus made available to match subsequences in the suitable positions relative to the entities. The quotation is usually considered the text span that cover the union of the FB, B and BA subsequences. In the investigation domain, targeted here, relations of interest tend to be realized between entities even at a very long distance in the text. For this reason, as we will see in Section 2.1, we followed an approach simpler than the one in [5]. Only the two FB and BA sequences are considered as originating subsequences. A single sequence is then obtained through direct juxtaposition of FB and BA over which the kernel computation is run. Sometimes it covers the entire sentence where the entities e_i and e_j are both quoted. However, when longer distances and multiple sentences are treated, the resulting kernel acts only on text fragments, that are more local to e_i and e_j. As a results, this kernel is oriented to capture the shallow (local) syntactic information implicit in the fragments. It will be hereafter referred as K_{Seq}.

As kernel composition allows to adopt several representations for an incoming text unit, we also exploited typical lexical representations of the source examples through a bag-of-word (BOW) approach. In this representation, every token in a text window including the two entities (using also the n $Fore$ words of the entity earlier in the text and the n $After$ words of the latter entity) is taken into account in the BOW. The resulting kernel, K_{BOW} hereafter, cannot capture some task specific aspects, e.g. the distance between the involved entities. It has thus been extended through special features, as discussed in the next section: the resulting alternative model will be referred as K_{XBow}.

In this way, lexical and syntagmatic spaces are modeled independently, via K_{BOW}/K_{XBow} and K_{Seq} respectively. The overall kernel is defined through the kernel combination: $K(X_1, X_2) = K_{XBow}(X_1, X_2) + K_{Seq}(X_1, X_2)^3$.

2.1   Feature Modeling for Investigative Relations

The adoption of an empirical perspective requires the availability of annotated examples of relationship instances as they are observed into incoming objects o (here the text units t_{ij}). Then every individual o is mapped into a suitable vectorial form x. This step is carried out through the extraction of a set of properties (i.e. features) from the source objects t_{ij}.

Every analysts' annotation is used to build the set of positive training examples for the relationship class $r \in \mathcal{R}$. Moreover, every positive instance for a relation, say r_k, is also a negative example for every relation r_l ($l \neq k$) that insists on the same entity pair of r_k. For example, every accepted instance of the KNOWS relation (between PP pairs) is also a negative instance for the IDENTIFIES relation (see Table 1 for a full description). In addition, negative training examples also stem from rejected cases. In order to build the full set of negative examples, we computed all possible entity pairs from a document that: (1) are not positive examples of any relation, and (2) obey to at least one relationship class in the domain schema (i.e. it is an entry in Table 1). This assumption states that every candidate quotation, suggested by at least one candidate entity pair, is a negative example when no annotation is available for it.

---

[3] A normalised version $K_{Norm}(X_1, X_2)$ is adopted for all the kernels K, where $K_{Norm}(X_1, X_2) = \frac{K(X_1, X_2)}{K(X_1, X_1)K(X_2, X_2)}$.

Notice that the above assumption makes the set of candidate pairs to proliferate in long documents. However, *inter-sentence* relations are very infrequent between very far sentences. In order to keep the set of candidate pairs manageable, thresholds to the maximal distance allowed between two entities e_i and e_j are imposed. The analysis of the annotated corpus showed that most of the entity pairs in valid relation instances generally occurred within a limited distance[4] The distribution of valid relations allowed to define a criteria (statistical filter hereafter) that filters out the (e_i, e_j) pairs whose distance is above a threshold. The optimal threshold has been estimated over a development set as the 90-th percentile that maximizes coverage while minimising the number of false instances introduced. As different relations produce different distributions different thresholds are adopted for each relationship class. The statistical filter is then clearly applied in the training (to gather useful negative examples) as well as in the test phase.

The complexity of the relation extraction task targeted in this project asks for a suitable (vector) description x_{ij} of individual examples t_{ij}. Features have to cover a variety of phenomena ranging from lexical information (e.g. expressing the main verbs denoting the target relations, such as *to_meet* for relation KNOWS) to grammatical constraints. Other, task specific features have been designed to better capture useful textual hints. In all the experiments the set of features described below has been adopted.

Lexical units. Words in texts are expressed through their surface representations (tokens) or through the corresponding lemmatised forms (lemmas).

Entity Types. In order to increase the generalization power of individual features, each textual mention to an entity (e.g. "*Mario*", "*Roma*") is substituted by the label of its corresponding class. For example in the excerpt "*Mario ha abitato a Roma per un periodo*[5]", the active tokens in the representation become {PP, *ha, abitato, a*, PL, *per, un, periodo* }.

Distance between mentions to entities. Although the token distance between the involved entities is used as a filter for candidate pairs, the distance is also useful to impose more or less stricter criteria on other features. So, discrete values are obtained as the 3 main percentiles (33%, 66%, 100%) of the distributions of distance values for each individual relations. Different three-valued labeling are obtained for the different relationship classes.

Punctuation. Punctuation in the *Fore*, *Between* and *After* portions of the involved textual units t_{ij} are all represented via special labels, accounting for the relative position of each punctuation mark with respect to the entities. For example, a comma in the *Fore* component of a textual unit (i.e. before the entity e_i appearing earlier in the text) is denoted by #,F, while #,B is reserved for

---

[4] Distance is measured in term of number of tokens.
[5] *Mario has been living in Rome for a while.*

commas appearing between the two entity mentions. Moreover, each feature is weighted according through its number of occurrences within the corresponding component (e.g. *Fore* vs. *Between*).

Ordering of mentions. This boolean feature OM denotes the property of the textual unit t_{ij} to instantiate a relation r_k in agreement with the order of this latter. For example, while the HANGS OUT relation is clearly orientated from people PP to places PL, the fragment *"A Roma l'incontro con Mario si protrasse sino a tarda notte*[6]*"* mentions the two entities in the reverse order: in this case, the feature OM assumes the value `false`.

3   Performance Evaluation

The industrial impact of the proposed SVM-based technology has been evaluated on real test collections, in coordination with the analyst teams. The overall objective of the experiments was to provide a comparative analysis of different learning algorithms, and measure the accuracy reachable. Some aspects more related to the applicability of REVEAL to the current operational investigative practices are discussed in the conclusive Section 4. This section first discusses the experimental set-up (Section 3.1), while the comparison of different learning algorithms over the employed test data is discussed in Section 3.2.

3.1   Experimental Set-Up

The experimental corpus, made of 86 documents, annotated by two teams of analysts, has been derived from two collections of public judicial acts related to the legal proceedings against the same large criminal enterprise. The corpus has been split into a 90% component (i.e. 79 documents) for training and the remaining 10% (7 documents) then used for testing[7]. The splitting has been applied by trying to preserve, in the test set, the same distribution of instances across relationship classes as those observed in the training data. Although manual annotations have been added for 15 different relationship classes, due to lack of evidence in the training data for some classes, the experimentation was focused on the seven relations reported in Table 1. Skewed distributions are observed, where some relations are much more common in documents like PP HANGS OUT AT A PL or PP KNOWS PP and other are very infrequent AN ASSET IS CONNECTED TO A PLACE. Some of the relations were not well represented in the training data but they have been selected for their high relevance for investigations. This allows also to verify the robustness of the REVEAL models. The experimental corpus is described in Table 2. It shows the overall number of instances available for training (column 2) and testing (column 3) over each individual relation: percentages are relative to the number of positive cases that

---

[6] *"In Rome, the meeting with Mario lasted 'til late night".*

[7] A distinct set of 10 documents has been used as a development set to compute the parameters, such as the statistical filters and the SVM settings.

Table 2. Experimental Data Set

Id	Relationship Class	Training instances (% of positives)	Test instances
r1	PP KNOWS PP	3985 (16.18%)	519
r2	PP IDENTIFIES PP	3985 (5%)	519
r3	PP HANGS OUT PL	2359 (14.83%)	229
r4	PP BELONGS TO CE	1717 (35.11%)	103
r5	CE INCLUDES CE	604 (20.19%)	10
r6	MC IS LINKED TO JP	62 (51.6%)	22
r7	MC IS LINKED TO PP	231 (42.85%)	39

have been used for training. Notice how the first two rows (relations KNOWS and IDENTIFIES) have the same number of cases: they in fact operate on the same number of candidate pairs, as their semantic signature (i.e. (PP × PP)) coincides.

3.2 Comparative Analysis

The first experiment aimed at evaluating the impact of different feature models across a set of learning strategies. In order to test the impact of the REVEAL models against some performance baselines, we adopted two well-known learning algorithms, i.e. C4.5 decision tree learner[9] and a NaiveBayes model to the same data sets[8]. Both systems have been run over the feature set characterising the K_{XBOW} kernel (i.e. bag-of-words extended with the domain features discussed in Section 2.1). Moreover, a simple baseline making random choices across the candidate pairs (filtered according to the 90-th percentile statistics), has been evaluated. All the algorithms were tested against the data shown in Table 2. The comparative evaluation is shown in Table 3 where performances of the different algorithms are shown. The last three rows represent models trained over the different kernels designed in REVEAL[9]. Although the precision of NaiveBayes is better than the bag-of-word model (i.e. the simplest SVM model), it achieves an overall lower F1 measure (0.39 vs. 0.45). The kernel-based models all show a good coverage with recall scores over 0.75. The two models proposed in REVEAL (i.e. K_{XBOW} and $K_{XBOW} + K_{Seq}$) are the best performing models.

The good results obtained through the different kernels, as shown by Table 3, inspired an impact analysis of the different models over the individual relations.

In Table 4 the F-measure scores as obtained for individual relations according to the REVEAL models are reported. Most of the relations obtained an excellent score, reaching in some case an F1 of 1. On some more complex relationship

---

[8] The Weka toolkit was employed also to optimize both algorithms over the development set.

[9] For the SVM learning, we used the SVMlightTK platform, available at: http://dit.unitn.it/ moschitt/Tree-Kernel.htm. The sequence kernel here supported is a special case of the tree kernel discussed in [10].

Table 3. Comparative evaluation among classification algorithms

Algorithm	P	R	F1	Acc
Random Choice	0.13	0.4	0.21	41%
Decision Tree	0.21	0.26	0.23	66%
NaiveBayes	0.36	0.48	0.39	63%
K_{BOW}	0.32	0.75	0.45	66%
K_{XBOW}	0.70	0.83	0.73	85%
$K_{XBOW} + K_{Seq}$	0.75	0.85	**0.75**	**88%**

Table 4. F-measure score of the SVM models over individual relationship classes

Id	Relationship Class	K_{XBOW}	$K_{XBOW} + K_{SK}$
r1	PP KNOWS PP	0.398	0.523
r2	PP IDENTIFIES PP	1	1
r3	PP HANGS OUT at a PL	0.40	0.684
r4	PP BELONGS to CE	0.66	0.747
r5	CE INCLUDES CE	1	1
r6	MC IS LINKED TO JP	0.70	0.70
r7	MC IS LINKED TO PP	1	1

classes, as PP KNOWS PP and PP HANGS OUT PL, the K_{XBOW} kernel achieves lower performances, basically due to the presence of dialectal or syntactically odd expressions. The combination of the two kernels, last column of Table 4, seems to overcome most of these problems. Notice that the weaker relation is r_1 (KNOWS) where also experts show a very high disagreement. It seems that although relatively shallow features are adopted, and no syntactic parsing is applied, the trained SVM seems to deal with most of the phenomena in an harmonic way with humans: relation detection exhibits a similar behavior where complex cases are hard for both. In particular, if the is $K_{XBOW} + K_{Seq}$ kernel

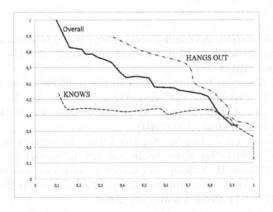

Fig. 1. Precision/Recall curve

is applied only to the 755 cases (that is the 65% of the overall test set) where full agreement is observed between teams of analysts, its F1 achieves the much better value of 0.82. As a final test, we computed the precision-recall curve for the REVEAL kernel $K_{XBOW} + K_{Seq}$, obtained by varying the SVM parameters. The curve in Fig. 1, defines the trade-off between recall and precision for two relationship classes and the micro-averaged results from all relations (*overall*). As apparent, the plot shows a regular shape and it suggests that parameter tuning can be effectively applied to capture the required trade-off between the suitable coverage and accuracy of the method.

4   Conclusive Remarks

From an industrial point of view the REVEAL project represented a relevant and successfully experience. The technology designed and tested in the project has been shown to be very effective. Several technological benefits are related to the specific modeling proposed in REVEAL, as discussed in Section 2. A wide range of experimental activities have been discussed in the paper (Section 3). As a results, the automation of the annotation step cannot be yet considered as a comprehensive solution, as performances are not always acceptable. However, a very attractive semi-automated solution has been enabled where the analyst role is to inspect the system suggestions to validate them. As validation is quite simpler than annotating "*from scratch*", a significant overall speed-up can be expected. During the project, the REVEAL processing time for a medium sized document of 15 pages has been estimated being about 13 minutes[10]. The current professionals are able to annotate the same amount of text in not less than 4 hours. Although this latter measure is surprisingly good, the speed-up achievable is about 18 times. Regarding future enhancement of our system, we would like to exploit advanced shallow semantic approaches such as predicate argument structures, e.g. [11,12,10] and combined syntactic and semantic kernels [13].

References

1. ASTREA (Information and communication for justice) coordinated by Italian Research Council/Research Institute on Judicial Systems (IRSIG-CNR), http://astrea.cineca.it/
2. Carreras, X., Marquez, L.: Introduction to the conll-2005 shared task: Semantic role labeling. In: Proc. of CoNLL, Ann Arbor, Michigan, pp. 152–164 (2005)
3. Zelenko, D., Aone, C., Richardella, A.: Kernel methods for relation extraction. J. Mach. Learn. Res. 3, 1083–1106 (2003)
4. Culotta, A., Sorensen, J.: Dependency tree kernels for relation extraction. In: Proceedings of ACL 2004, Barcelona, Spain, pp. 423–429 (2004)
5. Bunescu, R., Mooney, R.: Subsequence kernels for relation extraction. In: Weiss, Y., Schölkopf, B., Platt, J. (eds.) Advances in Neural Information Processing Systems 18, pp. 171–178. MIT Press, Cambridge (2006)

---

[10] A standard dual core Workstation has been employed for these measures.

6. Vapnik, V.: The Nature of Statistical Learning Theory. Springer, Heidelberg (1995)
7. Shawe-Taylor, J., Cristianini, N.: Kernel Methods for Pattern Analysis. Cambridge University Press, Cambridge (2004)
8. Lodhi, H., Saunders, C., Shawe-Taylor, J., Cristianini, N., Watkins, C.: Text classification using string kernels. Journal of Machine Learning Research 2 (2002)
9. Quinlan, R.: C4.5: Programs for Machine Learning. Morgan Kaufmann, San Francisco (1993)
10. Moschitti, A., Pighin, D., Basili, R.: Tree Kernels for Semantic Role Labeling. Computational Linguistics Special Issue on Semantic Role Labeling (3) (2008)
11. Moschitti, A., Cosmin, A.B.: A semantic kernel for predicate argument classification. In: CoNLL 2004, Boston, MA, USA (2004)
12. Giuglea, A.M., Moschitti, A.: Semantic Role Labeling via Framenet, Verbnet and Propbank. In: Proceedings of ACL 2006, Sydney, Australia (2006)
13. Bloehdorn, S., Moschitti, A.: Structure and semantics for expressive text kernels. In: Proc. of CIKM 2007 (2007)

Automatic Cluster Selection Using Index Driven Search Strategy

Denis Ferraretti[1,2], Giacomo Gamberoni[1,2], and Evelina Lamma[1]

[1] ENDIF-Dipartimento di Ingegneria, Università di Ferrara, Ferrara, Italy
{denis.ferraretti,giacomo.gamberoni,evelina.lamma}@unife.it
[2] intelliWARE snc, Ferrara, Italy
{denis,giacomo}@i-ware.it

Abstract. Clustering is the task of categorizing objects into different classes in an unsupervised way. Hierarchical clustering algorithms are usually very effective in detecting the dataset underlying structure. However, they do not create clusters, but compute only a hierarchical representation of the dataset. It is then desirable to define an automatic technique for cluster creation in hiearchical clustering algorithms.

To this purpose, in this paper we present an algorithm that finds the best clustering partition according to clustering validity indexes. In particular, our automatic approach performs a validity index-driven search through a clustering tree. The best partition is then selected cutting the tree in a non-horizontal way. The algorithm was implemented in a software tool and then tested on different datasets. The overall system makes then hierarchical clustering an automatic step, where no user interaction is needed in order to select clusters from a hierarchical cluster representation.

1 Introduction

Clustering is one of the best known problems in data mining. It is defined as the task of categorizing objects having several attributes into different classes such that the objects belonging to the same class are similar, and those that are broken down into different classes are not. Clustering is the subject of active research in several fields such as statistics, pattern recognition, machine learning and data mining. A wide variety of clustering algorithms have been proposed for different applications: the most common distinction is between partitioning and hierarchical clustering [1].

Partitioning algorithms create a "flat" decomposition of a dataset. Hierarchical clustering algorithms do not actually partition a data set into clusters, but compute a hierarchical model, which reflects its possibly clustering structure. The result of these algorithms is often represented by a dendrogram, i.e., a tree that iteratively splits a dataset into smaller subsets until each subset consists of only one object [2].

The first problem with these algorithms is that clusters are not explicit and have to be determined somehow from the representation.

R. Serra and R. Cucchiara (Eds.): AI*IA 2009, LNAI 5883, pp. 172–181, 2009.
© Springer-Verlag Berlin Heidelberg 2009

Moreover, clustering is mostly unsupervised process, thus the evaluation of the clustering algorithms is very important. Several clustering validity approaches have been developed [1].

In this paper we present an algorithm for the automatic extraction of clusters from a dataset. Starting from a dendrogram (produced by a standard hierarchical clustering algorithm), the technique uses clusters validity indexes to drive a search to the best cluster structure "cutting" the dendrogram with a *non-horizontal* border. In this way, it adds to hierarchical clustering a further step that can provide an effective cluster partition, optimizing a chosen evaluation index. We implement our algorithm in a software tool written in JAVA that computes hierarchical clustering and represents results in a dendrogram. It provides some cluster analysis tools like validity index computation, Color Mosaic and also gives the ability to automatically or manually cut the dendrogram in order to obtain a cluster partition.

The paper is organized as follows: general properties of clustering algorithms and related works are outlined in Section 2. Cluster validity indexes are introduced in Section 3. A detailed explanation of our algorithms for automatic clusters extraction is given in Section 4, while in Section 5 we present the experimental results over two different dataset. Finally, section 6 concludes the paper.

2 Clustering and Related Works

Hierarchical agglomerative clustering builds the hierarchy starting from the individual elements considered as single clusters, and progressively merges clusters according to a chosen similarity measure defined in features space [3]. The output of hierarchical clustering is a tree represented by a dendrogram, as represented in the upper part of the main window in Figure 3. The Color Mosaic (lower part of the main window in Figure 3) provides to the human expert an aid to represent of all the features of the whole dataset "at a glance".

The most standard way to define a partition from the tree built by a hierarchical clustering algorithm is to make an *horizontal cut* of the tree at a specified level. This is usually done by defining a parameter: either the number of desired classes, or the height of the cutting line. A more flexible approach is to allow the user to perform a *non-horizontal cut*. This approach can provide more opportunist cuttings: the user may want to have more details in some classes than in some others, or may want to group into the same class objects which appear to be unsimilar according to the clustering criterion [4]. In Figure 3 we can see an example in which a non-horizontal cut provides a partition that can not be obtained by an horizontal cut.

Starting from the root node, the user can divide the clusters going down through the tree structure, by selecting a node to "open", i.e. he can split that cluster in two sub-clusters. In this way, it is possible to choose the number of classes by "cutting" the tree at desired level. In the example (Figure 3), the dataset is split into five clusters (labeled from AAA to EEE). As a result, each identified cluster (black nodes in Figure 3) represents a set of examples with similar distribution of the features.

In literature, some methods for automatic clusters extraction from a hierarchical representation can be found on [5], [2] and [6].

In [5] the authors propose a method for reachability plots that is based on the steepness of the "dents" in a reachability plot. Unfortunately, this method requires an input parameter, which is difficult to understand and hard to determine.

In [2], the authors analyze the relation between hierarchical clustering algorithms that have different outputs, i.e. between the Single-Link method, which produces a dendrogram, and OPTICS, which produces a reachability plot. They develop methods to convert dendrograms and reachability plots into each other. Then they introduce a new technique to create a tree that contains only the significant clusters from a hierarchical representation as nodes.

In a third work, [6], several cluster evaluation techniques for gene expression data analysis are described. Normalisation and validity aggregation strategies are proposed to improve the prediction of the number of relevant clusters. The authors use K-means clustering algorithm and the work is tested only over a 2-classes datasets.

Another interesting and pioneering work is [4] where a *non-horizontal* dendrogram cut is proposed for the first time. Even if this last paper introduces the idea of a *non-horizontal* cut of the dendrogram, it does not provide any automatic procedure for this task. We then decided to extend and apply the concepts of automatic cluster extraction, presented in the former papers, in this particular tree cutting process.

3   Clusters Validation

One of the most important issues in cluster analysis is the evaluation of clustering results in order to find the partition (cluster configuration) that best fits the underlying data: this is the main goal of cluster validation.

Our system implements three evaluation indexes (Dunn, Davies-Bouldin and C-index), which assess cluster compactness and isolation. For each index we define a *global value*, referred to a chosen partition structure. These values are the standard evaluation measures from literature, but they provide only a rating for the whole clustering structure. In this paper we will consider only Dunn's Index, since it is one of the simplest to compute, and it did provide the best results in in our experiments.

The *Dunn's Index* [7] is based on the idea of identifying the cluster sets, that are compact and well separated. For any partition of clusters, where c_i represents the $i - th$ cluster of such partition, the Dunn's validation index, D, can be computed with the following formula:

$$D = \min_{1 \le i \le n} \left\{ \min_{1 \le j \le n} \left\{ \frac{\delta\left(C_i, C_j\right)}{\max_{1 \le k \le n}\left\{\Delta\left(C_k\right)\right\}} \right\} \right\}$$

where $\delta\left(C_i, C_j\right)$ is the distance between clusters C_i and C_j (inter-cluster distance[1]); $\max_{1 \le k \le n}\left\{\Delta\left(C_k\right)\right\}$ is the intra-cluster distance of cluster C_k, and n is the number of clusters.

---

[1] Inter-cluster distance is referred to two objects from different clusters, intra-cluster distance is referred to two objects from the same cluster.

difference with the old value. If this value is smaller than a fixed threshold, we stop the tree exploration.

By choosing a negative threshold the algorithm continue to search until the new clustering partition is significantly worse than the previous one (lower than ϵ), considering Dunn's Index validity measure. This threshold provides a simple method to avoid local maxima or flat zones.

5   Experimental Results

We test our technique over two different datasets from *UCI Machine Learning Repository*[10]: the Iris and the Synthetic Control Chart Time Series dataset. As a first step, we normalize all of the attributes with a linear adjustment in order to bring them in $[0.0, 1.0]$ range. Then, we use a hierarchical agglomerative clustering algorithm with Euclidean distance and complete linkage strategy. The choice of these distance and linkage strategy is driven by two simple considerations: first they are the most known and used techniques and second our tests gives best results only with these ones.

Every result is finally evaluated through the true instance-class assignment given by the dataset. Using *Expand-worst search* driven by Dunn's Index with $\epsilon = -0.005$ we obtained interesting results. Further tests with different distances and linkage strategies for clustering algorithms or the use of other validity indexes to drive the search of clustering configuration, do not have yielded the expected results for these dataset, also using *Go-to-best search* we do not obtain significant improvements.

In another work [11], we use Dunn, Davies-Bouldin and C-index in a combined solution to perform the driven search of cluster configuration. In that case we developed a prototype system, called I^2AM, that analyzes borehole log images using several image processing algorithms in order to extract numerical values for each characteristic and performs an automatic combined index-driven search over a hierarchical clustering representation. In [11] we also compare different search strategies and tree cutting mode.

To evaluate the improvement of our technique we also compare our hierarchical clustering results with clustering partitions given by K-means algorithm [12]. K-means is one of the simplest and best known unsupervised clustering learning algorithm. For each clustering solution we use the number of clusters as a parameter to run K-means and then we compute the information entropy.

5.1   Iris Dataset

This is perhaps the best known database in pattern recognition and clustering literature. It consists of 150 instances with four numerical attributes (sepal length in cm, sepal width in cm, petal length in cm and petal width in cm) and 3 classes (Iris Setosa, Iris Versicolour, Iris Virginica) with a distribution of 33.3% for each class.

Figure 1 shows the output of our system: the dendrogram with a Color Mosaic (on the left) and tables with global and specific validity index (top right) and

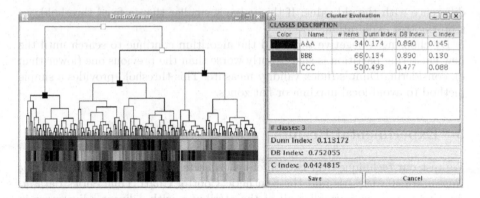

Fig. 1. The main window shows the dendrogram with the clustering configuration found by automatic exploration of Iris dataset. Other windows to the right show global and specific validity index measure for the selected cluster configuration and some clustering parameters.

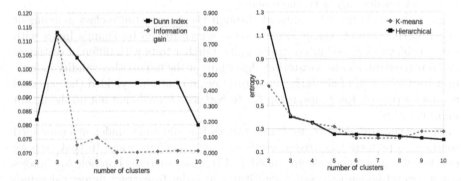

Fig. 2. Results for IRIS dataset. On the left: Plot of Dunn's Index and Information gain. On the right: Entropy comparison between K-means and hierarchical clustering.

clustering parameters (bottom right). The found solution counts three clusters outlined by black nodes in the dendrogram, classes are clearly visible in the Color Mosaic. It is important to note that in this case, the clusters selected from the *non-horizontal* cut are the same if we horizontally cut above the top black node.

In Figure 2 we report the values of the Dunn's Index and information gain versus the number of clusters given from different clustering configurations. For three clusters we have a local maximum, the δ value is less than ϵ and hence the algorithm stops at this partition. Also the information gain indicates that three is the right number of clusters: beyond this value the information gain is always near 0 and this means that entropy has no interesting improvement.

These results confirm that in this simple case our technique can discover the actual dataset partitioning. Looking to the Color Mosaic (see Figure 1) it is possible to note that found clusters are close to real classes. Information gain computation (Figure 2 on the left) shows that there are no better solutions near to the chosen one. Moreover, for three clusters, information entropy (Figure 2 on the right) of hierarchical clustering is a little bit smaller than the K-means information entropy and this reveals a slightly better homogeneity.

5.2 Synthetic Control Chart Time Series Dataset

This dataset contains 600 examples of control charts synthetically generated. There are six different classes of control charts: Normal, Cyclic, Increasing trend, Decreasing trend, Upward shift, Downward shift. Every chart has 60 time values, it means 60 numerical attributes. This is a good benchmark for time series clustering algorithms because Euclidean distance will not be able to achieve perfect accuracy. In particular the following pairs of classes will often be confused: Normal/Cyclic, Decreasing trend/Downward shift and Increasing trend/Upward shift. The right distance to use is the Derivative Dynamic Time Warping (DDTW, [13]). DDTW correctly found the six high-level groups, but Euclidean distance did not.

The resulting configuration given by our algorithm for this dataset (see Figure 3) shows five different clusters. This not exactly match the true dataset classification (six classes), indeed in the Color Mosaic it is possible to note that the first cluster includes instances from two different classes but other clusters are very homogeneous.

As we can see in Figure 4 our algorithm stops at five clusters although the right number of clusters should be six. This is because Dunn's Index for six clusters is lower than the index for five and this difference is lower than ϵ. The opening of other nodes using *Expand-worst search* leads to a growing number of clusters. We can observe that this will increase the Dunn's Index but information gain will remain always near 0. Also the information entropy for hierarchical clustering (see Figure 4 on the right) gives best results: near to the right number

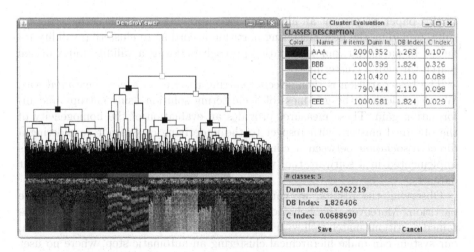

Fig. 3. The dendrogram with the clustering configuration found by automatic exploration of Synthetic Control Chart Time Series dataset

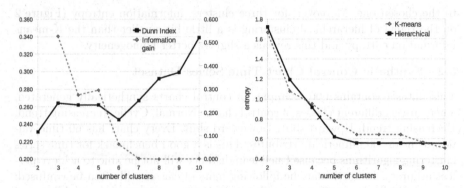

Fig. 4. Results for Synthetic Control Chart Time Series dataset. On the left: Plot of Dunn's Index and Information gain. On the right: Entropy comparison between K-means and hierarchical clustering.

of clusters (from 4 to 9) this measure is smaller than the K-means one. This means that the hierarchical techinque is able to find more homogeneous clustering solutions (according to the true instance-classes assignement). Observing these results we can make some considerations. First, the extracted clustering configuration is possible only if we allow a *non-horizontal* dendrogram cut. Indeed, in Figure 3, clusters CCC, DDD and EEE have a lower distance measure than the cluster BBB. This brings to the second important result: the Dunn-driven search improves the results even if the Euclidean distance is not suitable for this type of dataset. In this way we can achieve a good result, in terms of clusters homogeneity.

6   Conclusion

In this paper we present an algorithm for improving hierarchical clustering results, providing the best clustering partition according to clustering validity indexes. In particular, our automatic approach performs a validity index-driven search through a clustering tree.

The best partition is then selected cutting the tree in a *non-horizontal* way. We also calculate the goodness of a clustering solution using entropy and information gain. These measures provides an evaluation of the homogeneity of the obtained clusters with respect to the underlying classes not considering a direct association between a cluster and a specific class. The algorithm was implemented in a software tool written in JAVA and then tested on different datasets.

Observing the results, we can assess that our system provides a reliable partition. Moreover, the behavior of information gain confirms that the obtained partitions match well with the underlying structure of the datasets. Then, our system can make hierarchical clustering an automatic step, where no user interaction is needed in order to select clusters from a hierarchical cluster representation.

References

1. Kovacs, F., Legany, C., Babos, A.: Cluster validity measurement techniques. Technical report, Department of Automation and Applied Informatics, Budapest University of Technology and Economics, Budapest, Hungary (2002)
2. Sander, J., Qin, X., Lu, Z., Niu, N., Kovarsky, A.: Automatic extraction of clusters from hierarchical clustering representations. In: Whang, K.-Y., Jeon, J., Shim, K., Srivastava, J. (eds.) PAKDD 2003. LNCS (LNAI), vol. 2637, pp. 75–87. Springer, Heidelberg (2003)
3. Theodoridis, S., Koutroumbas, K.: Pattern Recognition, 3rd edn. Academic Press, London (2006)
4. Boudaillier, E., Hébrail, G.: Interactive interpretation of hierarchical clustering. In: Komorowski, J., Żytkow, J.M. (eds.) PKDD 1997. LNCS, vol. 1263, pp. 288–297. Springer, Heidelberg (1997)
5. Ankerst, M., Breunig, M., Kriegel, H.P., Sander, J.: OPTICS: Ordering points to identify the clustering structure. In: ACM SIGMOD International Conference on the Management of Data, pp. 49–60. ACM, New York (1999)
6. Bolshakova, N., Azuaje, F.: Cluster validation techniques for genome expression data. Signal Processing 83, 825–833 (2003)
7. Dunn, J.: Well separated clusters and optimal fuzzy partitions. Journal of Cybernetics, 95–104 (1974)
8. Quinlan, J.R.: Induction of decision trees. Machine Learning 1(1), 81–106 (1986)
9. Shannon, C.E.: A Mathematical Theory of Communication. CSLI Publications (1948)
10. Asuncion, A., Newman, D.: UCI machine learning repository (2007)
11. Ferraretti, D., Gamberoni, G., Lamma, E., DiCuia, R., Turolla, C.: A new tool for the petroleum industry based on image analysis and hierarchical clustering. In: 10th International Conference on Intelligent Data Engineering and Automated Learning (IDEAL 2009). Springer, Heidelberg (2009)
12. Macqueen, J.B.: Some methods for classification and analysis of multivariate observations. In: Procedings of the Fifth Berkeley Symposium on Math, Statistics, and Probability, vol. 1, pp. 281–297. University of California Press (1967)
13. Keogh, E.J., Pazzani, M.J.: Derivative dynamic time warping. In: First SIAM International Conference on Data Mining (SDM 2001) (2001)

The Necessity of Machine Learning and Epistemology in the Development of Categorization Theories: A Case Study in Prototype-Exemplar Debate

Francesco Gagliardi

Department of Philosophical and Epistemological Studies
University of Rome *La Sapienza*, Via Carlo Fea, 00161 Rome, Italy
francesco.gagliardi@libero.it

Abstract. In the present paper we discuss some aspects of the development of categorization theories concerning cognitive psychology and machine learning. We consider the thirty-year debate between *prototype-theory* and *exemplar-theory* in the studies of cognitive psychology regarding the categorization processes. We propose this debate is ill-posed, because it neglects some theoretical and empirical results of machine learning about the bias-variance theorem and the existence of some *instance-based* classifiers which can embed models subsuming both prototype and exemplar theories. Moreover this debate lies on a epistemological error of pursuing a, so called, *experimentum crucis*. Then we present how an interdisciplinary approach, based on synthetic method for cognitive modelling, can be useful to progress both the fields of cognitive psychology and machine learning.

Keywords: Machine Learning; Epistemology; Cognitive Psychology; Categorization Theories; Bias-Variance Dilemma; Instance-Based Learning.

1 Introduction

Categorization is the adaptive fundamental process by which we "cut" the physical and social reality. It permits us to understand and make predictions about objects and events in our world [14,16,19]. Therefore, categorization is a pervasive cognitive activity by which human mind divides the world in categories by building concepts that are mental representations of these categories.

The understanding of categorization processes of human mind is one of the most important and controversial intellectual challenges of cognitive science and artificial intelligence. In fact, categorization is a fundamental process for both human and artificial intelligence.

In the following section we recall briefly the main theories of categorization developed in cognitive psychology, introduce the prototypes-exemplars debate, and explain some methodological drawbacks of this research line. Then we relate machine-learning with categorization theory to overcome the aforementioned diatribe and illustrate how some classifier systems can be proposed as a categorization models which subsume both prototype-theory and exemplar-theory. Then we introduce some

R. Serra and R. Cucchiara (Eds.): AI*IA 2009, LNAI 5883, pp. 182–191, 2009.

observations on the necessity of a more interdisciplinary approach to the study of categorization.

2 Categorization in Cognitive Psychology

The main theories concerning the concepts [18,24,30], illustrated in the following are: the classical theory also known as Aristotelian, the prototypes theory and the exemplars theory.

Classical Theory. According to the classical theory a concept is defined by a set of characteristics, called *defining features*, which are necessary and sufficient conditions for its application: concepts are mentally represented as definition, or better *are* logical predicates.

Prototypes Theory. The first theory [27,28], which overcomes many of the problems related to categorization encountered by the classical theory, affirms that concepts are prototypes representing the typical characteristics of objects of a category rather than necessary and sufficient conditions.

According to the *prototypes theory* people tend to identify a category of objects and to reason about its members, by referring to a precise *typical* object of the category.

Exemplars Theory. A different point of view on concepts consists of considering them as a collection of stored exemplars in memory. This theory, know as *exemplar theory*, was proposed for the first time by Medin and Schaffer in 1978 [20].

It is totally different from the other theories, because it rejects the idea, common to the classical and prototypes theory, that humans have some kind of representation able to describe the whole category.

The Typicality View. The theories of prototypes and exemplars, jointly taken, constitute the so called *typicality view* on concepts. In fact, both theories, even if in contrast, are based on experimental evidences that as a whole they show the existence of a phenomenon of typicality in categorization processes (e.g. see *"Typicality as phenomenon"* in [24, pg.28]). This "phenomenon" cannot be fully explained by any of the theories developed till now: *"no theory has a ready explanation for all of the findings even within each specific topic"* [24, pg.5].

2.1 The Prototypes-Exemplars Debate and the *Naïve Epistemology* of an *Experimentum Crucis*

Prototypes and exemplars theories supersede the limitations and the experimental inadequacy of the classical theory, but when considered separately turn out to be incomplete and unsatisfactory [24, pg.4].

As matter of fact, to support one theory as absolutely correct against the other, leads to substantial difficulties: when considering categorization performed by human mind based only on prototypes, it should be claimed that the ability of remembering single observed objects does not influence our concepts construction, even in case of observation of atypical instances; while, for the exemplars based categorization, it should be affirmed a sort of inability of abstraction from observations and that, in a way, concepts do not exist because they are only a secondary effect of our mnemonic abilities. Nevertheless, in the past thirty years a lot of literature concerning experimental psychology focused on the comparison between prototypes theory and exemplar theory and carrying out experiments in order to demonstrate the correctness of one theory or the other one [24, pg.4].

For example we can consider the following two papers in conflict [23] and [35]. In the former prototypes theory is supported, while in the latter exemplars theory is supported, even if they make use of the same data set, the so called *"Medin and Schaffer 5-4 stimulus set"* [20]. This data set has being used extensively since thirty years in the cognitive psychology, even though it is composed of only nine instances belonging to two categories of five and four observations.

Many of these works are attempts to realize a sort of *experimentum crucis* which would definitively distinguish between the two competing theories. This research line is exposed to deep epistemological criticism for two distinct reasons: in general, *experimenta crucis* are impossible, or nearly, to be realized even in the exact sciences as physics[1], and in particular, as pointed out by Robert Merton [21,22] in the field of social sciences, many theories are *middle range theories,* that is, theories with a limited explicative range. In fact, as we will see below, the framing of categorization theories in the field of machine learning, shows that prototypes and exemplars theories are "by construction" incomplete and partial, thus it has no meaning to try to realize an *experimentum crucis* for them.

As matter of fact, the research line related to the diatribe of prototypes vs. exemplars appears to be a dead end, because it is fruitless and not decisive and perhaps it should be considered as an attempt to use some methodological practices, which are believed to belong to exact sciences. This kind of approach could be understood as *"physics envy"*[2], instead we prefer to consider it as a form of *"naïve epistemology"* because it is based more on a simple view on how sciences evolve than on an sort of envy.

3 Theories of Categorization and Machine Learning

The ultimate aim of the researches about categorization [24, pg.3] is the understanding of representations of categories that we build, the concepts, and by which we perform different cognitive tasks, such as recognition of new objects, inferences, communication and more.

The problem of *knowledge representation* is a historical problem of artificial intelligence and is still being widely discussed [6,12], especially in connection to the problem of cognitive plausibility of artificial systems, in other words, it is discussed about when these systems can be considered realistic models of human thinking and not merely mimetic.

In artificial intelligence one can define different areas and research lines, just in relation to the kind of knowledge representation used by artificial systems; and in particular, in the filed of machine learning, one can define different families of automatic classification systems on the basis of the kind of the used classes representations (e.g. rules, instances, decision trees, support vectors or other).

---

[1] Consider for example the *Duhem-Quine thesis.*

[2] The expression *"physics envy"* is attributed to the psychologist Sheldon H. 'Shep' White [26].

A common aspect of prototypes theory and exemplar theory is the idea that each category is represented by instances belonging to the class: in one case the instances are the prototypes abstracted from observations, and in the other case are the same previously observed instances.

In the area of *machine learning*, one of the learning methodologies for automatic classification known in literature is the so called *instance based learning*, [8,1,33] for which the classes, learnt by the classifier system, are represented by some instances saved in memory, which can be prototypes abstracted from the training set or can be exemplars directly observed. Classification is performed by relating a new instance, which class is unknown, to the instances saved in memory which classes are known.

Therefore, the field of machine learning, and in particular of instance-based learning, is the natural context where to study the theories of human categorization based on prototypes or exemplars, and to embrace the synthetic method [5,7] consisting in the realization of classifier systems which embody models of categorization.

Within *instance-based* learning framework it is possible to connect the characteristics of robustness and sensibility of a classifier system with categories representation based, respectively, on prototypes or exemplars. In fact, prototypes based classifiers, such as the *Nearest Prototype Classifier* (*NPC*) and the *Nearest Multiple-Prototype Classifier* (*NMPC*) [3], construct the representative instances of the class, called *prototypes*, as the barycentres of an observations subset. These systems obtain robust classifications, that is, not sensitive to noisy and atypical observations[3]. On the other way, classifiers based on exemplars, such as the *Nearest Neighbour Classifier* (*NNC*) and its well known generalization *k-NNC*, use as the set of representative instances the whole set of observations of classes, without any elaboration or abstraction. These systems, which are entirely based on the ability to save all observations in memory and obtain, especially the *NNC*, classifications extremely sensible and little robust[4].

In the family of *instance-based* systems the classifiers *NPC* and *NNC* represent the limit cases of maximum robustness and maximum sensibility respectively; they use types of classes representations that can be related to (see table 1) the theories of prototypes and of exemplars, respectively.

Table 1. Schematic synthesis of relations between categorization theories and instance-based classifier systems

		based on:	use:	maximize:		
Cognitive Psychology	Prototypes Theory	Abstraction	Prototypes	Robustness	Nearest Prototype Classifier	Machine Learning
	"Typicality" Theory				Hybrid Classifier	
	Exemplars Theory	Memorization	Exemplars	Sensibility	Nearest Neighbour Classifier	

---

[3] Intuitively, the robustness of these classifier systems can be understood by observing that the presence of outliers (due to noise or atypical instances of the class) has little or not influence on barycentre calculation.

[4] Intuitively the sensibility of these classifier systems is due to a sort of "data fidelity" or the classification of new instances is merely based on the comparison between new instances and the previous observed ones, without distinction between noise and correct observations (both typical and atypical).

As it is well known in machine learning a classifier system, whether natural or artificial, is the result of the trade-off between the two contrasting requisites of robustness and sensibility [9, chap.9].

This trade-off can be pursued by *regularization techniques* in the classification scheme, for example, by incorporating a score function to prevent overfitting, i.e., poor generalization performance. Regularization is a way of trading off bias and variance in the classifier model (see, for instance, [13]). The problem of overfitting and the use of regularization techniques is formally grounded on the *Bias-Variance Theorem* and on the *Bias-Variance dilemma* (e.g. [9, chap.9]).

The purpose is to restrain the complexity of the classifier, so that the decision boundary becomes smoother or fewer features are effectively utilized. The classifier can be regularized by tuning an additional parameter that deals with a complexity measure (e.g. in decision tree induction one can stop the growing of the tree early, according to a given criterium).

In instance-based classifiers a regularization (e.g. [32]) can be obtained with a compromise between the nearest prototype and nearest neighbour classifier, by tuning the learning algorithm which infers the representative instances, as, for example in the *Prototype-Exemplar Learning Classifier (PEL-C)* [11]. In this classifier, the number and the type of learnt representative instances is obtained with a regularization technique based on a *cross-validation* procedure [15, pg.108] [34, pg.149] to optimize the predictive accuracy.

The representations of the learnt classes have the lowest cross-validation error and they can be said to optimally regularize this classifier. This classifier obtain experimentally classes representations composed just of a mixture of prototypes and atypical exemplars.

Summing up, the introduced conceptual link between machine learning and the study of human categorization shows that a categorization entirely based on prototypes or on exemplars, cannot be absolutely proposed for both theoretical reasons and empirical evidences due to bias-variance theorem and to the existence of instance-based classifiers regularized by hybrid representations of classes.

Thus the thirty-year prototypes-exemplars diatribe is ill-posed from the beginning, and if we consider also that categorizations theories are *middle range* theories for which is unfeasible to realize an *experimentum crucis*, it is clear that it is absolutely groundless to assert the correctness of one of the two categorization theories against the other; a theory which subsumes both of them should be sought by the regularization techniques between prototypes and exemplars representations.

3.1 Some Computational Cognitive Models of Categorization

Other algorithms that make use, more or less deliberately, of regularization techniques are briefly illustrated.

The VAM (*Varying Abstraction Model*) [31] is a model belonging to instances-based classifiers which uses the *k-means* algorithm; it has been introduced to overcome the contraposition between prototypes and exemplars. This classifier can vary the abstraction of prototypes from one prototype for class, until to the "degenerate" prototypes which represent a subclass composed of one single instance; this type of prototype is called by the authors *pseudo-exemplar*.

Authors report that this model has better performances when instances abstraction is at an intermediate value. Although it seems authors are unaware of the fact this result is linked to bias-variance dilemma, it can be fully explained from our previous argumentations about trade-off of robustness and sensibility.

The rational model [2] and *the mixture model* [29] share with the VAM the idea of varying the level of abstraction of prototypes. All these are a sort of *NMPCs, Nearest Multiple-Prototype Classifier* (see above), and thus they have the interesting feature of being a generalization of the *NPC*. In machine learning there are also some hybrid classifiers which, in a way, further generalize these ones, such as the *T.R.A.C.E. (Total Recognition by Adaptive Classification Experiments)* [25] and the *PEL-C (Prototype-Exemplar Learning Classifier)*.

We think that the *PEL-C* is a very promising classifier for catching the "phenomenon" of typicality and to develop a "typicality theory" of categorization because it is able to describe the learnt categories by a mixture of prototypical instances with variable abstraction and atypical exemplars.

Therefore within the family of instance-based classifiers, we can develop systems that are computational models of categorization, such as *VAM*, the *mixture model* or the *PEL-C*, and which subsume both the prototypes and exemplars theories and, hence, they can help to realize a *theory of typicality* which would explain the phenomenon of typicality.

We do not deepen the comparison among these different models because our purpose is to show the feasibility of developing classifiers with instance-based representations, which can be proposed to overcome the contraposition between prototypes and exemplars, and not to state the best one nor to definitely propose a "typicality" theory.

In fact we suggest these classifiers can be the starting point of a new research line to develop theories of categorization soundly grounded on machine learning and synthetic method.

4 On Interdisciplinary Approach between Machine Learning and Cognitive Psychology

As we showed above, some drawbacks in the categorization research could be superseded by using machine learning knowledge and by embracing the synthetic method [5,7], as it would be required by the interdisciplinary nature of the categorization problem.

Machine learning can be a powerful instrument to formalize and to develop scientific theories in cognitive sciences, in fact cognitive psychology researches could have avoided thirty years of wasting time in an unfruitful diatribe, if they had not limited to a superficial use of mathematics for the development of cognitive theories, instead of a more foundational use of computational modelling of cognitive functions.

Unfortunately, in the other field we should considers that AI researchers mainly left *"the original vision of AI that was to understand the principle that support cognitive processing and to use them to construct computational systems with the same breath of abilities as humans"* [17, pg.4] and in particular the modern researches of machine learning and data mining *"have been concerned primarily with improving predictive accuracy rather than with acquiring cognitive structures that support intelligent behaviour, which was the original motivation for launching machine learning as a subfield of artificial intelligence"* [17, pg.2]

This is in accordance with the technological turn of AI, the *product-directed* one, which *"always aimed at constructing machines working perfectly in single specific sectors, without being interested in human general cognitive processes"*[5] [4, pg.49]; approach to be opposed to the *theory-direct* one, aimed at the comprehension of human cognitive processes. In this way, machine learning researches neglect possible theoretical involvements with cognitive sciences.

This lack of cooperation and integration between cognitive psychology and machine learning implies a waste of resource and energy for both.

In fact, besides the diatribe previously illustrated, we can observe that some psychological researchers reinvent models basically yet-known in machine learning community, (e.g. the *VAM* is a sort of *NMPC* and the obtained results are partially expected), and the few psychologists, who develop formal models of categorizations in a very sound way, do not obtain the diffusion that they should deserve (e.g. *the mixture model* [29] has citation count equal to zero in the *ACM portal*[6]).

Moreover, machine-learning researchers usually neglect results of cognitive psychology while the development of classifiers, according to the evidences of cognitive psychology, can be useful also in some applicative tasks of data mining and machine learning, as for example the knowledge discovery process.

This task can be defined as *"a non-trivial process of identifying valid, novel, potentially useful and ultimately understandable patterns from collections of data"* [10] and we think it can obtain much benefits from the development of classifiers with a cognitive plausible internal representation, because a psychologically realistic classes representation conducts more easily to achieve understandable patterns from data, which are meaningful for external and non-technical users.

We think a deeper cooperation between these fields can be a progress key for both because many of the problems dealt in cognitive psychology and artificial intelligence, as for example the problem of categorization, are exactly the same whether one considers natural or artificial systems.

5   Concluding Remarks

"Ideally in cognitive science, computational models and psychological experimentation go hand in hand" [30, pg.8], conversely, above we put in evidence how the "monodisciplinary" use of cognitive psychology and machine learning produces disappointing results, not only from the interdisciplinary viewpoint of cognitive science, as it is obvious, but also from the viewpoints of the aims of the single sectorial disciplines.

In fact, on one hand, cognitive psychology produced thirty years of an unfruitful prototypes-exemplars diatribe, which could be avoided if one had not limited oneself to a superficial use of mathematics for the development of cognitive theories, instead of a more foundational use of machine-learning.

---

[5] The original text is in Italian.
[6] See http://portal.acm.org/citation.cfm?id=605071 Accessed 21/may/2009.

On the other hand, in the field of machine learning and data mining, research mainly ignores the experimental evidences produced by the psychological research, neglecting possible theoretical involvements with cognitive sciences and hampering technological results themselves.

The advantages of a more interdisciplinary research are obvious if we consider that many of the problems dealt in cognitive psychology and artificial intelligence are the same whether one considers natural systems, as human minds, or artificial systems. Any distinction or disciplinary fragmentation cannot be other than the result of sociological dynamics which are harmful for the progress of scientific knowledge.

Hence, we should embrace the synthetic method, i.e. the incorporation of theories in computational systems, as a foundational tool to progress cognitive psychology research as well as artificial intelligence one.

The interdisciplinary nature of cognitive modelling is essential for cognitive science and artificial intelligence: because synthetic method is not a mathematical make-up for non-formal cognitive theories already developed, but it is a tool for developing them, as well as, it is not a waste-time activity in developing artificial intelligence systems, because cognitive theories can empower them.

Acknowledgments. I wish to thank Prof. R. Cordeschi for the valuable suggestions on a previous version of this work. Heartedly thanks are due to Dr. A. Brindisi for her support.

References

1. Aha, D.W., Kibler, D., Albert, M.K.: Instance-based learning algorithms. Machine Learning 6, 37–66 (1991)
2. Anderson, J.: The adaptive nature of human categorization. Psychological Review 98, 409–429 (1991)
3. Bezdek, J.C., Reichherzer, T.R., Lim, G.S., Attikiouzel, Y.: Multiple-prototype classifier design. IEEE Transactions on Systems, Man and Cybernetics, Part C 28(1), 67–79 (1998)
4. Cordeschi, R.: Indagini meccanicistiche sulla mente: la cibernetica e l'intelligenza artificiale. In: Somenzi, V., Cordeschi, R. (eds.) La filosofia degli automi. Origini dell'intelligenza artificiale (new edition). Bollati Boringhieri, Torino, pp. 19–61 (1994)
5. Cordeschi, R.: The Discovery of the Artificial. Behavior, Mind and Machines Before and Beyond Cybernetics. Kluwer, Dordrecht (2001)
6. Cordeschi, R.: Vecchi problemi filosofici per la nuova Intelligenza Artificiale. Networks. Rivista di Filosofia dell'Intelligenza Artificiale e Scienze Cognitive 1, 1–23 (2003), http://www.swif.uniba.it/lei/ai/networks/
7. Cordeschi, R.: Steps toward the synthetic method: symbolic information processing and self-organizing systems in early Artificial Intelligence. In: Husbands, P., Holland, O., Wheeler, M. (eds.) The Mechanical Mind in History, pp. 219–258. The MIT Press, Cambridge (2008)
8. Cover, T.M., Hart, P.E.: Nearest Pattern Classification. IEEE Trans. on Information Theory. IT 13(1), 21–27 (1967)
9. Duda, R., Hart, P., Stork, D.: Pattern Classification, 2nd edn. John Wiley & Sons, New York (2000)

10. Fayyad, U., Piatetsky-Shapiro, G., Smyth, P., Uturusamy, R.: Advances in Knowledge Discovery and Data Mining. The MIT Press, Cambridge (1996)
11. Gagliardi, F.: A Prototype-Exemplars Hybrid Cognitive Model of Phenomenon of Typicality in Categorization: A Case Study in Biological Classification. In: Love, B.C., McRae, K., Sloutsky, V.M. (eds.) Proceedings of the 30th Annual Conference of the Cognitive Science Society, pp. 1176–1181. Cognitive Science Society, Austin (2008), http://csjarchive.cogsci.rpi.edu/Proceedings/2008/pdfs/p1176.pdf
12. Gärdenfors, P.: Conceptual Spaces: the Geometry of Thought. The MIT Press, Cambridge (2000)
13. Geman, S., Bienenstock, E., Doursat, R.: Neural Networks and the Bias/Variance Dilemma. Neural Computation 4, 1–58 (1992)
14. Houde, O.: Categorization. In: Houde, O., Kayser, D., Koenig, O., Proust, J., Rastier, F. (eds.) Vocabulaire de sciences cognitives. Presses Universitaires de France (1998)
15. Henery, R.J.: Methods for Comparison. In: Michie, D., Spiegelhalter, D.J., Taylor, C.C. (eds.) Machine Learning, Neural and Statistical Classification, pp. 107–124. Prentice Hall, Englewood Cliffs (1994)
16. Kruschkea, J.K.: Categorization and Similarity Models. In: Smelser, N.J., Baltes, P.B. (eds.) International Encyclopedia Of The Social & Behavioral Sciences, pp. 1532–1535. Pergamon, Oxford (2001)
17. Langley, P., Laird, J.E.: Artificial intelligence and intelligent systems. Technical Report. Computational Learning Laboratory, CSLI, Stanford University, CA (2006), http://cll.stanford.edu/~langley/papers/aics.100.pdf
18. Medin, D.L.: Concepts and conceptual structure. American Psychologist 44, 1469–1481 (1989)
19. Medin, D.L., Aguilar, C.: Categorization. In: Wilson, R.A., Keil, F. (eds.) The MIT Encyclopedia of the Cognitive Sciences (MITECS), pp. 104–106. The MIT Press, Cambridge (1999)
20. Medin, D.L., Schaffer, M.M.: Context theory of classification learning. Psychological Review 85, 207–238 (1978)
21. Merton, R.K.: Social Theory and Social Structure. The Free Press, New York (1949,1968)
22. Merton, R.K.: Sociology of Science: Theoretical and Empirical Investigations. University of Chicago Press, Chicago (1973)
23. Minda, J.P., Smith, J.D.: Comparing prototype-based and exemplar-based accounts of category learning and attentional allocation. Journal of Experimental Psychology: Learning, Memory, Cognition 28, 275–292 (2002)
24. Murphy, G.L.: The big book of concepts. The MIT Press, Cambridge (2002)
25. Nieddu, L., Patrizi, G.: Formal methods in pattern recognition: A review. European Journal of Operational Research 120, 459–495 (2000)
26. Rogoff, B.: Apprenticeship in Thinking. Cognitive Development in Social Context. Oxford University Press, Oxford (1991)
27. Rosch, E.: Cognitive Representations of Semantic Categories. Journal of Experimental Psychology 104(3), 192–233 (1975)
28. Rosch, E., Mervis, C.B.: Family resemblance: Studies in the internal structure of categories. Cognitive Psychology 7, 573–605 (1975)
29. Rosseel, Y.: Mixture models of categorization. Journal of Mathematical Psychology 46, 178–210 (2002)
30. Thagard, P.: Mind: Introduction to cognitive science, 2nd edn. The MIT Press, Cambridge (2005)

31. Vanpaemel, W., Storms, G., Ons, B.: A Varying Abstraction Model for Categorization. In: Bara, B.G., Barsalou, L., Bucciarelli, M. (eds.) Proceeding of the XXVII Annual Conference of the Cognitive Science Society, pp. 2277–2282 (2005)

32. Veenman, C.J., Reinders, M.J.T.: The nearest subclass classifier: a compromise between the nearest mean and nearest neighbor classifier. IEEE Transactions on Pattern Analysis and Machine Intelligence 27(9), 1417–1429 (2005)

33. Wilson, D.R., Martinez, T.R.: Reduction techniques for exemplar-based learning algorithms. Machine Lerning 38(3), 257–268 (2000)

34. Witten, I.H., Frank, E.: Data Mining: Practical Machine Learning Tools and Techniques with Java Implementations, 2nd edn. Morgan Kaufmann, San Francisco (2005)

35. Zaki, S.R., Nosofsky, R.M., Stanton, R.D., Cohen, A.L.: Prototype and Exemplar Accounts of Category Learning and Attentional Allocation: A Reassessment. J. Exp. Psychol. Learn. Mem. Cognit. 29(6), 1160–1173 (2003)

A Lexicographic Encoding for Word Sense Disambiguation with Evolutionary Neural Networks

A. Azzini, C. da Costa Pereira, M. Dragoni, and A.G.B. Tettamanzi

Università degli Studi di Milano, Dipartimento di Tecnologie dell'Informazione
Via Bramante 65, 26013 Crema (CR), Italy
{antonia.azzini,mauro.dragoni,celia.pereira,andrea.tettamanzi}@unimi.it

Abstract. We propose a supervised approach to word sense disambiguation based on neural networks combined with evolutionary algorithms. Large tagged datasets for every sense of a polysemous word are considered, and used to evolve an optimized neural network that correctly disambiguates the sense of the given word considering the context in which it occurs.

A new distributed scheme based on a lexicographic encoding to represent the context in which a particular word occurs is proposed.

The viability of the approach has been demonstrated through experiments carried out on a representative set of polysemous words.

1 Introduction

Word Sense Disambiguation (WSD) (see a recent survey [8]) consists of assigning the most appropriate meaning to a polysemous word. The automatic word sense disambiguation process, in general, consists of two steps: (i) considering the possible senses of the given word; and (ii) assigning each occurrence of the word to its *appropriate* sense which depends on the context in which the word occurs. Representing the context of a word is then one of the most important steps in the automatic process of WSD. The more *effective* the representation of the context is, the more satisfactory the results of the WSD process are. Two kinds of context representations commonly used in connectionism are distributed [6] and localist schemes [3]. The major drawback of the latter is the high number of inputs needed to disambiguate a single sentence, because every input node has to be associated to every possible sense; in fact, this number increases staggeringly if one wants to disambiguate an entire text.

The original aspects of this work w.r.t. [1] are a novel representation used to encode the sentences, and new constraints on the network topologies. We propose (i) a new distributed encoding scheme based on a lexicographic encoding to represent the context of a particular word, and (ii) a supervised approach to word sense disambiguation based on neural networks (NNs) combined with evolutionary algorithms (EAs) [1]. We dispose of large tagged datasets describing the contexts in which every sense of a polysemous word occurs, and use them to

R. Serra and R. Cucchiara (Eds.): AI*IA 2009, LNAI 5883, pp. 192–201, 2009.

evolve an optimized NN that correctly disambiguates the sense of a word given its context.

We obtain a class of neural networks, each of them specialized in recognizing the correct sense of their corresponding word, one for each polysemous word in the dictionary.

To represent context, we took advantage of the lexicographic annotation that WordNet assigns to each word sense, which classifies it into one of forty-five categories based on syntactic category and logical groupings. The activations of the input neurons are obtained by summation of the activation patterns representing the words occurring in a given context, excluding the target word, after removing stop words and stemming the remaining words.

The viability of the approach has been demonstrated through experiments carried out on a representative set of polysemous words. Comparisons with the best entries of the Semeval-1/Senseval 4 competition have shown that the proposed approach is competitive with state-of-the-art WSD approaches.

The paper is organized as follows: Section 2 provides an overview of different kinds of supervised algorithms discussed in the literature; Section 3 presents a new distributed scheme based on a lexicographic encoding to represent the context in which a particular word occurs; Section 4 describes the evolutionary algorithm used to design neural network classifiers; Section 5 describes the experiments that have been carried out to test and validate the proposed approach and, finally, Section 6 concludes.

2   Supervised Approaches to WSD

One of the most successful research areas in WSD concerns the corpus-based approaches and, in particular, supervised approaches in which machine learning techniques are used to learn and classify senses from sense-tagged corpora [11]. Usually, in supervised approaches, the training dataset corresponds to a set of vectors whose features describe the relevant information about the examples they refer to [5]. The most popular classification approaches can be summarized in the following five groups, while a detailed description can be found in the literature [15].

- Methods based on probability which are generally used as classifiers by considering a set of probabilistic parameters that represent the distributions of different categories [4].
- Methods based on similarity, in which the sense disambiguation is obtained by considering the similarity between new cases and a set of vector examples, one for each sense, that serve as prototypes stored in a memory space [9].
- Methods based on rule selection are well described in the literature and known as Decision Trees and the Decision Lists [14].
- Methods based on Rule Combination which refer to algorithms that learn and combine several simple and imprecise rules into a single, highly accurate classifier, as the AdaBoost algorithm [10].

– Kernel-based Methods improve the classification process by allowing the learning of non-linear functions through a non-linear mapping of the input features, by using kernel functions [12].

Generally, when considering supervised learning methods, particular attention has to be given to a critical aspect, which regards the definition of the datasets involved in the training and testing processes. Indeed, the training set should be representative of the task that will be afterwards performed on other unseen data used to test and validate the implemented method. In this case, if only one corpus is considered to define the datasets, even if test and validation examples are different from those used for training, they are expected to be quite similar, influencing the accuracy performance of the WSD, that becomes a highly application-domain-dependent problem. This aspect has been considered in this work, and two different corpora have been used in the experiments, in order to avoid the problems generated by the use of a single corpus.

3   Lexicographic Encoding

A critical problem in supervised approaches to WSD is how to represent the context in which a word is used. In our case, such representation should be specifically targeted to its use with neural networks.

We consider a lexicographic encoding to represent the context in which a particular word occurs. To create a lexicographic representation of a word we use the lexicographic annotation that WordNet assigns to each word sense — each synset is classified into one of forty-five lexicographic categories based on syntactic category and logical groupings. An example of lexicographic categories is shown in Table 1.

Table 1. Example of Lexicographic Categories

Lexicographic Category	Category Description
noun.artifact	nouns denoting man-made objects
noun.location	nouns denoting spatial position
noun.process	nouns denoting natural processes

For each word that occurs in a sentence, every associated synset has been extracted and, for each synset, the related lexicographic information is considered. The context of a word w is then represented as a vector of the forty-five lexicographic categories, and the elements of such a vector correspond to the contribution of the instances of the other words in the sentence to the corresponding category. The so defined context is then given in input to the neural network.

Formally, the contribution $C_k(w)$, of an instance of word w to the k-th component of the context vector C is calculated as follows:

$$C_k(w) = \frac{N_k(w)}{N(w)} \tag{1}$$

where $N_k(w)$ is the number of synsets of w whose category is k, and $N(w)$ is the number of synsets of word w. As we can see, the contribution of a monosemous word is maximal, i.e., it is 1.0.

Let S be the sentence in which the word to be disambiguated w occurs. The k-th element of the vector context C of S, C_k, is given by:

$$C_k = \sum_{w \in S} C_k(w). \tag{2}$$

For example, starting from a sentence '*part aqueduct system*', where the target word, *tunnel*, has two senses, namely (1) "a passageway through or under something" and (2) "a hole made by an animal", the contribution to each input neuron $(C_1, \ldots, C_{45})$ is calculated as shown in Table 2, where the number in parenthesis is the number of instances of the lexicographic category in the sense list of each word.

Table 2. Input for the sentence "part aqueduct system"

Word	Lexicographic Category	Contribution	Word	Lexicographic Category	Contribution
part		(18 senses)	aqueduct		(1 sense)
	\<noun.act\>	0.110 (2)		\<noun.artifact\>	1.000
	\<noun.artifact\>	0.055 (1)			
	\<noun.body\>	0.055 (1)	system		(9 senses)
	\<noun.cognition\>	0.175 (3)		\<noun.artifact\>	0.111 (1)
	\<noun.communication\>	0.055 (1)		\<noun.attribute\>	0.111 (1)
	\<noun.location\>	0.055 (1)		\<noun.body\>	0.222 (2)
	\<noun.object\>	0.055 (1)		\<noun.cognition\>	0.334 (3)
	\<noun.possession\>	0.055 (1)		\<noun.group\>	0.111 (1)
	\<noun.relation\>	0.055 (1)		\<noun.substance\>	0.111 (1)
	\<verb.motion\>	0.110 (2)			
	\<verb.social\>	0.055 (1)			
	\<verb.contact\>	0.110 (2)			
	\<adv.all\>	0.055 (1)			

4   The Neuro Evolutionary Algorithm

We apply an evolutionary approach for NN design, previously validated on different benchmarks and real-world problems [2,1], which uses a variable-size representation of individuals. A population of classifiers, the individuals, is defined through Multi-Layer Perceptrons (MLPs), a type of feed-forward NNs. The evolutionary process is based on the joint optimization of NN structure and weights and uses the error Backpropagation (BP) algorithm to decode a *genotype* into a *phenotype* NN.

Individuals are defined with a pre-established input layer size for each network, set to the context vector size, while the output size is given by the number of senses of the target word. Different hidden layer sizes and different numbers of neurons for such layers are defined for each NN, according to two exponential distributions, in order to maintain diversity among all the individuals. The number of neurons in each hidden layer is constrained to be greater than or equal to the number of network outputs, in order to avoid hourglass structures, whose performance tends to be poor. Indeed, a layer with fewer neurons than the outputs destroys information which later cannot be recovered.

A detailed description of the evolutionary process is presented in [1], and here it is briefly reported. At each generation, a population consisting of the best $\lfloor n/2 \rfloor$ individuals is selected by truncation from a population of size n; the remaining NNs are then duplicated in order to replace those eliminated, and the population is randomly permuted. Elitism allows the survival of the best individual unchanged into the next generation and the solutions to get better over time. Then, for all individuals of the population the algorithm mutates the weights and the topology of the offsprings, trains the resulting network, calculates fitness, and saves the best individual and statistics about the entire evolutionary process.

The recombination opearator is not used in this approach, due to the detrimental effects on NNs [13]. Weights mutation perturbs the weights of the neurons before performing any structural mutation and applying the BP to train the network. All the weights and the corresponding biases are updated by using variance matrices and evolutionary strategies applied to the synapses of each NN, in order to allow a control parameter, like mutation variance, to self-adapt rather than changing their values by some deterministic algorithm. The topology mutation is implemented with four types of mutation by considering neurons and layer addition and elimination. The addition and the elimination of a layer and the insertion of a neuron are applied with three independent probabilities, while the elimination of a neuron is carried out only if the contribution of that neuron is negligible with respect to the overall network output.

Finally, the fitness of an individual is calculated based on the confusion matrix, as the difference between the number of output neurons and Trace, that corresponds to the sum of the diagonal elements of the row-wise normalized confusion matrix. Such a confusion matrix represents the conditional probabilities of the predicted outputs given the actual outputs. Following the commonly accepted practice of machine learning, the problem data are partitioned into three sets: training, test and validation set, used respectively to train, to stop the training, thus avoiding overfitting, and to assess the generalization capabilities of a network.

5   Experiments and Results

In this work, the genetic parameters p_{layer}^{+}, p_{layer}^{-}, and p_{neuron}^{+} are set to 0.05, the same values used in [1]. To validate the proposed approach, we have compared its performance to eight state-of-the-art WSD algorithms that ranked in the topmost positions at the last Senseval 4. The performance of the proposed encoding scheme is also compared with the performance obtained by using the positional encoding scheme presented in [1]. We selected the set of 30 words from Senseval 4's Task 17, listed in Table 3, for which detailed results are provided by the organizers, indicating the performance of the best eight systems, and compared the results obtained by the neuro-evolutionary algorithm to the results obtained by those best eight systems. For each word evaluation we carried out 10 runs of the evolutionary algorithm.

Since the rules for Task 17 allowed the use of any available resource for training, we used training and test datasets extracted from the IXA Group's web corpus for all words except *drug* and *part*. The word *drug* is monosemous in WordNet, while in Task 17 it was assigned two distinct senses; the word *part* is not included in the IXA Group's web corpus. Therefore, the training and test sets have been extracted from the training dataset provided by the Senseval 4 organizers.

For all words, the evolved ANNs have been scored on three validation sets:

1. a validation set extracted from the IXA Group's corpus with fine-grained senses taken from WordNet (IXA Fine);
2. a validation set extracted from the IXA Group's corpus with coarse-grained sense aggregation, where senses are taken from the OntoNotes [7] ontology like in the Senseval 4 Task 17 dataset (IXA Coarse);
3. the Senseval 4 Task 17 validation set, the same on which the entries in the evaluation had been scored, where senses are taken from the OntoNotes ontology.

The results of this experiment are shown in Table 3. Overall, they show that our approach, besides obtaining satisfactory results when validated with the same corpus on which it is trained, has a performance that is essentially comparable to state-of-the-art WSD systems.

To correctly interpret the data presented in Table 3, a few remarks are in order. No data is available for *drug* and *part* on the IXA Fine and IXA Coarse datasets.

For *rate*, our system has a suspiciously worse accuracy on the IXA Coarse dataset than on the IXA Fine dataset. This happens because one of the WordNet senses used in the IXA Fine dataset does not belong to any of the OntoNote senses according to which the IXA Coarse dataset is formed; therefore, records with that sense have been discarded — too bad that the evolved NN had an accuracy of 68.33% on the discarded sense!

Using the lexicographic encoding, the neuro-evolutionary approach lies within the benchmark range in 22 cases, and above range in 1 case out of 30; still, it should be noticed that approximately half of the below-range cases are actually quite close to the benchmark minimum. Such information is shown in the last three columns of Table 3.

Table 3 also reports the results obtained on the IXA validation set; the second column in the table contains the number of fine-grained senses of considered words while the third column provides the number of coarse-grained senses. The next three columns provide, respectively, the maximum, minimum, and average distance between senses for each given word, measured as the minimum number of ontology edges connecting two concepts in WordNet: the purpose of including this information is to give an idea of how close together or far apart the senses of a word are. The seventh and eighth columns of Table 3 give the accuracy, measured on the IXA validation set, of the NN disambiguator, obtained for each word over 10 runs of the neuro-evolutionary algorithm.

It should be noticed that not all senses differ semantically by the same amount. It can be observed that, for instance, the senses of *people* are quite close to each

Table 3. A summary of the results of applying the neuro-evolutionary approach to the disambiguation of 30 test words, with a comparison to the results obtained on Senseval 4 Benchmark

Word	# of Senses	# of Groups	Senses Dist. Max.	Min.	Avg.	Positional Fine Grained	Coarse Grained	Lexicographic Fine Grained	Coarse Grained	Benchmark Results	Max.	Min.
area	6	5	12	7	10	59.11	75.95	70.65	77.84	69.40	89.00	65.00
authority	7	6	18	3	12	57.00	57.00	59.49	59.49	70.00	86.00	33.00
base	19	12	22	4	13	33.44	34.27	55.19	56.93	68.40	80.00	40.00
bill	10	8	22	5	13	36.46	36.64	39.87	39.90	75.20	99.00	22.00
carrier	11	10	18	2	11	31.48	38.94	57.24	57.24	70.00	71.00	62.00
chance	5	4	15	3	10	65.38	65.38	91.49	91.49	50.00	73.00	20.00
condition	7	4	15	2	8	76.69	94.71	79.28	93.43	75.80	91.00	56.00
defense	11	8	18	2	10	36.65	37.91	46.14	46.79	30.00	57.00	29.00
development	8	3	15	3	9	49.51	81.51	60.92	90.97	64.30	100.00	62.00
drug	1*	2*	-	-	-	-	-	-	-	86.40	96.00	78.00
effect	6	4	12	6	9	66.94	75.16	82.99	84.58	75.90	97.00	77.00
exchange	11	6	17	2	10	25.88	39.89	35.98	54.02	73.30	92.00	79.00
future	3	3	13	9	10	91.91	91.91	97.49	97.49	86.90	98.00	83.00
hour	4	4	13	5	9	19.83	19.83	62.42	62.42	89.40	92.00	58.00
job	13	10	20	2	11	24.07	26.91	38.58	43.36	81.60	90.00	69.00
management	2	2	10	10	10	97.62	97.62	94.77	94.77	70.50	98.00	64.00
network	5	4	11	2	7	80.37	80.37	75.26	75.26	90.70	98.00	82.00
order	15	9	13	3	9	33.69	33.83	46.82	47.70	91.10	95.00	90.00
part	12	7	16	3	9	-	-	-	-	65.70	97.00	66.00
people	4	3	7	2	4	44.01	44.35	50.69	50.69	90.40	96.00	90.00
point	24	13	18	3	10	32.10	37.16	51.33	53.85	81.20	92.00	79.00
policy	3	2	16	11	14	80.34	85.47	86.71	82.91	100.00	97.00	64.00
position	16	6	14	2	8	21.01	22.90	40.37	50.25	38.60	78.00	53.00
power	9	4	16	4	9	40.26	71.15	62.02	77.70	65.20	92.00	74.00
president	6	3	19	2	11	61.20	64.79	77.31	79.25	71.60	98.00	85.00
rate	3	2	11	5	8	59.63	55.38	64.29	62.39	86.80	92.00	81.00
source	9	6	16	5	10	33.16	35.15	51.09	52.32	38.20	86.00	29.00
space	8	6	14	2	8	48.04	59.99	44.04	63.99	76.90	100.00	71.00
state	8	4	19	2	11	82.08	84.14	82.38	83.49	78.90	86.00	79.00
system	8	7	15	6	9	26.72	27.06	38.82	40.29	49.30	79.00	59.00

other. This means that confusion of senses may be expected, as the contexts in which semantically close meanings of a word are used may be very similar or even coincide. As a matter of fact, a quick inspection of the results suggests that, unsurprisingly, better accuracy is obtained for words whose senses are more apart.

To provide a more detailed, critical discussion of the results, it will be convenient to consider the graphical representation of the corresponding (row-wise normalized) confusion matrices.

In Figure 1 and in Figure 2, the confusion matrices for the 28 words considered are shown for the fine-grained disambiguation and for the coarse-grained disambiguation respectively. A visual comparison of the confusion matrices obtained for the same word confirms that a coarse-grained disambiguation reaches, as expected, better results. In most cases, the two disambiguation levels obtain the same results, it happens because the number of groups is equal to the number of senses.

For a few words, namely *area, defense, exchange, hour, job,* and *network,* both granularities yield qualitatively similar results, while for word *development* the coarse-grained disambiguation significantly improves the quality of results. For word *exchange,* for example, the EA classifies with satisfactory performances four of the eight senses of the word, listed in Table 3, respectively the first,

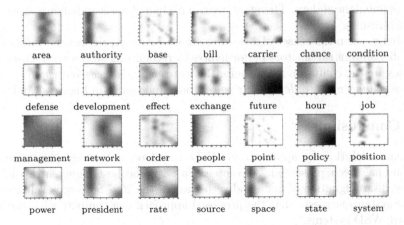

Fig. 1. Confusion matrices of the 28 Senseval 4 words, calculated for the IXA Fine validation set

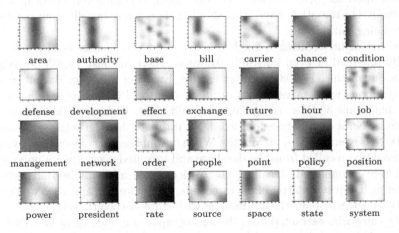

Fig. 2. Confusion matrices of the 28 Senseval 4 words, calculated for the IXA Coarse validation set

the second, the third, and the sixth. The fourth and the seventh senses are not classified so well, in that they tend to get confused with the third sense. This phenomenon may be explained by the high similarity among some of the word senses, as well as among all the sentences in the dataset associated to each of them. A similar situation happens also with the other words listed above.

A completely different case involves words like *authority*, *condition*, *state*, and *president*. For these words, the evolved NN takes the "shortcut" of always recognizing only one of the senses, without considering the others. The unbalance of the dataset for the two senses would seem a possible explanation. However, further experiments with artificially balanced datasets for the same words have shown similar results and, after all, the way fitness is calculated neutralizes the bias of unbalanced data. Therefore, we can exclude dataset unbalance as a cause of such results.

Finally, when the senses of a word are more far apart, the neuro-evolutionary classifier obtains satisfactory performances by recognizing all the word senses. Examples are the confusion matrices for the words *base, bill, carrier, chance, effect, future, management, order, policy, power, rate,* and *source* in Figures 1 and 2.

6   Conclusions

A neuro-evolutionary approach to WSD based on a lexicographic encoding scheme has been presented. A comparison with the best eight entries in the coarse disambiguation of noun and verbs task of the 2007 Senseval evaluation of WSD systems suggests that the proposed approach can compete with state-of-the-art WSD systems.

In order to put the results obtained from this work in the right perspective, one should consider that the classifiers were trained on a different dataset from the one used for benchmarking. On one hand, this puts an handicap on our classifiers. On the other hand, this should rule out any benefit from overfitting the training data, if any.

At first sight, creating a single NN for every ambiguous word might seem hardly practical or even infeasible. However, there are just 15,935 polysemous words out of the 117,798 WordNet entries. Evolving a NN for disambiguating a polysemous word takes, on an ordinary desktop PC, two hours on average. Assuming some 30 PCs are available day and night, 45 days would be enough to evolve a NN for each polysemous word. We estimate that the entire set of almost 16,000 NNs would occupy 30 Mbytes. When disambiguating a document, a stored NN would be recalled from the database and executed every time a polysemous word were encountered. Recalling a network can take a few milliseconds, whereas executing it is just a matter of microseconds. Therefore, the approach we propose can be considered realistic and feasible with state-of-the-art technology.

The categories extracted from the lexicographic annotations of WordNet may be regarded as a convenient loose approximation of (very high-level) semantic categories. Future work in this direction, therefore, will involve using some finer semantic categorization of the words occurring in the context of the target word. An evaluation of the overall disambiguation performance was out of the scope of this work. That would require evolving a classifier for every polisemous word, which is still in our future plan.

References

1. Azzini, A., Dragoni, M., da Costa Pereira, C., Tettamanzi, A.: Evolving neural networks for word sense disambiguation. In: Proc. of HIS 2008, Barcelona, Spain, September 10-12, pp. 332–337 (2008)
2. Azzini, A., Tettamanzi, A.: Evolving neural networks for static single-position automated trading. Journal of Artificial Evolution and Applications 2008(Article ID 184286), 1–17 (2008)

3. Cottrell, G.: A Connectionist Approach to Word Sense Disambiguation. Pitman, London (1989)
4. Escudero, G., Màrquez, L., Rigau, G.: An empirical study of the domain dependence of supervised word sense disambiguation systems. In: Proc. of the Joint SIGDAT conference on Empirical methods in natural language processing and very large corpora, Morristown, NJ, USA, pp. 172–180. ACL (2000)
5. Escudero, D.M., Màrquez, G.L., Rigau, G.: Supervised Corpus-Based Methods for WSD, pp. 167–207. Springer Netherlands, Heidelberg (2006)
6. Hinton, G., McClelland, J., Rumelhart, D.: Distributed representations. In: Parellel Distributed Processing: explorations in the microstructure of cognition. MIT Press, Cambridge (1986)
7. Hovy, E., Mitchell, M., Palmer, M., Ramshaw, L., Weischedel, R.: Ontonotes: The 90% solution. In: Proc. of the Human Language Technology Conf., New York, pp. 57–60 (2006)
8. Navigli, R.: Word sense disambiguation: A survey. ACM Comput. Surv. 41(2), 1–69 (2009)
9. Ng, H.: Exemplar-based word sense disambiguation: some recent improvements. In: Proc. of EMNLP 1997, pp. 208–213 (1997)
10. Schapire, R., Singer, Y.: Improved boosting algorithms using confidence-rated predictions. Machine Learning 37(3), 297–336 (1999)
11. Tratz, S., Sanfilippo, A., Gregory, M., Chappell, A., Posse, C., Whitney, P.: Pnnl: A supervised maximum entropy approach to word sense disambiguation. In: Proc. of Int. Workshop on Semantic Evaluations SemEval 2007, Prague, June 2007, pp. 264–267. Association of Computational Linguistics (2007)
12. Vapnik, V.: Statistical Learning Theory. John Wiley & Sons, Chichester (1998)
13. Yao, X., Liu, Y.: A new evolutionary system for evolving artificial neural networks. IEEE Transactions on Neural Networks 8(3), 694–713 (1997)
14. Yarowsky, D.: Hierarchical decision lists for word sense disambiguation. Computational Humanities 34(2), 179–186 (2000)
15. Zhou, X., Han, H.: Survey of word sense disambiguation approaches. In: Proc. of the Int. Artificial Intelligence Research Society Conference, Clearwater Beach, Florida, USA, pp. 307–313. AAAI Press, Menlo Park (2005)

Feature Membership Functions in Voronoi-Based Zoning

S. Impedovo, A. Ferrante, R. Modugno, and G. Pirlo

Dipartimento di Informatica, Università degli Studi di Bari, via Orabona 4, 70126, Bari
Centro "Rete Puglia", Università degli Studi di Bari, via G. Petroni 15/F.1, 70100 Bari

Abstract. Recently, the problem of zoning design has been considered as an optimization problem and the optimal zoning is found as the one which minimizes the value of the cost function associated to the classification. For the purpose, well-suited zoning representation techniques based on Voronoi Diagrams have been proposed and effective real-coded genetic algorithms have been used for optimization.

In this paper, starts from the consideration that whatever zoning method is considered, the role of feature membership function is crucial, since it determines the influence of a feature to each zone of the zoning method. Thus, in the paper the role of feature membership functions in Voronoi-based zoning methods is investigated. For the purpose, abstract-level, ranked-level and measurement-level membership functions are considered and their effectiveness is estimated under different Voronoi-based zoning methods.

The experimental tests, carried out in the field of hand-written numeral recognition, show that the best results are obtained when specific measurement-level membership functions are used.

1 Introduction

Zoning is a diffuse strategy for pattern classification. It allows extracting topological information from patterns, according to well-defined partitioning criteria of the pattern image. Precisely, given a pattern image B, a zoning $Z_M=\{z_1, z_2, ..., z_M\}$ of B is a partition of B into M sub-images, named zones, each one providing information related to a specific part of the pattern [1, 2]. Zoning has been used in many application fields. For instance, in the field of handwritten character recognition, zoning has been massively considered along with the use of geometrical features as a useful technique to handle the enormous variability of handwritten characters, due to different styles and personal variations in writing [2,3].

In the past, several zoning methods have been proposed based on standard partitioning criteria [1-5]. Figure 1 shows some standard zoning methods Z_{nxm} defined by using a (n x m) regular grid.

More recently a new technique for zoning design has been proposed in which zoning design is considered as an optimization problem, zoning description is based on Voronoi Diagrams and a real-coded genetic algorithm is used to find the optimal zoning. Although this technique provides better results than traditional zoning approaches, the choice of the feature membership function is still an open problem. Indeed, it is easy to verify that whatever zoning method is considered, the way in

R. Serra and R. Cucchiara (Eds.): AI*IA 2009, LNAI 5883, pp. 202–211, 2009.

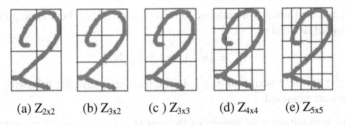

(a) Z_{2x2} (b) Z_{3x2} (c) Z_{3x3} (d) Z_{4x4} (e) Z_{5x5}

Fig. 1. Traditional Zoning Methods

which a feature influences each zone of the zoning method is a fundamental problem and the definition of the most profitable membership function strongly affects system performance [6].

Starting from this consideration this paper addresses the problem of feature membership function. Several functions are considered at abstract-level, ranked-level and measurement-level, and their effectiveness in Voronoi-based zoning systems is evaluated. The organization of the paper is as follows. In Section 2 the problem of zoning-based classification is introduced. Section 3 summarizes the Voronoi-based zoning design technique. Section 4 addresses the problem of feature membership functions and presents the membership function considered in this paper. Section 5 presents some experimental results. They have been carried out in the field of handwritten numeral recognition and shows that the best classification results are achieved by a measurement-level membership function based on exponential model.

2 Zoning-Based Classification

The classification of an unknown pattern x can be considered as a mapping D defined as [7, 8]:

$$D: S(x) \rightarrow \Omega \cup \{\mathbb{C}\} \qquad (1)$$

being

- S(x) the representation of x based on the features set $F=\{f_1,...,f_T\}$,
- $\Omega=\{C_1,..., C_K\}$ the set of class labels,
- $\mathbb{C}$ the class of rejections.

Of course, when zoning is considered, $S_F(x)$ [7, 8] can be described as

$$S(x) = \{ (f_i, z_j, w_{ij}) \}_{i=1,2,...Q , \; j=1,2,...M} \qquad (2)$$

where:

- (f_i, z_j, w_{ij}) means that feature f_i has been detected in pattern x, and the influence of f_i on zone z_j is equal to w_{ij};
- Q is the total number of features detected in x;
- M is the number of zones.

In this case pattern classification is performed in two steps. The first step concerns the learning phase in which feature distribution is determined. The second step concerns the classification of unknown patterns. Let $Z_M=\{z_1, z_2, ..., z_M\}$ be a zoning method, X^L the learning set defined as:

$$X^L = \{x^L_r \mid v=1,2,3...,N^L\}$$

and let $S(x^L_r)$ be the description of x^L_r. In the learning phase the following statistical distributions are obtained:

- the normalized total weight of the pair (f_i, z_j) for the learning patterns of the class C_k: (where n^k is the number of learning patterns belonging to the class C_k: $n^k=card\{x_r \in X^L | x_r \in C_k\}$) :

$$W^k_{ij} = \frac{1}{n^k} \sum_{x_r \in X^L, x_r \in C_k, (f_i, z_j, w_{ij}) \in S(x_r)} w_{ij} \quad , \tag{3}$$

being w_{ij} the weight indicating the influence of feature f_i to the zone z_j.

- the relevance S^k_{ij} of the normalized total weights for each pattern class:

$$S^k_{ij} = \frac{W^k_{ij}}{\sum_{k=1,2,...,K} W^k_{ij}} \quad , \tag{4}$$

where, of course, $S^k_{ij}=0$ if $\sum_{k=1,2,...,K} W^k_{ij} = 0$.

In the second step, i.e. the classification phase, the confidence value that an unknown input pattern x_t belongs to the class C_k, is given by the score [2]:

$$Score_{C_k}(x_t) = \sum_{q=1}^{Q^t} w_{i^t_q, j^t_q} \cdot S^k_{i^t_q, j^t_q} \quad , \tag{5}$$

where $S(x_t) = \{(f_{i^t_q}, z_{j^t_q}, w_{i^t_q j^t_q})\}$ is the description set of x_t (with $i^t_q \in \{1,2,...,T\}$ and $j^t_q \in \{1,2,...,M\}$, $\forall q=1,2,...,Q^t$).

Now, given

$$k^* = \underset{k \in \{1,2,...,K\}}{\arg \max} \; Score_{C_k}(x_t) \tag{6a}$$

$$k^{**} = \underset{\substack{k \in \{1,2,...,K\} \\ k \neq k^*}}{\arg \max} \; Score_{C_k}(x_t) \tag{6b}$$

the pattern x_t is classified as belonging to the class C_{k^*}, if and only if

$$Score_{C_{k^*}}(x_t) - Score_{C_{k^{**}}}(x_t) < \varepsilon \tag{7}$$

being ε a suitable threshold value.

3 Voronoi-Based Zoning Methods

Voronoi Diagrams have been recently used for zoning representation and design [8]. This technique considers zoning design as an optimization problem and the optimal zoning $Z^*_M = \{z^*_1, z^*_2, ..., z^*_M\}$ is found as the zoning for which the cost function $CF(Z_M)$ associated to classification is minimum [6]. In this paper the cost function is defined as [8]:

$$CF(Z_M) = \eta \cdot Err(Z_M) + Rej(Z_M) \qquad (8)$$

where:

- $Err(Z_M)$ is the substitution error rate;
- $Rej(Z_M)$ is the rejection rate;
- the coefficient η is the cost value associated to the treatment of an error with respect to a rejection (in our tests $\eta=5$).

More precisely, let B be a pattern image and $P=\{p_1, p_2,..., p_M\}$ a set of M distinct points in B. The *Voronoi Diagram* determined by P is the partition of B into M zones $\{z_1, z_2, ..., z_M\}$, with the property that each zone z_i corresponding to $p_i \in P$, contains all the points $\underline{p}$ for which it results:

$$ED\ (\underline{p}, p_i) < ED\ (\underline{p}, p_j)\ , \qquad \forall p_j \in P,\ p_j \neq p_i\ . \qquad (9)$$

Moreover, let $\underline{p}$ be a point for which it results:

$$ED(\underline{p}, p_i) = \min\{ED\ (\underline{p}, p_j) \mid p_j \in P\}, \qquad \forall i \in I,\ I \subseteq \{1\ 2,..., M\},$$

it is here assumed that:

$$\underline{p} \in z_{\min\{\ i\ \mid\ i \in I\ \}}. \qquad (10)$$

Therefore, when *Voronoi Diagrams* are used for zoning description, the zoning design optimization problem can be written as follows:

$$\textit{Find the set } \{p^*_1, p^*_2, ..., p^*_M\} \textit{ so that } CF(Z^*_M)=\min_{\{p_1, p_2, ..., p_M\}} CF(Z_M) \qquad (11)$$

with:

- $Z^*_M = \{z^*_1, z^*_2,..., z^*_M\}, z^*_j$ is the Voronoi region corresponding to p^*_j, $\forall j=1,2,..,M$;
- $Z_M = \{z_1, z_2,..., z_M\}$, z_j is the Voronoi region corresponding to p_j, $\forall j=1,2,..,M$.

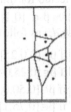

Fig. 2. Zoning representation by Voronoi Diagrams

For the optimization problem (11) a genetic algorithm has been considered [10]. The initial – population Pop=$\{\Phi_1, \Phi_2, ...,\Phi_i, ... ,\Phi_{Npop}\}$ is created by generating N_{pop} random individuals (N_{pop} even). Each individual is a vector $\Phi_i= \langle p_1,p_2,...,p_j,...,p_M \rangle$ (where each element p_j is a point defined as $p_j=(x_j,y_j)$) that corresponds to the zoning $Z_M=\{z_1, z_2, ..., z_M\}$, where each z_j is the Voronoi region corresponding to p_j, j=1,2,...,M.

Consequently, the fitness value of the individual $\Phi_i=\langle p_1,p_2,...,p_j,...,p_M \rangle$ is taken as the classification cost $CF(Z_M)$, obtained by (8), where $Z_M=\{z_1, z_2, ..., z_M\}$ and z_j is the Voronoi region corresponding to p_j, $\forall j=1,2,...,M$.

From the initial - population, the following genetic operations are used to generate the new populations of individuals:

1. *Individual Selection.* In the selection procedure $N_{pop}/2$ random pairs of individuals are selected, according to a *roulette-wheel* strategy [10].

2. *Crossover.* An arithmetic crossover is used to exchanges information between two selected individuals [10, 11]. In particular, let be

$$\langle p^a_1,p^a_2,...,p^a_{s-1}, p^a_s,...,p^a_M \rangle \tag{12a}$$

and

$$\langle p^b_1,p^b_2,...,p^b_{s-1}, p^b_s,...,p^b_M \rangle, \tag{12b}$$

two parent individuals, the two offspring individuals of the next generation are:

$$\langle \bar{p}^a_1,\bar{p}^a_2,...,\bar{p}^a_{s-1}, \bar{p}^a_s,...,\bar{p}^a_M \rangle \tag{13a}$$

and

$$\langle \bar{p}^b_1,\bar{p}^b_2,...,\bar{p}^b_{s-1}, \bar{p}^b_s,...,\bar{p}^b_M \rangle; \tag{13b}$$

where

$$\bar{p}^a_s =\alpha.p^a_s+(1-\alpha).p^b_s \tag{14a}$$

$$\bar{p}^b_s =\alpha.p^b_s+(1-\alpha).p^a_s ; \tag{14b}$$

being α a random value in the range [0,1].

3. *Mutation.* A non-uniform mutation operator is considered, based on the mutation probability *Mut_prob* [10,11]. In particular, let $\Phi_i=\langle p_1,p_2,...,p_M \rangle$ be an individual and $p_j=(x_j,y_j)$ an element of Φ_i selected for mutation. In the first step, the mutation operators generate a random value $\beta \in [0,1]$ according to a uniform distribution and select a direction vector $d=(d_x,d_y)$ as follows:

  □ $d = d_x^+ = (1,0)$ (East),   if $\beta \in [0,0.25]$;
  □ $d = d_x^- = (-1,0)$ (West),   if $\beta \in [0.25,0.50]$;
  □ $d = d_y^+ = (0,1)$ (North),   if $\beta \in [0.50,0.75]$;
  □ $d = d_y^- = (0,-1)$ (South),   if $\beta \in [0.75,1]$.

In the second step, the displacement for p_j is determined according to a Non-uniform Mutation. In this case the value of p_j becomes equal to p'_j, with [12, 13]:

$$p'_j = p_j + d \cdot \max\_displ \left(1 - \delta^{\left[1 - \frac{iter}{N^{iter}} \right]^b} \right), \qquad (15)$$

where:

- δ is a random value generated in the range [0, 1], according to a uniform distribution;
- max_displ is the maximum displacement allowed;
- $iter$ is the counter of the generations performed;
- N^{iter} is the maximum number of generations;
- b is a system parameter determining the degree of non-uniformity.

4. *Elitist Strategy.* From the N_{pop} individuals generated by the above operations, one individual is randomly removed and the individual with the minimum cost in the previous population is added to the current population [10, 13].

Steps from (1) to (4) are repeated until N^{iter} successive populations of individuals are generated. When the process stops, the optimal zoning is obtained by the best individual of the last-generated population.

Figure 2 shows an example of zoning method achieved by an optimization process. The optimal zoning Z^*_9 is shown and the Voronoi points are also reported.

4   Feature Membership Functions

When a zoning is considered, classification performance strongly depends on the effectiveness of $S_F(x)$. In fact, pattern description should be able to absorb as much as possible intra-class variability and maintaining inter-class differences, at the same time. For the purpose, whatever zoning method $Z_M = \{z_1, z_2, ..., z_M\}$ is considered, the feature-zone membership function is very important.

Of course, for each f_i detected at point pf_i, the influence values w_{ij} are generally determined on the basis of proximity conditions between the position of f_i and z_j, $j=1,2,...M$. In particular, let

- $Z_M = \{z_1, z_2, ..., z_M\}$ be a zoning method corresponding to the Voronoi points $P = \{pz_1, pz_2, ..., pz_M\}$, where z_j is the Voronoi region corresponding to the Voronoi point pz_j, $j=1,2,...,M$;
- d_{ij} be the distance between the position pf_i in which the feature f_i is positioned and pz_j, i.e. $d_{ij} = dist(pf_i, pz_j)$.

Under these assumptions let

$$j_1, j_2, ... j_k, j_{k+1}, ... j_k$$

be a set of indexes so that

- $i_k \in \{1, 2, ..., M\}$, $\forall k = 1, 2, ... K$;
- $i_{k1} \neq i_{k2}$, $\forall k_1, k_2 = 1, 2, ... K$, $k_1 \neq k_2$.

and for which it results

$$d_{ik1} \leq d_{ik2} \ , \ \forall k_1, k_2 = 1, 2, \dots K,$$

the following feature-zone membership functions are considered in this paper:

1. **Abstract-level membership functions:**
   - *The Winner-takes-all (WTA)* membership function. This is the standard membership function used in traditional zoning-based classification. In this case, for each f_i, we have:
     - $w_{ij}=1$ if $j=j_1$
     - $w_{ij}=0$ otherwise;

   - *The k-Nearest Zone (k-NZ)* membership function. This is a generalization of the WTA function. In this case, for each f_i, we have:
     - $w_{ij}=1$ if $j \in \{j_1, j_2, \dots j_k\}$
     - $w_{ij}=0$ otherwise;

2. **Ranked-level membership functions:**
   - *The Ranked-based (R)* membership function. In this case, for each f_i, we have:
     - $w_{ij}=M-j_k$ if $j=j_k$
     - $w_{ij}=0$ otherwise;

3. **Measurement-level membership functions:**
   - *The Measurement-based* membership function. In this case, for each f_i, three membership models have been used:

     - *Linear Weighting Model (LWM)*
       - $w_{ij}=1 \, / \, d_{ij}$
     - *Quadratic Weighting Model (QWM)*
       - $w_{ij}=1 \, / \, d_{ij}^{\,2}$.
     - *Exponential Weighting Model (EWM)*
       - $w_{ij}=1 \, / \, 10^{\,dij}$.

5   Experimental Results

For the experimental results the set of pattern classes $\Omega_1=\{0,1,2,3,4,5,6,7,8,9\}$ has been considered, concerning the set of 10 numeral digits. Precisely, 10000 patterns from the CEDAR database (BR directory) have been used according to a k-fold cross validation strategy (k=5) [14].

Concerning the feature set, the set $F_1=\{f_1,\dots,f_9\}$ (see Fig. 3) has been extracted from the pattern images [6, 15] after skeletonization carried out by the *Safe Point Thinning Algorithm* [16], where:

- ❏ f_1: holes;
- ❏ f_2: vertical-up cavities;
- ❏ f_3: vertical-down cavities;
- ❏ f_4: horizontal-right cavities;

- f_5: horizontal-left cavities;
- f_6: vertical-up end-points;
- f_7: vertical-down end-points;
- f_8: horizontal-right end-points;
- f_9: horizontal-left end-points.

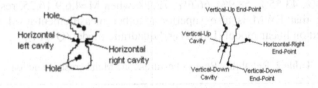

Fig. 3. Example of Features of the set F_1

Before conducting the experiments, the most suitable genetic operators and the free-parameter values of the optimization problem have been pre-estimated by performing some pilot tests [12, 13]. The selected parameters are: N_{Pop}=10, N^{iter}=1000, *Mut_prob*=0.05, *max_displ*=3, *b*=1.0. Figure 4 shows an example of how fitness improves with successive generations.

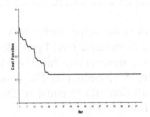

Fig. 4. Fitness improvement vs successive generations

Using these parameters, the optimal zonings Z^*_M for different number of zones (M=4, 6, 9, 16, 25) have been computed (for ε = 0.5). The performances of these zoning methods estimated on the database by a k-fold cross validation technique (k=5) is reported in Table 1 and 2.

Table 1. Recognition Rate vs Feature-Zone Membership Function

Zoning Method	Membership Functions						
	Abstract			Ranked	Measurement		
	WTA	2-NZ	3-NZ	R	LWM	QWM	EWM
Z^*_4	0,80	0,78	0,62	0,68	0,51	0,54	0,82
Z^*_6	0,87	0,83	0,77	0,62	0,51	0,56	0,89
Z^*_9	0,86	0,84	0,82	0,56	0,53	0,59	0,89
Z^*_{16}	0,88	0,86	0,86	0,54	0,52	0,62	0,90
Z^*_{25}	0,91	0,87	0,86	0,54	0,56	0,63	0,93

More precisely, Table 1 reports for each optimal zoning the recognition rates achieved by the various membership functions. The best results are always achieved by the measurement-level membership function when an exponential model (EWM) is used. In fact, EWM outperforms the best abstract-level membership function (WTA) of 2.5%, 2.2%, 3.4%, 2.2%, 2.1% when M=4,6,9,16,25, respectively. More-over, the improvement with respect to the ranked-level membership function (R) is equal to 20.6%, 43.55%, 58.9%, 66.6%, 72.2% when M=4,6,9,16,25, respectively. It worth noting that EWM is also superior to other measurement-level membership function based on linear model (LWM) and quadratic model (QWM).

Table 2. Reliability Rate vs Feature-Zone Membership Function

Zoning Method	Membership Functions						
	Abstract			Ranked	Measurement		
	WTA	2-NZ	3-NZ	R	LWM	QWM	EWM
Z^*_4	0,81	0,79	0,63	0,69	0,52	0,55	0,83
Z^*_6	0,88	0,84	0,78	0,63	0,52	0,57	0,91
Z^*_9	0,87	0,85	0,83	0,57	0,54	0,60	0,91
Z^*_{16}	0,89	0,87	0,87	0,55	0,57	0,63	0,91
Z^*_{25}	0,92	0,88	0,87	0,55	0,57	0,64	0,93

Table 2 reports the reliability rates achieved by the various membership functions, for each optimal zoning. Also in this case EWM provides the best results. EWM out-performs the best abstract-level membership function (WTA) of 2.5%, 3.4%, 4.6%, 2.2%, 1.0% when M=4,6,9,16,25, respectively. The improvement with respect to the ranked-level membership function (R) is equal to 20.1%, 44.4%, 59.6%, 65.4%, 69.0% when M=4,6,9,16,25, respectively.

5 Conclusion

This paper addresses the problem of the feature membership function in Voronoi-based zoning methods.

The experimental tests, obtained in the field of hand-written numeral recognition, have been achieved by using different membership functions. In particular, abstract-level, ranked-level and measurement-level membership functions have been considered.

The results demonstrate that measurement-level membership functions based on an exponential model are adequate to model the influence of a feature to each zone of the zoning method.

References

[1] Trier, O.D., Jain, A.K., Taxt, T.: Feature Extraction Methods For Character Recognition – A Survey. Pattern Recognition 29(4), 641–662 (1996)
[2] Plamondon, R., Srihari, S.N.: On-line and Off-line Handwriting Recognition: A comprehensive survey. IEEE Trans. Pattern Analysis and Machine Intelligence 22(1), 63–84 (2000)

[3] Suen, C.Y., Guo, J., Li, Z.C.: Analysis and Recognition of Alphanumeric Handprints by Parts. IEEE Trans. Systems, Man and Cybernetics 24(4), 614–630 (1994)

[4] Dimauro, G., Impedovo, S., Modugno, R., Pirlo, G.: Numeral recognition by weighting local decisions. In: Proc. ICDAR 2003, Edinburgh, UK, pp. 1070–1074 (August 2003)

[5] Baptista, G., Kulkarni, K.M.: A high accuracy algorithm for recognition of hand-written numerals. Pattern Recognition 4, 287–291 (1988)

[6] Lucchese, M.G., Impedovo, S., Pirlo, G.: Optimal Zoning Design by Genetic Algorithms. IEEE Trans. Sys. Men and Cybern. - Part A 36(5), 833–846 (2006)

[7] Kuncheva, L.I., Jain, L.C.: Designing Classifier Fusion Systems by Genetic Algorithms. IEEE Trans. Evolut. Comput. 4(4) (2000)

[8] Aurenhammer, F.: Voronoi Diagrams: A Survey of a Fundamental Geometric Data Structure. ACM Computing Surveys 3(3), 345–405 (1991)

[9] de Berg, M., Schwarzkopf, O., van Kreveld, M., Overmars, M.: Computational Geometry: Algorithms and Applications. Springer, Berlin (2000)

[10] Baeck, T.: Evolutionary Algorithms in Theory and Practice: Evolution Strategies, Evolution Programming, Genetic Algorithms. Oxford Univ. Press, New York (1996)

[11] Michalewicz, Z.: Genetic Algorithms + Data Structure=Evolution Programs. Springer, Berlin (1996)

[12] Grefenstette, J.J.: Optimization of control parameters for genetic algorithms. IEEE Trans. SMC 16(1), 122–128 (1986)

[13] Back, T., Fogel, D., Michalewicz, Z. (eds.): Handbook of Evolutionary Computation. Institute of Physics Publishing Ltd., Bristol and Oxford University Press, New York (1997)

[14] Hull, J.: A database for handwritten text recognition research. IEEE Trans. PAMI 16(5), 550–554 (1994)

[15] Heutte, L., Paquet, T., Moreau, J.V., Lecourtier, Y., Olivier, C.: A Structural / Statistical Features Based Vector for Handwritten Character Recognition. Pattern Recognition Letters 9, 629–641 (1998)

[16] Naccache, N.J., Shinghal, R.: SPTA: A proposed algorithm for thinning binary patterns. IEEE Trans. SMC 14(3), 409–418 (1994)

Optimal Planning with ACO

M. Baioletti, A. Milani, V. Poggioni, and F. Rossi

Dipartimento Matematica e Informatica
Università di Perugia, Italy
{baioletti,milani,poggioni,rossi}@dipmat.unipg.it

Abstract. In this paper a planning framework based on Ant Colony Optimization techniques is presented. Optimal planning is a very hard computational problem which has been coped with different methodologies. Approximate methods do not guarantee either optimality or completeness, but it has been proved that in many applications they are able to find very good, often optimal, solutions. Our proposal is to use an Ant Colony Optimization approach, based both on backward and forward search over the state space, using different pheromone models and heuristic functions in order to solve sequential optimization planning problems.

1 Introduction

Optimal sequential planning is a very hard task which consists on searching for solution plans with minimal length or minimal total cost. The cost of a plan is an important feature in many domains because high cost plans can be of little utility, almost like non executable plans. Several approaches to optimization planning have been recently proposed [9].

The main contribution of our work consists in investigating the application of the well known Ant Colony Optimization (ACO) meta–heuristic [6,5] to optimal propositional planning. ACO is a meta–heuristic inspired by the behaviour of natural ant colonies that has been successfully applied to many *Combinatorial Optimization* problems. The first application of ACO to planning has been proposed very recently [1]. The approach, although still preliminary, seems promising. Moreover it is known that stochastic approaches to planning can perform very well [7]. ACO seems suitable for planning because there is a strong analogy between the construction solution process of ACO and planning as progressive/regressive search in the state space. The basic idea is to use a probabilistic model of ants, in which each ant incrementally builds a solution to the problem by randomly selecting actions according to the pheromone values deposited by the previous generations of ants and to a heuristic function.

A first framework has been presented in [1], where the basis of our approach has been placed. In that work we proposed to solve optimal planning problems using an ACO approach to perform a forward heuristic search in the state space. From this first work we have developed the research following two lines [3,2]. At first we have deepened this model providing and testing several pheromone models, pheromone updating strategies and parameters tuning [1,3]. Then we have proposed a new approach based on the idea of *regression planning* where the search starts from the goal state and proceeds towards

R. Serra and R. Cucchiara (Eds.): AI*IA 2009, LNAI 5883, pp. 212–221, 2009.

the initial state [2]. The forward ACO model uses the *Fast Forward (FF)* heuristic [10], while in the backward version the heuristic h^2 in the h^m family [8] is used.

In this paper we summarize some new ideas regarding pheromone models and new experimental results. In particular we present the best results obtained so far that are reached using the *Action–Action* pheromone model and an improved implementation.

2   Ant Colony Optimization

Ant Colony Optimization (ACO) is a meta–heuristic used to solve combinatorial optimization problems, introduced since early 90s by Dorigo et al. [6], whose inspiration comes from the foraging behavior of natural ant colonies.

ACO uses a colony of artificial ants, which move in the search space and build solutions by composing discrete components. The construction of a solution is incremental: each ant randomly chooses a component to add to the partial solution built so far, according to the problem constraints. The random choice is biased by the pheromone value τ related to each component and by a heuristic function η. Both terms evaluate the appropriateness of each component. The probability that an ant will choose the component c is

$$p(c) = \frac{[\tau(c)]^\alpha [\eta(c)]^\beta}{\sum_x [\tau(x)]^\alpha [\eta(x)]^\beta} \tag{1}$$

where the sum on x ranges on all the components which can be chosen, and α and β are tuning parameters for pheromone and heuristic contributions.

The pheromone values represent a kind of memory shared by the ant colony and are subject to updating and evaporation. In particular, the pheromone can be updated at each construction step or for complete solutions (either all or the best ones) in the current iteration, possibly considering also the best solution of the previous iterations. The update phase is usually performed by adding to the pheromone values associated to components of the solution s an increment $\Delta(s)$ which depends on a solution quality function $F(s)$[1].

The pheromone evaporation has the purpose of avoiding a premature convergence towards not optimal solutions and it is simulated by multiplying it by a factor $1 - \rho$, where $0 < \rho < 1$ is the evaporation rate.

The simulation of the ant colony is iterated until a satisfactory solution is found, an unsuccessful condition is met or after a given number of iterations.

ACO has been used to solve several combinatorial optimization problems, reaching in many cases state of art performance, as shown in [6].

3   Optimal Propositional Planning

Automated planning is a well known AI problem deeply studied in the last years. Many different kinds of problems defined over several planning models have been proposed. Our aim is to solve *Optimal Propositional Planning* problems defined in terms of finding the shortest solution plan of a classical propositional planning problem.

---

[1] For minimization problem whose objective function is f, $F(s)$ is a decreasing function of f.

The standard reference model for planning is the Propositional STRIPS model [11], called also "Classical Planning". In this model the world states are described in terms of a finite set $\mathcal{F}$ of propositional variables: a state s is represented with a subset of $\mathcal{F}$, containing all the variables which are true in s.

A Planning Problem is a triple $(\mathcal{I}, \mathcal{G}, \mathcal{A})$, in which $\mathcal{I}$ is the initial state, $\mathcal{G}$ denotes the goal states, and $\mathcal{A}$ is a finite set of actions.

An action $a \in \mathcal{A}$ is described by a triple $(pre(a), add(a), del(a))$, where $pre(a)$, $add(a)$, $del(a) \subseteq \mathcal{F}$. $pre(a)$ is the set of preconditions: a is executable in a state s if and only if $pre(a) \subseteq s$. $add(a)$ and $del(a)$ are, respectively, the sets of positive (negative) effects: if an action a is executable in a state s, the state resulting after its execution is

$$Res(s, a) = (s \cup add(a)) \setminus del(a). \tag{2}$$

Otherwise $Res(s, a)$ is undefined.

A linear sequence of actions $(a_1, a_2, \ldots, a_n)$ is a plan for a problem $(\mathcal{I}, \mathcal{G}, \mathcal{A})$ if a_1 is executable in $\mathcal{I}$, a_2 is executable in $s_1 = Res(\mathcal{I}, a_1)$, a_3 is executable in $s_2 = Res(s_1, a_2)$, and so on. A plan $(a_1, a_2, \ldots, a_n)$ is a solution for $(\mathcal{I}, \mathcal{G}, \mathcal{A})$ if $\mathcal{G} \subseteq s_n$.

It is possible to formulate planning as a backward search in the state space, called *regression planning*. The search process manage a current set g of subgoals, initially coinciding with $\mathcal{G}$. An action a is compatible with g if $del(a) \cap g = \emptyset$ and in this case the subgoals before the execution of a are computed by regression

$$Regr(a, g) = (g \setminus add(a)) \cup pre(a). \tag{3}$$

In the regression planning a linear sequence of actions $(a_1, a_2, \ldots, a_n)$ is a plan for a problem $(\mathcal{I}, \mathcal{G}, \mathcal{A})$ if a_n is compatible with $\mathcal{G}$, a_{n-1} is compatible with $g_{n-1} = Regr(a_n, \mathcal{G})$, a_{n-2} is compatible with $g_{n-2} = Regr(a_n, g_{n-1})$, and so on. A plan $(a_1, a_2, \ldots, a_n)$ is a solution for $(\mathcal{I}, \mathcal{G}, \mathcal{A})$ if $g_0 \subseteq \mathcal{I}$.

4   Planning with Forward Ants

In this first approach, ants are used to perform a forward search through the state space. The ants build a plan starting from the initial state s_0 executing actions step by step. Each ant draws the next action to execute in the current state from the set of executable actions. The choice is made by taking into account the pheromone value $\tau(a)$ and the heuristic value $\eta(a)$ for each candidate action a with a formula similar to (1). Once an action a is selected, the current state is updated by means of the effects of a.

Choosing always only executable actions means that the entire plan is executable in the initial state. The construction phase stops when a solution is found (the goals are true in the current state), a dead end is encountered (no action is executable) or an upper bound L_{max} for the the length of action sequence is reached.

Pheromone Models. The first choice is to define solution components in terms of pheromone values. Components are composed by an action and further information in order to perform a more informed choice. Therefore the function $\tau(a, i)$ also depend on some other data i locally available to the ant.

However we have also studied a simple pheromone model, called *Action*, in which τ depends only on the action.

We have designed and implemented four different pheromone models.

In the *State–Action* model, τ also depends on the current state s. This is by far the most expensive pheromone model, because the number of possible components is exponential with respect to the problem size. On the other hand, the pheromone values have a straightforward interpretation as a policy (although in terms of a preference policy): $\tau(a, s)$ represents how much it is desirable (or better, it has been useful) to execute a in the state s.

In the *Level–Action* model τ also depends on the time step in which the action will be executed. This model is clearly more parsimonious then the state–action, but its interpretation is less clear. A drawback is that the desirability of an action at a time step t is unrelated to the values for close time steps, while if an action is desirable at time 2, i.e. at an early planning stage, it could also be desirable at time 1 and 3.

To solve this problem we introduce the *Fuzzy level–Action* model which is the fuzzified version of the previous model: the pheromone used for action a to be executed at time t can be seen as the weighted average of the pheromone values $\tau(a, t')$ for a and for $t' = t - W, \ldots, t + W$ computed with the level–action model. The weights and the time window W can be tuned by experimental results.

In the *Action–Action* model the pheromone also depends on the last executed action. This is the simplest way in which the action choice can be directly influenced by the previous choices. Using just the last choice is a sort of first order Markov property.

As we will discuss in the conclusions, this list of pheromone models is by no means exhaustive and many other pheromone models could be invented and tested.

Heuristic Function. The function $\eta(a)$ is computed as follows. First, we compute the state $s' = Res(s, a)$. Then the heuristic function $h_{FF}(s')$, defined in the FF planner [10], is computed as a distance between the state s' and the goals. Therefore we set

$$\eta(a) = \frac{1}{h_{FF}(s')}$$

The computation of $h_{FF}(s')$ can also give a further information, a set H of *helpful actions*, actions which can be considered as more promising. For the actions in H we use a larger value of η, defining

$$\eta(a) = \frac{1}{(1 - k)h_{FF}(s')}$$

where $k \in [0, 1]$.

Plan Evaluation. Plans are evaluated with respect to two different parameters. Let π a plan with length L and $\mathcal{S}_\pi = < s_0, s_1, \ldots, s_L >$ the sequence of all the states reached by π, then $h_{min}(\pi) = \min\{h_{FF}(s_i) : s_i \in \mathcal{S}_\pi\}$ and $t_{min}(\pi) = \min\{i : h_{FF}(s_i) = h_{min}(\pi)\}$

$h_{min}(\pi)$ is the distance (estimated by the heuristic function) between the goals and the state closest to the goals.

$t_{min}(\pi)$ is the time step at which the minimum distance is firstly obtained. Note that for a solution plan $h_{min}(\pi) = 0$ and $t_{min}(\pi)$ corresponds the length of π.

A possible synthesis of these two parameters is given by $P(\pi) = t_{min}(\pi) + Wh_{min}$ (π) where $W > 1$ is a weight which gives a large penality to non valid plans and then drives the search process towards valid plans.

Pheromone Updating. We have implemented a Rank Ant System pheromone update method [4], which operates as follows.

The updating process is performed only at the end of each iteration. First, the pheromone values are evaporated by a given rate ρ.

Then, the solutions are sorted in the decreasing order with respect to P.

Let $\pi_2, \ldots, \pi_\sigma$ the $\sigma - 1$ best solutions found in the current iteration and let π_1 the best solution found so far, then the pheromone value $\tau(c)$ for the component c is increased by the value $\Delta(c) = \sum_{i=1}^{\sigma}(\sigma + 1 - i)\Delta_i(c)$ where

$$\Delta_i(c) = \begin{cases} \frac{1}{P(\pi_i)} & \text{if } c \text{ takes part in } \pi_i \\ 0 & \text{otherwise} \end{cases}$$

The Algorithm. The optimization process continues for a given number N of iterations, in each of them n_a ants build plans with a maximum number of actions bound to L_{max}. c is the initial value for pheromone. The pseudo code of the resulting algorithm is shown in figure 1.

Algorithm 1. The algorithm ACOPlan-F

```
 1: π_best ← ∅
 2: InitPheromone(c)
 3: for g ← 1 to N do
 4:     for m ← 1 to n_a do
 5:         π_m ← ∅
 6:         state ←initial state of the problem
 7:         for i ← 1 to L_max and state is not final do
 8:             A_i ← feasible actions on state
 9:             a_k ← ChooseAction(A_i)
10:             extend π_m with a_k
11:             update state
12:         end for
13:     end for
14:     Update(π_best)
15:     Sort(π_1, ..., π_{n_a})
16:     UpdatePheromone(π_best, π_1, ..., π_{σ-1}, ρ)
17: end for
```

5   Planning with Backward Ants

The other possibility is to use backward ants. The construction phase operates in a backward direction. Each ant begins with the subgoals set g equals to goals and at each time step it chooses an action compatible with g and such that reaches at least one current subgoal, again according to the 1 formula. After having chosen an action a, the subgoals set is updated as $Regr(a, g)$.

In this way, the created sequences are always compatible with all the subgoals. The construction phase stops when a solution is found (g is a subset of the initial state), a dead end is encountered (no action is compatible with g can be found) or the upper bound L_{max} is reached.

Pheromone Models. The same considerations seen for the pheromone models of forward ants can also be reformulated for backward ants. We thus propose five different pheromone models. The models *Action, Level–Action, Fuzzy Level–Action* and *Action–Action* are similar to those defined for forward ants. The only difference is that the level number is counted from the goal, instead of the initial state.

While the *State–Action* model cannot be defined in this framework, as subgoals are not complete states, it is possible to define a new model, called *Goal–Action* in which τ also depends on the goal which is reached by a. This model has a nice interpretation in that $\tau(a, \gamma)$ quantifies how much it is promising to reach the goal γ by executing a.

Heuristic Function. The heuristic value for an action a is computed by means of the function h_{HSP}^2, described in [8], which estimates the distance between the initial state and the subgoals $g' = Regr(a, g)$.

Besides the fact that h_{HSP}^2 is suited for a backward search, it is important to note that most of the operations needed to compute it are performed only once, at the beginning of the search process. Moreover it is worth to notice that h_{HSP}^2 can be easily adapted to planning problems in which actions can have a non uniform cost.

Plan Evaluation and Pheromone Update. The plans are evaluated in a similar way as done for forward ants, except that h_{min} is defined in terms of the sequence $\mathcal{G}_\pi =< g_0, g_1, \ldots, g_L >$ of subgoal sets generated by π.

Also the pheromone update process is exactly the same. Note that, also in this case, $h_{min}(\pi) = 0$ if and only if π is a valid plan and thus, for valid plans, $P(\pi)$ reduces to the length of π.

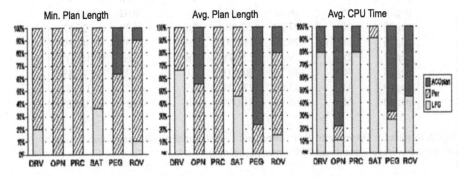

Fig. 1. Results collected for the domains *DriverLog* (DRV),*Openstacks* (OPN), *ParcPrinter* (PRC), *Satellite* (SAT), *PegSol* (PEG) and *Rover* (RVR)

	ACOplan			LPG		
Problem	Min Length	Avg. Length	Time	Min Length	Avg. Length	CPU Time
opn1	17	17	0.07	17	17	0.96
opn2	20	20	0.13	20	20	1.15
opn3	23	23	0.26	23	23	6.53
opn4	27	27	0.29	27	27.1	7.18
opn5	31	31	0.96	31	31	2.07
opn6	32	32	40.78	32	32.95	34.65
opn7	38	38	0.78	38	38	32.25
opn8	41	41	1.84	41	41.05	66.96
opn9	42	42	93.09	42.67	43.01	92.49

	ACOplan			LPG		
Problem	Min Length	Avg. Length	Time	Min Length	Avg. Length	CPU Time
prc1	8	8	0.01	8	8	0.01
prc2	15	15	0.01	15	15	0.02
prc3	22	22	0.02	22	22	0.01
prc4	29	29	0.10	29	29	0.03
prc5	36	36	0.25	36	36	0.03
prc6	43	43	0.61	43	43	0.03
prc7	50	50	5.88	50	50	0.14
prc8	57	57	12.22	57	57	0.16
prc9	64	64	23.71	64	64	0.17
prc10	71	71	40.61	71	71	0.08

	ACOplan			LPG		
Problem	Min Length	Avg. Length	Time	Min Length	Avg. Length	CPU Time
peg17	24	24	13.29	24	27.6	33.49
peg18	21	22	13.16	21	25.70	29.99
peg19	23	23.45	12.05	23	24.90	34.06
peg20	22	22.65	21.37	23	27.60	74.83
peg21	24	24.90	27.91	25	27.40	61.51
peg22	21	21.75	9.61	21	25.20	61.39
peg23	24	26.45	48.14	27	31.60	17.49
peg24	25	26.25	64.07	26	30.02	93.31
peg25	26	27.20	52.70	28	34.30	26.52
peg26	28	29.50	117.00	29	33.30	42.39
peg27	28	29.45	170.71	28	36.65	150.37
peg28	40	40.55	200.03	41	45.35	75.53
peg29	41	43.50	278.94	45	49.55	117.81
peg30	56	58.43	318.85	60	60	335.89

Fig. 2. Results for *Openstacks*, *Parcprinter* and *Pegsol* domains collecting solution lengths (minimum and average values) and CPU time (in seconds)

	ACOplan			LPG		
Problem	Min Length	Avg. Length	Time	Min Length	Avg. Length	CPU Time
drv1	7	7	0.01	7	7	0.03
drv2	19	19.15	6.49	19	19	0.53
drv3	12	12	0.03	12	12	0.08
drv4	16	16.15	9.90	16	16	0.08
drv5	18	18.2	39.81	18	18	2.28
drv6	11	11.7	21.37	11	11	0.85
drv7	13	13	0.27	13	13	0.08
drv8	22	23	39.52	22	22	9.25
drv9	22	22.1	24.03	22	22	0.54
drv10	17	17	8.22	17	17	0.54
drv11	19	19.25	14.16	19	19	23.06
drv12	37	42.1	615.58	35	35.25	237.28
drv13	26	26	39.81	26	26	4.55
drv14	30	32.4	1250.42	28	28	12.54
drv15	38	43.15	2504.10	32	32.80	499.78

	ACOplan			LPG		
Problem	Min Length	Avg. Length	Time	Min Length	Avg. Length	CPU Time
rvr6	36	36	3.33	36	36	2.32
rvr7	18	18	0.07	18	18	10.86
rvr8	26	26	0.28	26	26.1	158.48
rvr9	31	31	7.07	31	31	0.09
rvr10	35	35	9.13	35	35	133.03
rvr11	30	30.65	222.39	30	30	13.41
rvr12	19	19	0.16	19	19	0.17
rvr13	43	43.95	52.95	43	43.65	190.79
rvr14	28	28	22.07	28	28	0.94
rvr15	41	41.15	622.45	42	42.25	474.17
rvr16	41	41	65.37	41	41	116.97
rvr17	47	47	204.03	47	47	10.33
rvr18	41	41	333.82	40	41.85	434.88
rvr19	64	66.85	1445.33	66	68.7	519.62
rvr20	89	96.2	1409.46	88	92	480.21

	ACOplan			LPG		
Problem	Min Length	Avg. Length	Time	Min Length	Avg. Length	CPU Time
stl1	9	9	0.01	9	9	0.01
stl2	13	13	0.13	13	13	0.01
stl3	11	11	0.34	11	11	0.01
stl4	17	17	8.47	17	17	0.07
stl5	15	15	14.12	15	15	0.08
stl6	20	20	12.68	20	20	0.01
stl7	22	23	141.58	21	21	0.45
stl8	26	28.71	314.54	26	26	0.04
stl9	33	37.13	530.68	27	27	16.19
stl10	33	38.11	469.18	29	29	0.26
stl11	40	42.67	891.72	31	31	0.31

Fig. 3. Results for *Driverlog*, *Rovers* and *Satellite* domains collecting solution lengths (minimum and average values) and CPU time (in seconds)

The Algorithm. The algorithm, called *ACOPlan–B*, is very similar to *ACOPlan–F* and will not be reported. The only relevant difference is that the state is replaced by the subgoals set.

6 Experimental Results

We chose to compare ACOPlan with the LPG system [7]. The choice is twice motivated: LPG is a stochastic planner and it is very efficient from both a computational and a solution quality point of view. Moreover, when it runs with *-n* option, it gives solution plans with, in general, a number of actions very close to the optimum (often it can find solutions with the optimum number of actions). In this paper we present experimental results for ACOPlan–F over the domains *Driverlog, Rovers, Satellite, Openstacks, Parcprinter* and *Pegsol*. These domains have been chosen among the set of benchmarks because they offer a good variety and the corresponding results allow us interesting comments. In this test phase, the best results have been obtained with the *Action–Action* pheromone model and they are shown in the next tables.

ACOPlan–F has many parameters that have to be chosen. After systematic tests we decided to use this setting: 10 ants, 5000 iterations, $\alpha = 2$, $\beta = 5$, $\rho = 0.15$, $c = 1$, $k = 0.5$.

Since both the systems are not deterministic planners, the results collected here are the mean values obtained over 100 runs.

In Fig. 2 the results of tests over the domains from the last competition *Openstacks, Parcprinter* and *Pegsol* are shown, while in Fig. 3 we report the results of tests over the older domains *Driverlog, Rovers* and *Satellite*. To save space only the results for some problems for each domain are reported. All the tables have the same structure: in the first column problem names are listed; in the next columns the length of best solution plans, the average solution length and the average execution times respectively are reported for each planner.

As often reported in the literature, the performance of each planner depends on the domain at hand and, also in the same domain, the results could vary. In Fig.1 we summarize the results relative to the minimum solution length, the average solution length, and the average CPU time spent to extract the solution plan with minimum length respectively, for all the solved problems in the cited domains. The different colorations report the percentage of the cases in which ACOPlan is the winner, of the cases when LPG is the winner and of the cases ended in a draw (represented by *Par* in the graph).

With respect to ACOPlan–B, we have planned to perform a second track of tests, after having performed a tuning process for the parameters α, β, ρ, c, as done for forward ants.

7 Conclusions and Future Work

In this paper we have described a planning framework based on Ant Colony Optimization meta–heuristic to solve optimal sequential planning problems. The preliminary empirical tests have shown encouraging results and that this approach is a viable method for optimization in classical planning. For these reasons we are thinking to improve and extend this work in several directions.

We are planning to perform extensive sets of experiments in order to compare the different pheromone models and to contrast the difference between forward and backward ants. Another possibility could be to define other pheromone models and using different ACO techniques. The heuristic function used for forward ants should be replaced with a heuristic function which can handle action costs, following the approach proposed in the FF(h_a) system [9].

Finally we are considering to apply ACO techniques also to other types of planning. The extensions of classical planning which appears to be appealing for ACO are planning with numerical fluents and planning with preferences, being both optimization problems. These extensions are quite simple, although a major problem seems to be the choice of a suitable heuristic function, mainly for backward ants.

References

1. Baioletti, M., Milani, A., Poggioni, V., Rossi, F.: An ACO approach to planning. In: Proc of the 9th European Conference on Evolutionary Computation in Combinatorial Optimisation, EVOCOP 2009 (2009)
2. Baioletti, M., Milani, A., Poggioni, V., Rossi, F.: Ant search strategies for planning optimization. Accepted to International Conference on Planning and Scheduling, ICAPS 2009 (2009)
3. Baioletti, M., Milani, A., Poggioni, V., Rossi, F.: PlACO: Planning with Ants. In: Proc of The 22nd International FLAIRS Conference. AAAI Press, Menlo Park (2009)
4. Bullnheimer, B., Hartl, R.F., Strauss, C.: A New Rank Based Version of the Ant System - A Computational Study. Central European Journal for Operations Research and Economics 7, 25–38 (1999)
5. Dorigo, M., Gambardella, L.M.: Ant Colony System: A Cooperative Learning Approach to the Traveling Salesman Problem. IEEE Transactions on Evolutionary Computation 1(1), 53–66 (1997)
6. Dorigo, M., Stuetzle, T.: Ant Colony Optimization. MIT Press, Cambridge (2004)
7. Gerevini, A., Serina, I.: LPG: a Planner based on Local Search for Planning Graphs. In: Proceedings of the Sixth International Conference on Artificial Intelligence Planning and Scheduling (AIPS 2002). AAAI Press, Toulouse (2002)
8. Haslum, P., Bonet, B., Geffner, H.: New Admissible Heuristics for Domain-Independent Planning. In: Proc. of AAAI 2005, pp. 1163–1168 (2005)
9. Helmert, M., Do, M., Refanidis, I.: International Planning Competition IPC-2008, The Deterministic Part (2008), http://ipc.icaps-conference.org/
10. Hoffmann, J., Nebel, B.: The FF Planning System: Fast Plan Generation Through Heuristic Search. Journal of Artificial Intelligence Research 14, 253–302 (2001)
11. Nau, D., Ghallab, M., Traverso, P.: Automated Planning: Theory and Practice. Morgan Kaufmann, San Francisco (2004)

Navigation in Evolving Robots: Insight from Vertebrates
The Case of Geometric and Non-geometric Information

Michela Ponticorvo[1] and Orazio Miglino[1,2]

[1] Natural and Artificial Cognition Laboratory, University of Naples "Federico II",
Italy
[2] Laboratory of Autonomous Robotics and Artificial Life, Institute of Cognitive
Sciences and Technologies, National Research Council, Rome, Italy

Abstract. Mobile robots navigation is a broad topic, covering many different technologies and applications. It is possible to draw inspiration for robot navigation from vertebrates. Reviewing literature on vertebrates, it seems clear that they navigate by elaborating substantially two kinds of spatial information: geometric (environmental shape, distance from landmarks) and non geometric (colors, smells). In this paper we try to understand how these cues can be used by small populations of mobile robots in environments reproducing the main features of experimental settings used with vertebrates. The robots are controlled by neural networks, whose evolution determines robot navigation behaviour. We analyze how the artificial systems use these information, separately or jointly, and how is it possible to obtain mobile robots that exploit effectively geometric and non-geometric information to navigate in specific environments.

1   Introduction

Navigation is a very interesting subject for robotics study: as underlined by Franz and Mallot (2000) in their seminal review on the theme, many efforts have been dedicated to implement biological navigation behaviours in robots. These efforts have produced interesting results in reproducing insect navigation behaviour, but if we consider spatial abilities displayed by vertebrates it is evident that there is still a lot to be done. One possible source of inspiration may derive from vertebrates literature on spatial cognition.

In the present work we try to get insights from animals for robots' navigation starting from a very relevant issue in animal spatial navigation: the geometric and non-geometric information use. Behavioural and neurological evidence suggests that vertebrates have neuro-cognitive mechanisms that allow them to capture, represent and exploit information about the geometry of their environment. For example many different animals can locate a region within a larger space by using the distance between their current location and landmarks in the environment. This ability has been demonstrated in rats, birds, fish, chicks, primates and human beings. Non-geometric spatial information, such as the color

R. Serra and R. Cucchiara (Eds.): AI*IA 2009, LNAI 5883, pp. 222–231, 2009.
© Springer-Verlag Berlin Heidelberg 2009

of a landmark, smells, etc., is also a relevant source of information (for a review see Cheng and Newcombe, 2005). Neuro-anatomical studies have identified some of the neural structures involved.

In brief, studies have shown that a) rats exploit geometric information in the environment, ignoring contradictory non-geometric information (cfr. Gallistel, 1990); b) other organisms integrate geometric and non-geometric spatial information such as color (Sovrano et al., 2002 for fish *Xenotoca eiseni*) or featural cues (Chiandetti and Vallortigara, 2008 for young chicks); c) young children (under 5-6 years) fail to exploit non-geometric information in small spaces (Hermer and Spelke, 1996) but later learn to use non-geometric information consistently (Hermer-Vazquez et al., 2001). We observe a range of different behaviours: geometric primacy (in rats), integration of geometric and non-geometric information (in fish and young chicks), an acquired ability to integrate different classes of information (in children). In all cases geometric cues are of key importance. Use of geometry, unlike other sources of spatial information, is robust and is not affected by the presence of other cues (Cheng and Newcombe, 2005).

Gallistel attributes the ability to orient using geometric cues to a dedicated "brain module" in Fodor's sense of the term. In cases of conflict, this "geometric module", overrides other modules whose output is based on non-geometric information. At first sight, human behavior appears to contradict Gallistell's hypothesis - but in reality there is no contradiction. Until they are about six, children display "geometric primacy". But after that age, they can effectively integrate geometric with non-geometric information. Hermer and Spelke (1996), who have studied children's behaviour, suggest that this ability depends on language: mature use of language also appears around the age of six. According to these authors language makes it possible to link geometric and non-geometric information in a single "cognitive representation". If this argument is correct, animals such as rats, which have no superior language functions, are unable to progress beyond "geometric primacy": evolution forces them to rely on the automatic processing provided by the "geometric module". For human beings, Hermer and Spelke's argument has a certain plausibility. However, other experiments have produced data which it is hard to explain in this way. Sovrano et al. (2002) trained *Xenotoca eiseni* (a species of small fresh-water fish) to carry out Hermer and Spelke's task. The results show that fish in a rectangular tank with a colored wall orient using both geometric information (the shape of the aquarium) and non-geometric information (the color of the wall). In other words, geometric primacy does not apply to *Xenotoca eiseni*. Recent work has produced similar findings for mountain chickadees (Gray et al., 2005)that display non-geometric primacy. It seems unlikely that fish or birds use language to mediate between different forms of spatial information. Moreover Ratliff and Newcombe (2008) suggests that modularity is not the only possible explanation for human behaviour.

This debate, that is very lively in vertebrates' research community, in our opinion, can be exploited for mobile robots navigation. So, how can robots use geometric and non-geometric cues to navigate?

In what follows, we evolve mobile robots to use geometric and non-geometric cues for navigation: the variable role of geometric information in different species, according to Ratliff and Newcombe (2008) depends on the frequency with which organisms are exposed to different kinds of spatial information during their development and/or evolution. According to this hypothesis, any environment contains a certain proportion of geometric and a certain proportion of non-geometric information and organisms learn to respond to these information in varying degrees, depending on their salience. This means that if we manipulate the proportions of geometric and non-geometric information in the environment, we would observe different population of mobile robots displaying different navigation abilities. In environments providing mainly geometric information, we should obtain robots which are specialized in using this information (geometric primacy); if geometric information is less important, we should obtain robots that first learn to respond to the information they meet most frequently and later to other, rarer stimuli. Environments that contain both kinds of information in about the same proportion should facilitate the ability to respond simultaneously to both. In the study reported here, we use an Evolutionary Robotics (Nolfi and Floreano, 2000) approach to evolve populations of small mobile simulated robots with the ability to orient in a "rectangular open field box" - an environment already used to study the spatial behaviour of many different biological organisms.

In our own work, we accurately reproduced the same experimental conditions and observation techniques reported by Sovrano et al., in their study of redtail splitfins. By manipulating the frequency with which the robots come into contact with different classes of spatial information, we tried to obtain robots with different navigation abilities, that could exploit effectively the sources of information in the environment.

1.1   Animal Orientation in an Open-Field Box

To get insights from vertebrates navigation, let us start from the observed behaviour of fish, birds, rats and human beings on an well-known task often used in the study of spatial behaviour, the open-field box experimental setting. In the human version of the experiment a subject is placed in a rectangular room with white walls (Fig.1).

In one corner, the experimenter places a very inviting object (a reward). In the training phase the subject is allowed to see the reward, which the experimenter then hides. In the testing phase, the reward is buried. After a disorienting procedure the experimenter asks him/her to find the reward. If all the subject sees is the shape of the room, he/she can identify the reward zone and the rotationally-equivalent corner, but has no way of distinguishing between them. He/she will thus search for the reward in the wrong as well as in the correct corner. If one long wall is colored, for example it is blue, and if he/she uses this information, he/she will always be able to find the reward corner. This variation is known as the blue-wall task. Performing the task correctly requires subjects to integrate shape information with other non-geometric information such as the color of the wall. The blue wall task has been used with a number

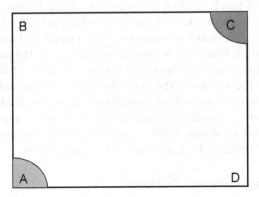

Fig. 1. Experimental setting. The organism is placed in a rectangular arena in which walls are painted white. Using no information except the shape of the box, the organisms has to locate a reward area (for example in corner (A)). From the organism's point of view the target area has the same geometric features (the same ratio between the long and the short side and the same sense (left vs. right) as the area in the opposite corner (C)). But if one of the walls is painted in a different color from the others this ambiguity disappears. In practice, the colored wall acts as a landmark.

of different species including rats, chicks, pigeons, fish *Xenotoca eiseni*, rhesus monkeys, and humans. Rats (Cheng, 1986) commit rotational errors even in the presence of olfactory, visual or tactile cues that should have enabled them to disambiguate the problem. Very young children in a small space (Hermer and Spelke, 1996) behaved in the same way as rats, not considering non-geometric information. By contrast, children over 6 years old were able to correctly solve the task. Moving from land to water, Sovrano and colleagues (2002) showed that *Xenotoca eiseni* also integrate geometric and non-geometric information. Gray et al. (2005) used the same task with mountain chickadees. These wild-caught birds show non-geometric primacy. When these birds are tested with conflicting information they follow the non-geometric cues.

2   Materials and Method

2.1   The Robot

In our experiments we used a software simulation of a Khepera robot, a round, 30mm tall micro-robot with a diameter of 55mm by the K-team group. It has eight infrared sensors that can detect obstacles within a range of 3cm. A round video camera (with a visual field of 360 degrees) is located on the upper part of the robot. The color camera has a pixel matrix of 500x582. To facilitate computation, we based all our calculations on the midline traversing the matrix. We also reduced the robot's visual field to 270 degrees, similar to the visual field of fish (private communication from V.A. Sovrano) and filtered the activity levels for robot pixels, so that all pixels tending towards white were coded with a value of '0' and all pixels tending towards blue or black received a code of '1'. Khepera moves using two wheels located on either side of its body, each controlled by a motor. The robot is equipped with a small on-board computer and a battery, facilitating relatively autonomous movement. We simulate evolution using software which captures the physical characteristics of the robot and the environment and then transfer the control system on the real robot on-board processor, according to a consolidated practice (Miglino et al., 1995). In our own experiment, we used a modified version of the "Evorobot" simulator developed by Nolfi (Nolfi and Floreano, 2000) - an environment created specifically for experiments with simulated populations of Khepera robots.

2.2   The Artificial Neural Network

The robot's control system consists of an Artificial Neural Network, a totally connected two-layer perceptron.

From a functional viewpoint, we use different kinds of input and output units. The sensor layer (18 units in all) consists of two bias units, 8 close-range proximity detectors, and 8 long range detectors of landmarks. The output layer consists of two motor units and one localization unit. The speed of each motor is proportional to the activation of one of the output units. The third output unit (the localization unit) temporarily halts the robot whenever its level of activation is higher than 0.5. We can thus interpret the activation of the localization unit as signalling the robot's perception that it is in a specific location.

2.3   The Experimental Setting

The experimental setting for our study was similar to the setting in which Sovrano et al.(2002) conducted their work. Each experiment took place in an arena with a "reward area" in one corner. The first environment we used is the complete "geometric + non-geometric arena" in which the long side opposite to the reward corner as well as the corners were colored blue (see Fig.1): in the arena where both geometric and non-geometric cues are available, if an organism uses no information but the shape of the box, the agent will commit a rotational error (choosing A or C). If it merges geometry and color information it will choose the correct corner A. Removing the blue wall from the complete arena, we obtain the "geometric arena" which is rectangular in shape with four white walls and angular landmarks. Instead removing the geometric information from the complete arena that is the rectangular shape of the arena and the angular landmarks we obtain a "non-geometric arena", square- shaped and with a blue wall. Both rectangular environments were 56.8 x 25.6 cm , while the square one was 56.8 x 56.8 cm. In all environments, a circular reward sector was located in the bottom-left corner. The reward sector had a radius of 8 cm. In the "geometric arena" the only available information is geometry, according to which we expect agent to choose the correct corner A as well as the rotationally equivalent corner C. In the "non-geometric arena" the blue wall disambiguates the task, so we expect the agent to choose the correct corner A.

2.4   The Genetic Algorithm

The robots were evolved using a genetic algorithm. At the beginning of the evolutionary process, we created the first generation of 100 robots. Each robot had a neural control system with random connection weights. We then tested the ability of each robot to localize the reward area in 100 different trials. At the beginning of a trial, the robot was positioned at the centre, facing in a random direction and was allowed to move around for 1500 computation cycles. Every time the robot reached and "identified" the target area (activation of the localization unit greater than 0.5) it stopped for 5 computation cycles (500 ms) and received one "reward point". The robot was assigned a final score consisting of the sum of reward points received during all the trials. After all robots were tested, the 80 robots with the lowest "reward score" were eliminated (truncation selection). During cloning, 35 per cent of connection weights were incremented by random values uniformly distributed in the interval [-1, +1]. These new neural control systems constituted a second robot generation. The testing/selection/reproduction cycle was iterated for 100 generations. To investigate the effects of differing frequencies of exposure to different classes of spatial information, we performed 11 experiments with different rates of exposure to "geometric arenas" and "colored arenas", keeping fixed the exposition at 10 per cent to the "blue wall task arena", that was our testing environment. This exposition was necessary to obtain robots that could behave correctly in different conditions. In our first experiment, all of the trials were conducted in the "colored wall arena". In each subsequent experiment, we increased the percentage of trials in the "geometric arena" by 10 per cent. Each experiment was replicated twenty times using different initial weights in each test. This procedure allowed us to test the statistical robustness of behavioural and evolutionary indicators.

3   Results

3.1   Comparison of the Two Tasks

There are differences in the computational complexity of the two tasks and in the time taken to evolve efficient solutions (here we consider the evolution with exposure to only one arena). Artificial Neural Networks associate input patterns (stimuli) with output patterns (behavioural responses). The greater the number of different response patterns the greater is the computational burden on the network. We can thus use the number of output patterns as a measure of the computational capability of the network. A study of the best individuals from each replication in each of the two arenas showed that robots which evolved in the geometric arena generated an average of 75.35 (SD = 31.5) different output patterns when tested in the same arena; robots which had evolved in the non-geometric arena produced an average of 53.50 (SD = 23.95) patterns. A comparison of means using the Student t test, showed that the difference was statistically significant $t(19) = 3.99$, $p = .0012$. In brief, the task in the non-geometric arena (in which disambiguating information is present) is computationally easier than the task in the geometric arena (where available information for the robot to work on is less complete). This hypothesis is confirmed by data on the speed of evolution. In the non-geometric arena, robots took an average of 73.20 generations (SD = 3.43) to achieve their maximum reward score; in the white arena they took longer: an average of 87.45 generations (SD = 2.77). The Student t test showed that the difference was statistically significant $t(19) = 2.88$, $p = .008$.

3.2 Evolution in Both Arenas

The blue wall task is not the simple sum of navigation in geometric arena plus navigating in the non-geometric one. It seems clear that merging these sources of information requires a careful balancing in their relative encoding. Therefore in this section, we present the results of 9 experiments in which individuals were exposed to both environments. As in our previous studies we replicated each experiment twenty times using different initial weights in each replication. In the first experiment, 10 % of trials were conducted in the geometric arena and 90 % in the non-geometric arena; in each of the following experiments we increased the proportion of trials in the geometric area by 10 % (to 20, 30 etc.). The test was composed by 100 trials in the geometric arena and 100 trials in the geometric + non-geometric arena.

Environmental Exposition and Performance. Figure 2 shows the percentage of robots with the ability to behave correctly in both arenas. We consider that robots behaved correctly if they do choose a corner in at least half of the trials and identify the right corner (or corners) at least n (number of correct corners/ total number of corners) + 1 times. We call these organisms "generalists". Results were obtained by monitoring the behaviour of robots which achieved the highest reward scores along generations in each of the 20 replications of each experiment. Of these, we identified 44.33 % of robots as "generalists", meaning that they behaved correctly in both arenas. As we can see from the chart, all experiments produce some individuals able to behave appropriately in both environments. However, the robots that achieve the highest scores are those which evolve for 70 % of the time in the white arena. In other words, to produce a good quantity of generalists, the level of exposure to the geometric arena must be slightly higher that the level of exposure to the non-geometric ones.

Environmental Exposition and Evolutionary Pathways. To gain insight into the dynamics underlying the emergence of "generalists", we identified the generation producing the "generalist" with the highest reward score. Starting from this peak point we studied the "developmental" pathways these robots had followed before acquiring the ability to perform correctly in both environments. To this end, we identified the generation at the mid-point between the first generation and the peak generation. In both cases, we examined all robots produced in this generation and compared the reward scores for each of the two environments with the scores recorded in the peak generation. If the difference between the two scores was statistically significant (the probability associated with t test

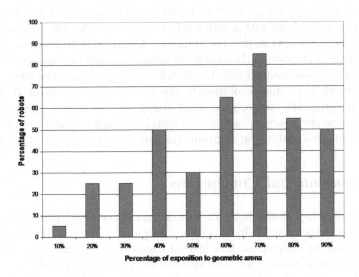

Fig. 2. Percentage of robots displaying optimal behaviour in both arenas

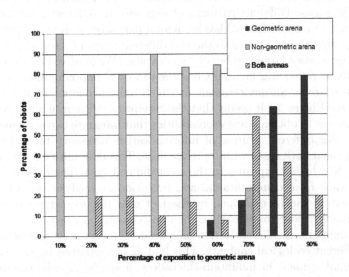

Fig. 3. Percentage of "specialist" and "generalist" individuals in each of the two arenas. The grey bar represents the percentage of robots that perform better in the non-geometric arena; the dark bar represents the percentage of robots that perform better in the geometric arena; the hatched bar represents the proportion of robots with similar performance in both arenas.

was below 1 %), the robot was considered to be a "specialist" in the environment in which it obtained the highest score. If not, the individual was considered to be a "generalist". Figure 3 illustrates the results of this analysis. For each series of experiments, we chart the number of geometric arena "specialists", the number of non-geometric arena "specialists" and the number of generalists. Unsurprisingly, individuals with a high level of exposure to the non-geometric arena tended to be specialists in this environment. As exposure to the non-geometric arena falls and exposure to the geometric arena increases the composition of the population changes dramatically. Geometric arena specialists first emerge when we conduct 60 % of their evolution in the geometric arena. With 70 % exposure to the geometric arena, generalists emerge as the largest group. With 80 % exposure to the geometric arena, this trend is completely reversed: geometric arena specialists are the largest group and non-geometric specialists vanish.

4   Discussion and Conclusions

Navigation is a key ability both for natural and artificial systems. In this paper we have tried to get insights from vertebrates spatial cognition: this approach has the advantage to understand how spatial cognition is organized in vertebrates to reproduce it in efficient artificial systems such as mobile robots.

How does navigation emerges depending on different environmental stimuli? The ability to behave correctly in both arenas emerged from very different evolutionary histories. Different patterns of exposure to different kinds of spatial information produced different developmental pathways. It is worth noting that simply exposing the organisms to the two different types of information was not enough to generate a large number of "generalists". We observed that the highest number of generalists was created when the robots were predominantly exposed to the white arena, an environment in which only geometric information was available (see Figure 3). It seems that the receipt for efficient navigating robots is the following: to obtain navigation abilities that integrate two separate spatial abilities (using different sources of information) is necessary to evolve robots presenting them both tasks with a precise dose. This dose must be balanced but must include a higher exposure to the most difficult task. The indication about difficulty come from computational and evolutionary analysis. In other words this observation can be explained by the fact that the geometric arena task is computationally more complex than the non-geometric arena task.

In our simulations different forms of spatial cognition arise through adaptation to different ecological contexts. This finding is supported by evidence from recent animal studies: in mountain-chickadees, a species of wild-caught birds, featural cues are more important than geometry. The authors hypothesize that this is because the birds "have little experience with salient right-angle cues, thus leading to reliance on featural over geometric information" (Gray et al., 2005, pag. 4). Also in fish (Brown et al., 2007) seems that rearing condition can affect relative dominance of cues in navigation. In the work reported here, this evidence provides the starting point for implementing navigation abilities in robots with varying level of exposure to different types of spatial information. Moreover the experimental set-ups they use become the test-bed for evolved robots. The ideal navigating robot should orient and move autonomously in real world, an ability that natural evolution has given to vertebrates: they are therefore a precious source of inspiration for mobile robot navigation: implementing a biologically plausible orientation system on a mobile robot platform may widen both robot navigation and biological orientation research.

In next studies these results must be complemented with an epigenetic approach including phylogenesis as well as ontogenesis. Moreover it is interesting to see how navigation is implemented in swarms of robotics.

References

1. Brown, A.A., Spetch, M.L., Hurd, P.L.: Growing in circles: rearing environment alter spatial navigation in fish. Psychological-Science 18, 569–573 (2007)
2. Cheng, K.: A purely geometric module in the rat's spatial representation. Cognition 23, 149–178 (1986)
3. Cheng, K., Newcombe, N.S.: Is there a geometric module for spatial orientation? Squaring theory and evidence. Psychonomic Bulletin and Review 12, 1–23 (2005)
4. Chiandetti, C., Vallortigara, G.: Is there an innate geometric module? Effects of experience with angular geometric cues on spatial re-orientation based on the shape of the environment. Animal Cognition 11, 139–146 (2008)
5. Franz, M.O., Mallot, H.A.: Biomimetic robot navigation. Robotics and Autonomous Systems 30, 133–153 (2000)
6. Gallistel, C.R.: The organization of learning. MIT Press, Cambridge (1990)
7. Gray, E.R., Bloomfield, L.L., Ferrey, A., Spetch, M.L., Sturdy, C.B.: Spatial encoding in mountain chickadees: Features overshadow geometry. Biology Letters 1, 314–317 (2005)
8. Hermer, L., Spelke, E.S.: Modularity and development: the case of spatial reorientation. Cognition 61, 195–232 (1996)
9. Hermer-Vazquez, L., Moffet, A., Munkholm, P.: Language, space, and the development of cognitive flexibility in humans: the case of two spatial memory tasks. Cognition 79, 263–299 (2001)
10. Miglino, O., Lund, H.H., Nolfi, S.: Evolving mobile robots in simulated and real environments. Artificial Life 4(2), 417–434 (1995)
11. Nolfi, S., Floreano, D.: Evolutionary Robotics. In: The Biology, Intelligence, and Technology of Self-Organizing Machines. MIT Press/Bradford Books MIT Press (2000)
12. Ratliff, K.R., Newcombe, N.S.: Is language necessary for human spatial reorientation? Reconsidering evidence from dual task paradigms. Cognitive Psychology 56, 142–163 (2008)
13. Sovrano, V.A., Bisazza, A., Vallortigara, G.: Modularity and spatial reorientation in a simple mind: Encoding of geometric and non-geometric properties of spatial environment by fish. Cognition 85, 51–59 (2002)

Social Tagging for Personalized Web Search

Claudio Biancalana[1,2]

[1] Department of Computer Science and Automation
Artificial Intelligence Laboratory
Roma Tre University
Via della Vasca Navale, 79, 00146 Rome, Italy
[2] Technical Scientific Committee
LAit S.p.A.
Via Adelaide Bono Cairoli, 68, 00145 Rome, Italy
claudio.biancalana@dia.uniroma3.it

Abstract. Social networks and collaborative tagging systems are rapidly gaining popularity as primary means for sorting and sharing data: users tag their bookmarks in order to simplify information dissemination and later lookup. Social Bookmarking services are useful in two important respects: first, they can allow an individual to remember the visited URLs, and second, tags can be made by the community to guide users towards valuable content. In this paper we focus on the latter use: we present a novel approach for personalized web search using query expansion. We further extend the family of well-known co-occurence matrix technique models by using a new way of exploring social tagging services. Our approach shows its strength particularly in the case of disambiguation of word contexts. We show how to design and implement such a system in practice and conduct several experiments. To the best of our knowledge this is the first study centered on using social bookmarking and tagging techniques for personalization of web search and its evaluation in a real-world scenario.

1   Introduction

The considerable quantitative increase in the amount of documents on the World Wide Web has led to a scenario in which disorganization gained the upper hand, due to the many different languages composing the documents, typically drafted by a huge number of authors on the Web. This fact leads to the need of supporting the user more efficiently in retrieving information on the web. Users easily find problems in retrieving information by means of a simple Boolean search to check the presence of the searched-for term in the web texts. Indeed, some web texts, consisting of terms that are often synonyms, or related to similar topics only, do not allow to conduct a proper search, and only take into consideration a few terms, which could be input by a user who is likely to have no or little experience in web searches. The Query Expansion (QE) technique fits in this disordered scenario to support the user in his/her search and allow them to widen the search domain, to include sets of words that are somehow linked to the frequency of the term the user specified in his/her query [1]. These may be simple synonyms or terms that are apparently not connected to syntactic meaning, but nevertheless linked to a context that is similar or identical to the one expressed by the original search provided

R. Serra and R. Cucchiara (Eds.): AI*IA 2009, LNAI 5883, pp. 232–242, 2009.

by the user [3]. Such information may be obtained in several ways, the main difference being the source used to obtain further information, which can be retrieved through the preferences explicitly indicated by the user, through the user's interaction with the system [6], through the incremental collection of information that links the query terms to document terms [5](for instance the search session logs) or by means of a simple syntactic analysis of the phrase forms that compose the documents [7],[10]. The approach presented in this paper has its origin from the calculation of co-occurrence matrices. We put forward an extension of the user model which makes use of information collected through Social Bookmarking services such as *del.icio.us*[1] and *StumbleUpon*[2]. The above websites allow us to store, organize, share and search bookmarks associated to Web pages, by the input of additional data (such as tags or short summaries), freely available to the whole community of users. The use of information with a social content, i.e., data based on the active participation of all involved users, is the subject of recent studies which underline both its positive and negative sides. Data reliability is sometimes spoiled by the introduction of erroneous or personal information, or by spamming phenomena [12]. Nevertheless, most of annotations turn out to be often consistent with the categories of the associated bookmarks [8]. The presence of a vast number of users, who agree in assigning a tag to a resource, has been shown to be a very reliable criterion [9]. Recently, the literature has offered several examples of interaction with the data retrieved by Social Bookmarking services. In particular we report an example [2] that has represented one of the most relevant input for the conception of our system: the Social Similarity Ranking (SSR), used to estimate the similarities between a search query and a Web page candidate for result; and the Social Page Ranking (SPR), which determines the rank of each page, based exclusively on the amount of input data associated to the page itself.

In our case, the two-dimensional matrix of co-occurrence is extended by inserting a third dimension containing metadata corresponding to tags assigned to the available resources on the Web, as inserted by the users of the most noticeable Social Bookmarking service, *del.icio.us*. The present paper is organized as follows: in Section 3 we illustrate the developed system. In Section 2 we present a theoretical debate on the reliability and reusability of the data collected through Social Bookmarking service. In Section 4 we describe the algorithms that distinguish the main innovation that we introduce. In Section 5 we present the experimentations carried out on the developed system, and the corresponding results; finally, we discuss our conclusions in section 6.

2 Data Reliability

The thriving of systems based on the active sharing of data is obviously an interesting topic as far as Information Retrieval and Natural Language Processing are concerned. The key elements the experts are focusing on are the following:

- is the information collected by users reliable?
- what are the best methods to exploit such a huge and heterogeneous amount of information?

---

[1] http://del.icio.us/
[2] http://www.stumbleupon.com/

Y. Yanbe et al. [12] consider the possibility of using social annotations as a new assessment indicator to appraise the popularity of a Web page, ousting the conventional ranking modality based on inbound and outbound links. SBSearch, the prototype system specifically developed for this occasion, tries to blend both approaches, revealing the main issues associated with Social Bookmarking:

- non negligible percentages of the entered bookmarks referring to other spheres (in particular the emotional one; for example the appreciation of the collected resource), thus undermining the actual credibility of the available data;
- the extreme dynamism of the Web (resources no longer available, domains changing sphere of interest) is a factor to be assessed duly before staking on a Social Bookmarking service, because the information turnover usually takes plenty of time;
- unfortunately, spamming is very frequent, and these phenomena must be observed constantly.

3   Social Query Expansion

To explore the potential of personalized query expansion by a social approach, we have developed an innovative search engine that can record and interpret users' behavior, in order to provide personalized search results, according to their interests. The whole procedure of personalization is completely transparent to the user, because it occurs in an implicit way based on of his/her choices, related to the terms in the submitted queries, and to the corresponding visited pages. The generation of a user profile occurs through the creation of a model, updated dynamically with the information derived from the searches (visited pages and corresponding search queries). The input queries are analyzed according to collected data, and if the comparison has a positive outcome (i.e., if the queries reflect the interests that the user has already shown in previous searches), then the system makes different Query Expansions, each one to a different semantic field, related to the terms inputed by the user, before carrying out the search. The final result is a page in which results are grouped in different blocks, each of them categorized through keywords, in such a way to facilitate the user in the choice of the result that is most coherent with his/her interests.

The system takes advantages of some resources freely available on the Web: *yahoo.com*, one of the most popular search engines. Results obtained in each search session are then shown to the user in such a way to underline the different categories of each group of results. The search of the tags associated to the pages visited by the user, is carried out by analyzing the information provided by two of the main sites of Social Bookmarking. In this case, the data collection occurs directly by parsing the html pages containing the necessary information.

In order to model the user visits, the system employs matrices based on co-occurrence at the page level: terms highly co-occurring with the issued keywords have been shown to increase precision when appended to the query. Many statistical measures have been developed to the best *term relationship* levels, either analyzing entire documents, lexical affinity relationship (i.e., pairs of closely related words contain exactly one of the initial query terms), etc.

The generic term t_x is in relation with all other n terms t_i (with $i = 1, \ldots, n$), according to a coefficient c_{xi} representing the co-occurrence measure between the two terms.

In a classical way, we can construct the correlation matrix using the HAL approach [4]: the co-occurence matrix is generated as follows: once a term is given, its co-occurrence is calculated with N terms to its right (ot its left). In particular, given a term t and considered the window of N terms to its right $f_t = \{w_1, \ldots, w_n\}$ we get $co - oc(t, w_i) = \frac{w_i}{i}, i = 1 \ldots, N$. A pair (a, b) is equal to pair (b, a): the co-occurence matrix is symmetrical. For each one of the training documents a co-occurence matrix is generated, whose lines are then normalized (on the maximum value). The matrices of the single document are then summed up, generating one single co-occurence matrix representing the entire corpus. The limit of this structure consists in the latent ambiguity of collected information: in presence of polysemy of the terms adopted by the user, the result of the query expansion risks to misunderstand the interests, leading to erroneous results. In order to overcome the above problem, in our system the classical model of co-occurrence matrix has been extended. The user model consists of a three-dimensional correlation matrix (see an example in figure 1). Each term of the matrix is linked to an intermediate level, containing the relative belonging classes, each accompanied by a relevance index. In this way, each term is *contextualized* before being linked to all the other terms present in the matrix, and led to well determined semantic categories, identified by tags.

In the example, the term *amazon*, if referred to the semantic class *nature*, shows high values of co-occurrence with the term *river*. Viceversa, if it is referred to the category *shopping*, is in strong relation with terms such a *books* and *buy*.

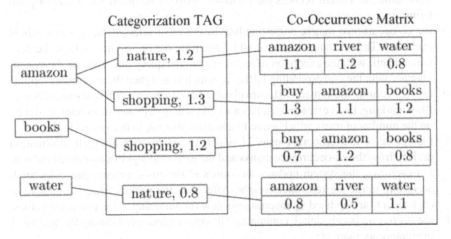

Fig. 1. An Example of Three-Dimensional correlation Matrix used for User Model

4   Algorithms

The system is based on two main algorithms: the first refers to the building and updating of the user model (discussed in subsection 4.1), the second to the query expansion (discussed in subsection 4.2). With reference to the pseudocode we notice that the co-occurrence matrix is represented by a map of maps for encoding knowledge and connecting this encoded knowledge to relevant information resources. Map of maps are organized around topics, which represent subjects of discourse; associations, representing relationships between the subjects; and occurrences, which connect the subjects to pertinent information resources.

4.1 Building and Updating the User Model

The building and updating of the user model are based on the pages chosen by the user during the searches. Starting with an empty model, every time the user clicks on a result after typing a search query, the system records the visited URL, together with the query originally used for the search.

Our system performs the analysis of the visited URLs in incremental way, according to the following algorithm:

1. a temporary map M is initialized, where it is possible to record the extracted data, before updating the pre-existent model (empty at first execution). The map keys are the encountered tags, the values are the relative two-dimensional co-occurrence matrices;
2. for each visited URL one obtains the corresponding html page, from which the textual information is extracted through a parser;
3. the list of terms is filtered in order to eliminate the stopwords, i.e., all those terms that are very frequent but irrelevant to the creation of the user model;
4. the list of terms undergoes a stemming by means of the Porter's algorithm. At the same time the system records the relations between stemmed terms and original terms;
5. the co-occurrence matrix corresponding to the most relevant k_{term} keywords is evaluated. The relevance is measured by counting the occurrences within the document itself, with the exception of terms used in the query (recorded by the system together with the corresponding URL), to which is assigned the maximum weight;
6. tags concerning the visited URLs are obtained, by accessing different sites of Social Bookmarking. Each extracted tag has a weight which depends on its relevance (i.e., on the number of users which agree to associate that tag to the visited URL);
7. the update of the temporary map M is performed, by exploiting all information derived from the co-occurrence matrix and the extracted tags in a combined fashion. For each tag_i the system updates the values of the co-occurrence just calculated, according to the tag relevance weight. After that, the vectors M_{tag_i,t_i}, relative to each term t_i are updated by inserting the new (or summing to the previous) values;
8. the set $terms$ is calculated, containing all terms encountered during the update of the temporary map M;
9. from the persistence layer one obtains a subset UM_{terms} of the user model in form of a three-dimensional matrix of co-occurrence, corresponding only to the terms contained in $terms$;
10. the matrix UM_{terms} is updated with the values of M. For each t_i belonging to $terms$, the set of keys ($tags$) is extracted from M, which points to values corresponding to t_i. For each tag_i belonging to $tags$, the vector M_{tag_i,t_i} is added to the

pre-existent vector UM_{t_i,tag_i}, updating the values for the terms already present, and inserting new values for the terms never encountered.

4.2 Query Expansion

Query expansion is performed starting from the original terms as inputted into the search engine, accessing the information collected in the user model. The result is a set of expanded queries, each of them associated to one or more tags. In such a way it is possible to present to the user different subgroups of results divided in categories. Exploiting the possibilities of submitting queries containing boolean logic of low level offered by *yahoo.com*, every expansion assumes the following form:

$$(t_{11} \; OR \; t_{12} \; OR \ldots OR \; t_{1x}) \; AND \; (t_{21} \; OR \ldots OR \; t_{2x})$$
$$AND \ldots AND \; (t_{y1} \; OR \ldots OR \; t_{yx})$$

where t_{yx} represents the generic term x corresponding to the stemmed root y, and the different terms coming from the same root undergo OR operation amongst them, because it is necessary that the result contains at least one of them. See examples in table 1.

Table 1. Example of multiple expansions

original query	categorization tags	expansions
amazon	**internet:**	buy **AND** (books **OR** book) **AND** amazon
amazon	**e-commerce, shopping:**	(rivers **OR** river) **AND** amazon

The algorithm of multiple expansion is the following:

1. let us suppose that the query Q is given, which is made of n terms q_i (with $i = 1, \ldots, n$). For each of them the system evaluates the corresponding stemmed term q_i', obtaining as a new result the new query Q';
2. for each term belonging to Q', from the three-dimensional co-occurrence matrix one extracts the corresponding two-dimensional vector q_i. Each of the extracted vectors may be thought as a map, whose keys are the tags associated to the terms q_i' (which possess a relevance factor) and the values of which are themselves *co-occurrence vectors* between q_i' and all the other encountered terms;
3. for each encountered tag the relevance factor is recalculated, adding up the single values of each occurrence of the same tag in all two-dimensional vectors. In this way, one produces a vector T in which tags are sorted according to the new relevance factor;
4. amongst all tags contained in T, only the higher k_{tag} are selected, on which the multiple expansion is performed;
5. for each selected tag t_i, one evaluates the vector sum_{t_i}, containing the sum of the co-occurrence values of the three-dimensional matrix, corresponding to all terms q_i' of the query Q';
6. for each vector sum_{t_i}, the most relevant terms k_{qe} (corrisponding to higer values) are selected. Combining the extracted terms with which if the query Q, a new query EQ' (made up of stemmed terms) is initialized;

7. for each expanded query EQ', the corresponding query EQ is calculated, through the substitution of stemmed terms with all the possible original terms recorded by the system, exploiting the boolean logic according to the scheme previously shown;
8. the query EQ obtained above, and the original tag t_i is inputed into the map M_{EQ}, in which the keys are expanded query and the values are sets of tags. If M_{EQ} already contains an expanded query identical to the input one, the tag t_i is added to the corresponding set of tags.

5   Tuning and Experimental Results

In this section we present the experimental results. The tuning phase consists of three main stages: *training, expansion test, tuning*.

During the first stage (*training*) the users (a pool of 12 independent referees) simulated an actual search session, centred on two topics which are chosen beforehand. The system records the various correspondences between the search query and the selected results. The users are asked to submit to the system one last query relative to at least one the chosen topics. In the second stage (*expansion test*) the system presents two lists of adjacent results, one obtained without modifying the query, and the other through Query Expansion, users evaluate which one corresponds better to their interests. Finally, in the third stage (*categorization test*) results obtained with the Query Expansion - which include categories - are shown, and the users are asked a correctness of the results by a questionnaire.

5.1   Settings

After data collection (*training* stage), we started by formulating a vast set of trial results, each time varying some parameters of the system. Once the tuning phase was over, we have independently chosen the set of most satisfying parameters, by submitting the referees' results. The main values relative to the terms candidate to the expansion are the following three:

- MAX EXPANSION TERMS
  is the maximum number of terms to be added to the original query, chosen to be equal to 3.
- MIN EXPANSION TERMS RELATIVE VALUE
  indicates the percentage threshold for selection of relevant terms after the first one, and it has a value of 0.9. In other words, if for instance the first candidate term has a relevance $r_1 = 2$, the system will judge as acceptable only the those terms with relevance of at least $r_2 = 0.9 \cdot 2 = 1.8$.
- LIMIT EXPANSION TERMS
  this parameter represents a constraint, depending on the number of term of the original query. Based on it, a query can be expanded with a number of terms equivalent, at most, to the number of original terms.

On the other hand, values relative to the tags are:

- MAX EXPANSION TAGS
  is the maximum number of tags selected for the multiple expansion, and has a value of 5.

– MIN EXPANSION TAGS RELATIVE VALUE
as in the case of the corresponding parameter relative to the terms candidate to the expansion, it is a percentage threshold for the selection of tags relevant to the expansion, fixed to a value of 0.7.

The comparison between different algorithms is made by using comparative performance values obtained from the algorithm under examination, on one single topic or the entire benchmark. Such performances are expressed in F_1-measures, so as to summarize, in one single measure, precision (π) and recall (ρ) values. As for the performance measures, we have precision, recall and F_1-measure (F_1):

$$\pi(t) = \frac{n_t}{50} \qquad \rho(t) = \frac{n_t}{N_t} \qquad F_1 = \frac{2 \times \pi \times \rho}{\pi + \rho}$$

where n_t stands for the number of returned links belonging to topic t, only the first 50 pages are taken in consideration for our tests, and N_t the overall number of test links belonging to topic t present in the index.

Table 2. The employed benchmark: statistics

Topic	Test links	Training links	Information Needs
Sports/Cycling/Human_Powered_Vehicles	15	5	yes
Computers/Home_Automation/Products_and_Manufacturers	27	7	yes
Business/Mining_and_Drilling/Consulting	74	18	yes
Games/Roleplaying/Developers_and_Publishers	52	14	yes
Business/Agriculture_and_Forestry/Fencing	100	27	yes
Shopping/Crafts/Paper	35	7	no
Arts/Performing_Arts/Magic	25	6	no
Science/Publications/Magazines_and_E-zines	26	7	no
Science/Social_Sciences/Linguistics	13	5	no
Recreation/Guns/Reloading	15	5	no
Tot.	382	101	

Table 3. Comparative F_1-Measurement

Topic	no QE	RF	*Our System*
Computers/Home_Automation/Products_and_Manufacturers	0.05	0.08	0.16
Sports/Cycling/Human_Powered_Vehicles	0.09	0.13	0.09
Games/Roleplaying/Developers_and_Publishers	0.10	0.18	0.18
Business/Mining_and_Drilling/Consulting	0.19	0.14	0.19
Business/Agriculture_and_Forestry/Fencing	0.05	0.14	0.57
Average	F_1	F_1	F_1
	0.10	0.13	**0.24**

5.2   The Employed Benchmark: The Open Directory Project

The Open Directory Project (ODP), also known as DMOZ[3], is a multilanguage directory of links belonging to the World Wide Web, namely a system to collect and classify links. The Open Directory Project has a hierarchic structure: the links are grouped into categories, also known as topics, and subcategories. It is therefore possible to identify a level-based organization within the hierarchy. An example of topics is *Top/Business/Forestry_and_Agriculture/Fencing*; excluding the level corresponding to Top, common to all topics, we have:

- *Level I: Business;*
- *Level II: Forestry and Agriculture;*
- *Level III: Fencing.*

Given the large quantity of links contained in ODP, we decided to consider only Level III links.

The pages corresponding to such links were downloaded from the World Wide Web, by using a parser; the textual information was taken from it, and then it was indexed by means of the Lucene indexing system[4]. Ten topics were then chosen from the Level III topics, five of which corresponding to the user's information needs, and five whose function was exclusively to generate noise in the creation of the user model. The links of each topics were then subdivided in a training set, corresponding to 25% of the links, and set of tests, corresponding to 75% of the links (see table 2).

5.3   Experimentation Methods

Once the user model is generated, it is possible to carry out real tests as follows. A query is built for each topic belonging to the user's information needs. The terms of the query are simply the terms that form the topic name. This query is then expanded according to the user model, and used to search for web pages within the created index, starting from all third-level links. The pages belonging to the training set of the considered topic are removed from the returned pages; only the first fifty are taken into consideration, which include the number of pages belonging to the topic under consideration. The index obtained with Lucene, starting from the third-level link of ODP, consists of 131,394 links belonging to 5,888 topics.

Personalization in this paper relies on the strength of the user community as it requires that search result documents have been tagged by users. For documents without bookmarks or tags, our personalization approach is not possible in practice because metadata about them is missing. An important task is therefore to analyze the expected availability of metadata for search result documents in real-world. For this reason it is not possible a direct comparison to other similar approach in the state of the art on the same benchmark.

Table 3 shows the results obtained by a system based on a traditional content-based user-modeling approach, where documents are represented in the Vector Space Model

---

[3] http://dmoz.org

[4] http://lucene.apache.org

and without Query Expansion (QE), in comparative terms. This system particularly focuses on the update of the user model by means of Relevance Feedback (RF) techniques [11], applied to the training pages content: for each category, the first ten keywords are taken from the corresponding training pages.

The keywords are obtained in terms of $tf \times idf$, and then used to expand the query. In our experimentation, our system obtained the best results, in terms of performance, on the reference benchmark.

6 Conclusions

In this research we developed an Information Retrieval system, based on Query Expansion and Personalization, that may help the user search for information on the Web, according to his/her information needs. The state of the art analysis of the main mechanisms of *personalized search* (in particular the query expansion) and the development of social bookmarking, were the starting points for the next realization of the system. As for personalization, the comparison amongst different methodologies in the literature allowed us a critical review, and at the same time it contributed to build a strong theoretical foundation. The study of the thematics linked to *Web 2.0*, and in particular the collaborative categorization, represented for us the initial inspiration for introducing the more original and innovative aspects which are distinctive of our system. Our choice to introduce the use of information derived from services of Social Bookmarking has been driven by the study of different sources, and it led to the ideation of an original approach, characterized by the user modeling by means of three-dimensional matrices of co-occurrence between terms. Experimental results were encouraging and confirmed the correlation with users' interests and the effective coherence and utility of their categorization.

References

1. Bai, J., Song, D., Bruza, P., Nie, J.-Y., Cao, G.: Query expansion using term relationships in language models for information retrieval. In: CIKM, pp. 688–695 (2005)
2. Bao, S., Xue, G., Wu, X., You, Y.: Optimizing web search using social annotations. In: WWW 2007: Proceedings of the 16th international conference on World Wide Web, pp. 501–510 (2007)
3. Burgess, C., Livesay, K., Lund, K.: Exploration in Context Space: Words, Sentences, Discourse. Discourse Processes 25(2&3), 211–257 (1999)
4. Burgess, C., Lund, K.: Hyperspace analog to language (hal): A general model of semantic representation. In: Proceedings of the annual meeting of the Psychonomic Society, vol. 12, pp. 177–210 (1995)
5. Gao, J., Nie, J.-Y., Wu, G., Cao, G.: Dependence language model for information retrieval. In: SIGIR 2004: Proceedings of the 27th annual international ACM SIGIR conference on Research and development in information retrieval, pp. 170–177. ACM Press, New York (2004)
6. Gasparetti, F., Micarelli, A.: Personalized search based on a memory retrieval theory. International Journal of Pattern Recognition and Artificial Intelligence (IJPRAI): Special Issue on Personalization Techniques for Recommender Systems and Intelligent User Interfaces 21(2), 207–224 (2007)

7. Teevan, J., Dumais, S.T., Horvitz, E.: Personalizing search via automated analysis of interests and activities. In: SIGIR 2005: Proceedings of the 28th annual international ACM SIGIR conference on Research and development in information retrieval, pp. 449–456. ACM Press, New York (2005)
8. Al-Khalifa, H.S., Davis, H.: Towards better understanding of folksonomic patterns. In: HT 2007: Proceedings of the 18th conference on Hypertext and hypermedia, pp. 163–166 (2007)
9. Halpin, H., Robu, V., Shepherd, H.: The complex dynamics of collaborative tagging. In: WWW 2007: Proceedings of the 16th international conference on World Wide Web, pp. 211–220 (2007)
10. Radlinski, F., Joachims, T.: Query chains: Learning to rank from implicit feedback. In: KDD 2005: Proceedings of the eleventh ACM SIGKDD international conference on Knowledge discovery in data mining, pp. 239–248 (2005)
11. Salton, G., Buckley, C.: Improving retrieval performance by relevance feedback, pp. 355–364. Morgan Kaufmann Publishers Inc., San Francisco (1997)
12. Yanbe, Y., Jatowt, A., Nakamura, S., Tanaka, K.: Can social bookmarking enhance search in the web? In: JCDL 2007: Proceedings of the 2007 conference on Digital libraries, pp. 107–116 (2007)

Partitioning Search Spaces of a Randomized Search

Antti E. J. Hyvärinen, Tommi Junttila, and Ilkka Niemelä

Helsinki University of Technology TKK
Department of Information and Computer Science
{Antti.Hyvarinen,Tommi.Junttila,Ilkka.Niemela}@tkk.fi

Abstract. This paper studies the following question: given an instance of the propositional satisfiability problem, a randomized satisfiability solver, and a cluster of n computers, what is the best way to use the computers to solve the instance? Two approaches, simple distribution and search space partitioning as well as their combinations are investigated both analytically and empirically. It is shown that the results depend heavily on the type of the problem (unsatisfiable, satisfiable with few solutions, and satisfiable with many solutions) as well as on how good the search space partitioning function is. In addition, the behavior of a real search space partitioning function is evaluated in the same framework. The results suggest that in practice one should combine the simple distribution and search space partitioning approaches.

1 Introduction

In this paper we develop distributed techniques for solving challenging instances of the propositional satisfiability problem (SAT). We are interested in using the best available SAT solvers as black-box subroutines or with little modification and in this way take advantage of the rapid development of SAT solver technology.

One of the interesting features in current state-of-the-art SAT solvers is that they use randomization and that their run times can vary significantly for a given instance. This opens up new opportunities for developing distributed solving techniques. The most straightforward idea is to employ a *simple distribution* approach where one just performs a number of independent runs using a randomized solver. This leads to surprising good speed-ups even when used in a grid environment with substantial communication and other delays [1]. The approach could be extended by applying particular restart strategies [2,3] or using an algorithm portfolio scheme [4,5]. Another key feature in modern SAT solvers is the use of conflict driven clause learning techniques. This feature can be exploited in the simple distribution approach and it has been shown that combining parallel learning schemes with a simple restart strategy leads to a powerful distributed SAT solving technique [6].

Another approach to developing parallel SAT solving techniques is based on *partitioning* the search space to multiple parts which can be handled in parallel. This can be achieved by constraint-based partitioning where the search space for a SAT instance $\mathcal{F}$ is split to n *derived* instances $\mathcal{F}_1, \ldots, \mathcal{F}_n$ by including additional constraints to $\mathcal{F}$. Typical implementation techniques include guiding paths [7,8,9] and scattering [10].

Both simple distribution and partitioning have their strengths. The former has led to surprisingly good performance but for really challenging SAT instances it provides no

R. Serra and R. Cucchiara (Eds.): AI*IA 2009, LNAI 5883, pp. 243–252, 2009.

mechanism for splitting the search to more manageable portions to be treated in parallel. Search space partitioning techniques offer an approach to achieving this. However, the interaction between partitioning and randomized SAT solvers is poorly understood and the paper aims to shed new light on this problem. It studies in detail combination of constraint-based partitioning and randomized SAT solvers, and provides an analysis on how an efficient and robust implementation can be achieved.

The rest of the paper is structured as follows. Section 2 reviews briefly relevant key characteristics of modern randomized SAT solvers and the simple distribution approach. Section 3 studies analytically the expected run time of a plain partitioning approach where a SAT instance is partitioned and then a randomized SAT solver is used to solve the resulting instances. The section provides fundamental results for two limiting cases, for ideal and void partitioning functions. Section 4 extends the study to a setting where simple distribution and partitioning are mixed. Section 5 provides an implementation of a randomized partitioning function and Section 6 verifies the results briefly using experiments, and conclusions are given in Section 7. An extended version of this work with more experiments and proofs for the propositions is available through the first author's home pages (http://www.tcs.hut.fi/~aehyvari/).

2   Randomization and Simple Distribution

Most modern SAT solvers apply search *restarts* and some form of *randomization* to avoid getting stuck at hard subproblems [11]. For instance, MiniSat [12] version 1.14 restarts the search periodically and makes two percent of its branching decisions pseudo-randomly. Despite restarts and randomness, the run times of a SAT solver on an instance $\mathcal{F}$ can vary significantly between some minimum t_{min} and maximum t_{max} (we assume that $t_{min} > 0$ and t_{max} is finite). Thus, we treat the run time of the solver on the instance as a random variable T and study the associated cumulative run-time distribution $q_T(t) = \Pr(T \leq t)$ (i.e. $q_T(t)$ is the probability that the instance is solved within t seconds) and its expected value $\mathbb{E}(T) = \int_{t_{min}}^{t_{max}} t q'(t) dt$. As an example, observe the run-time distribution $q(t)$ (approximated by one hundred sample runs) of an instance given in the left hand side plot of Fig. 1. Depending on the seed given to the pseudo-random number generator of MiniSat v1.14, the run time varies from less than a second to thousands of seconds.

This non-constant run time phenomenon can be exploited in a parallel environment by simply running n SAT solvers on the same instance $\mathcal{F}$ in parallel and terminating the search when one of the solvers reports the solution. We call this approach *Simple Distributed SAT solving* (SDSAT) and denote its run time by the random variable T_{sdsat}^n. The cumulative run time distribution is now improved from $q_T(t)$ of the sequential case to $q_{T_{sdsat}^n}(t) = 1 - (1 - q_T(t))^n$. This approach can be surprisingly efficient. As an example, for the instance in the left hand side plot of Fig. 1 the expected run-time in the sequential case is $\mathbb{E}(T) \approx 623s$ while for eight solvers $\mathbb{E}(T_{sdsat}^8) \approx 31s$ (that is, around 20 times less). For a more detailed analysis of running SDSAT in a parallel, distributed environment involving communication and other delays, see [1].

Although the SDSAT approach can reduce the expected time to solve an instance, it cannot reduce it below the minimum run time t_{min}. For an example, observe the sequential run time distribution $q(t)$ of another instance given in the right hand side plot

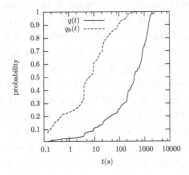 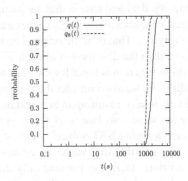

Fig. 1. The run time distributions of two instances for single (*the $q(t)$ plots*) and eight (*the $q_8(t)$ plots*) randomized SAT solvers

of Fig. 1; the variation of the run time is significantly smaller and the instance seems to have no short run times. Consequently, running eight SAT solvers in parallel does not reduce the expected run time significantly; in numbers, $\mathbb{E}(T) \approx 2,065s$ while for eight solvers $\mathbb{E}(T_{\text{sdsat}}^8) \approx 1,334s$ (i.e., only less than two times faster). Even more importantly, the *minimum run time stays the same irrespective of how many parallel solvers are employed*. As a summary, we can establish the following properties for the expected run time of the SDSAT approach:

Proposition 1. $t_{\min} \leq \mathbb{E}(T_{\text{sdsat}}^n) \leq \mathbb{E}(T)$ *for each* $n \geq 1$. *Furthermore,* $\mathbb{E}(T_{\text{sdsat}}^n) \to t_{\min}$ *when* $n \to \infty$.

As we have seen, SDSAT can allow super-linear speedup (meaning $\mathbb{E}(T_{\text{sdsat}}^n) < \mathbb{E}(T)/n$) for instances with strongly varying run time. However, as the maximum speedup obtainable with SDSAT is $\mathbb{E}(T)/t_{\min}$, this can only happen for "smallish" values of n and for more than $\mathbb{E}(T)/t_{\min}$ solvers the speedup is guaranteed to be sub-linear.

3 Partitioning

The basic idea in the form of partitioning we use in this paper is quite simple: given a SAT instance $\mathcal{F}$ and a positive integer n, use a *partitioning function* to compute a set $\mathcal{F}_1, \ldots, \mathcal{F}_n$ of *derived* SAT instances such that

$$\mathcal{F} \equiv \mathcal{F}_1 \vee \cdots \vee \mathcal{F}_n. \tag{1}$$

Now, in order to find whether $\mathcal{F}$ is satisfiable, we solve, in parallel, all $\mathcal{F}_1, \ldots, \mathcal{F}_n$ and deduce that $\mathcal{F}$ is satisfiable if at least one of $\mathcal{F}_1, \ldots, \mathcal{F}_n$ is. This method is called the *plain partitioning approach* in order to distinguish it from the composite approaches in Sect. 4. One way to implement partitioning functions is described in [10] (also see Sect. 5), where each $\mathcal{F}_i$ is obtained from $\mathcal{F}$ by conjoining it with a set of additional partitioning constraints.[1] In addition to the requirement (1), partitioning functions often ensure that the models of $\mathcal{F}_1, \ldots, \mathcal{F}_n$ are mutually disjoint.

---

[1] As explained in [13], guiding paths [7,8] can also be interpreted as partitioning constraints.

Intuitively, the ideal case is that the partitioning function can partition the instance $\mathcal{F}$ into n new instances $\mathcal{F}_1, \ldots, \mathcal{F}_n$ so that each new instance $\mathcal{F}_i$ is n times easier to solve than the original. That is, if the original instance $\mathcal{F}$ has the cumulative run time distribution $q_T(t)$, then the distribution of each $\mathcal{F}_i$ is $q_{T_i}(t) = q_T(nt)$. In this case we say that the partition function is *ideal* for the instance. As obtaining ideal partitioning functions can be difficult, we also consider the case of a *void* partitioning function where the partitioning fails totally, resulting in new instances which are as hard to solve as the original, i.e. have the same distribution $q_{T_i}(t) = q_T(t)$. This is a realistic scenario because modern, clause-learning SAT solvers, such as MiniSat, use sophisticated heuristics in the search: it is possible that values of certain variables are practically never considered. If the partition function constrains only these irrelevant variables, the difficulty of the instance does not decrease, and thus such a function is void.

In this section, we give an analytic study of the efficiency of the plain partitioning approach, under both ideal and void functions, when the fact that the SAT solver is randomized is taken into account. As the efficiency depends heavily on the satisfiability of the instance, we consider three cases: unsatisfiable instance, a satisfiable instance with many solutions, and a satisfiable instance with a unique solution. We have also simulated the plain partitioning approach on run time distributions of some real SAT instances; some results are given later in Sect. 4 after some composite approaches mixing simple distribution and plain partitioning have been described. A real partitioning function is considered in Sect. 5.

3.1 Unsatisfiable Instances

Assume that an unsatisfiable instance $\mathcal{F}$ is partitioned into n new instances $\mathcal{F}_1, \ldots, \mathcal{F}_n$ fulfilling Eq. (1). All new instances need to be shown unsatisfiable to deduce that $\mathcal{F}$ is unsatisfiable. When performed in parallel, this corresponds to waiting for the termination of the "unluckiest" run.

In the case of ideal partitioning function, each new instance $\mathcal{F}_i$ is n times easier to solve than the original $\mathcal{F}$, having run time distributions $q_{T_i}(t) = q_T(nt)$. We denote the random variable capturing the run time of the resulting plain partitioning approach under an ideal partitioning function by $T_{\text{part(ideal)}}^n$. As all the new instances have to be solved (in parallel), the corresponding run time distribution is $q_{T_{\text{part(ideal)}}^n}(t) = q(nt)^n$. Based on this, we have the following interesting results. First, ideal partitioning functions can provide *at most linear expected speed-up* on unsatisfiable instances:

Proposition 2. $\mathbb{E}(T_{\text{part(ideal)}}^n) \geq \mathbb{E}(T)/n$ for each $n \geq 1$.

In fact, it can be shown that linear speed-up can only be obtained on instances that have a constant run time distribution, i.e. when $t_{\min} = t_{\max}$. However, the expected run time is never worse than that of solving the original instance with one solver:

Proposition 3. $\mathbb{E}(T_{\text{part(ideal)}}^n) \leq \mathbb{E}(T)$ for each $n \geq 1$.

When the number n of SAT solvers run in parallel is increased, the expected run time $\mathbb{E}(T_{\text{part(ideal)}}^n)$ approaches $t_{\max}/n$, i.e., linear speed-up w.r.t. the *maximum* run time. Plain partitioning with ideal partitioning functions and simple distribution cannot be totally ordered; there are distributions for which $\mathbb{E}(T_{\text{sdsat}}^n) < \mathbb{E}(T_{\text{part(ideal)}}^n)$ and others for which

$\mathbb{E}(T^n_{\text{part(ideal)}}) < \mathbb{E}(T^n_{\text{sdsat}})$ when n is smallish. However, as $\mathbb{E}(T^n_{\text{part(ideal)}}) \leq t_{\max}/n$ and $\mathbb{E}(T^n_{\text{sdsat}}) \geq t_{\min}$, we have that $\mathbb{E}(T^n_{\text{part(ideal)}}) < \mathbb{E}(T^n_{\text{sdsat}})$ for sufficiently large n.

Let us next consider the case of a void partitioning function, i.e. the case when the partitioning fails so that the run time distribution $q_{T_i}(t)$ of each new instance $\mathcal{F}_i$ is equal to $q_T(t)$ of the original instance $\mathcal{F}$. We denote by $T^n_{\text{part(void)}}$ the run time of the resulting plain partitioning approach. As all $\mathcal{F}_i$ have to be solved, the run time distribution of $T^n_{\text{part(void)}}$ is $q_{T^n_{\text{part(void)}}}(t) = q_{T_i}(t)^n = q_T(t)^n$. From this it follows that for unsatisfiable instances *it is not possible to obtain any speedup with void functions*:

Proposition 4. $\mathbb{E}(T^n_{\text{part(void)}}) \geq \mathbb{E}(T)$ *for each* $n \geq 1$.

In fact, *the more resources one uses, the closer to the maximum run time one gets*: $\mathbb{E}(T^n_{\text{part(void)}}) \to t_{\max}$ when $n \to \infty$.

3.2 Satisfiable Instances with Many Solutions

We next consider the case when a satisfiable SAT instance $\mathcal{F}$ is partitioned into n new instances $\mathcal{F}_1, \ldots, \mathcal{F}_n$ fulfilling Eq. (1). In order to deduce that $\mathcal{F}$ is satisfiable, it is enough to show that *any* of the new instances is satisfiable. In this section we assume that each new instance $\mathcal{F}_i$ is satisfiable, postponing the case where only one is satisfiable to the next section.

Let us consider the case of ideal partitioning function first. Again, we denote the random variable describing the run time of the resulting plain partitioning approach by $T^n_{\text{part(ideal)}}$. As the probability that none of the n solvers has solved the associated new instance within time t is $(1-q_T(nt))^n$, run time distribution of $T^n_{\text{part(ideal)}}$ is $q_{T^n_{\text{part(ideal)}}}(t) = 1 - (1 - q_T(nt))^n$. Several interesting properties follow from this. First, with n parallel solvers, *the expected run time is n times smaller than that of the Simple Distributed SAT*: $\mathbb{E}(T^n_{\text{part(ideal)}}) = \mathbb{E}(T^n_{\text{sdsat}})/n$. Therefore, when compared to solving the original instance, we notice that *on satisfiable instances with many solutions we may expect at least linear speed-up*:

Proposition 5. $\mathbb{E}(T^n_{\text{part(ideal)}}) \leq \mathbb{E}(T)/n$ *for each* $n \geq 1$.

When the number n of parallel SAT solvers is increased, $\mathbb{E}(T^n_{\text{part(ideal)}})$ approaches $t_{\min}/n$. Thus, one can in principle obtain almost linear speed-up w.r.t. *the minimum run time*.

In the case of a void partitioning function the run time of each new instance is the same as that of the original. As each new instance is assumed to be satisfiable, solving any of them is enough to deduce the satisfiability of the original instance. Therefore, *for satisfiable instances with many solutions, the plain partitioning approach with a void partitioning function effectively reduces to Simple Distributed SAT*:

Proposition 6. $\mathbb{E}(T^n_{\text{part(void)}}) = \mathbb{E}(T^n_{\text{sdsat}})$.

3.3 Satisfiable Instances with One Solution

When a satisfiable instance $\mathcal{F}$ with only one satisfying truth assignment is partitioned into n new instances $\mathcal{F}_1, \ldots, \mathcal{F}_n$, it is likely that only one of the new instances is satisfiable while the others are unsatisfiable. Therefore, the satisfiable new instance has to be solved to deduce that $\mathcal{F}$ is satisfiable. The problem is that it is not known which of the new instances this is.

In the case of ideal partitioning function, the run time of the satisfiable new instance is n times smaller than that of the original instance. Therefore, if all the n new instances are solved in parallel and the solving is terminated as soon as the satisfiable new instance is solved, *linear speed-up is obtained with an ideal partitioning function*:

Proposition 7. $\mathbb{E}(T^n_{\text{part(ideal)}}) = \mathbb{E}(T)/n$.

In the case of a void partitioning function, the run time of the satisfiable new instance is the same as that of the original instance. Thus, *using a void partitioning function results neither in speed-up nor in loss of efficiency*:

Proposition 8. $\mathbb{E}(T^n_{\text{part(void)}}) = \mathbb{E}(T)$.

4 Composite Approaches

The analysis of the previous section shows that the plain partitioning approach can potentially obtain even super-linear speed-ups, whereas an improper, void implementation can by Prop. 4 result in worse expected run time than that of one solver. The two approaches presented here aim at being at least as efficient as solving the instance with one solver. Assume that we have resources to run n SAT solvers in parallel and consider the following approaches that mix simple distribution and plain partitioning.

- *Repeated Partitioning.* In this approach, we run in parallel $k = \lfloor \sqrt{n} \rfloor$ copies of the plain partitioning approach, each copy splitting the instance $\mathcal{F}$ into k new instances $\mathcal{F}_{i,1}, \ldots, \mathcal{F}_{i,k}$ and solving each $\mathcal{F}_{i,j}$ once. We denote the random variable describing the run time of this approach by $T^n_{\text{rep-part}}$.
- *Safe Partitioning.* This approach reverses the order of SDSAT and partitioning compared to repeated partitioning: the instance $\mathcal{F}$ is partitioned into $k = \lfloor \sqrt{n} \rfloor$ new instances $\mathcal{F}_1, \ldots, \mathcal{F}_k$ and each new instances $\mathcal{F}_i$ is solved with k SAT solvers in parallel. The run time of this is denoted by the random variable $T^n_{\text{safe-part}}$.

Unfortunately, when using repeated partitioning on an unsatisfiable instance and the partitioning function is void, the experiments show that $\mathbb{E}(T) \leq \mathbb{E}(T^n_{\text{rep-part}})$ and that $\mathbb{E}(T^n_{\text{rep-part}}) \rightarrow t_{\max}$ when $n \rightarrow \infty$. However, safe partitioning (i) is as good as repeated partitioning (i.e. $\mathbb{E}(T^n_{\text{safe-part}}) = \mathbb{E}(T^n_{\text{rep-part}})$) on satisfiable instances and we conjecture that it is at least as good (i.e. $\mathbb{E}(T^n_{\text{safe-part}}) \leq \mathbb{E}(T^n_{\text{rep-part}})$) on unsatisfiable ones when the same partition function is applied; (ii) is equal to SDSAT on satisfiable instances with many solutions when the partition function is void; and (iii) seems experimentally at least as fast as solving the original instance with one solver, i.e., $\mathbb{E}(T^n_{\text{safe-part}}) \leq \mathbb{E}(T)$ even when the instance is unsatisfiable and the partitioning function is void.

To illustrate the approaches and results presented in Sects. 2, 3, and 4, Fig. 2 shows the expected run times of different approaches when applied to the same instances as in Fig. 1. As the left hand side instance is satisfiable with many solutions, the "sdsat+others" line depicts the behavior of the SDSAT approach as well as all the considered partitioning approaches when a void partitioning function is applied. The instance at the right hand side is unsatisfiable.

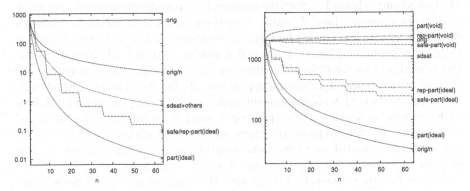

Fig. 2. Expected run times (in seconds) of different approaches on the instances in Fig. 1 when the number n of SAT solvers run in parallel is varied

5   Implementing a Randomized Partitioning Function

This section briefly describes how we implemented a partitioning function called *scattering* [10]. The implementation is based on MiniSat 1.14, constructs partitions having pairwise disjoint models, and is randomized so that the partitions differ depending on the random seed given as input to the function.

The function works in two phases. First, the function simply runs MiniSat as is to obtain heuristic values for the Boolean variables in the instance. The first phase ends when the instance is solved or a fixed time limit (currently 300 seconds) is reached. If the instance was not solved, the function enters the second phase, where the derived instances $\mathcal{F}_1, \ldots, \mathcal{F}_n$ are constructed from $\mathcal{F}$ by adding constraints. For each $1 \leq i \leq n - 1$, the scattering function uses the obtained heuristic values to select a conjunction of d_i literals, $C_i = l_{i,1} \wedge \ldots \wedge l_{i,d_i}$, and conjoins it to the derived instance $\mathcal{F}_i$. The function ensures that no derived instances share models by inserting the negation $(\neg l_{i,1} \vee \ldots \vee \neg l_{i,d_i})$ of the conjunction C_i in F_i to each derived instance $\mathcal{F}_j$ with $i < j \leq n$. Scattering is a partition function, since the last instance contains only the negations of the conjunctions corresponding to the "remaining" search space. Finally, the randomization of the scattering function follows naturally from that of MiniSat: the derived instances depend on the random seed passed to MiniSat.

6   Experimental Results on Partitioning

The following experiments study the behavior of the presented approaches both under ideal and real (scattering) partitioning functions on some real-world SAT instances. A summary of the results is presented in Table 1. The instances represent hard SAT formulas in the sense that their randomized run times often exceed one hour. Furthermore, `cube-11-h14-sat` is a satisfiable instance where the scattering function always resulted in a unique satisfiable derived instance, `dated-10-13-s` is a satisfiable instance where the scattering function always resulted in several satisfiable derived instances (unless solved by the function), and `APRoVE07-09` is an unsatisfiable instance.

The first column, labeled T, reports the sequential run-time distribution of the instances. The next two columns, labeled "simple distribution" in the table, report the results for Simple Distributed SAT solving for eight and 64 resources. The next four columns, labeled "ideal partitioning", report run-time distributions when an ideal partitioning function is used: first for the plain partitioning approach with eight and 64 resources, and then for safe and repeated partitioning approaches. The last four columns, labeled "scattering", report the run time distributions obtained when scattering (recall Sect. 5) is used as the partitioning function; first for plain partitioning with eight and 64 resources, and then for safe and repeated partitioning approaches with 64 resources. The rows of the table report the expected, minimum, median and maximum run times together with the first and third quartile (the values of t such that $q(t) \leq 0.25$ and $q(t) \leq 0.75$).

The run-time distributions for SDSAT and "ideal partitioning" approaches were obtained by solving the instance one hundred times with MiniSat 1.14. The resulting distributions were then used to compute the results analytically. None of the results include delays associated with parallel environments.

The distributions in columns T^8_{scatter} and T^{64}_{scatter} are obtained by running the plain partitioning approach with the scattering function fifty times using different random seeds. The resulting distribution was directly used to compute the values for the repeated partitioning approach ($T^{64}_{\text{rep-part}}$) under "scattering". To compute the results for the column $T^{64}_{\text{safe-part}}$ under "scattering", each derived instance was solved seven more times, thereby directly simulating an implementation of safe partitioning with scattering. The run times do not include the time required to run the scattering function. If the instance was solved while scattering, the run time is reported as zero.

The results in "simple distribution" columns show good scalability for dated-10--13-s and moderate scalability for other instances, as predicted by analytical results when $t_{\min}$ is close to $\mathbb{E}(T)$. The columns under "ideal partitioning" show that partitioning can in theory result in even better speed-up for these instances. Surprisingly, in the actual implementation (T^8_{scatter}, T^{64}_{scatter}) we see that plain scattering results in higher expected run-times than simple distribution for these instances. This reflects the difficulty of obtaining ideal partitioning functions.

Comparison of the "ideal partitioning" approaches confirms the discussion in Sect. 4. In particular, safe partitioning results in lower expected run time than repeated partitioning for unsatisfiable instances. However, the results under "scattering" show the opposite; repeated partitioning has consistently lower expected run time than safe partitioning. For example, observe the expected run times for safe and repeated partitioning approaches for the instance cube-11-h14-sat with unique satisfiable derived instance: in "ideal partitioning" they are equal, whereas the scattering-based safe partitioning is significantly worse than the scattering-based repeated partitioning approach. To study this, we computed run-time distributions for some of the satisfiable derived instances (not shown in the table), and it turns out that their expected run times varied between 109.1 and 4,773 seconds. Thus the hardness (expected run time) of a derived instance produced by scattering is also a random variable with possibly a very large range, and running the scattering function independently several times increases the probability of finding derived instances with low expected run times. This explains the good speed-up obtained by repeated partitioning when compared to safe partitioning.

Table 1. Comparing approaches for parallel search

	T	simple distrib.		ideal partitioning				scattering			
		T^8_{sdsat}	T^{64}_{sdsat}	$T^8_{\text{part(ideal)}}$	$T^{64}_{\text{part(ideal)}}$	$T^{64}_{\text{safe-part}}$	$T^{64}_{\text{rep-part}}$	T^8_{scatter}	T^{64}_{scatter}	$T^{64}_{\text{safe-part}}$	$T^{64}_{\text{rep-part}}$
				cube-11-h14-sat							
Exp	4,832	3,110	2,685	604,0	75.50	388.7	388.7	3,537	3,378	2,265	839.6
Min	2,629	2,629	2,629	328.6	41.07	328.6	328.6	117.5	69.49	13.93	117.5
Q_1	3,641	2,748	2,629	455.1	56.89	343.5	343.5	2,291	1,193	1,665	141.7
Med	4,661	3,009	2,640	572.7	72.83	376.1	376.1	3,459	3,473	2,381	338.2
Q_3	5,730	3,362	2,741	716.3	89.53	420.2	420.2	4,662	5,288	3,182	1,719
Max	10,050	10,050	10,050	1,256	157.0	1,256	1,256	11,500	7,199	4,405	11,500
				dated-10-13-s							
Exp	2,266	128.2	16.45	16.02	0.2570	2.056	2.056	261.2	317.3	21.22	0.2566
Min	10.09	10.09	10.09	1.262	0.1577	1.262	1.262	0	0	0	0
Q_1	283.2	28.00	10.09	3.499	0.1577	1.262	1.262	0	0	0	0
Med	784.7	109.4	10.58	13.67	0.1653	1.322	1.322	19.86	9.858	7.037	0
Q_3	2,093	181.4	17.63	22.68	0.2755	2.204	2.204	227.7	84.70	17.77	0
Max	37,930	37,930	37,930	4,741	529.6	4,741	4,741	6,449	10,095	172.0	6,449
				AProVE07-09							
Exp	4,016	2,361	1,685	759.7	117.7	392.5	586.7	2,598	1,719	1,632	1,559
Min	1,552	1,552	1,552	194.1	24.26	194.1	194.1	1,261	486.3	723.9	1,261
Q_1	3,086	2,033	1,552	668.0	106.2	352.3	554.2	1,757	1,122	1,271	1,466
Med	3,905	2,389	1,563	770.6	110.3	396.4	581.4	2,267	1,437	1,637	1,567
Q_3	4,732	2,666	1,736	843.7	129.8	414.3	606.4	3,550	1,894	1,912	1,694
Max	9,302	9,302	9,302	1,163	145.3	1,163	1,163	4,539	6,617	2,968	4,539

7   Conclusions

The paper investigates distributed techniques for solving challenging SAT instances and focuses on combining constraint-based search space partitioning with randomized SAT solving techniques. The paper studies first analytically the expected run time of a plain partitioning approach where a SAT instance is partitioned and then a randomized SAT solver is used to solve the resulting instances. Analytical results are derived for two limiting cases, for ideal and void partitioning functions. The investigation is then extended to a setting where simple distribution and partitioning are mixed. Finally the paper proposes a randomized partitioning function and compares the function against the ideal case.

The analytical results show that partitioning can potentially lead to catastrophic failures where an increase in computing resources leads to a decrease in solving efficiency for unsatisfiable instances. The empirical results show in part that a good implementation is usually able to avoid this failure, but plain partitioning can nevertheless be worse than an approach based on simple distributed SAT solving (SDSAT). Both problems are avoided in practice with safe and repeated partitioning. The experimental and analytical comparisons show an interesting relationship between the safe and repeated partitioning approaches, suggesting that an ideal partitioning function would profit from safe partitioning whereas randomness in the partitioning function can be better exploited with repeated partitioning.

Acknowledgments. The authors are grateful for the financial support of the Academy of Finland (projects 122399 and 112016), Helsinki Graduate School in Computer Science and Engineering, Jenny and Antti Wihuri Foundation, Emil Aaltosen Säätiö, Finnish Foundation for Technology Promotion, and Technology Industries of Finland Centennial Foundation.

References

1. Hyvärinen, A.E.J., Junttila, T.A., Niemelä, I.: Strategies for solving SAT in grids by randomized search. In: Autexier, S., Campbell, J., Rubio, J., Sorge, V., Suzuki, M., Wiedijk, F. (eds.) AISC 2008, Calculemus 2008, and MKM 2008. LNCS (LNAI), vol. 5144, pp. 125–140. Springer, Heidelberg (2008)
2. Luby, M., Sinclair, A., Zuckerman, D.: Optimal speedup of Las Vegas algorithms. Information Processing Letters 47(4), 173–180 (1993)
3. Luby, M., Ertel, W.: Optimal parallelization of Las Vegas algorithms. In: Enjalbert, P., Mayr, E.W., Wagner, K.W. (eds.) STACS 1994. LNCS, vol. 775, pp. 463–474. Springer, Heidelberg (1994)
4. Huberman, B.A., Lukose, R.M., Hogg, T.: An economics approach to hard computational problems. Science 275(5296), 51–54 (1997)
5. Gomes, C.P., Selman, B.: Algorithm portfolios. Artificial Intelligence 126(1–2), 43–62 (2001)
6. Hyvärinen, A.E.J., Junttila, T., Niemelä, I.: Incorporating clause learning in grid-based randomized SAT solving. Journal on Satisfiability, Boolean Modeling and Computation 6, 223–244 (2009)
7. Böhm, M., Speckenmeyer, E.: A fast parallel SAT-solver: Efficient workload balancing. Annals of Mathematics and Artificial Intelligence 17(4–3), 381–400 (1996)
8. Zhang, H., Bonacina, M., Hsiang, J.: PSATO: A distributed propositional prover and its application to quasigroup problems. Journal of Symbolic Computation 21(4), 543–560 (1996)
9. Jurkowiak, B., Li, C., Utard, G.: A parallelization scheme based on work stealing for a class of SAT solvers. Journal of Automated Reasoning 34(1), 73–101 (2005)
10. Hyvärinen, A.E.J., Junttila, T., Niemelä, I.: A distribution method for solving SAT in grids. In: Biere, A., Gomes, C.P. (eds.) SAT 2006. LNCS, vol. 4121, pp. 430–435. Springer, Heidelberg (2006)
11. Gomes, C.P., Selman, B., Crato, N., Kautz, H.A.: Heavy-tailed phenomena in satisfiability and constraint satisfaction problems. Journal of Automated Reasoning 24(1/2), 67–100 (2000)
12. Eén, N., Sörensson, N.: An extensible SAT-solver. In: Giunchiglia, E., Tacchella, A. (eds.) SAT 2003. LNCS, vol. 2919, pp. 502–518. Springer, Heidelberg (2004)
13. Hyvärinen, A.E.J.: Approaches to grid-based SAT solving. Research Report TKK-ICS-R16, Helsinki University of Technology (June 2009)

Improving Plan Quality in SAT-Based Planning

Enrico Giunchiglia and Marco Maratea

DIST, University of Genova, Viale F. Causa 15, Genova, Italy
{enrico,marco}@dist.unige.it

Abstract. Planning as Satisfiability (SAT) is the best approach for optimally (wrt makespan) solving classical planning problems. SAT-based planners, like SAT-PLAN, can thus return plans having minimal makespan guaranteed. However, the returned plan does not take into account plan quality issues introduced in the last two International Planning Competitions (IPCs): such issues include minimal-actions plans and plans with "soft" goals, where a metric has to be optimized over actions/goals. Recently, an approach to address such issues has been presented, in the framework of planning as satisfiability with preferences: by modifying the heuristic of the underlying SAT solver, the related system (called SATPLAN(P)) is guaranteed to return plans with minimal number of actions, or with maximal number of soft goals satisfied. But, besides such feature, it is well-known that introducing ordering in SAT heuristics can lead to significant degradation in performances. In this paper, we present a generate-and-test approach to tackle the problem of dealing with such optimization issues: without imposing any ordering, a (candidate optimal) plan is first generated, and then a constraint is added imposing that the new plan (if any) has to be "better" than the last computed, i.e., the plan quality is increased at each iteration. We implemented this idea in SATPLAN, and compared the resulting systems wrt SATPLAN(P) and SGPLAN on planning problems coming from IPCs. The analysis shows performance benefits for the new approach, in particular on planning problems with many preferences.

1 Introduction

Planning as Satisfiability (SAT) [1] is the best approach for optimally (wrt makespan) solving classical planning problems. The SAT-based planner SATPLAN [2] has been the winner in the deterministic track for optimal planners in the 4th International Planning Competition (IPC-4) [3] and co-winner in the IPC-5 [4] (together with another SAT-based planner, MAXPLAN [5]). SAT-based planners inherit from the approach the property that the returned plan has minimal makespan guaranteed. However, the returned plan does not take into account plan quality issues introduced in the last two International Planning Competitions (IPCs): such issues include minimal-actions plans and plans with "soft" goals, where a metric has to be optimized over actions/goals. Recently, an approach to tackle this problem has been presented, in the framework of planning as satisfiability with preferences [6,7]: given a (minimal) makespan, by imposing that the heuristic of the underlying SAT solver first select literals which correspond to "not to perform" actions, or "to satisfy" soft goals, the related system, called SAT-PLAN(P), is guaranteed to return plans with minimal number of actions, or with the maximal number of soft goals satisfied. But, besides such feature, and the fact that the

R. Serra and R. Cucchiara (Eds.): AI*IA 2009, LNAI 5883, pp. 253–263, 2009.

first computed plan is guaranteed to be "optimal", it is well-known that introducing ordering in SAT heuristics can lead, at least theoretically, to significant degradation in performances [8]: in SATPLAN(P), this phenomenon also happened experimentally on large planning problems with many actions.

In this paper, we present a different, generate-and-test, approach to tackle the problem of dealing with such optimization issues: without imposing any ordering, a (candidate optimal) plan is first generated, and then a constraint is added imposing that the new plan (if any) has to be "better" than the last computed, extending what has been done in different contexts in both CSP and SAT, e.g., [9,10,11], and also in OR, via "cuts". Thus, the plan quality is increased at each iteration of the algorithm. We implemented this idea in SATPLAN, and called SATPLAN-GNT the resulting system. We compared SATPLAN-GNT wrt SATPLAN(P) on both classical planning problems coming from IPCs, and on planning problems coming from the "SimplePreferences" track of the IPC-5, having all goals as "soft". The results of our analysis reveal that: (i) on planning problems with many preferences, SATPLAN-GNT performs better than SATPLAN(P); (ii) on planning problems with (relatively) few preferences, SATPLAN-GNT and SATPLAN(P) perform similarly; and (iii) SATPLAN(P) and SATPLAN-GNT are both overall competitive to SGPLAN on planning problems coming from the "SimplePreferences" track of the IPC-5, where SGPLAN was the clear winner.

2   Preliminaries

Let $\mathcal{F}$ and $\mathcal{A}$ be the set of *fluents* and *actions*, respectively. A *state* is an interpretation of the fluent signature. A *complex action* α is an interpretation of the action signature, and, intuitively, models the concurrent execution of the actions satisfied by α. A *planning problem* is a triple $\langle I, tr, G \rangle$ where

- I is a Boolean formula over $\mathcal{F}$ and represents the set of *initial states*;
- tr is a Boolean formula over $\mathcal{F} \cup \mathcal{A} \cup \mathcal{F}'$ where $\mathcal{F}' = \{f' : f \in \mathcal{F}\}$ is a copy of the fluent signature and represents the *transition relation* of the automaton describing how (complex) actions affect states (we assume $\mathcal{F} \cap \mathcal{F}' = \emptyset$);
- G is a Boolean formula over $\mathcal{F}$ and represents the set of *goal states*.

The above definition of planning problem differs from the traditional ones in which the description of actions' effects on a state is described in an high-level action language like STRIPS or PDDL. We used this formulation because the techniques we are going to describe are largely independent of the action language used, at least from a theoretical point of view. The only assumption that we make is that the description is deterministic: there is only one state satisfying I and for each state s and complex action α there is at most one interpretation extending $s \cup \alpha$ and satisfying tr. Consider a planning problem $\Pi = \langle I, tr, G \rangle$. In the following, for any integer i

- if F is a formula in the fluent signature, F_i is obtained from F by substituting each $f \in \mathcal{F}$ with f_i,
- tr_i is the formula obtained from tr by substituting each symbol $p \in \mathcal{F} \cup \mathcal{A}$ with p_{i-1} and each $f \in \mathcal{F}'$ with f_i.

If n is an integer, the *planning problem* Π *with makespan* n is the Boolean formula Π_n defined as $I_0 \wedge \bigwedge_{i=1}^{n} tr_i \wedge G_n, n \geq 0$, and a *plan* is an interpretation satisfying such formula.[1] For example, consider the planning problem of going to work from home for an husband and a wife. Assume that they can use the car or the bus or the bike, but one has to stay home with the child, this scenario can be formalized using two fluent variables *AtWorkH* and *AtWorkW*, and three action variables *Car*, *Bus* and *Bike*. The problem with makespan 1 can be expressed by the conjunction (here indicated with ",") of the formulas:

$$
\begin{aligned}
&\neg AtWorkH_0, \neg AtWorkW_0, \\
AtWorkH_1 &\equiv \neg AtWorkH_0 \equiv (Car_0 \vee Bus_0 \vee Bike_0), \\
AtWorkW_1 &\equiv \neg AtWorkW_0 \equiv (Car_0 \vee Bus_0 \vee Bike_0), \\
&\neg AtWorkH_1 \vee \neg AtWorkW_1,
\end{aligned}
\tag{1}
$$

in which the first two formulas correspond to the initial state, the third and the fourth to the transition relation, and the last indicates that exactly one goal can be reached, mimicking that goals are soft. The planning problem has many solutions, each corresponding to a non-empty subset of $\{Car_0, Bus_0, Bike_0\}$ for each fluent. Among those plans, in a minimal-actions plan a single action is performed. About the "soft" goals, the simple characterization in (1) can be extended to include "preferences", by means of weights associated to the violation of the goals, or an ordering on them.

3   Optimal Plans and Qualitative Preferences

In this section we formalize the definition of optimal plans in the framework of planning as satisfiability with preferences, first presented in [6]. The focus of our presentation is on the *qualitative* approach to the problem, and to preferences on literals. This is not a limitation, given that in [6] it is showed how to deal with quantitative preferences (where a preference in a pair $\langle P, c \rangle$ where P has the same meaning as before, and c is a function that maps literals to positive integer) and to (qualitative and quantitative) preferences on formulas, by means of a reduction to our framework of qualitative preferences on literals. Qualitative and quantitative approaches both received attention in planning, each having its pros and cons, as commented in, e.g., Sec. 2.3 of [4].

Let Π_n be a planning problem Π with makespan n. A *qualitative preference (for Π_n)* is a pair $\langle P, \prec \rangle$ where P is a set of literals (the preferences, in our case they will be built on action variables or soft goals) and $\prec$ is a partial order on P. The partial order can be extended to plans for Π_n. Consider a qualitative preference $\langle P, \prec \rangle$. Let π_1 and π_2 be two plans for Π_n. π_1 *is preferred to* π_2 *(wrt $\langle P, \prec \rangle$)* iff (i) they satisfy different sets of preferences, i.e., $\{p : p \in P, \pi_1 \models p\} \neq \{p : p \in P, \pi_2 \models p\}$, and ($ii$) for each preference p_2 satisfied by π_2 and not by π_1 there is another preference p_1 satisfied by π_1 and not by π_2 with $p_1 \prec p_2$. The second condition says that if π_1 does not satisfy a preference p_2 which is satisfied by π_2, then π_1 is preferred to π_2 only if there is a good reason for $\pi_1 \not\models p_2$, i.e., π_1 satisfies a "more preferred" preference (not satisfied by π_2). We write $\pi_1 \prec \pi_2$ to mean that π_1 is preferred to π_2. It is easy to see that $\prec$ defines a

---

[1] In the following, we continuously switch between plans and satisfying interpretations.

partial order on plans for Π_n wrt $\langle P, \prec \rangle$. A plan π is *optimal* for Π_n (wrt $\langle P, \prec \rangle$) if it is a minimal element of the partial order on plans for Π_n, i.e., if there is no plan π' for Π_n with $\pi' \prec \pi$ (wrt $\langle P, \prec \rangle$). As an example, consider the planning problem in (1); if we have the qualitative preference (in the following, we show only the action variables assigned to true in the optimal plan):

a. $\langle \{\neg Bike_0, \neg Bus_0, \neg Car_0\}, \emptyset \rangle$, i.e., the situation in which we prefer not to perform actions, and the preferences are equally important ($\prec = \emptyset$), there are three optimal, i.e., minimal-actions, plans, corresponding to $\{Bike_0\}$, $\{Bus_0\}$, $\{Car_0\}$.

b. $\langle \{\neg Bike_0, \neg Bus_0, \neg Car_0\}, \{\neg Bike_0 \prec \neg Car_0\} \rangle$, i.e., the situation in which again we prefer not to perform actions, and not to take the bike is preferred over not to take the car, then there are two optimal plans, i.e., $\{Bus_0\}$, $\{Car_0\}$.

c. $\langle \{AtWorkH, AtWorkW\}, \emptyset \rangle$, i.e., the situation in which we prefer to satisfy the (soft) goals, and the goals are equally important ($\prec = \emptyset$), there are 14 optimal plans, each corresponding to a non-empty subset of $\{Car_0, Bus_0, Bike_0\}$ for each soft goal.

d. $\langle \{AtWorkH, AtWorkW\}, \{AtWorkH \prec AtWorkW\} \rangle$, i.e., the situation in which again we prefer to satisfy the soft goals, and satisfying the first is preferred to the second, there are 7 optimal plans, each corresponding to a non-empty subset of $\{Car_0, Bus_0, Bike_0\}$ and with the fluent $AtWorkH$ true.

4   Computing Optimal Plans via Generate-and-Test

We have already discussed in the introduction that the algorithm in [6,7] has drawbacks related to imposing an ordering on preferences to be followed while branching: from [8], it is well-known that such ordering can significantly degrade the performances. Here we present a different, generate-and-test, approach for solving the problem of generating plans with minimal number of actions, or with maximal number of soft goal satisfied: given a (minimal) makespan, it first generates a (candidate optimal) plan, and then a constraint is added imposing that the new plan (if any) has to be "better" than the last computed, extending what has been done in both CSP and SAT, e.g., in [9,10,11], and also in OR, via "cuts". Thus, crucial for the above procedure is a condition which enables us to say which are the plans that are preferred (wrt $\langle P, \prec \rangle$) to a plan π: such a formula, where l is a literal, is

$$\left(\bigvee_{l:l \in P, l \notin \pi} l \right) \wedge \left(\bigwedge_{l':l' \in P, l' \in \pi} \left(\bigvee_{l:l \in P, l \notin \pi, l \prec l'} l \vee l' \right) \right). \tag{2}$$

which codifies conditions (i) and (ii) in the previous section. A plan π' is preferred to π wrt $\langle P, \prec \rangle$ iff π' satisfies (2), as stated by the following theorem.

Theorem 1. *Let π and π' be two plans. Let $\langle P, \prec \rangle$ be a qualitative preference. π' is preferred to π wrt $\langle P, \prec \rangle$ if and only if π' satisfies the preference formula in (2).*

Theorem 1 follows from (2) by construction. In Figure 1 we present the new solving procedures, in which:

- in QL-PLAN-GNT-2: for each $p \in P$, $v(p)$ is a newly introduced variable; $v(P)$ is the set of new variables, i.e., $\{v(p) : p \in P\}$; $v(\prec) = \prec'$ is the partial order on $v(P)$ defined by $v(p) \prec' v(p')$ iff $p \prec p'$; in QL-PLAN-GNT-1: $P' = P$ and $\prec' = \prec$.

- π is an *assignment* (or, equivalently, a (candidate) plan for Π_n), i.e., a consistent set of literals. An assignment π corresponds to the partial interpretation mapping to true the literals $l \in \pi$.
- $cnf(\varphi)$, where φ is a formula, is a set of clauses (i.e., set of sets of literals, with $\top$ denoting the empty set of clauses) such that: (i) for any interpretation π in the signature of $cnf(\varphi)$ such that $\pi \models cnf(\varphi)$ it is true also that $\pi' \models \varphi$, where π' is the interpretation π restricted to the signature of φ; and (ii) for any interpretation $\pi' \models \varphi$ there exists an interpretation π, $\pi \supseteq \pi'$, such that $\pi \models cnf(\varphi)$.
  There are well known methods for computing $cnf(\varphi)$ in linear time by introducing additional variables, e.g., [12].
- l is a literal and $\bar{l}$ is the complement of l;
- φ_l returns the set of clauses obtained from φ by (i) deleting the clauses $C \in \varphi$ with $l \in C$, and (ii) deleting $\bar{l}$ from the other clauses in φ;
- *Reason* returns the set of clauses corresponding to (2);
- *ChooseLiteral* returns an *unassigned* literal l (i.e., such that $\{l, \bar{l}\} \cap \pi = \emptyset$) in φ.

$\langle P, \prec \rangle := $ a qualitative preference; $\psi := \top$; $\pi_{opt} := \emptyset$

function QL-PLAN-GNT-1(Π, n)
1   **return** PREF-DLL$(cnf(\Pi_n), \emptyset, P, \prec)$

function QL-PLAN-GNT-2(Π, n)
2   **return** PREF-DLL$(cnf(I_0 \wedge \wedge_{i=1}^{n} tr_i \wedge_{p \in P} (v(p) \vee p)), \emptyset, v(P), v(\prec))$

function PREF-DLL$(\varphi \cup \psi, \pi, P', \prec')$
3   **if** $(\emptyset \in (\varphi \cup \psi)_\pi)$ **return** FALSE;
4   **if** $(\pi$ is total) $\pi_{opt} := \pi$; $\psi := Reason(\pi, P', \prec')$; **return** FALSE;
5   **if** $(\{l\} \in (\varphi \cup \psi)_\pi)$ **return** PREF-DLL$(\varphi \cup \psi, \pi \cup \{l\})$;
6   $l := ChooseLiteral(\varphi \cup \psi, \pi)$;
7   **return**   PREF-DLL$(\varphi \cup \psi, \pi \cup \{l\})$ **or**
                 PREF-DLL$(\varphi \cup \psi, \pi \cup \{\bar{l}\})$.

Fig. 1. The algorithms of SATPLAN-GNT

Note that in the algorithm we have not specified $\langle P, \prec \rangle$ to be a preference on literals: in fact, goals are usually formulas. Thus, when dealing with soft goals (QL-PLAN-GNT-2 algorithm), we add what are called "goal selectors", i.e., variables which are placeholders for the goals: preferences are then expressed over such set of variables, and in this way we are back to our setting of preferences on literals. Goal selectors have the same meaning of clause selectors in Max-SAT. When preferences are expressed over action variables (QL-PLAN-GNT-1 algorithm), adding such variables and modifying the ordering is not needed.

Thus, given a planning problem Π, a makespan n, and a qualitative preference $\langle P, \prec \rangle$, an optimal plan is computed by invoking PREF-DLL [11], a modified version of standard DLL for computing "optimal" solutions, on $cnf()$, i.e., the set of clauses corresponding to the planning problem Π with makespan n, possibly with the modified preferences. More in details, PREF-DLL is standard DLL except that when a new plan π is found (at line 4), it is set to be the (actual) optimal plan π_{opt} (initially set to $\emptyset$), a set of clauses corresponding to (2) are assigned to ψ, and FALSE is returned to continue the search looking for "better" plans. When PREF-DLL terminates, the last plan found is optimal (if one exists), i.e., no plan exists which is preferred to the actual π_{opt}, as stated by the following theorem.

Theorem 2. *Let Π be a planning problem, n the makespan, and $\langle P, \prec \rangle$ a qualitative preference. QL-PLAN-GNT-1 and QL-PLAN-GNT-2 terminate, and then π_{opt} is empty if Π does not have a plan of makespan n, and an optimal plan wrt $\langle P, \prec \rangle$, otherwise.*

Theorem 2 follows from the correctness of PREF-DLL (Theorem 2 in [11]) and the assumptions on *cnf*. Notice that each time a new π' is found at line 4 of Figure 1, which is better than the actual π_{opt}, ψ may be overwritten because we can discard the clauses added because of π since they are entailed by the new clauses added because of π', as stated by the following theorem.

Theorem 3. *Let $\langle P, \prec \rangle$ be a qualitative preference. Let $\pi_1, \pi_2, \ldots, \pi_k$ be the sequence of plans computed in QL-PLAN-GNT-1 or QL-PLAN-GNT-2 for a fixed makespan, and $\psi_1, \psi_2, \ldots, \psi_k$ be the corresponding formulas computed as in (2). For each i, $0 < i < k$, ψ_{i+1} entails ψ_i.*

As a consequence, in PREF-DLL, the formula ψ is overwritten as soon as a new model π is found (line 4). QL-PLAN-GNT-1 and QL-PLAN-GNT-2 are thus guaranteed to work in polynomial space in the size of the input planning problem, makespan and preference. Coming back to our original problems, if we want to compute plans with minimal number of actions, assuming $act(\Pi_n)$ is the set of variables in Π_n corresponding to action variables, it is enough to set $P := \{\overline{a} | a \in act(\Pi_n)\}$ and $\prec := \emptyset$, and calling the QL-PLAN-GNT-1 algorithm to obtain the expected result (in the qualitative case). If we are interested in computing plans with the maximal number of soft goals satisfied, given SG to be the set of soft goals of the problem, it is enough to set $P := \{v(p) | p \in SG\}$ and $\prec := \emptyset$, and calling the QL-PLAN-GNT-2 algorithm to obtain the expected result.

As an example of the behavior of the QL-PLAN-GNT-1 algorithm in Figure 1, consider the planning problem (1) (which thus corresponds to Π_n) and the preference a. at the end of the previous section, the search for a minimal-actions plan may proceed (depending on *ChooseLiteral*) as follows:

1. $\pi_1 = \{Car_0, Bike_0, Bus_0\}$, then $\psi_1 : \overline{Car_0} \vee \overline{Bike_0} \vee \overline{Bus_0}$, i.e., at least one action has to be assigned as in P. Assume the next plan is
2. $\pi_2 = \{Bike_0, Bus_0\}$, then $\psi_2 : (\overline{Bike_0} \vee \overline{Bus_0}) \wedge \overline{Car_0}$, which asks that at least one among $Bike_0$ and Bus_0 has to be assigned as in P, while Car_0 has to remain assigned as in P; assume now the next computed plan is
3. $\pi_3 = \{Bike_0\}$, then $\psi_3 : \overline{Bike_0} \wedge \overline{Car_0} \wedge \overline{Bus_0}$, which would be satisfied only by a plan where all actions are not performed: given this plan does not satisfy (the constraints related to the transition relation of) (1), π_3 is optimal. In PREF-DLL, this is achieved by means of no other plan is computed at line 4 of Figure 1, and the procedure eventually exits at line 3 with π_3 as π_{opt}.

Table 1. Results on domains coming from IPCs. $x(y)$ stands for y instances solved with x secs of mean CPU time.

	SATPLAN(P)(W)	SATPLAN-GNT(W)	SATPLAN(P)()	SATPLAN-GNT()
pipesworld-notankage	85.57(9)	110.37(9)	40.92(11)	100.21(13)
pipesworld-tankage	193.86(6)	217.72(7)	32.59(7)	97.96(8)
satellite	12.6(2)	7.34(2)	3.34(4)	226.04(4)
promela-optical	58.96(11)	108.2(13)	123.38(9)	18.59(13)
psr-small	34.82(47)	32.08(48)	11.85(44)	15(48)
depots	76.02(5)	43.84(5)	194.24(5)	123.08(9)
zenoTravel	10.44(8)	10.79(8)	64.41(9)	40.76(11)
freeCell	10.8(2)	8.91(2)	89.81(4)	15.19(3)
logistics	97.92(10)	5.77(13)	2.4(22)	78.37(25)
mprime	60.17(19)	60.24(19)	27.59(14)	12.58(19)
mystery	24.47(13)	28.29(15)	32.36(13)	11.69(15)
openstacks	–	–	717.31(4)	–
pathways	22.89(5)	13.96(5)	63.78(7)	5.79(7)
storage	16.67(9)	7.83(9)	42.1(12)	24.24(11)
TPP	0.08(5)	151.17(7)	0.14(8)	123.59(19)
elevator	18.99(15)	1.75(15)	54.08(30)	0.63(15)
rovers	83.41(6)	22.9(6)	79.31(8)	110.38(16)

5 Implementation and Experimental Evaluation

As we already said in the introduction, we used SATPLAN (ver. of Feb. 2006) as underlying planning system. SATPLAN is the reference SAT-based system. SATPLAN can only handle STRIPS domains. We extended SATPLAN in order to incorporate such ideas (i.e., to implement QL-PLAN-GNT-1/QL-PLAN-GNT-2 at each makespan of the SATPLAN's approach), and we called SATPLAN-GNT the overall system: it implements PREF-DLL in MINISAT which is also one of the solvers SATPLAN can use, and that we set as default for SATPLAN.[2]

Experiments for minimal-actions plans. We considered several STRIPS domains from the first five IPCs (the recent IPC-6 does not have basic STRIPS problems). Given $P :=$ $\{\overline{a}|a \in act(\Pi_n)\}$ defined above, we considered both the qualitative preference $\langle P, \emptyset \rangle$ and the quantitative preference $\langle P, c \rangle$ in which c is the constant function 1, i.e, the setting where an uniform cost is associated to "not to perform" each action: the related objective function is thus the minimization of the number of actions involved in the plan. We used Warners encoding [13] (denoted with "W") to reduce to qualitative preferences (with a non-empty partial order): it showed the best performances on planning problems in [6,7], and it is thus the same used in SATPLAN(P). In Table 1 there are the results of our analysis. The first column is the domain of problem, then SATPLAN(P)(W) and SATPLAN-GNT(W) (resp. SATPLAN(P)() and SATPLAN-GNT()) denote the systems working on the quantitative (resp. qualitative) case. Results are presented as in the

---

[2] SATPLAN's default solver is SIEGE: we run SATPLAN with SIEGE and MINISAT and we have seen no significant differences in performances in terms of both CPU time and plans quality.

Max-SAT Evaluations[3] by $x(y)$, where y is the number of solved instances within the time limit (900s on a Linux box equipped with a Pentium IV 3.2GHz processor and 1GB of RAM), and x is the mean solving time of solved instances (used to break ties). "−" means that no instance is solved within the time limit. The 4 systems solve the same (sub)set of instances in all domains, but for satellite and storage in the qualitative case. We can note that in the quantitative case SATPLAN-GNT(W) has an edge over SATPLAN(P)(W): it often solves more instances, and never less, while in the qualitative case results are mixed. In general, SATPLAN-GNT convergence to the optimal solution is effective, and only few interactions are needed: we performed a detailed analysis on selected benchmarks, and we noted that, in mean, only 2.5 iterations were needed, and the quality of the first solution was already very good. In the qualitative case, on same domains, splitting preferentially on action variables, without the burden introduced by the W-encoding, can efficiently lead to the optimal solution. Finally, note that often the differences in performances are in the order of one/few instances, or just in term of mean CPU time: this is in line with state-of-the-art results, given that often in optimization problems solving even one more, or just in a faster way, the available benchmarks can be a significant result (e.g., in the Max-SAT Evaluations, where often the domain winner is granted by only the mean CPU time).

Experiments with soft goals. We considered two type of problems. First, we evaluated SATPLAN-GNT on some of the instances from [6,7]. Such instances were created from the original STRIPS instances of the domain we mentioned above, but considering all goals as being "soft" (but with the constraint that at least one has to be satisfied, otherwise the empty plan is always a solution). We do not show results for this analysis, but we just summarize it. The vast majority of these benchmarks were already solved very efficiently by SATPLAN(P). We considered 10 problems (each from a different domain) where SATPLAN(P) took considerable time to solve, in the quantitative case: on such problems, SATPLAN-GNT took in mean around half a time wrt SATPLAN(P) to solve them. But, besides the fact that in the instances so far mentioned goals are precisely soft, i.e., they can be satisfied, or not, without affecting plan validity, such instances are not fully satisfactory because they are non-conflicting, i.e., all soft goals can be satisfied at the same time. For this reason, given that the case in which not all the goals can be satisfied (often called over-subscription planning) is practically very important, we also evaluated some domains from the "SimplePreferences" track of the IPC-5, which include the possibility to express and reason on conflicting soft goals. Given that such domains are non-STRIPS, and some ADL constructs are used, we have used the following compilation technique: the preferences (goals) in the IPC-5 problems are translated into preconditions of dummy actions, which achieve new dummy literals defining the new problem goals. Then, these new actions can be compiled into STRIPS actions by using an existing tool (we have used both Hoffmann's tool for compiling ADL actions into STRIPS actions, namely ADL2STRIPS, and a modification of the same tool used in IPC-5, based on LPG). In our analysis we have included the domains where all goals are soft (but conflicting in general, changing all weights associated to goals violation to be 1), and preferences are only expressed on goals, i.e., the Storage and Pathways domains. Results are presented as in the reports of the IPC-5, considering,

---

[3] See http://www.maxsat07.udl.es/ for the last.

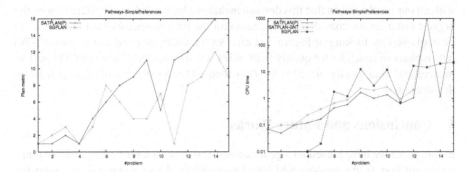

Fig. 2. Pathways domain, "SimplePreferences" track of IPC-5. Left: Plan metric, i.e., number of unsatisfied soft goals, for SATPLAN(P)(W) (and thus SATPLAN-GNT(W)) and SGPLAN. Right: CPU time for SATPLAN(P)(W), SATPLAN-GNT(W) and SGPLAN (in log scale).

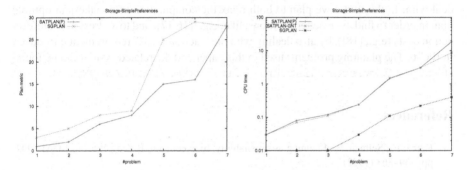

Fig. 3. Storage domain, "SimplePreferences" track of IPC-5. Left: Plan metric, i.e., number of unsatisfied soft goals, for SATPLAN(P)(W) (and thus SATPLAN-GNT(W)) and SGPLAN. Right: CPU time for SATPLAN(P)(W), SATPLAN-GNT(W) and SGPLAN (in log scale).

for each domain, both plan metric and CPU time to find the plan. In the analysis, we considered SATPLAN(P)(W) and SATPLAN-GNT(W) (given the metric is defined quantitatively in IPC-5 on soft goals) and, as a reference, SGPLAN, the clear winner of the "SimplePreferences" track at IPC-5. For the Pathways domain in Figure 2 we can note (Figure2 Right) that SATPLAN(P)(W) and SATPLAN-GNT(W) perform similarly (with SATPLAN(P)(W) being slightly better) and better than SGPLAN for non-easy (i.e., from problem #6, as numbered in the IPC-5) problems, but for two problems (#13 and #15) only solved by SGPLAN within the time limit. About the plan metric (Figure 2 Left), we can see that SGPLAN, overall, returns plans of slightly better quality, i.e., it can satisfy more soft goals. For the Storage domain, instead, in Figure 3 (Right) we can note that all systems solve all instances considered[4], with SGPLAN being around one order of magnitude faster than the other systems, which nonetheless solve each problem in less than 20s, with SATPLAN-GNT(W) being faster of around a factor of 2. This fact is in line

---

[4] We have considered all instances that the tools could compile. We are contacting the authors to be able to possibly compile bigger instances.

with all our results, given that this domain includes a high number of preferences on the biggest instances we considered. The reason for the performance gap wrt SGPLAN can be explained by looking at Figure 3 (Left): SATPLAN(P)(W) and SATPLAN-GNT(W) return plans of much better quality than SGPLAN. The tradeoff between CPU performances and plan quality of SATPLAN-GNT (and SATPLAN(P)) is effective, at least on this domain.

6   Conclusions and Future Works

In this paper we have presented a generate-and-test approach for finding optimal plans which, differently to a previous SAT-based approach (i) does not constrain the heuristic, (ii) works in polynomial space, and (iii) shows performance benefits. The most related approach is the one in [14] where the authors show how to extend GP-CSP in order to planning with preferences expressed as a TCP-net [15]. In the Boolean case, TCP-net can be expressed as Boolean formulas, and the problem they consider is the same we deal with. In the future we plan to both relax the computation of a makespan-optimal plan in order to find even better solutions, like, e.g., [16,17], and to address non-uniform action costs (e.g., [18]), by also dealing with the ":action-costs" requirement introduced in IPC-6. The planning problems used in this paper and the related system can be found at http://www.star.dist.unige.it/~marco/SATPLAN-GNT/.

References

1. Kautz, H., Selman, B.: Planning as satisfiability. In: Neumann, B. (ed.) Proc. of ECAI 1992, pp. 359–363 (1992)
2. Kautz, H., Selman, B.: Unifying SAT-based and graph-based planning. In: Dean, T. (ed.) Proc. of IJCAI 1999, pp. 318–325. Morgan Kaufmann, San Francisco (1999)
3. Hoffmann, J., Edelkamp, S.: The deterministic part of IPC-4: An overview. Journal of Artificial Intelligence Research 24, 519–579 (2005)
4. Gerevini, A., Haslum, P., Long, D., Saetti, A., Dimopoulos, Y.: Deterministic planning in the 5th IPC: PDDL3 and experimental evaluation of the planners. Artificial Intelligence 173(5-6), 619–668 (2009)
5. Xing, Z., Chen, Y., Zhang, W.: Maxplan: Optimal planning by decomposed satisfiability and backward reduction. In: Proc. of 5th IPC, ICAPS 2006, pp. 53–55 (2006)
6. Giunchiglia, E., Maratea, M.: Planning as satisfiability with preferences. In: Proc. of AAAI 2007, pp. 987–992. AAAI Press, Menlo Park (2007)
7. Giunchiglia, E., Maratea, M.: SAT-based planning with minimal-#actions plans and "soft" goals. In: Proc. of AI*IA 2007, pp. 422–433 (2007)
8. Järvisalo, M., Junttila, T.A., Niemelä, I.: Unrestricted vs restricted cut in a tableau method for boolean circuits. Annals of Mathematics and Artificial Intelligence 44(4), 373–399 (2005)
9. Castell, T., Cayrol, C., Cayrol, M., Berre, D.L.: Using the Davis and Putnam procedure for an efficient computation of preferred models. In: Proc. of ECAI 1996, pp. 350–354 (1996)
10. Gavanelli, M.: An algorithm for multi-criteria optimization in CSPs. In: Proc. of ECAI 2002, pp. 136–140. IOS Press, Amsterdam (2002)
11. DiRosa, E., Giunchiglia, E., Maratea, M.: A new approach for solving satisfiability problems with qualitative preferences. In: Proc. of ECAI 2008, pp. 510–514. IOS Press, Amsterdam (2008)

12. Tseitin, G.: On the complexity of proofs in propositional logics. Seminars in Mathematics 8 (1970)
13. Warners, J.P.: A linear-time transformation of linear inequalities into CNF. Information Processing Letters 68(2), 63–69 (1998)
14. Brafman, R.I., Chernyavsky, Y.: Planning with goal preferences and constraints. In: Proc. of ICAPS 2005, pp. 182–191. AAAI Press, Menlo Park (2005)
15. Boutilier, C., Brafman, R.I., Domshlak, C., Hoos, H.H., Poole, D.: CP-nets: A tool for representing and reasoning with conditional ceteris paribus preference statements. Journal of Artificial Intelligence Research 21, 135–191 (2004)
16. Büttner, M., Rintanen, J.: Satisfiability planning with constraints on the number of actions. In: Proc. of ICAPS 2005, pp. 292–299. AAAI Press, Menlo Park (2005)
17. Chen, Y., Lv, Q., Huang, R.: Plan-A: A cost-optimal planner based on SAT-constrained optimization. In: Proc. of 6th IPC, ICAPS 2008 (2008)
18. Keyder, E., Geffner, H.: Heuristics for planning with action costs revisited. In: Proc. of ECAI 2008, pp. 588–592. IOS Press, Amsterdam (2008)

A New Approach to Iterative Deepening Multiobjective A*

J. Coego, L. Mandow, and J.L. Pérez de la Cruz*

Dpto. Lenguajes y Ciencias de la Computación, Universidad de Málaga 29071,
Málaga, Spain
{jcoego,lawrence,perez}@lcc.uma.es

Abstract. Multiobjective search is a generalization of the Shortest Path
Problem where several (usually conflicting) criteria are optimized simultaneously. The paper presents an extension of the single-objective IDA*
search algorithm to the multiobjective case. The new algorithm is illustrated with an example, and formal proofs are presented on its termination, completeness, and admissibility. The algorithm is evaluated
over a set of random tree search problems, and is found to be more efficient than IDMOA*, a previous extension of IDA* to the multiobjective
case.

1 Introduction

Heuristic search in Shortest Path Problems is a central field of study in Artificial Intelligence. The Multiobjective Shortest Path Problem (MSPP) is an
extension of the Shortest Path Problem with practical applications in different
domains, like circuit partitioning [1], operator scheduling, channel routing [2], or
domain independent planning [3]. Multiobjective problems require the evaluation of several different and frequently conflicting objectives for each alternative.
These problems rarely have a single optimal solution. Most frequently, a set of
non-dominated (Pareto-optimal) solutions can be found, each one presenting a
particular trade-off between the objectives under consideration. The number of
non-dominated solutions in MSPP is known to grow exponentially with solution
depth in the worst case [4]. Fortunately, several classes of interesting multiobjective problems do not exhibit this worst-case behavior [5].

This paper deals with depth-first search strategies, which present the advantage of memory requirements linear with the depth of the solution. However, in
certain cases these algorithms may involve the consideration of an exponentially
larger set of paths when compared to best-first algorithms. Algorithms IDA* [6],
and RBFS [7] are members of this class. Previous multiobjective depth-first algorithms include IDMOA* [8], and MOMA*0 [2], the multiobjective extensions
of IDA* and RBFS respectively. This paper presents a natural extension of the

---

* Work partially funded by/Trabajo parcialmente financiado por: Consejería de Inno-
vación, Ciencia y Empresa. Junta de Andalucía (España) - P07-TIC-03018.

R. Serra and R. Cucchiara (Eds.): AI*IA 2009, LNAI 5883, pp. 264–273, 2009.

IDA* algorithm to the multiobjective case. The new algorithm is illustrated with an example, and formal proofs on its termination, completeness and admissibility are presented. Experimental tests over a set of random trees are presented. The algorithm is found to perform more efficiently than IDMOA*, a previous extension of IDA* to the multiobjective case.

The structure of the paper is as follows. Section 2 introduces some common terminology useful to understand multiobjective problems. Section 3 describes the new algorithm, and presents an illustrative example and several important formal properties. Section 4 describes the experimental setup, and analyzes the time requirements of the algorithms considered. Finally, some conclusions and future work are outlined.

2   Multiobjective Shortest Path Problems

Let us consider two q-dimensional vectors $v, v' \in \mathbb{R}^q$. A partial order relation $\prec$ denominated *dominance* is defined as follows, $v \prec v'$ iff $\forall i (1 \leq i \leq q)\ v_i \leq v'_i$ and $v \neq v'$. Given two q-dimensional vectors v and v' (where $q > 1$), it is not always possible to say that one is better than the other. For example in a bidimensional cost space vector $(2, 3)$ dominates $(2, 4)$, but no dominance relation exists between $(2, 3)$ and $(3, 2)$. Following this, an *indifference* relation $(v \sim v')$ is defined as v neither dominates, nor is dominated by, v'.

Given a set of vectors X, we shall define $nd(X)$ as the set of non-dominated vectors in X, i. e., $nd(X) = \{x \in X \mid \nexists y \in X \quad y \prec x\}$ Let G be a locally finite labeled directed graph $G = (N, A, c)$ with N nodes and A arcs (n, n') labeled with positive vectors $c(n, n') \in \mathbb{R}^q$, where $c(n, n') = (c_1, \ldots, c_q)$ being c_i the cost associated to the ith objective. The cost of a path is defined as the sum of the costs of its arcs; obviously, this cost is a q-dimensional vector. Let $g(n)$ denote the cost of the path stored in the search tree from the start node to n, $H(n)$ the set of non-dominated heuristic vectors of node n that estimate the cost of a solution from node n to a goal node and $F(n)$ the set of non-dominated evaluations of node n, computed as $f(n) = h(n) + g(n)$, where $h(n) \in H(n)$. A multiobjective search problem in G is stated as follows:

Given a start node $s \in N$ and a set of goal nodes $\Gamma \subseteq N$, find the set of all *non-dominated* paths P in G, i. e., the set of all paths P such that (i) P goes from s to a node in Γ; (ii) the cost of P is non dominated by the costs of any other path satisfying (i). Such set is called Γ^* and C^* is the set of all costs of non-dominated solution paths.

3   Algorithm PIDMOA*

Iterative deepening search proceeds by a sequence of depth-first searches or *iterations*. Before each iteration, a *threshold* is set and search is discontinued when this threshold is reached in each expanded path. The idea of iterative deepening was applied to heuristic search by Korf in [9] (algorithm IDA*). The main idea is that the threshold for iteration $i + 1$ is set as the minimum scalar $f(n)$ value of the nodes n at which search was discontinued in iteration i. Later this idea was generalized to multiobjective search and in this way algorithm IDMOA* was defined by Harikumar and Kumar in [8].

IDMOA* focuses on a single objective at a time, so it also maintains a scalar threshold to test generated nodes in order to discontinue search. This implies that tests against the current threshold will be carried out quickly. Initially IDMOA* focuses on the first objective, applying iterative deepening until it gets the set of non-dominated solutions that have the smallest value for the first objective. The same scheme applies to the remaining objectives, with the exception that each of them will also have an upper limit, given by the maximum value of such objectives in all solutions found so far. IDMOA* is proven to be admissible, that is, it finds the whole set of non-dominated solutions.

The main drawback of this algorithm is that while setting an objective, it does not process or take into account values that appear during the expansions for the rest of pending objectives. Besides this, although the dominance tests with the threshold vector remain simple, IDMOA* must include extra tests to delete non-dominated solutions possibly added to the solution set in prior steps.

These drawbacks of IDMOA* lead us to present a new approach to multiobjective iterative deepening. The resulting algorithm is called PIDMOA* (*Pareto Front Iterative Deepening Multiobjective A**).Contrary to IDMOA*, PIDMOA* takes into account all objectives *simultaneously*. That means that in each iteration we consider a set *Threshold* of non-dominated vectors and search is discountinued at any node n such that its vector valued cost is dominated by a vector in *Threshold*.

The arguments of PIDMOA* are a graph G, a start node s and a set of goal nodes Γ. It maintains a set SOLUTION of found solutions (initialized to $\varnothing$) and a set of thresholds *Threshold* (initialized to the subset of non-dominated heuristic vectors of the start node). Each solution is a pair $(\gamma, \boldsymbol{f}(\gamma))$ where γ is a solution node and $\boldsymbol{f}(\gamma) \in F(\gamma)$ is the value of a solution path to γ. Succesive iterations are defined by the actualizations of the set of cost vectors *Threshold*. When the set is empty, the algorithm terminates.

Actualizations of *Threshold* are done by performing a depth-first search DFS starting from s. During this search, when reaching a node n the following tests are done: (i) if n is fully dominated by previously found solutions, n is discarded and search is discontinued at n; (ii) if $\boldsymbol{f}(n)$ is greater than some current threshold, it is accumulated as a threshold for the next iteration and search is discontinued at n; (iii) if n is a goal node, n is added to SOLUTION; (iv) if none of the above conditions hold, search is continued in a depth-first fashion.

3.1   Example

Figure 1 shows a sample tree search problem, where each arc is labelled with a vector cost (2 objectives), and each node is labelled with an heuristic vector estimate. Search starts at the root node. PIDMOA* would set its threshold initially to $\{(5,5)\}$, i.e. the heuristic estimate of the root. In its 1st iteration, all nodes with estimates non-dominated by the Threshold would be explored in a depth first fashion. The set of explored nodes and the Threshold of iteration 1 are shown in figure 2. At the 2nd iteration, the Threshold would is updated to $\{(7,5),(5,7)\}$, i.e. the non-dominated set of all path costs that exceeded the previous threshold at iteration 1. The new threshold, and the set of explored nodes in the 2nd iteration is shown in figure 3. Subsequent iterations perform in a similar fashion and are shown in figures 4-5. A solution is found at iteration 3 (cost $(10,5)$). All vectors dominated by this solution are discarded from further thresholds. Search is also discontinued at nodes with cost dominated by solutions already found. A 2nd solution is found at iteration 4 (cost $(5,10)$), and the threshold becomes empty. Solutions are found at iteration 4, and the Threshold becomes empty. Next we can see the pseudocode of the algorithm.

PIDMOA* (G, s, Γ)
SOLUTION = $\varnothing$; Threshold = $\{\boldsymbol{h}(s)\}$
WHILE Threshold $\neq \varnothing$
      Threshold = DFS (s, Threshold)
return (SOLUTION)

DFS (node, threshold_par)
nodomvectorsf = $\{\boldsymbol{f}(node) \in F(node) \mid (\nexists(\gamma, P^*(\gamma)) \in SOLUTION \mid P^*(\gamma) \preceq \boldsymbol{f}(node))\}$
IF (nodomvectorsf = $\varnothing$) THEN return $\varnothing$;
domvectorsf = $\{\boldsymbol{f}(node) \in F(node) \mid (\exists \boldsymbol{c} \in threshold\_par \mid \boldsymbol{c} \prec \boldsymbol{f}(node))\}$
IF (domvectorsf $\neq \varnothing$) THEN return domvectorsf;
IF (node $\in \Gamma$) THEN
      SOLUTION = SOLUTION $\cup$ $\{(node, \boldsymbol{f}(node))\}$
ELSE
      threshold_dfs = $\varnothing$
      successors = expand_node (node)
      FOR each n in successors DO
            threshold_dfs = threshold_dfs $\cup$ DFS(n, threshold_par)
            delete dominated vectors within threshold_dfs;
      return (threshold_dfs)

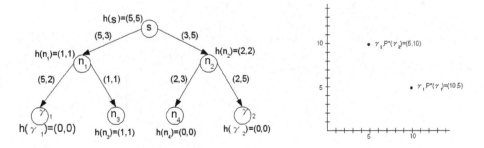

Fig. 1. Multiobjective problem

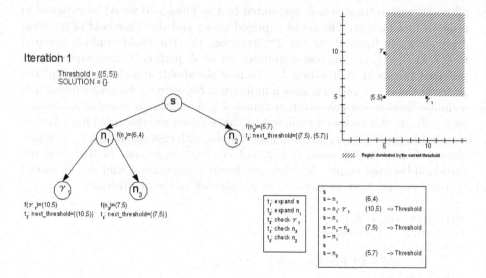

Fig. 2. 1st iteration of PIDMOA*

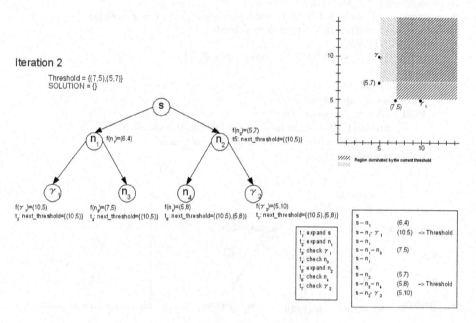

Fig. 3. 2nd iteration of PIDMOA*

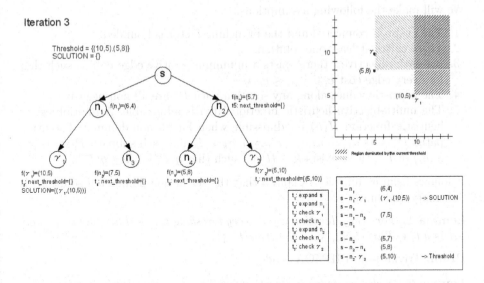

Fig. 4. 3rd iteration of PIDMOA*

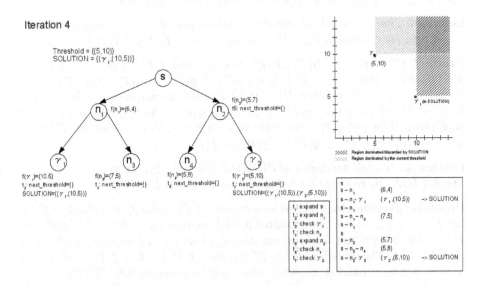

Fig. 5. 4th iteration of PIDMOA*

3.2   Properties of PIDMOA*

We will make the following assumptions:

1. The graph is connected and the branching factor is bounded.
2. There exists at least one solution.
3. For each objective i there exists a minimum positive edge cost ε_i such that for every edge cost c, $0 < \varepsilon_i \le c_i \forall i$.
4. The heuristic values along any solution path P^* are all non-negative.
5. The multiobjective heuristic function H(n) is admissible. A multiobjective heuristic function $H(n)$ is admissible when for all non-dominated solution paths $P^* = (s, n_1, ..., n_i, n_{i+1}, ..., \gamma_k)$, $\gamma_k \in \Gamma$ and each subpath $P_i^* = (s, n_1, ..., n_i)$ of P^*, there exists $h \in H(n_i)$ such that $g(P_i^*) + h \preceq g(P^*)$.

It follows from assumptions 2 and 3 that the set of non-dominated solutions is nonempty and finite.

Lemma 1. *For every iteration i and every threshold $t_{i+1} \in Threshold_{i+1}$, there exists a $t_i \in Threshold_i$ such that $t_i \prec t_{i+1}$.*

Proof: Trivial from PIDMOA* code.

Lemma 2. *Nodes expanded in iteration i will be also expanded in iteration $i+1$.*

Proof: If n was expanded in interation i, then for all $t_i \in Threshold_i$, $f(n) \sim t_i$ or $f(n) \preceq t_i$. From Lemma 1, we have that $t_i \prec t_{i+1}$, hence $f(n) \sim t_{i+1}$, or $f(n) \preceq t_i \prec t_{i+1}$, hence $f(n) \prec t_{i+1}$. That is, for all $t_{i+1} \in Threshold_{i+1}$, $f(n) \sim t_{i+1}$ or $f(n) \prec t_{i+1}$, so n will be expanded in iteration $i + 1$.

Lemma 3. *For every non-dominated solution path $P^* = (s = n_0, n_1, \ldots, n_i, \ldots, \gamma)$ and every subpath $P_i^* = (s = n_0, n_1, \ldots, n_i)$, $f(P_i^*) \preceq f(P^*)$.*

Proof: $f(P_i^*) = g(P_i^*) + h(P_i^*)$, $h(P_i^*) \in H(n_i)$ and $f(P^*) = g(P^*) + h(P^*) = g(P^*) + h(\gamma)$. But $h(\gamma) = 0$, hence $f(P^*) = g(P^*) = g(P_i^*) + g(P_r^*)$ where $P_r^* = (n_i, n_{i+1}, \ldots, \gamma)$. Since $H(n)$ is admissible (assumption 5), there exists $h(n_i) \in H(n_i)$ such that $g(P_i^*) + h(n_i) \preceq g(P^*)$. Substracting $g(P_i^*)$ from both terms, $h(n_i) \preceq g(P_r^*)$. So it exists an $h(n_i) \in H(n_i)$ such that $f(P_i^*) = h(n_i) + g(P_i^*) \preceq g(P_i^*) + g(P_r^*) = g(P^*) = f(P^*)$.

Lemma 4. *In each iteration i of PIDMOA*, for all $\gamma \in \Gamma$ such that $(\gamma, f(\gamma)) \notin SOLUTION$, there exists a $t \in Threshold_i$ such that $t \prec f(\gamma)$.*

Proof: Let us assume that there does not exist $t \in Threshold_i$ such that $t \prec f(\gamma)$. Let $P^* = (s, n_1, \ldots, \gamma)$. From Lemma 3, we have that for all $P_i^* \subset P^*$, $f(P_i^*) \preceq f(P^*)$. Since γ was not found in iteration i, for all $t \in Threshold_i$, $f(P^*) \sim t$ or $f(P^*) \preceq t$. Therefore for all $t \in Threshold_i$, $f(P_i^*) \sim t$ or $f(P_i^*) \preceq t$. Every $n_i \in P^*$ will be expanded, except γ, which will be added to SOLUTION.

Lemma 5. *If there exists $P^*(\gamma_j) = (s, \ldots, \gamma_j)$, $\gamma_j \in \Gamma$ such that $(\gamma_j, f(P_{\gamma_j}^*)) \in SOLUTION$, then it does not exist $P^*(\gamma_i) = (s, \ldots, \gamma_i)$, $\gamma_i \in \Gamma$ such that $f(P_{\gamma_i}^*) \prec f(P_{\gamma_j}^*)$. That is, for all $(\gamma_j, f(P_{\gamma_j}^*)) \in SOLUTION$, $(\gamma_j, f(P_{\gamma_j}^*)) \in C^*$; all the solutions found by PIDMOA* are non-dominated solutions.*

Proof: Let us assume that there exists $\gamma_i \in \Gamma$ such that $\boldsymbol{f}(P^*_{\gamma_i}) \prec \boldsymbol{f}(P^*_{\gamma_j})$. If $(\gamma_i, \boldsymbol{f}(P^*_{\gamma_i})) \in$ SOLUTION, then $(\gamma_j, \boldsymbol{f}(P^*_{\gamma_j}))$ will never join SOLUTION because of the first test in PIDMOA* code (associated to variable *nodomvectorsf*). On the other hand, if $(\gamma_i, \boldsymbol{f}(P^*_{\gamma_i})) \notin$ SOLUTION, since $(\gamma_j, \boldsymbol{f}(P^*_{\gamma_j})) \in$ SOLUTION, then (a) for all $\boldsymbol{t} \in Threshold_i$, $P^*_{\gamma_j} \sim \boldsymbol{t}$, or $P^*_{\gamma_j} \prec \boldsymbol{t}$. But $(\gamma_i, \boldsymbol{f}(P^*_{\gamma_i})) \notin$ SOLUTION, therefore there exists $\boldsymbol{t} \in Threshold_i$ such that $\boldsymbol{t} \prec \boldsymbol{f}(P^*_{\gamma_i})$. So, there exists $\boldsymbol{t} \in Threshold_i$ such that $\boldsymbol{t} \prec \boldsymbol{f}(P^*_{\gamma_i}) \prec \boldsymbol{f}(P^*_{\gamma_j})$, which contradicts (a).

Theorem 1. *When PIDMOA* finishes, SOLUTION = C^*, that is, every non dominated solution will be found by PIDMOA*.*

Proof: Let us assume that there exists $(\gamma_x, \boldsymbol{f}(P^*_{\gamma_x})) \in C^*$ such that $(\gamma_x, \boldsymbol{f}(P^*_{\gamma_x})) \notin SOLUTION$. From Lemma 4, there exists $\boldsymbol{t} \in Treshold_n$ such that $\boldsymbol{t} \prec \boldsymbol{f}(P^*_{\gamma_x})$. Since $(\gamma_x, \boldsymbol{f}(P^*_{\gamma_x})) \in C^*$, $\boldsymbol{f}(P^*_{\gamma_x})$ will be returned from DFS call and $Threshold_n$ will not be empty and hence PIDMOA* will not finish.

Theorem 2. *PIDMOA* finishes and at its termination SOLUTION = C^*.*

Proof: By theorem 1, it suffices to prove that PIDMOA* always finishes. Let $c_{max} = (v_1, \ldots, v_k)$, where $v_i = \max\{y_i\}$ and y_i is the i-th objective value for every $\boldsymbol{f}(P^*_{\gamma_x})$ included in SOLUTION. Then for all $(\gamma_i, \boldsymbol{f}(P^*_{\gamma_i})) \in SOLUTION$ it holds that $\boldsymbol{f}(P^*_{\gamma_i}) \prec c_{max}$. By assumptions 1 and 3, each expanded path will reach a cost of c_{max} in at most $\lceil \frac{\max\{v_i\}}{\max\{\varepsilon_i\}} \rceil$ steps. At step $\lceil \frac{\max\{v_i\}}{\max\{\varepsilon_i\}} \rceil$, each expanded node will verify that there exists $(\gamma, \boldsymbol{f}(P^*_\gamma)) \in SOLUTION$ such that $\boldsymbol{f}(P^*_\gamma) \prec \boldsymbol{f}(node)$ and, as a result of this, the set of thresholds for the next iteration will be empty, so PIDMOA* will finish.

4 Empirical Evaluation

Several rounds of empirical evaluations were carried out to compare the performance of IDMOA* and PIDMOA*. Complete binary trees of increasing depth were generated up to depth 20. Arcs were labeled with integer random costs in the range [1,5] for two different objectives. Approximately 10% of the leaf nodes were designated as goal nodes in each tree. Both algorithms were run without heuristic (blind search), and using the difference between tree depth and node depth as an optimistic heuristic estimate for all nodes and both objective functions (heuristic search). This experimental setup is easy to reproduce. All results presented below are averaged for three different problems for each tree depth.

Figure 6a shows the time requirements of both algorithms. All of them show a sharp increase in time requirements, in accordance with the exponential growth rate of the explored graphs. Therefore these results can be better appreciated using a logarithmic scale for the time axis (figure 6b). Three important conclusions can be drawn. In the first place, the time requirements of BIDMOA are more than *an order of magnitude* smaller than those of IDMOA for the hardest problems considered (depth 20). Furthermore, the growth rate of time requirements with tree depth appears to be *larger* for IDMOA than for BIDMOA. Finally,

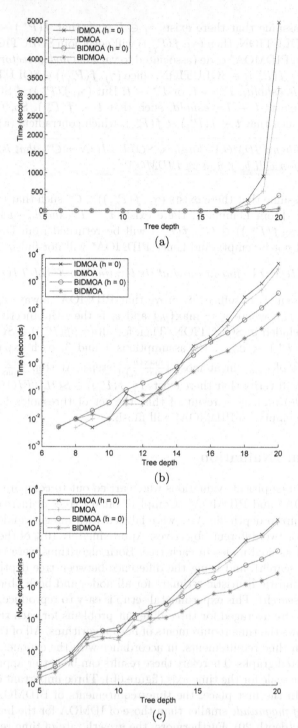

Fig. 6. (a) Time requirements with linear time scale; (b) Time requirements with logarithmic time scale; (c) Node expansions with logarithmic node scale

BIDMOA makes *better use of heuristic information*, achieving a comparatively larger reduction in time requirements when compared to blind search. An explanation of this behavior can be found in figure 6c, which shows the number of node expansions performed by the algorithms (logarithmic scale was used for the vertical axis). Analogous behavior of the algorithms was observed in additional test sets with larger cost ranges (the largest range tested was [1,100]).

5 Conclusions and Future Work

The paper presents PIDMOA*, a natural extension of the heuristic search algorithm IDA* to the multiobjective case. The algorithm is shown to be optimal and complete under reasonable asumptions. PIDMOA* optimizes all objectives simultaneously, and performs iterative deepening using dominance checks over a set of vector-valued bounds to discontinue search at each deepening stage. This is in contrast with IDMOA*, that sequentially optimizes each objective, and uses scalar bounds for each deepening stage. Experimental tests over randomly generated binary trees with two objectives show that the deepening strategy of PIDMOA* clearly outperforms IDMOA* in time requirements and node expansions, and makes better use of heuristic information.

Future work includes deeper formal and experimental analyses of the performance of PIDMOA*, and comparison to other multiobjective search algorithms.

References

1. Harikumar, S., Kumar, S.: Multiobjective search based algorithms for circuit partitioning problem for acceleration of logic simulation. In: Tenth Int. Conf. on VLSI Design, pp. 239–242 (1997)
2. Dasgupta, P., Chakrabarti, P.P., DeSarkar, S.C.: Multiobjective Heuristic Search. Vieweg, Braunschweig/Wiesbaden (1999)
3. Refanidis, I., Vlahavas, I.: Multiobjective heuristic state-space search. Artificial Intelligence 145, 1–32 (2003)
4. Hansen, P.: Bicriterion path problems. Lecture Notes in Economics and Mathematical Systems, vol. 177, pp. 109–127. Springer, Heidelberg (1979)
5. Müller-Hannemann, M., Weihe, K.: On the cardinality of the pareto set in bicriteria shortest path problems. Annals OR 147(1), 269–286 (2006)
6. Korf, R.E.: Iterative-deepening A*: an optimal admissible tree search. In: Proc. of the IX Int. Joint Conf. on Artificial Intelligence (IJCAI 1985), pp. 1034–1036 (1985)
7. Korf, R.E.: Linear-space best-first search. Artificial Intelligence 62, 41–78 (1993)
8. Harikumar, S., Kumar, S.: Iterative deepening multiobjective A*. Information Processing Letters 58, 11–15 (1996)
9. Korf, R.E.: Depth first iterative deepening: an optimal admissible tree search. Artificial Intelligence 27, 97–109 (1985)

High Performing Algorithms for MAP and Conditional Inference in Markov Logic

Marenglen Biba, Stefano Ferilli, and Floriana Esposito

Department of Computer Science
University of Bari, Via E. Orabona, 4, 70125, Bari, Italy
biba@di.uniba.it, ferilli@di.uniba.it, esposito@di.uniba.it

Abstract. Markov Logic (ML) combines Markov networks (MNs) and first-order logic by attaching weights to first-order formulas and using these as templates for features of MNs. However, MAP and conditional inference in ML are hard computational tasks. This paper presents two algorithms for these tasks based on the Iterated Robust Tabu Search (IRoTS) metaheuristic. The first algorithm performs MAP inference by performing a biased sampling of the set of local optima. Extensive experiments show that it improves over the state-of-the-art algorithm in terms of solution quality and inference times. The second algorithm combines IRoTS with simulated annealing for conditional inference and we show through experiments that it is faster than the current state-of-the-art algorithm maintaining the same inference quality.

1 Introduction

Many real-world applications of AI require both probability and first-order logic (FOL) to deal with uncertainty and structural complexity. Logical approaches successfully handle complexity, while statistical ones successfully handle uncertainty. Recently, in the field of statistical relational learning (SRL), several models for combining logical and statistical approaches have been proposed [3]. Most of these approaches combine probabilistic graphical models (PGMs) with subsets of FOL. This paper focuses on Markov Logic (ML) [11], a powerful representation that has finite FOL and PGMs as special cases. It extends FOL by attaching weights to formulas and using these as templates for MNs.

Maximum *a posteriori* (MAP) inference in MNs means finding the most likely state of output variables given the state of the input variables and this problem is NP-hard. Since for ML, the MAP state is the state that maximizes the sum of the weights of the satisfied ground clauses, it can be found by weighted MAX-SAT solvers. The authors in [14] use MaxWalkSAT [13] to find the MAP state. This paper presents a novel algorithm that exploits Iterated Local Search (ILS) [7] and Robust Tabu Search (RoTS) [16] metaheuristics. Experiments in real-world domains show that it outperforms the state-of-the-art algorithm for MAP inference in ML, in terms of solutions quality and inference running times.

R. Serra and R. Cucchiara (Eds.): AI*IA 2009, LNAI 5883, pp. 274–283, 2009.
© Springer-Verlag Berlin Heidelberg 2009

Conditional inference in graphical models involves computing the distribution of query variables given evidence and it was shown to be #P-complete [12]. An inference algorithm for ML is that of [10] where the authors combine ideas from satisfiability (SAT) and Markov Chain Monte Carlo (MCMC) methods. It uses SampleSAT [17] as a subroutine. In this paper we propose the novel algorithm SampleIRoTS based on the ILS and RoTS metaheuristics, that interleaves RoTS steps with simulated annealing ones in an ILS. SampleIRoTS is then plugged in the novel algorithm MC-IRoTS. Experimental evaluation shows that on a large number of inference tasks, MC-IRoTS performs faster than the state-of-the-art algorithm while maintaining the same quality of predicted probabilities.

We begin by briefly introducing ML, then describe the algorithms with the relative experiments and then conclude.

2   Markov Logic and Probabilistic Inference in MLNs

A MN models the joint distribution of a set of variables $X = (X_1, X_2, \ldots, X_n) \in \chi$ [2]. It is composed of an undirected graph G and a set of potential functions. The graph has a node for each variable, and the model has a potential function ϕ_k for each clique in the graph. The joint distribution represented by a MN is given by: $P(X = x) = \frac{1}{Z} \prod_k \phi_k(x_{\{k\}})$ where $x_{\{k\}}$ is the state of the kth clique. Z, known as the *partition function*, is given by: $Z = \sum_{x \in \chi} \prod_k \phi_k(x_{\{k\}})$. A Markov Logic Network [11] L is a set of pairs $(F_i; w_i)$, where F_i is a formula in FOL and w_i is a real number. Together with a finite set of constants $C = \{c_1, c_2, \ldots, c_p\}$ it defines a MN $M_{L;C}$ as follows: 1. $M_{L;C}$ contains one binary node for each possible grounding of each predicate appearing in L. 2. $M_{L;C}$ contains one feature for each possible grounding of each formula F_i in L. The weight of the feature is w_i associated with F_i in L. There is an edge between two nodes iff the corresponding ground predicates appear together in at least one grounding of one formula in L. The probability distribution over possible worlds x specified by $M_{L;C}$ is given by $P(X = x) = \frac{1}{Z} \exp(\sum_{i=1}^{F} w_i n_i(x))$ where F is the number of formulas in the MLN and $n_i(x)$ is the number of true groundings of F_i in x.

In [9] was shown how the problem of finding the most likely state of a Bayesian network given evidence can be solved by reduction to weighted SAT. For MLNs, MAP inference reduces to finding a truth assignment that maximizes the sum of weights of satisfied clauses. The state-of-the-art algorithm for MAP inference in ML proposed in [14] is based on MaxWalkSat. MCMC methods are the most popular approach to approximate inference [4]. However, inference must also handle deterministic dependencies and MCMC methods give poor results for deterministic or near deterministic ones. On the other side, SAT procedures cannot be applied to probabilistic dependencies. An approach to deal with both is that of [10] which uses SampleSAT [17] in a MCMC algorithm to uniformly sample the SAT solutions. SAT solvers find solutions very fast but sample highly non-uniformly, whereas MCMC methods may take exponential time to converge. The authors in [17] proposed a strategy combining random walk with Metropolis transitions. Experimental results in [10] show that MC-SAT greatly outperforms Gibbs sampling.

Algorithm 1. Iterated Robust Tabu Search

Input: C: set of weighted clauses in CNF, BestScore: current best score)
CL_C = Random initialization of truth values for atoms in C;
$CL_S = LocalSearch_{RoTS}(CL_C)$;
BestAssignment = CL_S; BestScore = Score(CL_S);
repeat
   $CL'_C = Perturb_{RoTS}(BestAssignment)$;
   $CL'_S = LocalSearch_{RoTS}(CL'_C)$;
   if Score(CL'_S) $\geq$ BestScore **then**
      BestScore = Score(CL'_S)
   end if
   BestAssignment = accept(BestAssignment,CL'_S);
until k consecutive steps have not produced improvement
Return BestAssignment

3   The IRoTS Algorithm

SAT is the problem of deciding whether a given formula has a model. Here we focus on function-free FOL with the domain closure assumption. Propositionalization is the process of replacing a FOL knowledge base (KB) by an equivalent propositional one. A FOL KB is satisfiable iff the equivalent propositional KB is satisfiable. Thus, inference for FOL can be performed by propositionalization followed by SAT testing.

Among the best performing metaheuristics for MAX-SAT are Tabu Search (TS) [5] and Iterated Local Search (ILS) [7]. A highly competitive algorithm is ILS-YI [18] that uses local search based on 2- and 3-flip neighbourhoods. Another competitive algorithm is Robust Tabu Search (RoTS) [16], a special case of TS. In each step RoTS flips a non-tabu variable that achieves a maximal improvement in the total weight of unsatisfied clauses and declares it tabu for the next tt steps. tt is called the tabu tenure. An exception to this "tabu" rule is made if a more recently flipped variable achieves an improvement over the best solution so far (*aspiration*). Furthermore, whenever a variable has not been flipped for a long time, it is forced to be flipped. This implements a form of long-term memory and helps prevent stagnation of search. Finally, instead of using a fixed tabu tenure, every n iterations the parameter tt is randomly chosen from an interval $[tt_{min}, tt_{max}]$ according to a uniform distribution.

The original version of IRoTS for MAX-SAT was proposed in [15] showing that it is highly competitive on many MAX-SAT instances. On weighted and unweighted Uniform Random 3-SAT instances, IRoTS performs significantly better than state-of-the-art algorithms. Algorithm 1 starts by randomly initializing the truth values of the atoms. Then it uses RoTS to reach a local optimum CL_S. At this point, a perturbation method based on RoTS is applied leading to CL'_C and then again RoTS is applied to reach another local optimum CL'_S. The *accept* function decides whether the search must continue from the previous or last found local optimum CL'_S. In *Perturb* we use a fixed number of RoTS steps $9n/10$ with tabu tenure $n/2$ where n is the number of atoms. Regarding $LocalSearch_{RoTS}$, it performs RoTS steps until no improvement for n^2/d steps (we call d threshold ratio) with a tabu tenure $n/10 + 4$. The IRoTS that we develop here is different from [15] in that we do not dynamically adapt the tabu

Algorithm 2. The MC-IRoTS algorithm

```
MC-IRoTS (clauses, numSamples)
x^(0) ← Satisfy(hard clauses)
for i = 1 to numSamples do
    M ← φ
    for all c_k ∈ clauses satisfied by x^(i 1) do
        With probability 1 − e^{w_k} add c_k to M
    end for
    Sample x^(i) ∼ Unif_{SAT(M)}
end for
```

tenure and do not use a probabilistic choice in *accept* (we always accept the best solution found so far). Moreover, we use $d = 1$, whereas the authors in [15] use $d = 4$.

4   The MC-IRoTS Algorithm

The most widely used MCMC method is Gibbs sampling which proceeds by sampling each variable in turn given its Markov blanket. In order to generate samples from the correct distribution, it is sufficient that the Markov chain satisfy ergodicity and detailed balance. In the presence of strong dependencies, changes to the state of a variable given its neighbors become unlikely, and convergence of the estimates becomes very slow. In the limit of deterministic dependencies, ergodicity breaks down. Another widely used approach relies on auxiliary variables to capture the dependencies. This technique is also known as *slice sampling* [1]. The question whether random walk SAT procedures can be used to sample uniformly or near-uniformly from the space of satisfying assignments, was first dealt with in [17]. It was shown that random walk procedures can reach the full set of solutions of complex KBs. Moreover, by interleaving random walk steps with Metropolis transitions, it was shown that sampling becomes near-uniform. At near-zero temperature, simulated annealing (SA) samples solutions uniformly, but will take too long to find them. WalkSAT finds solutions very fast, but samples them non-uniformly. SampleSAT [17] samples solutions near-uniformly and highly efficiently by, at each iteration, performing a WalkSAT step with probability p and a simulated annealing step with probability $1 - p$. Here the idea is to combine IRoTS with SA. The novel algorithm performs with probability p a RoTS step and with probability $1 - p$ a SA step. We call this algorithm SampleIRoTS and the expectation is that it will be faster than SampleSAT. The goal is to exploit SampleIRoTS in an inference algorithm for ML and compare it with the state-of-the-art algorithm. The basic idea of how to use SampleSAT in an inference algorithm, was first proposed in [10]. The MC-IRoTS algorithm that we propose here, applies slice sampling to ML by using SampleIRoTS to sample a new state given the auxiliary variables. Algorithm 2 gives pseudo-code for MC-IRoTS similar to the one in [10]. $Unif_{SAT(M)}$ is the uniform distribution over the set SAT(M). At each step, hard clauses are selected with probability 1, and all sampled states satisfy them. As shown in [10], this kind of algorithm, generates a Markov chain which satisfies ergodicity and detailed balance.

Table 1. Inference results for predicate advisedBy with MLNs learned first with 500 iterations of PSCG (table above) and then by running PSCG for 10 hours (table below

fold	IROTS		MWS-T		MWS-TR			
	Cost	Time	Cost	Time	Cost	Time	preds	clauses
ai	93103.7	56.65	92393.5	62.98	92512.9	60.27	4760	185849
graphics	72221.8	26.91	72245.1	28.88	71659.8	46.30	3843	136392
language	32398.1	1.03	32668.2	1.08	32380.1	1.06	840	15762
systems	117144.0	125.71	118416.0	134.35	118629.0	192.16	5328	218918
theory	71726.1	17.85	71727.9	19.30	71873.1	34.03	2499	73600
average	77318.7	45.63	77490.1	49.32	77411.0	66.76	-	-
fold	Cost	Time	Cost	Time	Cost	Time	preds	clauses
ai	98513.5	71.34	99737.8	78.31	99876.4	74.53	4760	185849
graphics	28007.9	26.3	28074.8	28.02	28005.2	43.06	3843	136392
language	10985.8	1.48	11070.8	1.57	10711.3	1.52	840	15762
systems	73154.6	55.06	73471.8	59.92	73642.9	57.98	5328	218918
theory	90979.1	25.53	89517.9	26.73	89462.7	49.06	2499	73600
average	60328.2	35.94	60374.6	38.91	60339.7	45.23	-	-

5   Experiments

Regarding IRoTS we want to answer these questions: **(Q1)** Does IRoTS improve over the state-of-the-art algorithm for MLNs in terms of solutions quality? **(Q2)** Does IRoTS performance depend on the particular configuration of clauses' weights? **(Q3)** Does IRoTS performance depend on particular features of the dataset, i.e., number of ground clauses and predicates? **(Q4)** In case IRoTS finds better solutions than the state-of-the-art algorithm, what is the performance in terms of running times? **(Q5)** What is the performance of IRoTS for huge relational domains with hundreds of thousands of ground predicates and clauses? To perform MAP inference we need MLN models and evidence data. MLNs can be hand-coded or learned from training data. Since the goal here is to perform inference for complex models where it is hard to find the MAP state, we generated complex models from real-world data and tested IRoTS against the current state-of-the-art algorithm of [14]. We took as a dataset, the UW-CSE dataset and the MLN hand-coded model that comes together with this dataset. For the first experiment we learned weights using the algorithm PSCG [8] giving advisedBy as non-evidence predicate. We trained the algorithm for 500 iterations on each area of the dataset. After learning the MLNs, we performed MAP inference with query predicate advisedBy.

In order to equally compare IRoTS with MaxWalkSAT, we compared IRoTS with the tabu version of MaxWalkSAT and by using the same number of search steps for both algorithms. For IRoTS, parameters $d = 1$ and $k = 3$. We observed that on the language and theory folds the iterations were very fast and 3 steps without improvement were too few. Thus we used $k = 10$ only for these areas, and $k = 3$ for the rest. Anyway, for IRoTS we counted the overall number of flips (including perturbation) and we used the same number for MaxWalkSAT with tabu (MWS-T). The tabu tenure for MWS-T was set to the default of alchemy [6] (equal to 5). The results are reported in Table 1 where for each algorithm we report the cost of false clauses of the final solution and running time in minutes.

Table 2. Inference results for query predicates advisedBy (table above) and tempAdvisedBy (table below) on a model learned with both advisedBy and tempAdvisedBy as non-evidence predicates

fold	IROTS Cost	IROTS Time	MWS-T Cost	MWS-T Time	MWS-TR Cost	MWS-TR Time	preds	clauses
ai	50.85	2.15	50.85	13.55	50.85	10.30	4760	185762
graphics	62.10	2.10	62.10	8.40	62.10	6.30	3843	136297
language	9.75	0.12	9.75	0.32	9.75	0.27	840	15711
systems	52.96	11.20	52.96	80.71	52.96	60.07	5328	218820
theory	57.23	0.15	57.23	2.08	57.23	1.63	2499	73540
average	46.58	3.14	46.58	21.01	46.58	15.71	-	-
fold	Cost	Time	Cost	Time	Cost	Time	preds	clauses
ai	16112.70	93.89	16669.30	101.98	16309.00	143.29	4760	185672
graphics	12872.90	30.63	13153.10	32.81	12765.50	33.03	3843	136244
language	2238.57	0.73	2196.23	0.75	2024.31	0.68	840	15706
systems	19722.50	61.07	20352.70	65.65	19938.30	95.62	5328	218727
theory	7388.90	47.85	7668.17	55.60	7600.41	34.50	4892	261078
average	11667.11	46.83	12007.90	51.36	11727.50	61.42	-	-

Since IRoTS uses the perturbation procedure to escape local optima, it would be fair to compare IRoTS with a version of MWS-T that uses a similar mechanism to jump in a different region of the search space. Thus we compared IRoTS also with MWS-TR (Tabu&Restarts) with a number of ten restarts and number of flips for each iteration equal to 1/10th of the overall number of flips. In this way, the equality of comparison is maintained in order to perform the same number of flips for all algorithms. Moreover, for MWS-T we used the default tabu tenure of alchemy that is five, but it would be more interesting to compare IRoTS with MWS-TR with the same tabu tenure as IRoTS, i.e., $n/10 + 4$. Thus we used this tabu tenure for MWS-TR.

As it can be seen IRoTS on average finds solutions of higher quality than the other algorithms. IRoTS is also faster even though the number of search steps is the same. Thus, questions **(Q1)** and **(Q4)** can be answered affirmatively. However, we want to be sure that IRoTS performance towards the other algorithms does not depend on the particular weights of the model. For this reason we decided to generate other MLNs on the same dataset but with different weights. We did this by using again PSCG and running it for 10 hours instead of 500 iterations for each training set. This guarantees that the MLNs generated be different in terms of the clauses' weights. Inference results regarding these MLNs are reported in Table 1. As can be seen, again IRoTS performs better, thus question **(Q2)** can be answered affirmatively, since for the same number of groundings but with different clauses' weights, IRoTS finds better solutions.

A crucial question is whether IRoTS performance depends on the number of groundings. It is important to maintain the same performance for any number of groundings. Therefore we decided to consider an additional query predicate. We learned the weights using PSCG (50 iterations) but this time considering as non-evidence predicates two predicates: advisedBy and tempAdvisedBy. The final MLNs learned should predict the probability for all groundings of both predicates. We report experiments for each predicate in turn and finally for an inference task with both query predicates. This way we will have a different

Table 3. Results for query predicates advisedBy and tempAdvisedBy in a single inference task (table above) and three query predicates taughtBy, advisedBy and tempAdvisedBy (table below) using a model learned with advisedBy and tempAdvisedBy as non-evidence predicates

fold	IROTS		MWS-T		MWS-TR			
	Cost	Time	Cost	Time	Cost	Time	preds	clauses
ai	13367.10	294.01	15659.60	278.3	14610.00	331.79	9384	680351
graphics	11511.40	166.78	12622.90	160.78	11994.40	150.17	7564	495227
language	1996.73	7.1	2054.05	8.71	2004.55	8.1	1624	52491
systems	16823.70	362.78	19283.50	339.2	18484.60	337.44	10512	804425
theory	6845.18	80.65	7545.73	104.3	7112.81	103.05	4900	261890
average	10108.82	182.26	11433.16	178.26	10841.27	186.11	-	-
fold	Cost	Time	Cost	Time	Cost	Time	preds	clauses
ai	509738	52.97	536670	49.22	538508	50.52	23664	1894428
graphics	338790	40.33	338554	35.58	339004	32.62	23485	1510794
language	37034	1.18	43756.2	1.17	42886.7	1.14	5152	157944
systems	494128	48.58	619380	38.58	604742	36.62	26136	2045461
theory	175668	15.33	214252	11.5	210758	12.62	14504	794484
average	311071.6	31.68	350522.44	27.21	347179.74	26.7	-	-

number of groundings compared to the previous experiments. Results for each query predicate are reported in Table 2. For advisedBy, the algorithms find the same solution but IRoTS is much faster, while for tempAdvisedBy IRoTS performs much better than MWS-T and is more accurate than MWS-TR. Regarding running times IRoTS is clearly faster.

Inference results with both query predicates in a single inference are shown in Table 3. IRoTS is clearly better than the other algorithms. The difference in solutions quality is more evident towards MWS-T with an improvement of approximately 12% in the solutions quality. MWS-TR is competitive with IRoTS but looses on average 7% in terms of solutions quality towards IRoTS. Regarding running times IRoTS is slightly slower than the MWS-T and slightly faster than MWS-TR. However, the differences are not significant compared to the overall running times.

All results clearly answer questions **(Q2)** and **(Q3)**. We used different MLNs with different weights, but the better performance of IRoTS is not affected by clauses' weights. Moreover, we generated MLNs that together with evidence data produce a different number of groundings, but IRoTS performance does not change with the number of groundings. Regarding question **(Q4)**, IRoTS is in general faster. In only one case, IRoTS is slightly slower than MWS-T but finds much better solutions. Thus question **(Q4)** can be answered stating that even though it finds better solutions, IRoTS does not spend more time and in general is faster than both MWS-T and MWS-TR. Results in Table 3 answer question **(Q5)** since inference consists in a huge number of groundings. However, to fully answer **(Q5)**, we generated MLNs with an additional predicate such that the number of groundings could be extremely high. We used for this the *taughtBy* predicate that has three arguments, thus a huge KB must be solved for MAP inference. Results with three predicates are reported in Table 3 and IRoTS again performs better. The number of groundings is very high and in one fold it reaches over 2 million. This is common for SRL domains where grounding

Table 4. Inference results in the CORA domain

		MC-IRoTS			MC-SAT		
preds	clauses	CLL	AUC	Time	CLL	AUC	Time
1849	77701	-0.043± 0.003	0.901	22.71	-0.043± 0.003	0.901	26.95
67081	1171457	-0.248± 0.003	0.092	444.91	-0.247± 0.003	0.094	435.10
59536	113798	-1.686± 0.003	0.059	68.41	-1.714± 0.003	0.059	104.82
59536	118828	-0.170± 0.002	0.158	64.63	-0.146± 0.001	0.156	93.69
1681	2724901	-1.427± 0.010	0.050	1291.30	-1.447± 0.010	0.055	1479.75
9409	912673	-2.011± 0.007	0.083	207.22	-1.990± 0.007	0.090	200.76
71289	142311	-0.079± 0.000	0.815	363.59	-0.079± 0.000	0.813	385.50
69169	69169	-0.044± 0.000	0.907	50.91	-0.044± 0.000	0.907	56.72
59536	59536	-0.057± 0.001	0.797	42.86	-0.057± 0.001	0.799	45.53
1849	79507	-0.158± 0.011	0.333	28.93	-0.154± 0.011	0.348	29.96
3844	234546	-0.056± 0.005	0.432	72.71	-0.057± 0.005	0.434	82.76
1681	68921	-0.085± 0.009	0.452	24.22	-0.085± 0.009	0.447	25.28
8836	821842	-0.083± 0.002	0.324	275.11	-0.084± 0.002	0.319	294.25
6084	350298	-0.139± 0.005	0.099	66.56	-0.137± 0.005	0.116	70.10
1849	82216	-0.159± 0.010	0.333	28.16	-0.162± 0.010	0.343	33.81
71289	142311	-0.124± 0.001	0.406	94.61	-0.124± 0.001	0.410	112.59
59536	62051	-0.315± 0.004	0.283	43.31	-0.315± 0.004	0.283	49.03
3844	121086	-0.625± 0.024	0.076	34.65	-0.651± 0.024	0.069	35.92
9409	14065	-0.246± 0.008	0.108	5.86	-0.242± 0.008	0.110	6.51
11025	1157625	-0.101± 0.003	0.219	428.42	-0.100± 0.003	0.228	448.42
-	-	**-0.393± 0.006**	**0.346**	**182.95**	**-0.394± 0.006**	**0.349**	**200.87**

Table 5. Results for predicate advisedBy based on MLNs generated with 500 iterations of PSCG (table above) and by running PSCG for 10 hours (table below)

			MC-IRoTS			MC-SAT		
fold	preds	clauses	CLL	AUC	Time	CLL	AUC	Time
ai	4760	185849	-0.031±0.005	0.043	59.72	-0.033±0.005	0.008	69.66
graphics	3843	136392	-0.023±0.005	0.005	45.28	-0.023±0.005	0.005	53.35
language	840	15762	-0.049±0.016	0.011	3.5	-0.049±0.016	0.011	3.63
systems	5328	218918	-0.026±0.005	0.074	62.45	-0.028±0.005	0.006	80.92
theory	2499	73600	-0.028±0.007	0.101	24.93	-0.029±0.007	0.007	28.68
average	-	-	-0.031±0.008	0.047	39.18	-0.032±0.008	0.007	47.25
fold	preds	clauses	CLL	AUC	Time	CLL	AUC	Time
ai	4760	185849	-0.033±0.005	0.008	59.47	-0.029±0.005	0.156	70.4
graphics	3843	136392	-0.023±0.005	0.005	45.85	-0.023±0.005	0.005	54.23
language	840	15762	-0.049±0.016	0.011	3.61	-0.049±0.016	0.011	3.68
systems	5328	218918	-0.027±0.005	0.006	71.99	-0.027±0.005	0.006	87.35
theory	2499	73600	-0.029±0.007	0.007	24.33	-0.029±0.007	0.007	28.18
' average	-	-	-0.032±0.008	0.007	41.05	-0.031±0.008	0.037	48.77

of FOL clauses is combinatorial. The results however show that IRoTS is slightly slower than the other two algorithms, even though IRoTS finds better solutions. This can be due to the fact that IRoTS does not get stuck in local optima (as early as the other algorithms) but continues searching, thus spending more time. Thus question **(Q5)** can be answered affirmatively in that for tasks with a huge number of ground predicates and clauses, IRoTS is superior in terms of solutions quality.

For MC-IRoTS, we want to answer these questions: **(Q1)** Does MC-IRoTS improve over the state-of-the-art algorithm in terms of running time? **(Q2)** What is the performance of MC-IRoTS compared to the state-of-the-art algorithm in terms of quality of probabilities produced? We first learned MLNs using the CORA dataset and then performed inference with query predicates sameBib,

Table 6. Results for advisedBy (table above) and tempAdvisedBy (table below) based on MLNs learned with both advisedBy and tempAdvisedBy as non-evidence predicates

fold	preds	clauses	MC-IRoTS			MC-SAT		
			CLL	AUC	Time	CLL	AUC	Time
ai	4760	185762	-0.028±0.004	0.066	66.8	-0.027±0.004	0.102	76.9
graphics	3843	136297	-0.026±0.004	0.017	51.1	-0.024±0.004	0.004	57.63
language	840	15711	-0.043±0.012	0.233	4	-0.037±0.011	0.221	4.1
systems	5328	218820	-0.026±0.004	0.067	78.58	-0.029±0.004	0.004	89.5
theory	2499	73540	-0.022±0.005	0.307	27.99	-0.025±0.005	0.204	31.19
average	-	-	-0.029±0.006	0.138	45.69	-0.028±0.006	0.107	51.86
fold	preds	clauses	CLL	AUC	Time	CLL	AUC	Time
ai	4760	185672	-0.007±0.002	0.030	59.01	-0.007±0.002	0.032	71.46
graphics	3843	136244	-0.004±0.001	0.174	43.96	-0.006±0.002	0.042	53.44
language	840	15706	-0.002±0.002	1.000	3.26	-0.004±0.004	0.008	3.42
systems	5328	218727	-0.008±0.002	0.008	69.51	-0.007±0.002	0.019	84.54
theory	2499	73513	-0.013±0.004	0.005	23.9	-0.014±0.005	0.003	28.95
average	-	-	-0.007±0.002	0.243	39.93	-0.008±0.003	0.021	48.36

Table 7. Results for two predicates advisedBy and tempAdvisedBy (table above) and three predicates taughtBy, advisedBy, tempAdvisedBy in a single inference task (table below)

fold	preds	clauses	MC-IRoTS			MC-SAT		
			CLL	AUC	Time	CLL	AUC	Time
ai	9384	680351	-0.020±0.003	0.028	210.58	-0.019±0.003	0.004	243.07
graphics	7564	495227	-0.019±0.003	0.003	153.00	-0.019±0.003	0.002	183.7
language	1624	52491	-0.029±0.007	0.004	22.34	-0.029±0.008	0.004	17.56
systems	10512	804425	-0.018±0.002	0.027	305.73	-0.019±0.002	0.003	286.46
theory	4900	261890	-0.020±0.003	0.005	83.72	-0.023±0.004	0.004	97.61
average	-	-	-0.021±0.004	0.013	155.07	-0.022±0.004	0.004	165.68
fold	preds	clauses	CLL	AUC	Time	CLL	AUC	Time
ai	23664	1894428	-0.053±0.003	0.007	559.43	-0.047±0.002	0.009	672.84
graphics	23485	1510794	-0.012±0.001	0.001	434.36	-0.012±0.001	0.001	556.14
language	5152	157944	-0.022±0.004	0.005	44.08	-0.012±0.001	0.001	56.63
systems	26136	2045461	-0.017±0.001	0.003	604.5	-0.017±0.001	0.007	744.24
theory	14504	794484	-0.017±0.002	0.043	228.35	-0.020±0.002	0.003	289.32
average	-	-	-0.024±0.002	0.012	374.14	-0.024±0.002	0.005	463.83

sameAuthor, sameVenue and sameTitle. Results are reported in Table 4 where each algorithm was run with 1000 samples. We generated different models in order to have different ratios of ground predicates and ground clauses. This would help better evaluate the algorithms over a wide range of inference scenarios. As results show, MC-IRoTS improves over MC-SAT in terms of overall running time. In order to provide further experimental evidence, we performed inference also on the UW-CSE dataset by using the MLNs learned previously for MAP inference. The conditional inference results with these MLNs are reported in Tables 5, 6, 7. As results show, MC-IRoTS improves in terms of running time, while preserving almost the same quality in terms of CLL and AUC.

6   Conclusions

Markov Logic is a powerful combination of first-order logic and probabilistic graphical models. This paper proposes two high performing algorithms based on the iterated local search and tabu search metaheuristics for MAP and conditional inference in Markov Logic. Experiments show that the algorithms outperform the current state-of-the-art algorithms in terms of solutions quality and running times. As future work, we plan to dynamically adapt the algorithms parameters and perform multiple runs with statistical analysis of performances.

References

1. Damien, P., Wakefield, J., Walker, S.: Gibbs sampling for bayesian non-conjugate and hierarchical models by auxiliary variables. Journal of the Royal Statistical Society B 61, 2 (1999)
2. Della Pietra, S., Pietra, V.D., Laferty, J.: Inducing features of random fields. IEEE TPAMI 19, 380–392 (1997)
3. Getoor, L., Taskar, B.: Introduction to Statistical Relational Learning. MIT, Cambridge (2007)
4. Gilks, W.R., Richardson, S., Spiegelhalter, D.J.: Markov Chain Monte Carlo in Practice. Chapman and Hall, Boca Raton (1996)
5. Glover, F., Laguna, M.: Tabu Search. Kluwer Academic Publishers, Dordrecht (1997)
6. Kok, S., Singla, P., Richardson, M., Domingos, P.: The alchemy system for statistical relational ai. Tech. rep., CSE-UW, Seattle, WA (2005)
7. Loureno, H.R., Martin, O., Stutzle, T.: Iterated local search. In: Handbook of Metaheuristics, pp. 321–353. Kluwer Academic Publishers, USA (2002)
8. Lowd, D., Domingos, P.: Efficient weight learning for markov logic networks. In: Kok, J.N., Koronacki, J., Lopez de Mantaras, R., Matwin, S., Mladenič, D., Skowron, A. (eds.) PKDD 2007. LNCS (LNAI), vol. 4702, pp. 200–211. Springer, Heidelberg (2007)
9. Park, J.D.: Using weighted max-sat engines to solve mpe. In: AAAI, pp. 682–687 (2005)
10. Poon, H., Domingos, P.: Sound and efficient inference with probabilistic and deterministic dependencies. In: Proc. 21st AAAI, pp. 458–463 (2006)
11. Richardson, M., Domingos, P.: Markov logic networks. Machine Learning 62, 107–236 (2006)
12. Roth, D.: On the hardness of approximate reasoning. Artificial Intelligence 82, 273–302 (1996)
13. Selman, B., Kautz, H., Cohen, B.: Local search strategies for satisfiability testing. In: Cliques, Coloring, and Satisfiability: 2nd DIMACS, pp. 521–532 (1996)
14. Singla, P., Domingos, P.: Discriminative training of markov logic networks. In: Proc. 20th AAAI, pp. 868–873. AAAI Press, Menlo Park (2005)
15. Smyth, K., Hoos, H., Sttzle, T.: Iterated robust tabu search for max-sat. In: Canadian Conference on AI, pp. 129–144 (2003)
16. Taillard, E.D.: Robust taboo search for the quadratic assignment problem. Parallel Computing 17, 443–455 (1991)
17. Wei, W., Erenrich, J., Selman, B.: Towards efficient sampling: Exploiting random walk strategies. In: Proc. 19th Nat'l Conf. on AI (AAAI) (2004)
18. Yagiura, M., Ibaraki, T.: Efficient 2 and 3-flip neighborhood search algorithms for the max sat:experimental evaluation. Journal of Heuristics 7(5), 423–442 (2001)

A Robust Geometric Model for Argument Classification

Cristina Giannone, Danilo Croce, Roberto Basili, and Diego De Cao

University of Roma Tor Vergata, Roma, Italy
{giannone,croce,basili,decao}@info.uniroma2.it

Abstract. Argument classification is the task of assigning semantic roles to syntactic structures in natural language sentences. Supervised learning techniques for frame semantics have been recently shown to benefit from rich sets of syntactic features. However argument classification is also highly dependent on the semantics of the involved lexicals. Empirical studies have shown that domain dependence of lexical information causes large performance drops in outside domain tests. In this paper a distributional approach is proposed to improve the robustness of the learning model against out-of-domain lexical phenomena.

1 Introduction

Many natural language processing tasks, such as information extraction, question answering or dialogue, are tightly bound to semantic interpretations of arbitrary texts. Several paradigms have been proposed in order to define meta-models for such semantic structures, one eminent example being *frame semantics* [1], explored since late 70's. In frame semantics linguistic predicates and semantic roles are defined in terms of *frames*, that are predicate-argument structures modeling prototypical real world situations. A frame is evoked in a sentence through the occurrence of a specific set of *lexical unit* (LU), i.e. words (such as nouns or verbs) that linguistically express the situation of the frame. A noticeable contribution of a frame is the prediction of a set of prototypical semantic roles, i.e. semantic arguments, called *Frame Elements* (FE), that characterize the participants to the underlying event, irrespectively from individual lexical units. For example the following sentences evoke the KILLING frame, through the LU *kill*. In the examples some of the FEs of KILLING are emphasized (i.e VICTIM, KILLER and CAUSE).

$$[Rebels \ _{\text{KILLER}}] \ \textbf{killed} \ [the \ president \ _{\text{VICTIM}}] \quad (1)$$

$$[The \ blast \ _{\text{CAUSE}}] \ \textbf{killed} \ [everyone \ on \ deck \ _{\text{VICTIM}}] instantly, \ with \ the \ ... \quad (2)$$

In sentence (1), for example, *Rebels* is the grammatical subject of the verb *kill* and this implies the role KILLER, while *the president* is the direct object expressing the VICTIM role.

The task of automatic recognizing predicates and FEs in sentences is called *Semantic Role Labeling* (SRL) [2]. It is a prominent paradigm in automatic NL

R. Serra and R. Cucchiara (Eds.): AI*IA 2009, LNAI 5883, pp. 284–293, 2009.

analysis for its applicability to a wide range of NLP tasks ([3]). Recent works [4], [5] have also shown that syntactic constraints, acting on dependencies between the target word and its arguments, are necessary for accurate semantic role labelling (*SRL*) models. State-of-art approaches to SRL are supervised machine learning methods, such as Support Vector Machine (SVM), based on the observable syntactic structures of sentences (e.g. [6], [7]). Unfortunately, syntax alone is in general not sufficient to fully capture the proper argument semantics. In frame semantics, the linking between predicate words and their arguments can be very complex and the involved syntactic structure may show a huge variability even for the same verb. Moreover, the relationship between grammatical dependencies and the corresponding semantic roles is ambiguous. It is worth noticing that methods based on rich sets of syntactic features show a significant decrease in performance when applied to data of a different domain from the training one (i.e. in an *out-of-domain* fashion) [7]. This is alarming about the entire learning paradigm as the expected linguistic generalization seems not to have been achieved.

Sentences 1 and 2 are clear examples of this aspect, raising a number of issues. First, very similar syntactic structures suggest different thematic structures. In the above sentences, the grammatical subject of the predicate verb expresses the two different roles, KILLER and CAUSE respectively. This difference shows that syntax alone is not a sufficient explanation, as the inherent semantics of the role filler plays here a crucial role. In sentences 1 and 2 the grammatical subjects *Rebels* and *blast* have a strikingly different semantics and this is the main cause for the difference in the selection of the two different proto-agentive roles, KILLER and CAUSE respectively.

Two research directions here are interesting. First, minimizing the dependency on syntactic features is needed. Augmenting the capacity of a feature space to represent a target hypothesis function also corresponds to an higher risk of overfitting the data. In other words, the more complex the features, the narrower is the set of linguistic phenomena that will be properly treated outside the training corpus. Second, we want to increase the sensitivity to lexical information, that is critical in the role assignment task. In FrameNet, semantic types (such as SENTIENT for the frame elements KILLER) are used to constraint such roles and a learning model should also account for this. In this work we will study both aspects: (1) a simple classification approach is adopted, where the number of features that model individual semantic roles is drastically reduced with respect to previous works ([7]) and (2) an unsupervised process to acquire lexical information and use it as role preference criteria is applied. As a consequence only two independent features are considered:

- the *semantic head* (*h*) of the role, i.e. the word that carries the semantics of the syntactic argument. In sentence 2 the FE VICTIM is realised in the argument *everyone on deck*, whose semantic head is *everyone*.
- the direct *dependency relation* (*r*) connecting the semantic head (and consequently the represented role) to predicate words. In 2, the semantic head *everyone* is connected to the LU *kill* through the direct object (OBJ) relation.

In particular, the semantic heads are modeled as geometrical vectors within a *word space* [8], [9]. While the grammatical dependencies can be directly observed from the parsing output, lexical semantic information is derived from the annotations provided by the FrameNet corpus. Examples of role fillers are gathered from it and regions of the word space where they are represented are taken as models of individual semantic roles. Preferences for the interpretation of syntactic arguments in terms of semantic role are modeled as distances in the word space with respect to the known examples. In Section 2, we discuss previous related work. A weakly supervised approach based on the capability of the word space to model the semantics of argument fillers is then defined in Section 3. The evaluation on the argument classification step of a SRL task is then reported in Section 4. The discussion in Section 5 will comment on the promising results that confirm our claims.

2   Related Work

Semantic Roles are a traditional topic in computational semantics ([10]). Computational approaches to the recognition and analysis of thematic roles followed two main lines. In the traditional approach, role assignment is seen as the side effect of a theorem proving process triggered by semantic grammars and lexicons. Lexicalized approaches here are employed to infer roles through principles based on verbal argument selection ([10]). The theoretical development here has allowed a corresponding growth of large scale lexical resources, like FrameNet or Propbank. These developments made also available large sets of semantically annotated sentences that triggered example-driven machine learning methods trained over them. The idea is to avoid the manual construction of huge lexical repositories, but exploit learning methods for inducing complex recognition models through suitable generalization algorithms. Bayesian methods have been early employed giving rise to accurate role labeling systems [2]. More recently discriminative models, such as Support Vector Machines, have been studied and they reach state-of-art performances through suitable combinations of lexical and grammatical features ([11,6]). A survey of these models applied to the FrameNet role labeling task is discussed in [12]. While most of the above approaches are strictly supervised, as they fully depend on the quality of the source training material, unsupervised methods have also been attempted recently. In [13] a method to infer verb argument preferences from small sets of annotated examples is presented. Although lower performance levels are obtained with respect to supervised algorithms, this approach has much smaller training requirements. The basic idea is to exploit similarity between labeled dependency graphs characterizing a small set of seeding examples and a larger number of candidate novel sentences is obtained. The adopted similarity stems from grammatical constraints (i.e. similar dependency subgraphs) and semantic similarity between the involved argument head words (e.g. nouns or adjectives). Accordingly the system is able to retrieve (and rank) all sentences similar to annotated examples that are candidates for inheriting the same labeling. In this way, through several iterations the method increases the initially small set of examples and incrementally gathers novel instances. The novel issues raised by this work are related to the adoption of a weakly supervised perspective on the SRL task, and the enforcing of head semantics in an explicit way within the learning algorithm. In [13] in fact, in line with previous explorations (such as [9], [14] or [15]), semantic similarity is also computed in an unsupervised fashion through the adoption of a semantic word space. The novelty here is that implicit selectional preferences for role assignment are induced as distances within a vector space built over an unlabeled corpus.

Word (or semantic) spaces have been largely employed as models of similarity or relevance judgement in information retrieval, and have been recently devised also for NLP tasks. In general, a semantic space treats word contexts as sets of unordered words, whereas the vector representation takes place within a normed space. Applications of such technology have been studied for word sense disambiguation (e.g. [16]), acquisition of selectional preferences [14] as well as lexical unit induction [15]. Much of these works have been applied to unsupervised settings. In our work the semantic similarity triggers a lazy learning inference method based on the existing set of annotated examples observed in FrameNet. The purpose here is to evaluate the generalization capability in out-of-domain scenarios, where the targeted performance is measured over a corpus not used for training. As roles are general notions, linguistic generalization across resources is a strong need for effective SRL learning.

3   A Geometrical Model of Frame Elements

As the assignment of semantic roles is strictly related to the lexical meaning of argument heads, we will adopt a distributional perspective. The intuition is that the meaning of a word can be described by the set of textual contexts in which it appears (*Distributional Hypothesis* [17]). In distributional models, words are represented through vectors built over the observable contexts. Similar vectors suggest semantically related words. In view of modeling argument classification, we will extract the set H^{A_k} of semantic heads h (e.g. *Rebels* as in (1)) for a given Frame Element A^k (e.g. KILLER) and represent them as vectors $\overrightarrow{h}$. In such a space, a generic notion of *semantic relatedness* is modeled: two words close in the space are likely to be either in a paradigmatic or syntagmatic relation [8], [18]. In order to define vectors $\overrightarrow{h}$, an adjacency matrix M is used, whose rows describe words and whose columns describe the corpus contexts. Then an algebraic method, known as Latent Semantic Analysis (LSA) [19], is applied to M to acquire meaningful representations $\overrightarrow{h}$. LSA exploits the linear transformation called Singular Value Decomposition (SVD). It is a decomposition process that produces an approximation of the original matrix M, capturing semantic dependencies between context vectors, i.e. M's columns. The matrix M is transformed in the product of three matrices: U, S, and V such that $M = USV^T$. In SVD, the original space is replaced by a lower dimensional one M_l by truncating M to its first l dimensions. Each dimension is in fact a derived concept and truncation means neglecting the least meaningful information embodied by M. M_l captures the same statistical information of M in a new l-dimensional space, where each dimension is a linear combination of some original features (i.e. contexts). These newly derived features may be thought of as artificial concepts, each one representing an emerging meaning component as the linear combination of many different words (or contexts). As for the duality principle of LSA, words and documents are mapped by the SVD (through U and V respectively) into the same l-dimensional space.

The similarity between two argument heads h_1 and h_2 can thus be computed as the distance between $\overrightarrow{h_1}$ and $\overrightarrow{h_2}$, and is proportional to the number of the most informative contexts shared between them. In order to model a specific A^k, we first partition role fillers according to the different syntactic relations r that link them to the corresponding LU. Let $H_r^{A^k}$ be the subset of H^{A_k} of such semantic heads h. As a dependency based representation for r is used[1], from sentences (1) and (2) we get $H_{\mathrm{OBJ}}^{\mathrm{VICTIM}} = \{president, everyone\}$.

As the LSA vectors $\overrightarrow{h}$ are available for every semantic head h, a vector representation $\overrightarrow{A^k}$ for the role A^k can be obtained. Intuitively, it could be computed as the geometric centroid of the vectors of all its semantic heads. However, the rich nature of frame elements A^k implies that very different types of contexts may express the same situation. For example, *hitman* and *father* could be semantic heads for the role KILLER, but they are also likely to appear in very different contexts. Their vectors are thus likely to be distant in the space. As more than one region of the semantic space may act as a good geometrical representation for one frame element, a clustering phase is applied first. It is carried out by an adaptive clustering algorithm, based on k-means [20,21], applied over the sets $H_r^{A^k}$. Given a syntactic relation r, let $C_r^{A^k}$ denote the set of clusters derived for $H_r^{A^k}$: these are characterized by high internal similarity in the semantic space. Each cluster $c \in C_r^{A^k}$ is represented by a vector $\overrightarrow{c}$, computed as the geometric centroid of all its semantic heads $h \in c$. For a frame F, the above approach defines a geometric model of all the involved frame elements A^k: it consists of all centroids $\overrightarrow{c}$ with $c \subseteq H_r^{A^k}$. They represent A^k through sets of its similar role fillers, as they can be observed in FrameNet. Table 1 represents clusters for $H_{\mathrm{SBJ}}^{A^k}$ of the KILLING frame.

As we are targeting here only argument classification, in the remainder of this paper, we assume that for a given a sentence s, the evoking word for the correct frame F is already known. A sentence s can thus be modeled as the following set of role-relation pairs $s = \{(r_1, h_1), ..., (r_n, h_n)\}$ where the heads h_i are in the syntactic relation r_i in s with the underlying lexical unit of F.

For every head h_i, the vector $\overrightarrow{h_i}$ allows to estimate its similarity with the different potential roles A^k. Given the observed syntactic relation r_i in s, we select all clusters $c \in C_{r_i}^{A^k}$. The similarity between the j-th cluster for the FE A^k, i.e. $c_{kj} \in C_r^{A^k}$, and h_i is the usual *cosine similarity*.

---

[1] In [7] dependency-based systems are shown to achieve the best performances for argument classification in SRL.

Table 1. Clusters of semantic heads in the subject (SBJ) position for the frame KILLING

Role	Semantic head clusters
CAUSE	c_1: {blast, hurricane}
	c_2: {gunman, soldier, terrorist}
KILLER	c_3: {traveller, village}
	c_4: {creature, deer, dinosaur, eagle, herring}
VICTIM	c_5: {candidate, liberal, politician, president}

In this step, a threshold τ is adopted in order to select only those word clusters D_{r,h_i} that are good representations for h_i, so that

$$D_{r,h_i} = \bigcup_k \{c_{kj} \in C_r^{A^k} | sim(h_i, c_{kj}) \geq \tau\}$$

By applying a *k-nearest neighbours* (k-NN) strategy to D_{r,h_i}, we first rank clusters c_{kj} according to their similarity scores and then select a fixed number of the m most similar ones to h_i. The resulting set $D_{r,h_i}^{(m)}$ includes at most m good clusters. The probability of the role A^k to be the valid frame element for h_i can then be estimated through voting, i.e. as the ratio between the number of the nearest cluster in $D_{r,h_i}^{(m)}$ that are models for A^k (i.e. members of $C_r^{A^k}$) and the overall number of clusters:

$$prob(A^k | h_i, r) = \frac{C_r^{A^k} \cap D_{r,h_i}^{(m)}}{|D_{r,h_i}^{(m)}|} \tag{3}$$

In figure 1 the clusters described in table 1 for some semantic roles of the KILLING frame are shown. The candidate head $h_i = pirate$ is also projected in the space and then the $m = 4$ most similar clusters are selected. Although the closest cluster represents the CAUSE frame element, the k-NN approach results in the following preferences according to Eq. 3: $prob(\text{KILLER}|h_i, \text{SBJ}) = 3/4$, $prob(\text{CAUSE}|h_i, \text{SBJ}) = 1/4$, $prob(\text{VICTIM}|h_i, \text{SBJ}) = 0$.

Once Equation 3 is available, a reranking phase is carried out for deriving the best sequence of frame elements $A^{(k_1,...,k_n)}$ for the entire s, where k_j is a choice function within the role set of the underlying frame. A sequence $A^{(k_1,...,k_n)}$ is *valid* for s if a role appears only *once* in the sequence, i.e. $\forall i \neq j$ $A^{k_i} \neq A^{k_j}$ in $A^{(k_1,...,k_n)}$. So, given a *valid* sequence $A^{(k_1,...,k_n)}$ for a sentence s, its probability is given by:

$$prob(A^{k_1,...,k_n} | s) = \prod_{i=1}^n prob(A_i^k | r_i, h_i)$$

The output argument sequence A^* is the one maximizing the above probability:

$$A^* = \operatorname*{argmax}_{A^{(k_1,...,k_n)}} prob(A^{k_1,...,k_n} | s) \tag{4}$$

This problem is a *multiple assignment problem* where the most likely semantic role sequence $A^* = A^{(k_1,...,k_n)}$ for all heads of the sentence must be found. The

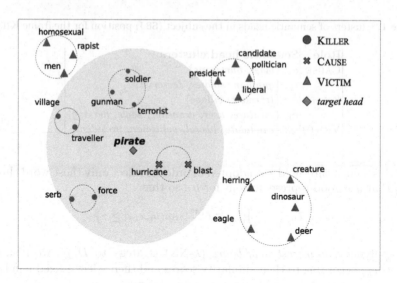

Fig. 1. The k-NN approach to classify $h_i = pirate$ over the $m = 4$ best clusters for the KILLING frame

Hungarian Algorithm [22] can thus be used to solve this optimization problem in polynomial time and produce the targeted role labeling.

4   Empirical Evaluation

The aim of evaluation is to compare the effectiveness of the presented model in the argument classification task. Different out-of-domain tests are carried out to compare the accuracy of our simple model with respect to different training conditions. The geometric model for semantic roles has been induced from three corpora: BNC, Europarl and TREC-2002 Vol.2. BNC is the corpus underlying FrameNet and consists of about 100 million words for English. Each context corresponds to a sentence and a resulting set of about 362,000 contexts has been obtained. The source vector representing an individual word defines the $tf.idf$ scores of the word in each sentence. The SVD reduction has been thus run over the resulting 4,530,000 × 362,000 matrix, with a dimension cut of $k = 300$. The same process is applied to the other corpora. Europarl is the European Parliament (EP) corpus made of about one million of sentences: these, here, are taken as contexts, and the LSA dimensionality cut was of $k = 200$. TREC finally includes about 110 million words in 230,401 documents: entire documents are here used as contexts and a dimensionality cut of $k = 250$ was applied. The above corpora are text collections of different domains, characterised by quite different linguistic phenomena. While the BNC corpus includes heterogeneous texts (e.g. novels, news, dialogues), EP reports on the European Parliament session and is dominated by direct speeches while TREC is made by news.

The training set used to derive distributional models of frame elements in every test is made of the subset of annotated sentences in FrameNet after leaving aside the test data set. It consists of 134,697 predicates whose sentences exhibit 271,560 arguments. This corresponds to the same data set used in [7], where the LTH parser [5] provided the dependency relations between role fillers and predicates. In order to detect the semantic head of FE fillers, the graph is visited to find the dependency path connecting the semantic head with the target lexical unit. Unfortunately in nearly 5% of the cases, we are not able to find a valid path. In these cases, clusters are selected in the semantic space independently from the syntactic relation: clusters built from examples of any relation are retained as useful candidates. Finally, in order to better generalize and reduce sparseness, we processed all the corpora (and the training data as well) with the Stanford Named Entity Recognizer [23]. Recognized proper nouns are replaced with 4 specific tokens (i.e. category labels): PERSON, LOCATION, ORGANIZATION and MISCELLANEOUS.

In all tests, valid clusters are defined with $\tau \geq 0.1$, while the value $m = 13$ is used to build every $D_{r,h_i}^{(m)}$ set of clusters. The achievable argument classification accuracy has been always tested on the FrameNet labeled corpus: a set of 14,952 predicates and 30,173 arguments also employed in [7] has been used as the test set. In the tests, gold-standard arguments are provided and their classification is then obtained through Eq.4. Notice that although the training FrameNet examples are kept fixed in the three experiments, the distributional models used to compute Eq. 4 are quite different. The models $C_r^{A^k}$ for the arguments A^k are in fact derived through clustering in the different LSA spaces where semantic similarities independent of the FrameNet examples are computed. While the test over the BNC corpus can be thus considered as an *in-domain test* (seemingly to what the supervised SRL training approach assumes), the other two experiments are to be considered out-of-domain tests.

Table 2. Argument classification accuracy

Corpus	Verbs	Nouns	Adj	All
BNC	79.64%	77.87%	**85.75%**	79.81%
Europarl	**79.73%**	**78.40%**	85.37%	**79.99%**
TREC	79.07%	75.30%	85.37%	78.59%

The results are shown in Table 2, where columns refer to the syntactic categories of the involved lexical units. The first row represents the BNC (i.e. in-domain) test where the achieved accuracy is 79.81%. This result compares with the state-of-art accuracy of 89% discussed in [7]: this latter is based on a complex set of syntactic features used for the SVM training. Given the relatively simple learning strategy of our model, this 10% gap is the one usually known between unsupervised and supervised systems, with no major weakness due to the simple syntactic features adopted (i.e. the relation r).

One important aspect is that the performance of the state-of-art model reported in [7] decreases to 71.1% in an out-of-domain test. A drop of about 18% is thus observed and this seems to be justified by the lexical sensitivity of the SVM training that is heavily domain dependent. Interestingly the accuracy reported in Table 2 by the use of different semantic spaces (the Europarl and TREC corpora) do not show any performance drop. Although lexical information is here acquired from sources different from the training corpus (i.e. BNC), domain dependence seems not to represent a major problem. We see that in the Europarl corpus, an even better accuracy is achieved with respect to the BNC. The major reason is that the sentence based model is quite effective in capturing paradigmatic relations (as also observed in [24]). In TREC a document based context model is used, and it seems to suffer from the noise due to long documents, characterized by multiple topics: in this case, a variety of topical similarities emerge and they seem not to contribute to the characterization of roles.

5   Conclusion

In this paper a flexible approach to the argument classification problem in SRL has been presented. It makes use of a semantic space to characterize frame elements introduced by individual dependency relations. The major design choices of our approach are the adoption of simple syntactic constraints to avoid overfitting and a portable model for the lexical semantics of argument heads. The results obtained are very interesting. The two features of our model, i.e. dependency relations among lexical units and distributional semantic similarity for argument fillers are quite effective. As a result robustness in out-of-domain test is strengthened. Unlike other supervised SRL approaches where the domain dependence of lexical features produces a strong performance drop (as argued in [11]), no reduction in accuracy is observed for our model. The method proposed seems widely applicable. It does not require large training data sets in the domain of application, as it can generalize lexical data outside of the labeled corpus. Efficiency is also a significant advantage. Whenever the parsing information as well as the LSA outcome are made available to the system, the training time of our model is about 40 minutes, while testing is below 20 minutes for the entire test set: labeling can thus proceed at an average rate of 700 sentences per minute. More work is required to assess the method on different training/test corpora for which differences in the syntactic style are significant. Moreover, the learning curve of our method is worth of a deeper analysis, as we need to investigate the training conditions under which these results can be reproduced.

References

1. Fillmore, C.J.: Frames and the semantics of understanding. Quaderni di Semantica 4(2), 222–254 (1985)
2. Gildea, D., Jurafsky, D.: Automatic labeling of semantic roles. Computational Linguistics 28, 245–288 (2002)
3. Shen, D., Lapata, M.: Using semantic roles to improve question answering. In: Proc. of EMNLP-CoNLL, Prague, pp. 12–21 (2007)

4. Punyakanok, V., Roth, D., Yih, W.: The necessity of syntactic parsing for s.antic role labeling. In: Proceedings of IJCAI 2005 (2005)
5. Johansson, R., Nugues, P.: Semantic structure extraction using nonprojective dependency trees. In: Proc. of SemEval 2007, Prague, June 23-24 (2007)
6. Moschitti, A., Pighin, D., Basili, R.: Tree kernels for semantic role labeling. Computational Linguistics 34(2), 193–224 (2008)
7. Johansson, R., Nugues, P.: The effect of syntactic representation on semantic role labeling. In: Proceedings of COLING, Manchester, UK, August 18-22 (2008)
8. Sahlgren, M.: The Word-Space Model. PhD thesis, Stockholm University (2006)
9. Padó, S., Lapata, M.: Dependency-based construction of semantic space models. Computational Linguistics 33(2), 161–199 (2007)
10. Dowty, D.: Thematic proto-roles and argument selection. Language 6(3) (1991)
11. Pradhan, S.S., Ward, W., Martin, J.H.: Towards robust semantic role labeling. Computational Linguistics 34(2), 289–310 (2008)
12. Baker, C.F., Ellsworth, M., Erk, K.: Semeval-2007 task 19: Frame s.antic structure extraction. In: Proc. of SemEval, Prague, pp. 99–104 (2007)
13. Fürstenau, H., Lapata, M.: Semi-supervised semantic role labeling. In: Proceedings of the 12th Conference of the European Chapter of the ACL (EACL 2009), Athens, Greece. Association for Computational Linguistics, March 2009, pp. 220–228 (2009)
14. Erk, K.: A simple, similarity-based model for selectional preferences. In: Proc. of the 45th Annual Meeting of the Association of Computational Linguistics, Prague. Association for Computational Linguistics, June 2007, pp. 216–223 (2007)
15. Pennacchiotti, M., De Cao, D., Basili, R., Croce, D., Roth, M.: Automatic induction of framenet lexical units. In: Proc. of EMNLP 2008 Hawaii (2008)
16. Schütze, H.: Automatic word sense discrimination. Computational Linguistics 24(1), 97–124 (1998)
17. Harris, Z.: Distributional structure. In: Katz, J.J., Fodor, J.A. (eds.) The Philosophy of Linguistics. Oxford University Press, New York (1964)
18. Pado, S.: Cross-Lingual Annotation Projection Models for Role-Semantic Information. PhD thesis, Saarland University (2007)
19. Landauer, T., Dumais, S.: A solution to plato's probl.: The latent semantic analysis theory of acquisition, induction and representation of knowledge. Psychological Review 104, 211–240 (1997)
20. Heyer, L.J., Kruglyak, S., Yooseph, S.: Exploring expression data: Identification and analysis of coexpressed genes. Genome Research (9), 1106–1115 (1999)
21. Basili, R., De Cao, D., Marocco, P., Pennacchiotti, M.: Learning selectional preferences for entailment or paraphrasing rules. In: Proc. of RANLP 2007 (2007)
22. Kuhn, H.W.: The hungarian method for the assignment probl. Naval Research Logistics Quarterly 2, 83–97 (1955)
23. Finkel, J.R., Grenager, T., Manning, C.D.: Incorporating non-local information into information extraction systems by gibbs sampling. In: ACL. The Association for Computer Linguistics (2005)
24. Pado, S., Lapata, M.: Dependency-based construction of semantic space models. Computational Linguistics 33 (2007)

Probabilistic Ontology Learner in Semantic Turkey

Francesca Fallucchi, Noemi Scarpato,
Armando Stellato, and Fabio Massimo Zanzotto

Disp, University "Tor Vergata" Rome, Italy
{fallucchi,scarpato,stellato,zanzotto}@info.uniroma2.it

Abstract. In this paper we present the Semantic Turkey Ontology Learner (ST-OL), an incremental ontology learning system, that follows two main ideas: (1) putting final users in the learning loop; (2) using a probabilistic ontology learning model that exploits transitive relations for inducing better extraction models.

1 Introduction

Ontologies and knowledge repositories are important components in Knowledge Representation (KR) and Natural Language Processing (NLP) applications. Yet, to be effectively used, ontologies and knowledge repositories have to be large or, at least, adapted to specific domains. Even huge knowledge repositories such as WordNet [1] are extremely poor when used in specific domains such as the medical domain (see [2]). Studying methods and building systems for automatically creating, adapting, or extending existing knowledge repositories using domain texts is a very important and active area.

A large variety of methods have been proposed: ontology learning methods [3,4,5] in KR as well as knowledge harvesting methods in NLP such as [6,7]. These learning methods use variants of the distributional hypothesis [8] or exploit some induced lexical-syntactic patterns (originally used in [9]). The task is generally seen as a classification (e.g., [10,11]) or a clustering (e.g., [4]) problem. This allows the use of machine learning or probabilistic models.

Models for automatic creating knowledge repositories generally exploit existing structured knowledge such as existing thesauri. Methods based on the Hearst's work [6] use existing pairs of words in a given semantic relation to extract patterns from corpora. These patterns are then used to induce novel pairs of words that are in the same semantic relation. For example, the pair of words *Bush* and *New Haven* are known examples of the semantic relation *has_born_in*. These can be used to extract from corpora that *is the birthplace of* is a good pattern to induce other instances of the above relation. Yet, these models hardly exploit the formal properties of the target relation. Then, these models do not properly exploit information that can be indirectly derived for existing data.

Some semantic relations such as hyperonymy and part-of have an extremely important property that is transitiveness. Exploiting this property along with

R. Serra and R. Cucchiara (Eds.): AI*IA 2009, LNAI 5883, pp. 294–303, 2009.

existing knowledge repositories during the discovering phase may help in building better knowledge extraction and structuring models. Such an idea is explored in [11,12].

Automatic models for extracting ontological knowledge from texts do not have the performance needed to extend existing ontologies with a high degree of accuracy. Then, resulting automatically expanded ontologies can become totally useless. Generally, systems for augmenting ontologies extracting information from texts foresee a manual validation for assessing the quality of ontology expansion. Yet, these systems do not use the manual validation for refining the information extraction model that proposes novel ontological information. The idea here is to prefer methods that can use decisions of final users to incrementally refine the model for extracting ontological information from texts, i.e., each decision of final users is exploited in refining the parameters of the extraction model. Including these new examples as training for a machine helps in augmenting the performances of the automatic extractor as shown in [13].

In this paper we present the Semantic Turkey Ontology Learner (ST-OL), an incremental ontology learning system, that follows two main ideas: (1) putting final users in the learning loop; (2) using a probabilistic ontology learning model that exploits transitive relations for inducing better extraction models.

The paper is organized as follows. We firstly review the related work (Sec. 2). We present the ideas behind our new ontology learning system (Sec. 3). We then introduce the system that follows the above principles (Sec. 4). Finally, we draw some conclusions (Sec. 5).

2   Related Work

Exploiting the above (and other) algorithms and techniques for inducing ontological structures from texts, different approaches have been devised, followed and applied regarding how to properly exploit the learned objects and translate them into real ontologies through dedicated editing tools. This is an aspect which is not trivially confined to importing induced data inside an existing (or empty) ontology, but identifies iterative processes that could beneficiate of properly assessed interaction steps with the user, giving life to novel ways of interpreting ontology development.

One of most notable examples of integration between ontology learning systems and ontology development frameworks are offered by Text-to-Onto [14], an ontology learning module for the KAON tool suite, which discovers conceptual structures from different kind of sources (ranging from free text to semi structured information sources such as dictionaries, legacy ontologies and databases) using knowledge acquisition and machine learning techniques; OntoLT [15], is a Protégé [16] plug-in able to extract concepts (classes) and relations (Protégé slots or Protégé OWL properties) from linguistically annotated text collections. It provides mapping rules, defined by use of a precondition language, that allow for a mapping between extracted linguistic entities and classes/slots.

An outdated overview of this kind of integrated tools (which is part of a complete survey on ontology learning methods and techniques) can be found in the public Deliverable 1.5 [17] of the OntoWeb project.

A more recent examples is offered by the Text2Onto [13] plug-in for the Neon toolkit [18], a renewed version of Text-To-Onto with improvements featuring ont-model independence (a *Probabilistic Ontology Model* is adopted as a replacement for any definite target ontology language), better user interaction and incremental learning. Lastly, in [19] the authors define a web browser extensions based on the Semantic Turkey Knowledge Acquisition Framework [20], offering two distinct learning modules: a relation extractor based on a light-weight and fast-to-perform version of algorithms for relation extraction defined in [7], and an ontology population module for harvesting data from html tables.

Most of the above define supervised cyclic *develop and refine* processes controlled by domain experts.

3   Incremental Ontology Learning

To efficiently set-up an incremental model for ontology learning, we need to address two issues:

- an efficient way to interact with final users
- an incremental learning model

The rest of the section shows how we obtain this using existing models and existing systems. We start from present the concept of incremental ontology learning (Sec. 3.1). Secondly, we describe the used ontology editor (Sec. 3.2). Finally, we introduce the used ontology learning methodology (Sec. 3.3).

3.1   The Concept

The incremental ontology learning process we want to model leverages on the positive interaction between an automatic model for *ontology learning* and the final users. We obtain this positive interaction using one additional component: an *ontology editor*. The overall process is organized in two phases: (1) the *initialization step* and (2) the *learning loop*. In the *initialization step*, the user selects the initial ontology and selects the corpus. The system uses these two elements to generate the first model for learning ontological information from documents. In the *learning loop*, the machine learning component extract a ranked list of pairs (*candidate_concept,superconcept*). The user selects, among the first k pairs, the correct ones to be added to the ontology. We can then use these choices to generate both positive and negative training examples for the ontology learning component. When the new ontology extraction model is learnt from the corpus, the updated ontology, and the growing *non-ontology*, the process restarts from the beginning of the loop.

Given a selected corpus C, the initial ontology O_0, and the generic ontology O_i at the iteration i, we can see the incremental learning process as the sequence of the resulting ontologies $O_0 \ldots O_n$. The *transition* function leverage on the ontology learning model M and the interaction with the user UV. This function can be represented as follows:

$$M_C(O_i, \overline{O}_i) = \widehat{O}_{i+1} \overset{UV}{\rightsquigarrow} (O_{i+1}, \overline{O}_{i+1}) \tag{1}$$

where M_C is the model learnt from the corpus, O_i is the ontology at the step i and $\overline{O}_i$ are the negative choices of the users at the same step. This model outputs a ranked list of possible updates of the ontology $\widehat{O}_{i+1}$. The user validation UV on the first k possibilities produces the updated ontology O_{i+1} and the updated *non-ontology* $\overline{O}_{i+1}$. At the initial step, the process has O_0 and $\overline{O}_0 = \emptyset$. The *ontology learner* produces the model $M_C(O_i, \overline{O}_i)$ building feature vectors representing the contexts of the corpus C where we can find pairs of pairs (*candidate_concept,superconcept*). These pairs are extracted from the ontology O_i and the *non-ontology* $\overline{O}_i$.

3.2   Semantic Turkey

Semantic Turkey is a Knowledge Management and Acquisition system developed by the Artificial Intelligence Group of the University of Rome, Tor Vergata. Semantic Turkey (ST, from now on) had been initially developed [21] as a web browser extension (currently implemented for the popular Web Browser Mozilla Firefox) for *Semantic Bookmarking*, that is, the process of *eliciting* information from (web) documents, to *acquire* new knowledge and *represent* it through representation standards, while *keeping reference* to its original information sources.

Semantic Bookmarks differ from their traditional cousins in that they abandon the purely partitive semantics of traditional links&folders bookmarking, and promote a new paradigm, aiming at "a clear separation between (acquired) knowledge data (the WHAT) and their associated information sources (the WHERE)". In practice, the user is able to select portions of text from web pages accessed from the browser, and to annotate them in a (user defined) ontology. A neat separation is maintained between ontological resources created through annotation, and the annotations themselves. This way, the user can easily organize its knowledge (by establishing relationships between ontology objects, categorizing them, better defining them through attributes etc...), while keeping multiple bookmarks in a separated space, pointing to ontology resources and carrying with them all information related to the taken annotations (such as the page where the annotation has been taken, its title, the text which was referring to the created/referenced ontology resource etc...). Easy-to-perform drag-and-drop operations were thought to optimize user interaction, by concentrating the different actions accompanying the creation of both the ontological resources and their related annotations in a few mouse clicks.

ST lately evolved [20] in a complete Knowledge Management and Acquisition System based on Semantic Web technologies: by introducing full support for ontology editing and by improving functionalities for annotation&creation, ST explored a new dimension which has no predecessor in the field of Ontology Development or Semantic Annotation, and is unique to the process of building new knowledge while exploring the web to acquire it. ST new objective was thus reducing the impedance mismatch between domain experts and knowledge investigators on the one side, and knowledge engineers on the other, by providing them with a unifying platform for acquiring, building up, reorganizing and refining knowledge. It is upon this framework that the ontology learning module that we introduce here has been implemented and integrated.

3.3   Probabilistic Ontology Learner

We use the Probabilistic Ontology Learning (POL) [11,12] to expand existing ontologies with new facts. In POL is possible to take into consideration both corpus-extracted evidences and the structure of the generated ontology. In the probabilistic formulation [11], the task of learning ontologies from a corpus is seen as a maximum likelihood problem. The ontology is seen as a set O of assertions R over pairs $R_{i,j}$. In particular we will consider the *is-a* relation. In this case if $R_{i,j}$ is in O, i is a concept and j is one of its generalization (i.e., the direct or the indirect generalization). For example, $R_{dog,animal} \in O$ describes that *dog* is an *animal* according to the ontology O.

The main probabilities are then: (1) the prior probability $P(R_{i,j} \in O)$ of an assertion $R_{i,j}$ to belong to the ontology O and (2) the posterior probability $P(R_{i,j} \in O | \vec{e}_{i,j})$ of an assertion $R_{i,j}$ to belong to the ontology O given a set of evidences $\vec{e}_{i,j}$ derived from the corpus. These evidences are derived from the contexts where the pair (i, j) is found in the corpus. The vector $\vec{e}_{i,j}$ is a feature vector associated with a pair (i, j). For example, a feature may describe how many times i and j are seen in patterns like "i as j" or "i is a j". These among many other features are indicators of an Is-a relation between i and j (see [6]).

Given a set of evidences E over all the relevant word pairs, in [11,12], the probabilistic ontology learning task is defined as the problem of finding an ontology $\hat{O}$ that maximizes the probability of having the evidences E, i.e.:

$$\hat{O} = \arg\max_{O} P(E|O)$$

In the original model [11,12], this maximization problem is solved with a local search. In the incremental ontology learning model we propose, this maximization function is solved using also the information coming from final users.

In the user-less model, what is maximized at each step is the ratio between the likelihood $P(E|O')$ and the likelihood $P(E|O)$ where $O' = O \cup N$ and N are the relations added at each step. This ratio is called multiplicative change $\Delta(N)$ and is defined as follows:

$$\Delta(N) = P(E|O')/P(E|O) \qquad (2)$$

The last important fact is that it is possible to demonstrate that

$$\Delta(R_{i,j}) = k \cdot \frac{P(R_{i,j} \in O | \vec{e}_{i,j})}{1 - P(R_{i,j} \in O | \vec{e}_{i,j})} =$$
$$= k \cdot odds(R_{i,j})$$

where k is a constant (see [11]) that will be neglected in the maximization process. We calculate the *odds* using the logistic regression.

Given the two stochastic variables Y and X, we can define as p the probability of Y to be 1 given that X=x, i.e.:

$$p = P(Y = 1 | X = x)$$

The distribution of the variable Y is a Bernoulli distribution, i.e.:

$$Y \sim Bernoulli(p)$$

Given the definition of the *logit* as:

$$logit(p) = \ln \left(\frac{p}{1-p} \right) \qquad (3)$$

and given the fact that Y is a Bernoulli distribution, the logistic regression foresees that the logit is a linear combination of the values of the regressors, i.e.,

$$logit(p) = \beta_0 + \beta_1 x_1 + ... + \beta_k x_k \qquad (4)$$

where $\beta_0, \beta_1, ..., \beta_k$ are called *regression coefficients* of the variables $x_1, ..., x_k$ respectively.

The remaining problem is how to estimate the regression coefficients. This estimation is done using the maximal likelihood estimation to prepare a set of linear equations using the above *logit* definition and, then, solving a linear problem using pseudo-inverse matrix [12]. We will called the logit vector with l, then we estimate the regression coefficients as following

$$\widehat{\beta} = X^T l \qquad (5)$$

In the user-oriented incremental ontology learning we propose, the above maximization is done including final users in the loop. In our task we do not find the ontology that maximizes the likelihood of having the evidences E. We calculate the probabilities step by step. Then we present an ordered set of choices to final users that will make the final decision on what to use on the next iteration. The order set is obtained using the logit function as it is equivalent to the order given by the probabilities. For this reason, in the following we will operate directly on the logit rather than on the probabilities. It is possible to calculate the logit vector to the i-th iteration using both equations (3) and (5) and obtained

$$XX^+ l_i = \widehat{l}_{i+1} \overset{UV}{\rightsquigarrow} l_{i+1} \qquad (6)$$

In each iteration, we calculate the logit vector using the logit vector of the previous iteration. After then the logit vector is changed in the user validation (UV). When the user accepts a new relation its probability is set to 0.99 while when the user discards a relation, its probability is set to 0.01. The matrix XX^+ is constant for each iteration. Here we have found a matrix XX^+ that is the constant model M_C of the equation (1). The matrix XX^+ only depends on the corpus C and not on the initial ontology. The logit vector l represents both the current ontology O_i and the negative ontology $\overline{O_i}$ as it includes the logit of both probabilities, i.e., 0.99 and 0.01.

4   Semantic Turkey-Ontology Learner (ST-OL)

The model described in previous section has been implemented and integrated in a Semantic Turkey extension called ST Ontology Learner (ST-OL). ST-OL provides a graphical user interface and a human-computer interaction work-flow supporting the incremental learning loop of our learning theory. If the user has loaded an ontology in ST, he can to improve it by adding new classes and new instances using ST-OL. The interaction process is achieved through the following steps:

- an *initialization phase* where the user selects the initial ontology O and the bunch of documents C where to extract new knowledge
- an *iterative phase* where the user launch the learning and validates the proposals of ST-OL

Thus, starting from the initial ontology O and a bunch of documents C, he has the possibility to use an incremental ontology learning model.

For the *initialization phase* (c.f., Sec. 3.1), the User Interface (UI) of ST-OL allows users to select the initial set of documents C (corpus), and to send both the ontology O and the corpus C to the learning module. To start this stage of the process, the user selects *"Initialize POL"* on the ST-OL panel (see Fig. 1). The probabilistic ontology learner analyzes the corpus, finds the contexts for each ontological pair, computes the first extraction model, and, finally, proposes the pairs that are in is-a relation. This first analysis is the more expensive one as it computes the matrix XX^+. Yet, this computation is done only once in the iterative process.

Once this initialization finishes, the *iterative phase* starts. ST-OL enables the button labeled *"Proposed Ontology"*. The effect of this button is to show the

Fig. 1. Initial Ontology extended with the pairs proposed by the POL System

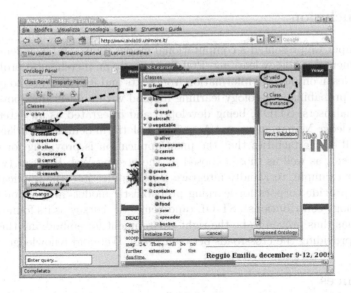

Fig. 2. Manual validation of new resources added to the ontology

initial ontology extended with the pairs proposed by POL. Figure 1 shows an example of an enriched initial ontology.

The main goal of ST-OL is to help in focusing the attention to the good added information. The user has the possibility of selecting the pairs he wants to add among the proposed pairs. To drive the attention towards the good pairs, we use different brightness of red for the different probabilities. More intense tonalities of red represent higher probabilities.

In order to focus only on possibly good pairs, ST-OL only shows pairs above a threshold τ of probabilities. For example, in Fig. 1, the relation, i.e., the pair, between "truck" and "container" is more probable than the relation between "spreader" and "container". Then different red tones are used. At this point, the user can accept or reject the information. After acceptance, the new information is stored in the ST ontological repository and can be browsed as usual. through the ontology panel on the Firefox sidebar. Fig. 2 shows what happened when the user accepted two proposed pairs: "mango" as instance of "fruit" and "pepper" as subclass of "vegetable".

The above activity enables the incremental model as it builds an upgraded probability vector. When the user accepts a new pair, ST-OL updates its probability to 0.99. When the user discards the pair, its probability is set to 0.01. These new values are used for the next iteration of the leaning process. After some manual evaluation, the user can decide to update the proposed ontology. Given the probabilistic ontology learning model presented in 3.3, this new evaluation is just a simple multiplication of the existing matrix XX^+ and the new vector. To force the recompilation, the user can use the *"Proposed Ontology"* button.

5   Conclusion

In this paper we presented a computational model POL and a system ST-OL for incremental ontology learning. POL is basically an incremental probabilistic model to learn ontological information from texts and it is designed to positively exploit a probabilistic ontology learning method within a learning loop that includes final users. ST-OL, being developed and integrated as an extension for the Knowledge Management and Acquisition platform Semantic Turkey, has inherited all of the facilities that the main application is providing for ontology development, as well as those exposed by the hosting Web Browser (which enabled, for example, to rapidly integrate a web spider into the application and use it to provide corpora for learning probabilistic models and/or for inducing new ontology contributions). ST-OL (and Semantic Turkey as its founding technology) has thus proven to be the right environment for embodying this kind of process, providing the crossroads between Users, Web and Knowledge.

References

1. Miller, G.A.: WordNet: A lexical database for English. Communications of the ACM 38(11), 39–41 (1995)
2. Toumouth, A., Lehireche, A., Widdows, D., Malki, M.: Adapting wordnet to the medical domain using lexicosyntactic patterns in the ohsumed corpus. In: AICCSA 2006: Proceedings of the IEEE International Conference on Computer Systems and Applications, pp. 1029–1036. IEEE Computer Society, Los Alamitos (2006)
3. Medche, A.: Ontology Learning for the Semantic Web. Engineering and Computer Science, vol. 665. Kluwer International, Dordrecht (2002)
4. Cimiano, P., Hotho, A., Staab, S.: Learning concept hierarchies from text corpora using formal concept analysis. Journal of Artificial Intelligence research 24, 305–339 (2005)
5. Navigli, R., Velardi, P.: Learning domain ontologies from document warehouses and dedicated web sites. Comput. Linguist. 30(2), 151–179 (2004)
6. Hearst, M.A.: Automatic acquisition of hyponyms from large text corpora. In: Proceedings of the 15th International Conference on Computational Linguistics (CoLing 1992), Nantes, France (1992)
7. Pantel, P., Pennacchiotti, M.: Espresso: Leveraging generic patterns for automatically harvesting semantic relations. In: Proceedings of the 21st International Conference on Computational Linguistics and 44th Annual Meeting of the Association for Computational Linguistics, Sydney, Australia. Association for Computational Linguistics, pp. 113–120 (July 2006)
8. Harris, Z.: Distributional structure. In: Katz, J.J., Fodor, J.A. (eds.) The Philosophy of Linguistics, Oxford University Press, Oxford (1964)
9. Robison, H.R.: Computer-detectable semantic structures. Information Storage and Retrieval 6(3), 273–288 (1970)
10. Pekar, V., Staab, S.: Taxonomy learning: factoring the structure of a taxonomy into a semantic classification decision. In: Proceedings of the Nineteenth Conference on Computational Linguistics, vol. 2, pp. 786–792 (2002)
11. Snow, R., Jurafsky, D., Ng, A.Y.: Semantic taxonomy induction from heterogenous evidence. ACL, 801–808 (2006)

12. Fallucchi, F., Zanzotto, F.M.: SVD feature selection for probabilistic taxonomy learning. In: Proceedings of the Workshop on Geometrical Models of Natural Language Semantics, Athens, Greece. Association for Computational Linguistics, pp. 66–73 ((March 2009)

13. Cimiano, P., Vlker, J.: Text2onto - a framework for ontology learning and datadriven change discovery. In: Montoyo, A., Muñoz, R., Métais, E. (eds.) NLDB 2005. LNCS, vol. 3513, pp. 227–238. Springer, Heidelberg (2005)

14. Maedche, A., Volz, R.: Icdm workshop on integrating data mining and knowledge management. In: The Text-To-Onto Ontology Extraction and Maintenance Environment, San Jose, California, USA (2001)

15. Buitelaar, P., Olejnik, D., Sintek, M.: A protégé plug-in for ontology extraction from text based on linguistic analysis. In: Bussler, C.J., et al. (eds.) ESWS 2004. LNCS, vol. 3053, pp. 31–44. Springer, Heidelberg (2004)

16. Gennari, J., Musen, M., Fergerson, R., Grosso, W., Crubzy, M., Eriksson, H.: The evolution of protégé-2000: An environment for knowledge-based systems development. International Journal of Human-Computer Studies 58(1), 89123 (2003)

17. Gómez-Pérez, A., Manzano-Macho, D.: Deliverable 1.5: A survey of ontology learning methods and techniques. Technical report (May 2003)

18. Haase, P., Lewen, H., Studer, R., Tran, D.T., Erdmann, M., d'Aquin, M., Motta, E.: The neon ontology engineering toolkit. In. In: WWW 2008 Developers Track (April 2008)

19. Bagni, D., Cappella, M., Pazienza, M.T., Pennacchiotti, M., Stellato, A.: Harvesting relational and structured knowledge for ontology building in the wPro architecture. In: Basili, R., Pazienza, M.T. (eds.) AI*IA 2007. LNCS (LNAI), vol. 4733, pp. 157–169. Springer, Heidelberg (2007)

20. Griesi, D., Pazienza, M.T., Stellato, A.: Semantic turkey - a semantic bookmarking tool (system description). In: Franconi, E., Kifer, M., May, W. (eds.) ESWC 2007. LNCS, vol. 4519, pp. 779–788. Springer, Heidelberg (2007)

21. Griesi, D., Pazienza, M.T., Stellato, A.: Gobbleing over the web with semantic turkey. In: Semantic Web Applications and Perspectives, 3rd Italian Semantic Web Workshop (SWAP2006), Scuola Normale Superiore, Pisa, Italy, December 18-20 (2006)

Semantic Annotation
of Legal Modificatory Provisions

Leonardo Lesmo, Alessandro Mazzei, and Daniele P. Radicioni

Dipartimento di Informatica, Università di Torino
C.so Svizzera, 185 10149, Torino, Italy
{lesmo,mazzei,radicion}@di.unito.it

Abstract. We consider the task of the automatic semantic annotation of the normative modifications enclosed in Italian legal texts. The system is based on a deep syntactic perser coupled with shallow semantic analysis based on frames.

1 Introduction

Semantic annotation, or semantic mark-up, is commonly acknowledged as a costly and complex matter. Much work has been carried out in various fields, such as Information Extraction, to the ends of automatically extracting notable information from texts [1]. For automatic annotation some advances have been obtained over restricted -more regular- domains, such as the *legal* field. For example, systems have been built that automatically identify and classify structural portions of legal documents and their intra- and inter-references [4,14]; other investigations are being carried out to produce semantic analysis [15,16]. Various initiatives have been established at the national and international levels to devise XML standards for describing legal sources and schemas to identify legal documents [10]. Unfortunately, the *human* annotation process is expensive and error-prone, so that such efforts will be viable only in conjunction with tools to extract in automatic fashion the *structural* and *semantic* data from legal texts. For instance, the Italian initiative NormeInRete (NIR)[1] allows using a text editor to mark up in a semi-automated fashion structural partition and normative references. However, the task of automatic annotation of semantic data is still an open issue.

In particular documents containing *modificatory provisions* are relevant in this perspective, in that modificatory provisions (also called legal or normative modifications) are explicitly concerned with describing how some other legal text –or other part of the current one– has to be modified. In this paper we describe a system for the automatic semantic annotation of these legal documents. Indeed a modificatory provision is basically a *change* made to one or more clauses within a text, or to the entire text. Handling modificatory provisions requires special attention, since they affect the entire normative system. Despite we are

---

[1] http://www.normeinrete.it/

R. Serra and R. Cucchiara (Eds.): AI*IA 2009, LNAI 5883, pp. 304–313, 2009.
© Springer-Verlag Berlin Heidelberg 2009

presently concerned with individual modificatory provisions, we stress that any modification made to a single norm may in principle affect other norms, as well. Let us consider, e.g., that if we differently characterize the definition of "patria potestà" (*parental authority*) defined in a given normative source, this modification affects all the norms referring to that definition. It should be considered, in this regard, that the uncertainty on the effects of normative modifications would undermine the certainty of the law, thereby making it hard to clearly understand what is the law, or which one of several versions of a provision counts as law. Automating the process to semantically annotate modificatory clauses and provisions would be of great help in simplifying the legal system and in consolidating texts of law. *Consolidated text* is the updated version of a normative text, the version embodying the changes: consolidating and consolidated texts create a sort of *legal network* of legal references [12].

Annotating a modificatory provision implies *1*) individuating the type of the modification (e.g., deletion, replacement), the text being modified (another law, decree, etc.), the portion of such document affected by the modification (be it a structural part, as "article 22" or a text fragment, as "the words 'six months'...''); *2*) generating a set of metadata that compactly describe such modification. This second step is important since adding semantic metadata allows for intelligent search and update of legal documents. In order to automatically execute these two steps, we need to apply a number of natural language processing technologies (NLP). In this paper we describe an NLP-based system –which is currently under active development– intended at automatically annotating with semantic metadata Italian legal texts containing modificatory provisions. This work is part of a joint research with the Modelling Legal Informatics Resource Group of CIRSFID (University of Bologna) that provided the taxonomy (i.e. a non-formal lightweight ontology of modificatory provisions that we currently use. The system relies on "deep" syntactic analysis [9] and "shallow" semantic interpretation of natural language: in this way we are able to enhance the legal texts with XML semantic meta-data. Our work has two main strengths. From a theoretical point of view, we show that the combination of deep syntax and shallow semantics is appropriate in specialized domains, such as the legal texts containing modificatory provisions, in which the language is more controlled. From a practical perspective, we provide human annotators with a prototype that can greatly speed-up the annotation of semantic meta-data in normative documents.

2   Related Works

A number of recent works are related in several ways with our project. Similar to our approach, the system described in [2] encodes the meaning of modificatory provisions with semantic frames. However this system is designed for assisting human users in the drafting process rather than in the annotation process, like in our case. The system SALEM [7] has some similarity with ours: it automatically annotates the modificatory provisions of NIR documents by using *shallow* syntactic parsing and a rule-based strategy to fill the semantic frames. One major difference between our project and SALEM is that our approach relies on a deep syntactic parser, whilst the system SALEM adopts a shallow syntactic analysis (produced by a chunk parser). In principle, one would argue that SALEM is more robust, while our system is more precise. For example, deep syntactic analysis enables to handle coordination or relative clauses. Additionally, SALEM produces analysis of general provisions, whilst our project is focussed on modificatory provision. The work in [11] is aimed at extracting enough information to build a question answering system on judicial opinions: it uses a deep syntactic parser to build a full syntactic description of the legal sentences. This parser produces constituency structure rather than dependency structure, as in our case. Moreover, in contrast to our approach, McCarty uses a logic language –i.e. a deep semantic structure–, to represent the entire semantics of the sentences, rather than using semantic frames. From a theoretical viewpoint, our approach (deep syntax and shallow semantics) can be though of as an intermediate one between SALEM (shallow syntax and shallow semantics) and that described by McCarty (deep syntax and deep semantics).

3   The Semantic Annotation Process

Our system relies on a rich taxonomy of modificatory provisions that has been carried out in some related works [5]. However, we presently consider only the changes affecting the provision text or form, and namely *integrations, substitutions* and *deletions*.

The annotation of modificatory provisions is a three steps process. In the first step we retrieve the possible location of a modificatory provision within the document, and we simplify the input sentences, so to prune text fragments that do not convey relevant information (*input preprocessing*). In the second step we perform the syntactic analysis of the retrieved sentences; in the third step we semantically annotate the retrieved provisions through a *tree matching* approach. We briefly illustrate the first two steps and then focus on the annotation phase.

3.1   Preprocessing and Syntactic Analysis

Extracting Modificatory Provisions. The input to the system is encoded in the *NormeInRete* (NIR)[2] standard format for Italian Legal Text. The NIR format encodes the structural elements used to mark up the main partitions of legal texts, as well as its atomic parts (such as articles, paragraphs, subparagraphs, and lettered and numbered items) and any non-structured text fragment. Additionally, the NIR standard includes in its Document Type Definitions a part dedicated to modifications, to implement this model in XML. Based on the XML structure, we retain the text excerpts contained between the tags ⟨corpo⟩[3], that is where the modifications may be found. We then replace the text tagged by

---

[2] http://www.normeinrete.it/sito_area3-ap_stan_rappresentazione_xml.htm
[3] *Body* in English.

$\langle rif \rangle^4$ and $\langle virgolette \rangle^5$, individuating the position where a modification occurs and a quoted text fragment (see below) with the respective *IDs*. For example, given the XML encoding of the sentence "All'articolo 40 [, comma 1, della legge 28 dicembre 2005, n. 262,] le parole: 'sei mesi' sono sostituite dalle seguenti: 'dodici mesi'." (*At article 40 [...] the words: 'six months' are replaced with the following: 'twelve months'.*), we obtain the rewritten sentence "All'RIF9, le parole: VIR1 sono sostituite dalle seguenti: VIR2." (see Fig. 2-*a* and 2-*c*). This sentence, which is much simpler to analyze with no loss of information, is then given in input to the parser.

Syntactic Analysis. The Turin University Parser (TUP) is a rule-based parser that returns the syntactic structure of sentences in dependency format. Dependencies are binary relations(e.g. *subject-relation*) between a dominant word (*the head*, e.g. the verb) and a dominated word (*the dependent*, e.g. the noun-subject) [9]. After two preliminary steps (the *morphological analysis* and *part of speech tagging*, necessary to recover the lemma and the part of speech (PoS) tag of the words), the sequence of words goes through three phases: *chunking*, syntactic analysis of the *coordination*, and *verbal subcategorization*. The output of the parser is the dependency tree of the input sentence (Fig. 1).

Since modificatory provisions often span over different adjacent phrases (e.g., verbal, nominal, adjectival, prepositional, adverbial phrases), it is necessary to put together such fragments in order to collect all pieces of information. To do so we implemented a *discourse manager*. Moreover, modificatory provisions can span over multiple sentences, which considerably increases the difficulty of the syntactic analysis and of the semantic interpretation, as well. Also punctuation (such as colon, semicolon and period) may determine the need to consider multiple sentences. In particular, the discourse manager interacts with the parser in two phases. Firstly each phrase is analyzed by the parser to discover the type of the phrase; secondly, the discourse manager puts them together according to some schema (e.g., *prepositional phrase + verbal phrase*) sending the resulting phrase to the parser. For example, the colon is commonly employed in locutions such as "at the article 127: the words 'six months' are replaced by the words: 'twelve months' ". Initially, the discourse manager sends the text fragments separately to the parser. Then, considering the roots of the resulting parse trees, the discourse manager puts them together according to the schema *prepositional phrase + verbal phrase + noun phrase* and sends them to the parser again.

3.2 Semantic Analysis

The semantic interpreter is a *rule-based* algorithm. The rules handle two sorts of information: the parse trees, and the domain knowledge encoding the legal modifications taxonomy.

To consider the main traits of modificatory provisions and how they are extracted, let us consider the modification in Fig. 2. To qualify a modification, it is

---

[4] Short form for the Italian word *riferimento*, *reference* in English.

[5] Short form for the Italian word *virgolette*, *quotes* in English.

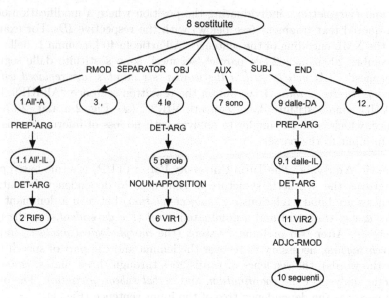

Fig. 1. Syntactic analysis of the sentence "All'RIF9, le parole: VIR1 sono sostituite dalle seguenti: VIR2", (*At RIF9 the words: VIR1 are replaced with the following: VIR2*)

necessary to extract information about the following points. *i*) the modification *type* : we presently consider only *integration*, *substitution* and *deletion*, but in the taxonomy there are many other classes; *ii*) the *position* can be specified as a sequence of words (e.g., "before the words: 'six months' "). Alternatively, the *position* can be specified relative to the document (e.g., "at article 40"); *iii*) *quoted text*, that can be of two types. Legal experts call the textual fragment being inserted/added the *novella* of a modificatory provision; while they call *novellando* the text fragment affected by the normative modification. For the sake of clarity, we will borrow that terminology. The *quoted text* identifies both *novella* and *novellando*. Such information is then used for generating the set of metadata that describe a given modification (Figure 2-*b*). However, the structure of normative modifications can be different. In facts, it could be present only the *novellando* (e.g., "[in the end of article 40] the words: 'six months' are deleted"), or only the *novella* (e.g., "[at the end of article 40] the words 'to be enforced' are added").

After having described the meaningful elements of normative modifications, let us consider how the modificatory provisions components are encoded within the system. Modifications are represented by means of *semantic frames*, composed by *slots* [1], such as *legalCategory*, *referenceDocument*, *modifyingText* and *modifiedText*. Frames are associated to the legal categories: for instance, the verbs belonging to the *legalCategory substitution* such as *replace*, *substitute*, *change*, *modify*, etc. have all the same slots. In this way we can add further verbs to the legal categories by taking advantage of their shared semantic frames. *Integration*, *substitution* and *deletion* can differ as regards the slots corresponding

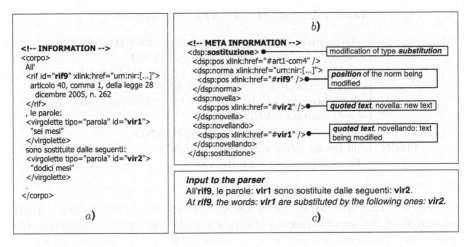

Fig. 2. Frame *a)*: XML encoding the sentence (extracted from the tag ⟨corpo⟩) "All'articolo 40 [, comma 1, della legge 28 dicembre 2005, n. 262,] le parole: 'sei mesi' sono sostituite dalle seguenti: 'dodici mesi'." (*At article 40 [...] the words: 'six months' are replaced with the following: 'twelve months'.*). Frame *b)*: the meta-description of the modification. Frame *c)*: the content of the tag ⟨corpo⟩, as it is given in input to the parser after the rewriting of constants *rif* (i.e., *reference*) and *virgolette* (quotes).

to the *quoted text*. For example, *deletion* has a *novellando* (in this case, the slot *modifiedText* should be filled) but no *novella*, and accordingly no *modifyingText* slot. Viceversa, both *substitution* and *integration* can have either *novella* or *novellando* or both of them.

A main rule is charged to test whether the root node of the syntactic tree is a verb, and if it belongs to the modificatory provisions taxonomy. For example, given the root verb, we take the verb lemma *inserire* (*insert*), we search for it in the knowledge base, and find that it is a possible instantiation of the *legalCategory integration*, together with the verbs *aggiungere* (*add*), *integrare* (*incorporate*), etc.. In this case we have a fundamental cue that the sentence being analyzed contains a modificatory provision. Also, based on the taxonomy, we are informed about the *legalCategory* of the modification at hand. The next step is in inspecting the dependents of the verb looking for the *position* of the modification, and for novella and novellando (*modifyingText* and *modifiedText*, respectively) arguments.

In this setting, filling a modification frame corresponds to searching a parse tree that fits to slots of a given type, and then to finding an appropriate mapping between (tree) dependents and (frame) slots. To perform tree matching, the other rules test the content of the verb arguments and the verb modifiers to fill the slots of current frame. In particular, the rules are charged to discover whether in the syntactic arguments like *subject*, *object* or in any modifier a constant such as *RIF* or *VIR* are present. For example, the syntactic tree corresponding to the *substitution* modification in Fig. 1 has as root node the verb "sostituire"

Table 1. Example of tests on the verb dependents

IF	- the subtree attached to the verb by a *RMOD* label (that is, *modifier*) contains a RIF_1 constant; **AND** - the subtree attached to the verb by a *SUBJ* label (that is *subject*) contains a VIR_1 constant; **AND** - the subtree attached to the verb by a *OBJ* label (that is, *object*) contains a VIR_2 constant
THEN	- fill *referenceDocument* with he RIF_1 **AND** - fill *modifiedText* with the VIR_2 **AND** - fill *modifyingText* with the VIR_1

(*substitute*). This verb is present in the knowledge base, which causes a frame associated to the *legalCategory: substitution* to be instantiated. Subsequently, a set of tests are executed on the verb dependents –i.e., the children nodes– to fill the appropriate slots (Table 1). The rule maps a syntactic pattern onto a set of semantic slots, performing a test on the subtrees rooted in the *subject*, in the *object* and in the *modifiers* (Fig. 3).

Further rules are designed to account for complex linguistic constructions, such as the case of coordination. Let us consider the following sentence: *All'articolo 1, comma 2, sono soppresse le lettere d) ed f); [At the article 1, comma 2, the letters d) and f) are deleted;]*. It is conveniently converted into: *All'RIF16, sono soppresse le RIF34) ed RIF35), [At RIF16, the RIF34) and RIF35) are deleted]*. The syntactic parser recognizes coordination and marks the edge between the conjuncts. The semantic frame corresponding to deletion of *RIF34* is filled similarly to the case of the substitution; additionally, the semantic interpreter recognizes the presence of a conjunct. In facts, the reference *RIF34* has a descendant node *RIF35* in a subtree that is connected by the *COORD* labeled edge.

Cases involving 2 conjuncts, such as *RIF34* and *RIF35*, are rather straightforward because we are handling homogeneous objects (both are *RIFs*), and because there is no ambiguity about the fact that *RIF34* and *RIF35* are coordinated. To extend the coverage, here are some further sorts of coordination types:

– Al *RIF22* le parole: *VIR4* e *VIR7* sostituiscono le parole: *VIR8* e *VIR11*, rispettivamente; (rough translation: At the *RIF22* the words: *VIR4* and *VIR7* substitute the words: *VIR8* and *VIR11*, respectively;);

– Al *RIF1* le parole *VIR1* sostituiscono *VIR4* e *VIR5* e al *RIF2* le *VIR12* sostituiscono *VIR13, VIR14* e *VIR15* (rough translation: At *RIF1*, the words *VIR1* replace the *VIR4* and *VIR5*, and at *RIF2* the *VIR12* replaces the *VIR13, VIR14* and *VIR15*).

As a general strategy, to handle such cases we bound the combinatorial explosion menacing the computation by assuming that coordinates can only be homogeneous (that is, a *VIR* cannot be coordinated to a *RIF*).

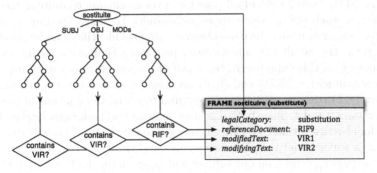

Fig. 3. Three tests are performed on a parse tree, and the appropriate slots are filled

4   Experimentation

From a practical viewpoint, our research is intended to assist human annotators in individuating legal modificatory provisions and qualifying them with as many details as possible. At the current stage of development we deal with modificatory provisions of either *integration, substitution* or *deletion* type. For the experimentation we used a dataset composed of 180 files, containing overall 11,944 XML *corpo* elements (see Section 3.1) and 2,306 modificatory provisions (namely, 809 integrations, 894 substitutions and 603 deletions) hand-annotated by the CIRSFID legal experts. To assess the performance of the system we record the percentage of modificatory provisions correctly annotated. Each correctly annotated modificatory provision consists of the tuple ⟨*type, position, novella, novellando*⟩, where *type* is one in {*integration, substitution, and deletion*}, *position* is the constant identifying the position into a given document where the modification occurs, and *novella* and *novellando* are both excerpts of quoted text (see Section 3.2). Our system obtains 83.0% *precision* and 71.7% *recall*. In particular, the figures we obtain for the *recall* on *integration, substitution* and *deletion* are 77.8%, 77.4% and 55.1%, respectively.

The overall result gives an estimation of the robustness of the approach, and of the implemented system as well: as far as we know, no experimentation has been conducted on datasets large as the present one. Provided that the present experimentation is performed with a preliminary release of the system, in the annotation of *integrations* and *substitutions* we obtain encouraging results, while only a poor accuracy is achieved in the case of *deletions*.

Most errors fall into one of the following classes: 1. preprocessing errors, such as misspelled words, or errors in the XML annotation; 2. discourse manager errors, occurring when chains of complex phrases are met; 3. parser errors, e.g., too complex syntactic structures, or unknown words (such as the verb *anteporre, to place sth. before*); 4. semantic interpretation errors, in cases in which the semantic interpreter is not able to extract the relevant information. By manual inspection we noted that errors in class 1 are currently by far the most frequent ones. In particular, the system skipped a large number of *corpo* elements (namely, 1,772 out of 11, 944). Then 14.8% of all possible texts fragments containing modifications do not reach the discourse manager module. To assess the impact of such kind of errors, we consider the percentage of correctly identified modificatory provisions w.r.t. the number of modificatory provisions that are actually analyzed by the parser. In this experiment, the *recall* of the system raises to 82.0%. In detail, we obtain 85.1%, 85.7% and 70.5% on *integration*, *substitution* and *deletion*, respectively. Further investigation is required to refine the analysis of the problems in classes 2 and 4. Since both discourse manager and semantic interpretation rely on hand-written rules, we plan to enlarge the rule bases. This will allow us to recognize a wider variety of phrase sequences and tree patterns, respectively. As regards as parsing errors, on one side we will improve the TUP parser, by tuning it on unknown syntactic constructions emerging from data.

5   Conclusions and Future Work

We have considered the problem of the automatic semantic annotation of modificatory provisions. We have delivered a robust NLP system that can speed-up the process of semantic metadata annotation of legal documents. We have illustrated the extraction of modificatory provisions as a three-steps process, where we first retrieve the relevant excerpts of text, parse them, and map the resulting parse trees onto the appropriate semantic frame. Finally, we have described a hand-crafted dataset used as a gold-standard for experimenting with the implemented system; we have then discussed the results obtained and reviewed some types of errors, pointing out possible improvements.

Since a key aspect of our approach is in the semantics conveyed by verbs, it will be necessary to compare and connect our work to the major international Projects focussed on analyzing verbs semantics, such as VerbNet [8], PropBank [13] and FrameNet [3]. Finally, another line of future work will involve extending the considered modificatory provisions to time-related modifications (e.g., we will consider *changes of efficacy* and *changes of force* [5]), that will require consider some formalisms for handling events and temporal expressions [6].

Acknowledgments

We wish to thank Raffaella Brighi and Monica Palmirani from the Modelling Legal Informatics Resource Group of CIRSFID (University of Bologna) for their work on the taxonomy currently in use. This work has been partially supported by the Converging Technologies Project ICT4LAW funded by Regione Piemonte.

References

1. Appelt, D.E., Israel, D.J.: Introduction to information extraction technology. In: Tutorial at IJCAI 1999 (1999)
2. Arnold-Moore, T.: Automatic generation of amendment legislation. In: Proceedings of the International Conference on Artificial Intelligence and Law (ICAIL), pp. 56–62 (1997)

3. Baker, C.F., Fillmore, C.J., Lowe, J.B.: The Berkeley FrameNet Project. In: Proceedings of the 17th International Conference on Computational Linguistics, Morristown, NJ, USA, pp. 86–90. ACL (1998)
4. Biagioli, C., Francesconi, E., Spinosa, P., Taddei, M.: The NIR project: Standards and tools for legislative drafting and legal document web publication. In: Proceedings of ICAIL Workshop on e-Government:Modelling Norms and Concepts as Key Issues, pp. 69–78 (2003)
5. Brighi, R., Lesmo, L., Mazzei, A., Palmirani, M., Radicioni, P.: Towards Semantic Interpretation of Legal Modifications through Deep Syntactic Analysis. In: Jurix 2008: The 21st Annual Conference, Frontiers in Artificial Intelligence and Applications. IOS Press, Amsterdam (2008)
6. Bunt, H., Overbeeke, C.: Towards Formal Interpretation of Semantic Annotation. In: Proceedings of the Sixth International Language Resources and Evaluation (LREC 2008), Marrakech, Morocco (May 2008)
7. Cherubini, M., Giardiello, G., Marchi, S., Montemagni, S., Spinosa, P., Venturi, G.: NLP-based metadata annotation of textual amendments. In: Proc. of Workshop on Legislative XML 2008, Jurix 2008 (2008)
8. Kipper, K., Korhonen, A., Ryant, N., Palmer, M.: Extensive Classifications of English Verbs. In: Proceedings of the 12th EURALEX International Congress, Turin, Italy (September 2006)
9. Lesmo, L.: The Rule-Based Parser of the NLP Group of the University of Torino. Intelligenza Artificiale 2(4), 46–47 (2007)
10. Lupo, C., Vitali, F., Francesconi, E., Palmirani, M., Winkels, R., de Maat, E., Boer, A., Mascellani, P.: General XML format(s) for legal Sources - ESTRELLA European Project. Deliverable 3.1, Faculty of Law, University of Amsterdam, Amsterdam, The Netherlands (2007)
11. McCarty, L.T.: Deep semantic interpretations of legal texts. In: ICAIL 2007: Proceedings of the 11th International Conference on Artificial Intelligence and Law, pp. 217–224. ACM Press, New York (2007)
12. Pagallo, U., Ruffo, G.: P2p systems in legal networks: another small world case. In: Proceedings of the 11th International Conference on Artificial Intelligence and Law, pp. 287–288. ACM Press, New York (2007)
13. Palmer, M., Gildea, D., Kingsbury, P.: The Proposition Bank: An Annotated Corpus of Semantic Roles. Computational Linguistics 31(1), 71–106 (2005)
14. Palmirani, M., Brighi, R.: An XML Editor for Legal Information Management. In: Traunmüller, R. (ed.) EGOV 2003. LNCS, vol. 2739, pp. 421–429. Springer, Heidelberg (2003)
15. Saias, J., Quaresma, P.: A Methodology to Create Legal Ontologies in a Logic Programming Based Web Information Retrieval System. Artificial Intelligence and Law 12(4), 397–417 (2004)
16. Soria, C., Bartolini, R., Lenci, A., Montemagni, S., Pirrelli, V.: Automatic Extraction of Semantics in Law Documents. In: Biagioli, C., Francesconi, E., Sartor, G. (eds.) Proceedings of the V Legislative XML Workshop, pp. 253–266. European Press Academic Publishing (2007)

Towards Extensible Textual Entailment Engines: The EDITS Package

Matteo Negri[1], Milen Kouylekov[1], Bernardo Magnini[1], Yashar Mehdad[1,2],
and Elena Cabrio[1,2]

[1] FBK-Irst, Trento, Italy
[2] University of Trento, Italy
{negri,kouylekov,magnini,mehdad,cabrio}@fbk.eu

Abstract. This paper presents the first release of EDITS, an open-source software package for recognizing Textual Entailment developed by FBK-irst. The main contributions of EDITS consist in: *i)* providing a basic framework for a distance-based approach to the task, *ii)* providing a highly customizable environment to experiment with different algorithms, *iii)* allowing for easy extensions and integrations with new algorithms and resources. System's main features are described, together with experiments over different datasets showing its potential in terms of tuning and adaptation capabilities.

1 Introduction

Textual Entailment (TE) has been proposed as a unifying generic framework for modeling language variability and semantic inference in different Natural Language Processing (NLP) tasks. The TE recognition (RTE) task [6] consists in deciding, given two text fragments (respectively called *Text* - *T*, and *Hypothesis* - *H*), whether the meaning of *H* is entailed (can be inferred) from the meaning of *T*. For instance, given the following *T-H* pairs:

T1: *Euro-Scandinavian media cheer Denmark v Sweden draw.*
H1: *Denmark and Sweden tie.*

T2: *Oracle had fought to keep the forms from being released.*
H2: *Oracle released a confidential document.*

an RTE system should assign the entailment judgement "YES" (*i.e.* an entailment relation holds between *T* and *H*) to *T1/H1*, and "NO" to *T2/H2*.

The RTE problem is relevant for many different areas of text processing research, since it represents the core of the semantic-oriented inferences involved in a variety of practical NLP applications. These include Question Answering (where systems have to identify, in large collections of documents, text portions that entail the correct answer), Information Retrieval (where query concepts have to be entailed by the relevant retrieved document), Information Extraction (where different textual realizations of the same relation should entail each

R. Serra and R. Cucchiara (Eds.): AI*IA 2009, LNAI 5883, pp. 314–323, 2009.
© Springer-Verlag Berlin Heidelberg 2009

other), Document Summarization (where redundant text portions in the source documents should be entailed by the output summary), and Machine Translation (where a correct translation should entail a reference gold standard translation and vice-versa). In the last few years, research on TE received a strong boost by the PASCAL Recognizing Textual Entailment Challenge [7]. In the four editions of the evaluation campaign, an increasing number of participants have been presented with increasingly complex definitions of the task. The released datasets (consisting in lists of T-H pairs to be judged as entailing or not) reflect the long-term objective of creating more and more natural evaluation settings.

Current approaches to TE recognition range from the use of Machine Learning techniques (using lexical, syntactic, and semantic features, as in [1]), to deep semantic oriented approaches (involving syntactic matching, logical inferences, and the use of theorem provers, as in [15]), similarity techniques (computing similarity scores between T and H both at lexical and syntactic levels, as in [10]), or different combinations of these approaches [2]. Each approach has its own advantages and disadvantages: while deep approaches are more effective to capture the more complex linguistic phenomena and model the language variability problem, shallow techniques are less resource demanding and more portable across domains and languages.

This paper presents EDITS (Edit Distance Textual Entailment Suite), a freely available open-source software tool for recognizing textual entailment. The main motivation underlying our work on EDITS is that, given the complexity of the TE task, due both to the linguistic phenomena that are involved, and to the variability of the application scenarios, TE engines actually need high capabilities to be adapted and optimized. We show that the distance-based approach implemented in EDITS has been successfully used in three experimental settings, each focusing different aspects of the problem, and different features of the system:

- the use of specialized entailment engines, which can be trained on specialized datasets and then organized in a way that the relative distances obtained on single datasets can be linearly combined;
- the automatic estimation of the costs for single edit operations: this is relevant for the portability of the system, since performance are optimized on the peculiarities of specific datasets;
- the adaptation to an application task (*i.e.* Question Answering), where optimized performance is obtained by means of application-oriented configurations.

The paper is structured as follows: Section 2 describes the main features of the package, its core components, and the workflow. Sections 3 and 4 respectively provide empirical results achieved in the design and combination of specialized entailment engines, and the automatic estimation of costs for edit operations over RTE data. Section 5 reports on experiments on entailment-based question interpretation, showing the potential of EDITS in terms of tuning and adaptation capabilities. Section 6 concludes the paper with the planned improvements for the future releases of the system.

2 EDITS: System Overview

The first release of the EDITS (1.0) package (available under GNU Lesser General Public Licence - LGPL) can be downloaded from *http://edits.fbk.eu*[1].

The package has been developed according to the two-way classification task proposed by the PASCAL-RTE evaluation campaigns: given as input a *T-H* pair, the returned output is represented by one of the two entailment judgments, either "YES" or "NO", supported by a confidence score.

EDITS implements a distance-based approach for recognizing textual entailment, which assumes that the distance between *T* and *H* is a characteristic that separates the positive *T-H* pairs, for which the entailment relation holds, from the negative pairs, for which the entailment relation does not hold. More specifically, EDITS is based on edit distance algorithms, and computes the *T-H* distance as the overall cost of the edit operations (*i.e.* insertion, deletion and substitution) that are necessary to transform *T* into *H*.

The edit distance approach used in EDITS builds on three basic components, which can be configured and easily extended by the user:

- An **Edit distance algorithm**, which calculates the set of edit operations that transform *T* into *H*. In the current release of the package, EDITS provides distance algorithms at three levels: *i) String Edit Distance*, where the three edit operations are defined over sequences of characters, *ii) Token Edit Distance*, where edit operations are defined over sequences of tokens of *T* and *H*, and *iii) Tree Edit Distance*, where edit operations are defined over single nodes of a syntactic representation of *T* and H[2];
- A **Cost scheme**, which defines the cost associated to each edit operation involving an element of *T* and an element of *H*. Such cost can be either a fixed value set by the user, or any function that returns a numerical value;
- Optional sets of **rules**, both *entailment rules* and *contradiction rules*, providing specific knowledge (*e.g.* lexical, syntactic, semantic) about the allowed transformations between portions of *T* and *H*. Each rule has a left hand side (an element of *T*) and a right hand side (an element of *H*), associated to a probability which indicates if the left hand side entails or contradicts the right hand side. Rules can be manually defined, or they can be extracted from any external resource available (*e.g.* WordNet, corpora, Wikipedia), and stored in XML files called *Rule Repositories*.

Each module, and its corresponding parameters, can be configured by the user through the XML EDITS Configuration File (ECF). A basic configuration file

---

[1] A complete system's description, installation and usage information are also available at the same address.

[2] EDITS 1.0 provides an implementation of the Levenshtein distance algorithm [11] (for String Edit Distance), a token-based version of the same algorithm (for Token Edit Distance), and the Zhang-Shasha algorithm [16] (for Tree Edit Distance).

includes at least one distance algorithm and one cost scheme, while rule repositories can be optional. EDITS provides a general framework which allows, through the ECF, to combine in different ways the existing algorithms/cost schemes, or replace them with new ones implemented by the user.

EDITS can work at different levels of complexity, depending on the linguistic analysis carried out over T and H. An internal representation format, called *ETAF* (EDITS Text Annotation Format) is defined such that both linguistic processors and semantic resources can be easily used within EDITS, resulting in a flexible, modular and extensible approach to TE. The format is used both for representing the input (T-H pairs), as well as for representing entailment and contradiction rules. ETAF allows to represent texts at three different levels of annotation: *i)* as simple strings; *ii)* as sequences of tokens with their associated morpho-syntactic properties; *iii)* as syntactic trees with structural relations among nodes. The three levels are considered in increasing complexity, and levels of higher complexity assume the levels of lower complexity. For example, texts annotated at syntactic level assume that the same texts are also tokenized. The lower the complexity level, the more language-independent the resulting configuration. For instance, while working at the level of strings allows for handling datasets in any language, working at the level of syntactic representations will require the availability of language-specific tools (*i.e.* a syntactic parser).

Given a certain configuration of its three basic components, EDITS can be trained over a specific RTE dataset in order to optimize its performance. In the training phase EDITS produces a *distance model* for the dataset, which consists in a distance threshold S (with $0 < S < K$) that best separates the positive and negative examples in the training data. During the test phase EDITS applies the threshold S, so that pairs resulting in a distance below S are classified as "YES", while pairs above S are classified as "NO". Given the edit distance $ED(T, H)$ for a T-H pair, a normalized entailment score is finally calculated by EDITS using the following formula:

$$entailment(T, H) = \frac{ED(T, H)}{(ED(T, \_) + ED(\_, H))} \tag{1}$$

where $ED(T, H)$ is the function that calculates the edit distance between T and H, and $(ED(T, \_) + ED(\_, H))$ is the distance equivalent to the cost of inserting the entire text of H and deleting the entire text of T. The entailment score has a range from 0 (when T is identical to H), to 1 (when T is completely different from H).

Once a distance model is available, EDITS can be run over a RTE test set. Besides the entailment judgment (*i.e.* "YES"/"NO"), for each pair the system provides the entailment score calculated by the algorithm, and the confidence score of the entailment assignment (*i.e.* the distance between the entailment score and the threshold S calculated at a training stage).

Figure 1 shows EDITS basic components and workflow. A detailed description of the edit distance approach implemented in EDITS can be found in [10].

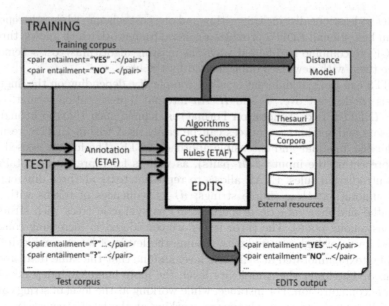

Fig. 1. EDITS workflow

3   Combining Specialized Entailment Engines

EDITS peculiarity of being easily extensible and adaptable to different tasks and approaches has been firstly explored through our participation at the RTE4 Challenge [5]. While most of the systems submitted to RTE face the problem of language variability with omnicomprehensive approaches, without trying to address the problem of the interaction between the linguistic phenomena underlying the entailment relation, our approach uses different independent Entailment Engines (EEs), each of which is specialized to deal with a specific aspect of language variability.

To experiment with the feasibility of the proposed approach, two simple entailment engines based on EDITS have been implemented, and organized to provide a linear combination of their respective distances on T-H pairs. In each engine (corresponding to a different configuration of EDITS), the cost of edit operations has been defined according to the specific linguistic phenomenon it should cope with. The cost schemes of the different specialized engines are designed in order not to intersect. In each EE specific costs are set for a phenomenon x, while for all the aspects considered by other modules costs are set to 0.

The first module that has been implemented (EDITSneg) is specialized to deal with negation phenomena, and works on Negative Polarity Items (NPIs, direct licensors of negation and antonyms). EDITSneg uses the Token Edit Distance algorithm to determine the less costly sequence of edit operations to transform T into H. As far as the cost scheme is concerned, specific costs are assigned to edit operations involving a NPI. The underlying intuition is that assigning higher costs to edit operations that involve NPIs should prevent the system from assigning positive entailment to a T-H pair in which one of the two fragments contradicts the other. In this scheme, all other words but the negative polarity items have a zero cost of insertion, deletion and substitution. To make experiments on negation, an artificial balanced dataset has been built. It is composed of 66 T-H pairs (each containing one NPI either in T or in H): 32 positive (with entailment) pairs and 34 negative (no entailment) pairs. Performances achieved by EDITSneg on this specialized dataset are satisfying (77% Acc), although phenomena like negation are easier to circumscribe and isolate than others.

The second module deals with lexical similarity. In EDITSlex, the cost scheme sets specific costs for edit operations considering WordNet similarities among tokens and the Token Edit Distance algorithm is used. A set of entailment rules has been generated, exploiting the Adapted Lesk algorithm (Extended Gloss Overlaps) from the WordNet similarity package [14] to set the probability of entailment among tokens. For RTE4, EDITlex is not set as an independent module, but it is integrated in a more general system that considers all but the negation phenomena (EDITSall-but-neg), plus WordNet similarities. In the cost scheme of EDITSall-but-neg, the cost of the insertion of a token is equal to the length (*i.e.* the number of words) of T, and the deletion of a token from T is equal to the length of H. The substitution cost is set as the sum of the insertion and the deletion of the tokens, if they are not equal. This means that the algorithm would prefer to delete and insert tokens rather than substituting them. Setting the insertion and deletion costs respectively to the length of T and H is motivated by the fact that a shorter text T should not be preferred over a longer one T' while computing their overall mapping costs with the hypothesis H. Setting the costs to fixed values would in fact penalize longer texts (due to the larger amount of deletions needed) even though they are very similar to H.

4   Learning Costs for the Edit Operations

Towards extending EDITS taking advantage of its flexibility and adaptability, we implemented a stochastic method based on Particle Swarm Optimization (PSO) [8] to automatically estimate and optimize the cost of each edit operation by maximizing the accuracy gained over the RTE training data[12]. For this purpose, we integrated EDITS and JSwarm-PSO[3], obtaining a flexible framework for experiments aiming at an improvement of our best configuration used to participate in the RTE-4 challenge (RTE-Conf in Table 1).

As far as the distance algorithm is concerned, RTE-Conf uses Tree Edit Distance. Concerning the cost scheme, RTE-Conf uses a list of the most common 500 English words as a stop word list, and assigns a deletion cost equal to 0 if a deleted word w from T is a stop word. Otherwise, the deletion cost is set to the number of words in H multiplied by w's length (number of characters). Similarly, the cost of inserting a word w in H is set to 0 if w is a stop word, and to the number of words in T multiplied by w's length otherwise. The cost of

---

[3] http://jswarm-pso.sourceforge.net/

Table 1. The accuracy on RTE datasets based on the RTE-Conf and RTE-Extended configurations

Model	Data set			
	RTE-4	RTE-3	RTE-2	RTE-1
RTE-Conf	57.6	59.37	57.75	55.5
RTE-Extended	59.5	62.4	59.87	58.62
Avg_Results	57.3	64.5	59.8	55.9

substituting two words is the Levenshtein distance (*i.e.* the edit distance calculated at the level of characters) between their lemmas, multiplied by the number of words in T, plus number of words in H. All these rules are parametrized with parameters with a fixed value set to 1. The actual goal of applying the PSO optimization approach is to change the value of these parameters in order to obtain the optimal scheme.

Our results are summarized in Table 1, which shows that the accuracy gained by applying the automatically optimized cost scheme (RTE-Extended) outperforms the RTE-Conf results over all the RTE test sets. In addition, it's also worth mentioning that in most of the cases RTE-Extended outperforms the average results achieved by participants in each edition of the challenge (Avg_Results row in Table 1). The analysis of the estimated optimal cost of each operation allowed to draw several conclusions, which have practical implications and are potentially useful for further research. Among these, for instance, we observed that the optimal cost of deletion was empirically estimated as 0, which means that deleting a node from the dependency tree of T does not affect the quality of the results. Our experiments illustrate that, with a little effort, EDITS can be extended in order to improve the general performance of the entailment system, also by capturing hidden phenomena that could not be easily explored otherwise.

5   Using EDITS in a QA Application

Besides working on RTE data, the flexibility of EDITS allows for effective configurations targeting specific problems and applications in other text processing areas. For instance, an approach to queston analysis for QA over structured data has been recently proposed in [13] and [9]. The addressed problem is to discover entailment relations between an input question Q (the *text* in the TE terminology) and a set of relational patterns (the *hypotheses*) representing possible lexicalizations of the relations of interest in a given domain. The assumption is that, if an entailment relation holds between Q and a pattern p associated to a relation R_i, then R_i is among the relations expressed in Q.

The task has been defined as a classification problem, where a question Q annotated with entities has to be assigned to all the relations $R_1,...,R_n$ it expresses, selected from a predefined set R. For instance, given the question *"What can I see [DATE: today] at cinema [SITE: Astra]?"*, the following relations represent the expected output of the system:

R1: HASMOVIESITE(MOVIE:?, SITE: "Astra")
R2: HASDATE(MOVIE:?, DATE: "today")

The classification is carried out using EDITS to compare the question Q against a set of textual patterns stored in a *Pattern Repository (P)*. P contains n sets of relational patterns. Each set represents a possible lexicalization of one relation R_i in R, and is associated to a SPARQL query fragment. Given the question Q, EDITS attempts to verify, for each relation R_i, if an entailment relation holds between Q and at least one of R_i's patterns. If so, the relation is added to the output of the system, and the corresponding SPARQL query fragment is used to access the target database. In case no relation is found, this is interpreted as evidence that the question is out of domain.

Adopting the configuration used for our RTE participation (RTE-Conf in Table 1), [13] and [9] report positive results over a dataset of English questions extracted from the QALL-ME benchmark[4], a multilingual corpus of annotated spoken requests in the domain of cultural events in a town (*e.g.* cinema, theatre, exhibitions).

Building on these positive results, further experiments with the same approach have been carried out on a customized configuration targeting specific issues posed by Italian questions in the *CINEMA* domain[5]. In such scenario, due to the unavailability of a reliable parser for the Italian language, we opted for a configuration based on the Token Distance Algorithm described in Section 2.

As far as the cost scheme is concerned, the customized configuration builds on the following observations:

1. In recognizing textual entailment the *relevance* of the terms involved in the edit operations is an important factor in determining the overall cost of transforming T into H. This is even more important in a restricted domain, where inserting or deleting domain-specific terms has a crucial impact on the transformation.

2. Often, as also emerged from the experiments reported in Section 4, the adoption of higher insertion costs as a negative impact. This is particularly true in our application scenario, which is characterized by *texts* (questions) that are usually much longer than the *hypotheses* (relational patterns[6]). In such specific scenario, high deletion costs have an undesired impact on the overall cost of the transformation of T into H. Often, in fact, very short (though precise) patterns are not entailed by long questions due to the high number of deletions performed.

---

[4] The QALL-ME benchmark has been developed within the EU funded QALL-ME project (http://qallme.fbk.eu), [3].

[5] Our dataset consists of 472 Italian questions (386 for training, 86 for test) marked as containing one or more relations chosen from a set of 14 binary relations.

[6] While the average lenght of the questions in our dataset is 12.6 words, the average length of the manually created patterns is 4.2 words.

In light of these considerations, a number of experiments has been carried out with different cost schemes assigning: *i)* lower costs to the *deletion* operation, and *ii)* higher *insertion* costs to terms that are relevant in the selected domain. In the resulting best configuration, deletion costs have been set to 0, and insertion costs for a term w have been set to the Inverse Document Frequency of w (IDF(W), calculated over the patterns for each relation).

As far as evaluation is concerned, since successful answer retrieval depends on capturing *all and only* the relations expressed in the question[7], we considered the proportion of test questions for which the system recognized all and only the relations marked in the gold standard. Such proportion (the *exact matches*) has been adopted as a more suitable metric for the QA scenario instead of standard Prec/Rec/F1 scores, or the overall Accuracy of the entailment assignment.

The best configuration achieved noticeable performance improvements over a simpler configuration used for comparison (Token Edit Distance, with insertion and deletion costs respectively set to the length of T and the length of H), with an *exact matches* increase from 24.4% (21 questions out of 86) to 76.7% (66 questions out of 86). These results further demonstrate the potentialities of EDITS in terms of customization for specific datasets.

6   Conclusions and Future Work

In this paper we presented the first release of EDITS, an open-source software package for recognizing Textual Entailment developed by FBK-irst. First, the main features and functionalities of the tool have been described. Then, to show the potential of EDITS in terms of tuning and adaptation capabilities, we reported on several experiments over different datasets and application scenarios. Such experiments addressed: the design and combination of specialized entailment engines, the automatic estimation and optimization of the costs for edit operations, and the customization of the tool for an entailment-based question interpretation task.

The planned work for the next releases of the package includes: *i)* the extension of the set of available distance algorithms, *ii)* the implementation of functionalities that allow for combining the output of specialized entailment engines into a single entailment score, *iii)* the development of a graphical user interface, to speed-up and further simplify the use of the system by non-expert users; *iv)* the creation of a comprehensive documentation of the internal structure of the system (algorithms, interfaces between modules), that will allow developers to extend it with customized modules and additional functionalities.

Acknowledgments. This work has been partially supported by the EU-funded project QALL-ME (FP6 IST-033860).

---

[7] Unrecognized relations would determine underspecified queries (often leading to redundant answers), while spuriously recognized relations would determine overspecified queries (leading to answer extraction failures).

References

1. Agichtein, E., Askew, W., Liu, Y.: Combining Lexical and Semantic Features for Recognizing Textual Entailment. In: Proceedings of the Text Analysis Conference TAC 2008, Gaithersburg, Maryland (2006)
2. Bos, J., Markert, K.: Combining Shallow and Deep NLP Methods for Recognizing Textual Entailment. In: Quiñonero-Candela, J., Dagan, I., Magnini, B., d'Alché-Buc, F. (eds.) MLCW 2005. LNCS (LNAI), vol. 3944, pp. 217–230. Springer, Heidelberg (2006)
3. Cabrio, E., Kouylekov, M., Magnini, B., Negri, M., Hasler, L., Orasan, C., Tomas, D., Vicedo, J.L., Neumann, G., Weber, C.: The QALL-ME Benchmark: a Multilingual Resource of Annotated Spoken Requests for Question Answering. In: Proceedings of LREC 2008, Marrakech, Morocco (2008)
4. Cabrio, E.: Specialized Entailment Engines: Approaching Linguistic Aspects of Textual Entailment. In: Proceedings of the NLDB conference, Saarbruecken, Germany (to appear, 2009)
5. Cabrio, E., Kouylekov, M., Magnini, B.: Combining Specialized Entailment Engines for RTE-4. In: Proceedings of TAC 2008, 4th PASCAL Challenges Workshop on Recognising Textual Entailment (2008)
6. Dagan, I., Glickman, O.: Probabilistic Textual Entailment: Generic Applied Modeling of Language Variability. In: Proceedings of the PASCAL Workshop on Learning Methods for Text Understanding and Mining, Grenoble, France (2004)
7. Dagan, I., Glickman, O., Magnini, B.: The PASCAL Recognising Textual Entailment Challenge. In: Quiñonero-Candela, J., Dagan, I., Magnini, B., d'Alché-Buc, F. (eds.) MLCW 2005. LNCS (LNAI), vol. 3944, pp. 177–190. Springer, Heidelberg (2006)
8. Eberhart, R.C., Shi, Y., Kennedy, J.: Swarm Intelligence. Morgan Kaufmann, San Francisco (2001)
9. Gretter, R., Kouylekov, M., Negri, M.: Dealing with Spoken Requests in a Multimodal Question Answering System. In: Dochev, D., Pistore, M., Traverso, P. (eds.) AIMSA 2008. LNCS (LNAI), vol. 5253, pp. 93–102. Springer, Heidelberg (2008)
10. Kouylekov, M., Magnini, B.: Tree Edit Distance for Recognizing Textual Entailment. In: Proceedings of the International Conference Recent Advances in Natural Language Processing (RANLP), Borovets, Bulgaria (2005)
11. Levenshtein, V.I.: Binary codes capable of correcting deletions, insertions, and reversals. Doklady Akademii Nauk SSSR 163(4) (1965)
12. Mehdad, Y.: Automatic Cost Estimation for Tree Edit Distance Using Particle Swarm Optimization. In: Proceedings of ACL 2009 (to appear, 2009)
13. Negri, M., Magnini, B., Kouylekov, M.: Detecting Expected Answer Relations through Textual Entailment. In: Gelbukh, A. (ed.) CICLing 2008. LNCS, vol. 4919, pp. 532–543. Springer, Heidelberg (2008)
14. Pedersen, T., Patwardhan, S., Michelizzi, J.: Word-Net:Similarity - Measuring the Relatedness of Concepts. In: Proceedings of Fifth Annual Meeting of the North American Chapter of the Association for Computational Linguistics (NACL 2004), Boston, MA (2004)
15. Tatu, M., Moldovan, D.: A Logic-based Semantic Approach to Recognizing Textual Entailment. In: Proceedings of COLINGACL 2006, Poster Sessions, Sydney, Australia (2006)
16. Zhang, K., Shasha, D.: Fast Algorithm for the Unit Cost Editing Distance Between Trees. Journal of Algorithms 11 (1990)

"Language Is the Skin of My Thought": Integrating Wikipedia and AI to Support a Guillotine Player*

Pasquale Lops, Pierpaolo Basile, Marco de Gemmis, and Giovanni Semeraro

Department of Computer Science, University of Bari "Aldo Moro", Italy
{lops,basilepp,degemmis,semeraro}@di.uniba.it

Abstract. This paper describes OTTHO (On the Tip of my THOught), a system designed for solving a language game, called *Guillotine*, which demands knowledge covering a broad range of topics, such as movies, politics, literature, history, proverbs, and popular culture. The rule of the game is simple: the player observes five words, generally unrelated to each other, and in one minute she has to provide a sixth word, semantically connected to the others. The system exploits several knowledge sources, such as a dictionary, a set of proverbs, and Wikipedia to realize a *knowledge infusion* process. The paper describes the process of modeling these sources and the reasoning mechanism to find the solution of the game. The main motivation for designing an artificial player for Guillotine is the challenge of providing the machine with the cultural and linguistic background knowledge which makes it similar to a human being, with the ability of interpreting natural language documents and reasoning on their content. Experiments carried out showed promising results. Our feeling is that the presented approach has a great potential for other more practical applications besides solving a language game.

1 Introduction

One of the most challenging fields in which Artificial Intelligence techniques have been exploited is represented by games. In terms of knowledge prerequisites it is possible to clearly distinguish closed-world games from open-world ones. The former provide the agent with all the knowledge necessary for playing the game, while for the latter no fixed sets of rules are sufficient to define the game play. An interesting case of open-world games is represented by *language games*, in which word (and phrase) meanings play an important role.

This paper describes a system that supports players of a language game called *Guillotine*. The game is broadcast in an evening show called *L'eredita'* by RAI, the Italian National Broadcasting Service, and involves a single player, who is given a set of five words, called *clues*, each linked in some way to a specific word that represents the unique solution of the game. In order to get a clue

---

* The first part of the title of the paper takes its inspiration from a sentence contained in the book "Power Politics" by Arundhati Roy (2002).

R. Serra and R. Cucchiara (Eds.): AI*IA 2009, LNAI 5883, pp. 324–333, 2009.

at a time, the player must choose between two different proposed words: one is correct, the other one is wrong. Each time she chooses the wrong word, the prize money is divided by half (the reason for the name *Guillotine*). The five words are generally unrelated to each other, but each of them is strongly related to the word representing the solution. Once the five clues are given, the player has one minute to guess the right answer. An example of the game follows: Given the five words *doppio, carta, soldi, pasta, regalo*, the solution is *pacco*, because "Pacco, doppio pacco e contropaccotto" is the title of an Italian movie, "pacco di carte" stands for "pack of cards", "pacco di soldi" means "wad" (of money), "pacco di pasta" stands for "parcel of pasta" and "pacco regalo" is the Italian phrase for gift-wrap. The player needs different types of knowledge, such as cultural and linguistic ones, to reason on and find the correct word. This paper presents OTTHO (On the Tip of my THOught), an information seeking system that supports a Guillotine player during the final stage of the game, when the five words are already available and the player has just one minute to guess the right answer. This stage poses a grand challenge for a computing system, because of the variety of subject matter, and above all because the clues given to the player may have subtle meanings and embody complexities at which humans excel and computers traditionally do not.

This paper extends the work reported in [12], since it enriches the knowledge base of the system through facts extracted from an encyclopedic knowledge source (Wikipedia). The paper is structured as follows. Related work are analyzed in Section 2. Section 3 describes the architecture of OTTHO, while details about the modeling of different types of knowledge sources are in Section 4. The reasoning mechanism for finding the solution of the game is briefly presented in Section 5 and experiments are in Section 6. Conclusions are in Section 7.

2   Related Work

The literature classifies games related to language in *word games*, in which word meanings are not important, and *language games*, in which word meanings play an important role. An example of language game is *Who Wants to be a Millionaire?*, in which the player should prove to have common sense knowledge or knowledge on popular culture to answer to a series of multiple-choice trivial questions. It has been shown that a system able to mine answers from the web plays the game about as well as people [8]. *Jeopardy!* is an American quiz show featuring trivia in topics such as history, literature, the arts, pop culture and science. Jeopardy! has a unique answer-and-question format in which contestants are presented with clues in the form of answers, and must phrase their responses in the form of a question. Recently, IBM announced that a question answering system called Watson will be delivered soon, which will try to challenge Jeopardy! human players. Another popular language game is solving crossword-puzzles. The first experience reported in literature is *Proverb* [10], that exploits large libraries of clues and solutions to past crossword puzzles. Differently from Proverb, OTTHO does not learn from previously solved games, due to the variety of subject matter and clues ambiguity. *WebCrow* [1] is the first solver for

Italian crosswords using the Web as knowledge base. Similarly to WebCrow, OTTHO exploits unstructured Web-based knowledge sources, such as a dictionary, an encyclopedia and a list of proverbs, etc. Even though techniques for text processing have some points in common, OTTHO differs in modeling the knowledge as a set of co-occurrences of terms, because the evidence of a strong relationship between words is the key factor for finding a set of candidate solutions. Furthermore, OTTHO exploits a specific strategy for representing the encyclopedic knowledge contained in Wikipedia, by relying on the WordSpace model [11], that represents concepts as vectors in a high-dimensional space, so that vectors representing concepts with related meanings are close in that space. Like WebCrow and the "Millionaire" system, OTTHO is language independent, in the sense that both the strategy adopted for modeling the knowledge sources and the reasoning mechanism are not influenced by the language of documents available in the sources.

3   OTTHO System Architecture

In order to *train* the artificial player for the Guillotine game, it is necessary to: 1) identify the *cultural* and *linguistic* background knowledge sources, and 2) model them in a way that some reasoning mechanisms can be applied. We analyzed hundreds of examples of the game, in particular how the clues were correlated to the solution, and we were able to identify the following *classes* of knowledge sources, ranked according to the frequency with which they were helpful in finding the solution of the game (Section 6):

1) Encyclopedia: the description of an article contains the solution;
2) Dictionary: the word representing the solution is contained in the description of a lemma or in some example phrases using that lemma;
3) Proverbs and Aphorisms: short pieces of text in which the solution is found very close to the clues;
4) Acronyms: the solution is one term of the acronym;
5) Titles of movies, songs, books, poetries.

These classes of sources need to be organized and processed in order to extract and model relationships between words, and to define a reasoning mechanism that, given the clues, is able to select the correct solution of the game among a set of candidate words. The system architecture is depicted in Figure 1. It is extremely modular and allows to plug in additional modules for enriching the background knowledge base without modifying the reasoning mechanism.

4   Modeling the Background Knowledge Sources

The modeling process must face the problem of the different characteristics of knowledge sources, resulting in a set of different heuristics for building the whole model on which to apply the reasoning mechanism. Since we are interested in

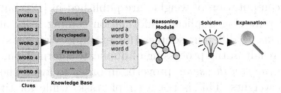

Fig. 1. OTTHO System Architecture

finding relationships existing between words, we decided to model each knowledge source using the correlations existing between terms occurring in that specific source.Indeed, we used a *term-term matrix* containing terms occurring in the modeled knowledge source in which each cell contains the weight representing the degree of correlation between the term on the row and the one on the column. The computation of weights is different for each type of knowledge source.

4.1 Modeling a Dictionary

We used the on-line De Mauro Paravia Italian dictionary containing 160,000 lemmas (old.demauroparavia.it/). We obtained a lemma-term matrix containing weights representing the relationship between a lemma and terms used to describe it. Because of the general lemma-definition organization of entries in the dictionary, we can fairly claim that the model is language-independent. Each Web page describing a lemma has been processed in order to extract the most relevant information useful for computing weights in the matrix. The text is processed in order to skip the HTML tags, even if the formatting information is preserved in order to give higher weights to terms formatted using bold or italic font. Stopwords are eliminated and abbreviations used in the definition of the lemma are expanded. Weights in the matrix are computed using a strategy based on a TF-IDF scheme, and normalized with respect to the length of the definition in which the term occurs and the length of the dictionary. A detailed description of the heuristics for modeling the dictionary is reported in [12].

4.2 Modeling Proverbs

As for the dictionary, a TF-IDF strategy has been used for defining the weights in the term-term matrix modeling the knowledge source of proverbs – 1, 600 proverbs gathered from the web at (http://web.tiscali.it/proverbiitaliani and http://giavelli.interfree.it/proverbi_ita.html). The following heuristics for improving the weight computation have been exploited:

- Terms in a proverb are correlated with all other terms in the same proverb;
- The closer the terms, the stronger their correlation, and the shorter the sentence, the stronger the correlation between terms;
- The greater the number of co-occurrences of two terms, the weaker their correlation.

Details on the computation of weights are published in [12], but a strategy for identifying and removing duplicate information coming from different sources follows. Indeed, using different sources (for example Web sites) for extracting proverbs might result in the processing of duplicate information. In this case we need to remove *roughly the same* proverbs in order to avoid the assignment of incorrect (fake) weights. This is not a simple task: while for titles of songs or books an exact match is enough, this could be not effective for proverbs, since the same proverb might be expressed using *roughly the same*, but *not the same* words (we call them semi-duplicate). In order to identify semi-duplicates we rely on the work by Broder [2]. Given two documents A and B he defines the mathematical notion of *resemblance r(A,B)* that seems to capture well the informal notion of *"roughly the same"*, not well captured by standard distances defined on strings, such as Hamming or Levenshtein. The *resemblance r(A,B)* of A and B is a number between 0 and 1 that expresses a degree of a set intersection. Given a document D, we define as *shingle* a sequence of contiguous tokens occurring in D. $S(D, w)$ denotes the *w-shingling* associated with D, defined as the *bag* of all its shingles of size w. For instance, the 2-shingling for $D=$[a, rose, is, a, rose] (tokens separated by comma) is $S(D, 2) = \{$[a,rose], [rose,is], [is, a], [a,rose]$\}$. The resemblance of two documents A and B is defined, depending on w, as:

$$r_w(A, B) = \frac{|S(A, w) \cap S(B, w)|}{|S(A, w) \cup S(B, w)|} \tag{1}$$

In order to tune the shingle size for short documents like proverbs, experiments were carried out on the collected proverbs, and we found better resemblance results by setting $w = 2$. The resemblance threshold for deeming two proverbs as semi-duplicates was set to 0.3. For $P_1=$[well, begun, is, half, done] and $P_2=$[well, begun, is, half, ended], it can be easily verified that $r_2(P_1, P_2) = 3/5$.

4.3 Modeling Wikipedia

Exploiting Wikipedia as a knowledge source has several advantages, such as its constant development by the community, the availability in several languages, its high accuracy [5] and the availability of generic and specialized knowledge for several domains. The process of modeling Wikipedia is different from the one adopted for proverbs and dictionary, due to the huge amount of information to be processed. We adopted a more scalable approach for processing Wikipedia entries, by using models for representing concepts through vectors in a high dimensional space, such as the *Semantic Vectors* or WordSpace models [11]. The core idea behind semantic vectors is that words and concepts are represented by points in a mathematical space, and this representation is learned from text in such a way that concepts with similar or related meanings are near to one another in that space (geometric metaphor of meaning). The semantic similarity between concepts can be represented as proximity in a n-dimensional space. Therefore, the main feature of the geometric metaphor of meaning is not that meanings can be represented as locations in a semantic space, but rather that similarity between word meanings can be expressed in spatial terms, as proximity in a high-dimensional space. One of the great virtues of semantic vectors is that they can be built using entirely unsupervised distributional analysis of free text. In addition, they make very few language-specific assumptions (tokenized text is just needed). The basis of semantic vectors model is the *distributional hypothesis* [6], according to which the meaning of a word is determined by the rules of its use in the context of ordinary and concrete language behavior. This means that words are semantically similar to the extent that they share *contexts* (surrounding words). If 'beer' and 'wine' frequently occur in the same context, say after 'drink', the hypothesis states that they are semantically related or similar. Co-occurrence is defined with respect to a context, for example a window of fixed length, or a document. Co-occurring words can be stored into matrices where the rows represent the terms and the columns represent contexts. Each row corresponds to a vector representation of a word. The strength of semantic association between words can be computed by using cosine similarity. This kind of techniques need to handle the potentially very high dimensionality of vectors. The solution is usually represented by *dimensionality reduction* techniques that allow representing high-dimensional data in a lower-dimensional space without losing information. *Latent Semantic Analysis (LSA)* [9] collects the text data in a words-by-documents co-occurrence matrix, that is then decomposed with singular-value decomposition (SVD) into smaller matrices, by capturing latent semantic structures in the text data. The main drawback of SVD is the scalability. Another approach is *Random Indexing* (RI) [7], that targets the problem of dimensionality reduction by removing the need for the huge co-occurrence matrix. RI represents a novel way of conceptualizing the construction of context vectors. Instead of collecting co-occurrences in a co-occurrence matrix at first and then extracting context vectors from it, RI incrementally accumulates context vectors, which can be later assembled into a co-occurrence matrix, if needed. In our work, we adopt the Semantic Vectors package [13] for creating semantic vectors from the Wikipedia corpus. The package supplies tools for indexing a collection of documents by means of the RI strategy, and relies on the Apache Lucene APIs. In order to exploit the Semantic Vectors model in our system, we downloaded the Italian Wikipedia XML dump (updated on May 2008), which was indexed by using Apache Lucene. Finally, the Semantic Vectors package was run on the Lucene index. After these steps, we obtained a model in which related concepts occur near to each other in that space. For example, if we search for terms near the word *pasta* we obtain: *spaghetti* and *semola* (bram). Each related term has a score measuring the correlation level.

5   The Reasoning Mechanism

Modeling several knowledge sources realizes a sort of "knowledge infusion" process into the system, in order to create a memory of world facts and linguistic knowledge. An algorithm for retrieving the most appropriate *pieces of knowledge* associated with the clues is needed to replicate the cognitive mechanism of a human being in the most faithful way. We adopt a *spreading activation model* [3], which has been used in other areas of Computer Science such as Information Retrieval [4]. The pure spreading activation model consists of a network data structure of nodes interconnected by links, that may be labeled and/or weighted and usually have directions. The processing is initiated by labeling a set of *source nodes* with activation weights, and proceeds by iteratively propagating that activation to other nodes linked to the source nodes. For each iteration, a termination condition is checked in order to end the search process over the network. We adopt this model as the reasoning mechanism of OTTHO. In the network for Guillotine, nodes represent words while links denote associations between words, obtained from the knowledge sources modeled as described in Section 4. The idea is that, like for the human beings, the reasoning process in OTTHO starts by looking at the clues. Thus, spreading in the Guillotine network is triggered by those words. The activation of clues causes words with related meanings (as modeled in the knowledge sources) to become active. At the end of the weight propagation process, the most "active" words represent good candidates to be the solution of the game. The Spreading Activation Network (SAN) for a run of the game is built by including initially source nodes $N_1, \ldots, N_5$, each one labeled with the word k_i given as a clue. Then, the SAN is populated by running m expansion phases. In the first expansion phase, for each clue k_i the most related words and corresponding correlation weights are extracted from the knowledge sources. We take as the most related words for k_i only those k_j with correlation weights higher than 0.1. For each k_j obtained by querying the knowledge sources, one node labeled with k_j is added to the SAN and linked to the source node k_i. The edge is directed from k_i to the k_j and labeled with the corresponding correlation weight. In the next expansion phase, the same process continues for the nodes added during the previous phase, and so on. In order to avoid overgrowing, we set $m = 4$. The spreading activation strategy consists of iterations (*pulses*). Given a SAN populated by nodes $N_1, \ldots, N_T$, each node N_i has an associated activation value at iteration p. A threshold F determines if a node is fired, that is whether it can spread its activation value over the SAN. At each iteration, every *fired* node propagates its activation to its neighbors as a function of both its current activation value and the weights of the edges that connect it with its neighbors. The spreading process is triggered by the clues, which are fired by setting their activation value to 1. Then, in the first pulse, clues spread the activation value to their neighbors. Only fired neighbors will spread their activation values in the next pulse and so on, until there are no fired nodes. The final result of the process is a list of the most *active* nodes (words). A thorough description of the spreading strategy is available in [12].

6   Experimental Evaluation

The goal of the experiments is twofold: 1) to measure the number of games solved by the system; 2) to evaluate whether the use of the encyclopedic knowledge source allows to solve more games than using just the dictionary and proverbs. A game is *solved* when the solution occurs in the Candidate Solution List (CSL) produced by OTTHO by selecting all the nodes N_i whose activation values are greater than a fixed threshold λ. Selected nodes are ranked according to their activation levels (in descending order), and the labels of the first k nodes are included in the CSL. Hereafter, CSL_k^λ denotes the CSL built with constraints k and λ. The reason for adopting this evaluation method is that ranking nodes might be a good strategy for determining some *candidate* solutions, but more sophisticated techniques are needed to select a *unique* answer among them, as the rules of the game require. At present, the component that picks up the answer from the CSL, namely the SOLVER, is not yet complete, therefore the evaluation is limited to the best k candidate solutions in the CSL. The dataset is a collection of $N = 50$ games attempted during the show by human players. For evaluating the system, we adopted two performance measures: Absolute Accuracy (AA) and Relative Accuracy (RA). Let N_k^λ be the number of games for which the correct solution is found in CSL_k^λ:

$$AA = N_k^\lambda/N \tag{2}$$

RA takes into account whether a game can be *potentially* solved by OTTHO. A game is *solvable* when the solution is found among words related to clues, obtained by searching the clues in the modeled knowledge sources $KS_1, \ldots, KS_n$.

Let $\#SOLVABLE_{KS_1,\ldots,KS_n}$ be the number of *solvable* games in the dataset, RA is the fraction of $\#SOLVABLE_{KS_1,\ldots,KS_n}$ successfully solved:

$$RA = N_k^\lambda/\#SOLVABLE_{KS_1,\ldots,KS_n} \tag{3}$$

Notice that, even if a game is solvable, there is no guarantee that the system will find the solution. For example, if the solution node for a solvable game is connected just to a single clue, it is likely that its activation level will receive less "energy" than other nodes connected to more clues. Thus, there is the risk that the activation level of the solution does not overcome the λ threshold. To give a more precise idea of the difficulty of the games available in the dataset, we analyzed also the *coverage* of the clues. A clue is *covered* by the available knowledge sources when an entry for that clue is found in at least one of them. The coverage for a game is defined as the number of covered clues (0-5). The higher the coverage, the higher is the likelihood that OTTHO will find the solution. Out of 250 clues in the dataset, 212 (84,8%) are covered by the dictionary, 111 (44,4%) by the proverbs, 246 (98,4%) by Wikipedia. Actually, almost all clues are covered by Wikipedia. This is a potentially positive result, since it means that the spreading algorithm will be able to reason on more clues than using only dictionary and proverbs. Experimental results reported in Table 1 confirm this feeling since $\#SOLVABLE_{Dictionary,Proverbs,Wikipedia}$ is the highest value observed. Results of coverage analysis are reported in Table 2. Values in columns from 2 to 5 are the number of games whose coverage is reported in the first column, taking into account each single knowledge source or all the available sources. The remaining columns show $N_k^{\lambda=0.1}$, with different values for k. We notice that OTTHO is able to solve games with higher coverage, therefore results reported in Table 2 are encouraging. Results of accuracy, as defined by equations (2) and

Table 1. Solvable games using different knowledge sources

Knowledge source(s)	Solvable games
Dictionary	36 (72%)
Proverbs	3 (6%)
Dictionary, Wikipedia	39 (78%)
Dictionary, Proverbs, Wikipedia	42 (84%)

Table 2. Coverage analysis

Coverage	Dictionary	Proverbs	Wikipedia	All sources	$N^{\lambda=0.1}_{k=10}$	$N^{\lambda=0.1}_{k=15}$	$N^{\lambda=0.1}_{k=100}$	$N^{\lambda=0.1}_{k=150}$
1	0	7	0	0	0	0	0	0
2	2	18	0	0	0	0	0	0
3	4	12	1	0	0	0	0	0
4	19	8	2	3	0	0	2	3
5	24	0	47	47	1	4	18	23
					1	4	20	26

Table 3. Accuracy of the system

	$N^{\lambda=0.1}_{k=10}$	$N^{\lambda=0.1}_{k=15}$	$N^{\lambda=0.1}_{k=100}$	$N^{\lambda=0.1}_{k=150}$
AA	2.0% (8.0%)	8.0% (10.0%)	40.0% (52.0%)	52.0% (68.0%)
RA	2.4% (10.0%)	9.5% (12.0%)	47.6% (62.0%)	61.9% (90.0%)

(3), are reported in Table 3. The accuracy of OTTHO using just the dictionary and the proverbs is also reported in parentheses. When strict constraints are set on the CSLs ($k = 10, k = 15$), the system accuracy is not satisfactory at all. Better results are obtained by allowing an extended list of candidate solutions (last two columns in Table 3). It seems that the integration of Wikipedia hurts performance in terms of accuracy. By the way the system is able to solve more games when the encyclopedic source is used and a larger set of clues is covered. The first conclusion of the experiments is that it is necessary to work more on the reasoning strategy for selecting the words to be the best candidates. Further experiments for tuning the (several) parameters of the reasoning mechanism (λ, pulses, activation value of a node, etc.) should be carried out. The most promising strategy for the development of the SOLVER seems to be the processing of large lists of candidate solutions, without setting severe constraints on them. The SOLVER should select the unique answer which "better explains" the input clues, thus it will discard solutions not able to explain the clues, rather than discard *a priori* candidate solutions with low activation values.

7   Conclusions

In this paper we described OTTHO, an information seeking system that supports a Guillotine player during the final stage of the game, consisting of guessing a hidden word semantically related to five words given as clues. We have defined the knowledge base for representing the *cultural* and *linguistic* background of the player and implemented a reasoning mechanism through a spreading activation algorithm able to retrieve the most appropriate *pieces of knowledge* necessary to solve the game. The paper relies on a previous work, in which we processed a dictionary and a set of proverbs as knowledge sources, and analyzes the effect of integrating Wikipedia. Experiments show that Wikipedia improves the coverage of the clues, but accuracy is not high. We are convinced that the proposed approach has a great potential even for more practical applications besides solving a language game. The system could be used for implementing an alternative paradigm for *associative information retrieval* [4], in which an initial indexing phase of documents can *spread* further *hidden* terms for retrieving related documents. For example, given the words *sabbia* (sand), *cemento* (cement), *rifiuti* (garbage), *dominio* (domain) and *impero* (empire), one of the solutions suggested by OTTHO is *ecomafia*, that is quite surprising, but (unfortunately) strongly connected to the previous words and can be used to find related documents.

References

1. Angelini, G., Ernandes, M., Gori, M.: Solving Italian Crosswords Using the Web. In: Bandini, S., Manzoni, S. (eds.) AI*IA 2005. LNCS (LNAI), vol. 3673, pp. 393–405. Springer, Heidelberg (2005)
2. Broder, A.Z.: On the Resemblance and Containment of Documents. In: Proc.of Compression and Complexity of Sequences, pp. 21–29. IEEE Comp.Society, Los Alamitos (1997)
3. Collins, A.M., Loftus, E.F.: A Spreading Activation Theory of Semantic Processing. Psychological Review 82(6), 407–428 (1975)
4. Crestani, F.: Application of Spreading Activation Techniques in Information Retrieval. Artificial Intelligence 11(6), 453–482 (1997)
5. Giles, J.: Internet Encyclopaedias Go Head to Head. Nature 438, 900–901 (2005)
6. Harris, Z.: Mathematical Structures of Language. Interscience, New York (1968)
7. Kanerva, P.: Sparse Distributed Memory. MIT Press, Cambridge (1988)
8. Lam, S.K., Pennock, D.M., Cosley, D., Lawrence, S.: 1 Billion Pages = 1 Million Dollars? Mining the Web to Play Who Wants to be a Millionaire. In: Proc. of the 2003 Conference on Uncertainty in Artificial Intelligence, pp. 337–345 (2003)
9. Landauer, T.K., Dumais, S.T.: A Solution to Plato's Problem: The Latent Semantic Analysis Theory of Acquisition, Induction, and Representation of Knowledge. Psychological Review 104(2), 211–240 (1997)
10. Littman, M.L., Keim, G.A., Shazeer, N.: A Probabilistic Approach to Solving Crossword Puzzles. Artificial Intelligence 134, 23–55 (2002)
11. Sahlgren, M.: The Word-Space Model: Using distributional analysis to represent syntagmatic and paradigmatic relations between words in high-dimensional vector spaces. PhD thesis, Stockholm University, Department of Linguistics (2006)
12. Semeraro, G., Lops, P., Basile, P., de Gemmis, M.: On the Tip of my Thought: Playing the Guillotine Game. In: IJCAI 2009, Proceedings of the 21st International Joint Conference on Artificial Intelligence, pp. 1543–1548. Morgan Kaufmann, San Francisco (2009)
13. Widdows, D., Ferraro, K.: Semantic Vectors: A Scalable Open Source Package and Online Technology Management Application. In: Proc. of the 6th International Conference on Language Resources and Evaluation, LREC 2008 (2008)

Analyzing Interactive QA Dialogues Using Logistic Regression Models

Manuel Kirschner[1], Raffaella Bernardi[1], Marco Baroni[2], and Le Thanh Dinh[1,3]

[1] KRDB, Faculty of Computer Science, Free University of Bozen-Bolzano, Italy
{kirschner,bernardi}@inf.unibz.it

[2] Center for Mind/Brain Sciences, University of Trento, Italy
marco.baroni@unitn.it

[3] Institute of Formal and Applied Linguistics
Faculty of Mathematics and Physics, Charles University in Prague, Czech Republic
lethanh.dinh@stud-inf.unibz.it

Abstract. With traditional Question Answering (QA) systems having reached nearly satisfactory performance, an emerging challenge is the development of successful Interactive Question Answering (IQA) systems. Important IQA subtasks are the identification of a dialogue-dependent typology of Follow Up Questions (FU Qs), automatic detection of the identified types, and the development of different context fusion strategies for each type. In this paper, we show how a system relying on shallow cues to similarity between utterances in a narrow dialogue context and other simple information sources, embedded in a machine learning framework, can improve FU Q answering performance by implicitly detecting different FU Q types and learning different context fusion strategies to help re-ranking their candidate answers.

1 Introduction

It is widely acknowledged that answering Follow Up Questions (FU Qs), viz., questions uttered after some other interaction, is a different task than answering isolated questions. Hence, Interactive Question Answering (IQA) systems have to tackle different challenges than Question Answering (QA) systems. The latter can rely only on the question to extract the relevant keywords. The former should take the previous interactions into consideration and achieve some form of context fusion, i.e., identify the information in the previous interactions that are relevant for processing the FU Q and answering it properly [1]. A first crucial context-dependent distinction is among topic shift and topic continuation FU Qs. These types of FU Qs might require different processing strategies. Hence, important sub-tasks within the IQA community are the identification of a typology of questions, automatic detection of the identified types, and the development of different context fusion strategies for each type [2,3,1].

In this paper, we aim to show how a system based on shallow cues to similarity between utterances in a narrow dialogue context and other simple information

R. Serra and R. Cucchiara (Eds.): AI*IA 2009, LNAI 5883, pp. 334–344, 2009.
© Springer-Verlag Berlin Heidelberg 2009

sources, embedded in a machine learning framework, can improve FU Q answering performance, and how such system can also implicitly detect different FU Q types and learn different answer ranking strategies to cope with them.

A further innovative aspect of our work is that instead of using the artificial Text Retrieval Conference (TREC) data most IQA systems are evaluated on (TREC 2001, TREC 2004), we train and test our system on real user questions collected via an on-line chatter Bot on a closed domain.

Section 2 places our proposal in the broader picture of IQA research. We describe our dialogue corpus in Section 3, and we introduce our general modeling framework and the features we use in Section 4. Versions of the model that do and do not take context into account are evaluated in Section 5. We conclude with an error analysis that points at directions for future improvements in Section 6.

2   Related Work

The importance of evaluating IQA against real user questions and the need to consider preceding system answers has already been emphasized [2,3,4,1]. In these earlier studies, dialogue corpora were collected via Wizard of Oz experiments and/or by giving specific tasks to the users. The corpus of dialogues we deal with consists of real logs in which the users were chatting with a Bot to obtain information in a help-desk scenario.

[4] and [5] look for salient transitions among utterances in a given context. To this end, they exploit deep semantic analyses to detect Argument-Predicate structure [4] and Centering Theories features [5]. Both [4] and [2] take Argument-Predicate structures as the base of semantic networks used to model context interaction and guide context fusion. In [2], the system relies also on deep and chunk reasoning on an external ontology. In this paper, we avoid any form of deep analysis of this sort.

Fine grained typologies of questions have been suggested [2,3,5], and different processing strategies have been proposed for the identified types. We consider the basic distinction between topic shift and topic continuation, and we propose a generalized linear model framework in which this distinction is automatically detected and used to improve the answering performance.

Our work is closely related to [1], that presents a question classifier that detects topic shifts and topic continuations by exploiting utterance similarity measures. However, we go two steps further by not requiring training data annotated for question type, and directly using the question classification cues to improve answer re-ranking performance.

Similarly to other work in QA [6,7], we use corpus-based similarity measures, with the important innovation that we extend them to similarity with previous utterances in the context. Finally, there is a large literature on using supervised machine learning for various aspects of QA, including question re-ranking [8]. Again, as far as we know we are the first to propose a supervised classifier that takes the previous dialogue into account for answering FU Qs.

3 Data

Most IQA systems have been trained and evaluated over the TREC (2001, 2004) data-sets, that consist of several sessions of related questions, the first of which sets the topic. Hence, there are no topic shifts, apart from the artificial ones at the first question of each session, if one considers the whole set as a single interaction [1]. Furthermore, there are no answers to rely on, and the questions were collected by TREC evaluators, i.e. they are not questions asked by users genuinely interested in the interaction. Hence, their nature is rather artificial. To overcome these limitations, we have been collecting a corpus of human-machine interactions with a Bot that provides information about a university library, picking a canned-text answer from a set of 484 information statements produced by the librarians (thus, the system faces the task that is often called *answer/passage re-ranking* in the QA literature).

The corpus consists of 139 4-turn snippets of human-machine interactions. We limit ourselves to four turns since there is evidence [2,9] that in most cases the previous two turns (Q_1 and A_1) contain enough information to process the FU Q (Q_2) and select its answer (A_2). Moreover, this makes our classifier well suited for practical applications, as it only relies on cues extracted from a fixed window of 3 utterances (Q_1, A_1 and Q_2). We refer to these 3 utterances as "the context". Note that A_1 correctly answers Q_1 in all snippets kept for this study.

Contexts in the corpus were manually classified by one of the authors as either topic shifts or topic continuations (57 and 82 cases, respectively). Below we report two sample topic continuation and one topic shift snippet examples. In a closed domain, the difference between the two types is rather fine-grained. We have considered topic continuations all those cases where the FU Qs are natural continuations of the conversation. The interactions are "task-related" information seeking dialogues, hence if two questions pertain to tasks where one is a typical development of the other, they are counted as "topic continuations" (example 1, where the actions are "borrow" and "search"). Questions that are about an entity introduced in Q_1 or A_1 (example 2) are also topic continuations. Shift/continuation annotation is a delicate task, making unsupervised, implicit detection of context (as in the experiments below) particularly desirable.

Example 1: Topic continuation (related tasks)
User. How many items can I borrow at once?   (Q_1)
System. You may borrow up to 40 items at time, independently of the type of
     medium.   (A_1)
User. Is there a help for searching books?   (Q_2)
System. To find books I'd recommend searching in our on-line catalogue (OPAC).
     (A_2)

Example 2: Topic continuation (entities in A_1)
User. How can I borrow a book?   (Q_1)
System. You can either get the item yourself or request it via our delivery
     service.   (A_1)

User. What are the requirements for it? (Q_2)
System. If you want an item to be sent to another Library branch simply click
   on the blue ORDER button to reserve it [...] (A_2)

Example 3: Topic shift
User. Where can I find a public phone? (Q_1)
System. Public phones are situated in the main entrance area. (A_1)
User. I have lost the Library Card. What can I do? (Q_2)
System. If you've lost your Library Card please tell my colleagues at the infor-
   mation desk. They will lock your account. (A_2)

Like in TREC data, the topic continuation FU Qs can contain ellipses, e.g., Q_1:
Where can I find design books? Q_2: and dvd?. Differently from TREC, both Q_1
and Q_2 could be just keywords, may contain noisy information such as typos or
bad grammar, and could be very similar: either the user is trying to refine the
question (the answer is correct but not what they want to know) or the topic
is further explored by moving the focus of attention on a new related entity or
a new related action: Q_1: *Could you recommend me some book? Q_2: Could you
recommend me some novel?.* These kinds of interactions seem typical of real user
data and they have been noticed also in other corpora of this type [2,1].

4   Model

Our goal is, given a FU Q (Q_2 in our dialogue snippets), to pick the best answer
from the fixed A_2 candidate set, by assigning a score to each candidate, and
ranking them by this score. Different context types might require different answer
picking strategies. Thus, we specify both A_2 *(identification) features*, aiming at
selecting the correct A_2 among candidates, and *context (identification) features*,
that aim at characterizing the context. The A_2 identification features measure
the similarity between an utterance in the context (e.g., Q_2) and a candidate
A_2. Context features measure the similarity between pairs of utterances in the
context (e.g., Q_1 and Q_2). They do not provide direct information about A_2,
but might cue a special context (say, an instance of topic shift) where we should
pay more attention to different A_2 identification features (say, less attention to
the relation between Q_2 and A_2, and more to the one between A_1 and A_2).

   We implement these ideas by estimating a generalized linear model from training data to predict the probability that a certain A_2 is correct given the context.
In this model, we enter A_2 features as main effects, and context features in interactions with the former, allowing for differential weight assignment to the same
A_2 features depending on the values of the context features.

4.1 A_2 Features

Most of our A_2 features measure the similarity between a context utterance and
A_2. The intuition is that the correct A_2 is similar to the context.

Lexical Similarity (*lexsim*): If two utterances (e.g., Q_2 and A_2) share some terms, they are similar; the more *discriminative* the terms they share, the more similar the utterances. Implemented by representing the utterances as vectors with the words they contain as dimensions. The value of each dimension is the *tf.idf* [10] of the corresponding word in the general ukWaC corpus,[1] calculated as:

$$\text{tf.idf}(w) = \sqrt{count(w)} \sqrt{\log \frac{|D|}{|D_w|}}$$

where *count(w)* returns the number of occurrences of the word in the corpus, $|D|$ is the number of corpus documents, and $|D_w|$ the number of documents containing the word. Weighting by *tf.idf* favours more discriminative terms, that occur in a restricted number of documents (e.g., *library* is less frequent but more discriminative than *long*). Similarity is quantified by the cosine of the angle between the vectors representing the two utterances being compared:

$$cos(\boldsymbol{u}_1, \boldsymbol{u}_2) = \frac{\boldsymbol{u}_1 \cdot \boldsymbol{u}_2}{||\boldsymbol{u}_1||\,||\boldsymbol{u}_2||}$$

Distributional Similarity (*distsim*): Two utterances are similar not only if they share the same terms, but also if they share similar terms (e.g., *book* and *journal*). Term similarity is estimated on the ukWaC corpus, by representing each content word (noun, verb, adjective) as a vector that records its corpus co-occurrence with other content words within a 5-word span. Raw co-occurrence counts are weighted by pointwise mutual information, that dampens the impact of frequent words [10]:

$$\text{mi}(w_1, w_2) = \log_2 \frac{p(w_1 \& w_2)}{p(w_1)\,p(w_2)}$$

where $p(w_1 \& w_2)$ is estimated by the proportion of $w_1 \& w_2$ co-occurrences over the total co-occurrence count for all pairs, and $p(w_*)$ from the marginal frequencies. Distributionally similar words, such as *book* and *journal* will have similar vectors [11]. An utterance is represented by the sum of the normalized distributional vectors of the words it contains. Similarity between utterances is again quantified by the cosine of their vector representations. We tried a few variants (20-word spans, raw frequency or log-likelihood ratio instead of mutual information, max similarity between nouns or verbs instead of summed vectors), but the resulting models were either highly correlated to the one we are reporting here, or they performed much worse in preliminary experiments. We leave it to further work to devise more sophisticated ways to measure the overall distributional similarity among utterances [12].

Semantic similarity (*semsim*): We try to capture the same intuition that similar utterances contain similar words, but we measure similarity using Word-

---

[1] http://wacky.sslmit.unibo.it/

Net [13].[2] We experimented with most of the WordNet similarity measures that were used by [1], settling for the Lin measure, that gave the best results across the board:

$$\text{linsim}(w_1, w_2) = \frac{2 \; ic(lcs(w_1, w_2))}{ic(w_1) + ic(w_2)}$$

where $lcs(w_1, w_2)$ is the lowest common subsumer of synsets containing the two words in the WordNet hierarchy (we pick the synsets maximizing the score), and $ic(x)$ (the information content) is given by $-\log(p(x))$. Probabilities are estimated from the sense-annotated SemCor corpus coming with the Word-Net::Similarity package. Following [1], the similarity between two utterances is determined by matching the words so as to maximize pairwise similarities, while normalizing for sentence length.

Action sequence (*action*): The binary action feature indicates whether two turns are associated with the same action, and thus represent an action sequence. For identifying the action associated with each A_2, we hand-annotated each of the 484 answer candidates with one of 25 relevant actions (borrowing, delivering, etc.). The action(s) associated with the other turns (Q_1, A_1, Q_2) are automatically assigned by looking for strings that match words that we think represent one of the 25 actions.

For each feature type, we compute its value for both the $A_1.A_2$ and $Q_2.A_2$ interplays: we will refer to them below as the *far* and *near* features, respectively (far and near in terms of distance of the compared utterance from A_2. We ignore $Q_1.A_2$ features for now). By crossing the measure types and the considered interplays, we obtain 8 A_2 features (*far.lexsim*, *near.lexsim*, *far.distsim*, etc.).

4.2   Context Features

Topic shifts across turns have been generally recognized as the main contextual factor affecting the relative role of context in FU Q answering [1]. If Q_2 continues the previous topic, then the previous context should still be relevant to A_2. If the topic shifted, the A_2 selection strategy should focus on the most recent turn only (i.e., Q_2 itself).

In order to verify that adding topic continuation information to the model does indeed help A_2 prediction, we used our manual coding of contexts for whether they contain a topic shift or not (*topshift*). This feature is of limited practical utility in a real life system: topic change or persistence should be detected by automated means.

A simple way to capture the notion of topic continuity is in terms of similarity between Q_2 and each of the preceding utterances (the less similar Q_2 is to Q_1 or A_1, the more likely it is that the topic shifted). Thus, the same utterance similarity measures that, when used to compare other utterances to A_2, serve as A_2 identification features, can be treated as continuous approximations to a

---

[2] We use the WordNet::Similarity package:
   http://wn-similarity.sourceforge.net/

topic shift when applied to the $Q_1.Q_2$ and $A_1.Q_2$ interplays. Since we defined 3 similarity measures (lexsim, distsim and semsim), we obtain 6 more context identification features ($Q_1.Q_2.lexsim$, $A_1.Q_2.lexsim$, $Q_1.Q_2.distsim$, etc.).

We tried various strategies to combine the topic approximation cues into composite measures (e.g., by defining "profiles" in terms of high and low $Q_1.Q_2$ and $A_1.Q_2$ scores, or by defining multiple interaction terms), but they did not improve on the simpler context features, and we do not report their performance here.

4.3   Logistic Regression

Logistic regression models (LRMs) are generalized linear models that describe the relationship between features (independent variables) and a binary outcome [14]. Our logistic regression equations, which specify the probability for a particular answer candidate A_2 being correct, depending on the learned intercept β_0, the other β coefficients (representing the contribution of each feature to the total answer correctness score), and the feature values $x_1, \ldots, x_k$ (which themselves depend on a combination of Q_1, A_1, Q_1 or A_2) have the form:

$$\text{Prob}\{\text{answerCorrect}\} = \frac{1}{1 + \exp(-X\hat{\beta})}$$

$$\text{where } X\hat{\beta} = \beta_0 + (\beta_1 x_1 + \cdots + \beta_k x_k)$$

Context typology is implicitly modeled by *interaction* terms, given by the product of an A_2 feature and a context feature (when we enter an interaction term with a context feature, we also always introduce the corresponding main effect). An interaction term provides an extra β to assign a differential weight to an A_2 feature depending on the value(s) of a context feature. In the simplest case of interaction with a binary 0-1 feature, the interaction β weight is only added when the binary feature has the 1-value.

We estimate the model parameters (the beta coefficients $\beta_1, \ldots, \beta_k$) using maximum likelihood estimation. Moreover, we put each model we construct under trial by using an iterative backward elimination procedure that takes off all those terms whose removal does not cause a significant drop in goodness-of-fit. All the results we report below are obtained with models that underwent this trimming procedure.

5   Evaluation

We match each of the 139 contexts (Q_1, A_1 and Q_2 sequences) in our dialogue corpus with each of the 484 A_2s in our pre-canned answer repository. Since the corpus had been pre-annotated for what is the (single) correct A_2 for each context, this produces 483 negative examples and 1 positive example for each context, that can be used to estimate a LRM.[3] We rank the $Q_1.A_1.Q_2.A_2$ 4-tuples constructed in this way in terms of the probability they are assigned by

---

[3] Our experiments with random sampling of the majority class (i.e., the negative training examples) did not improve model performance.

the estimated LRM, and we look at the rank of the *correct* A_2 for each context. We use "leave-one-out" cross validation by predicting the ranks of the A_2s given each context with a model that was trained on the remaining 138 contexts. The lower the average rank of the correct A_2 across contexts, the better the model predictions. In the tables below, we report these average ranks. When we report statistical results about the relative quality of models, these are based on paired Mann-Whitney tests across the 139 ranks. Statistical significance claims are based on the $p < 0.05$ level, after correction for multiple comparisons.

We will first look at models that only look at the relation between context utterances and A_2 ("main effects only" models), and then at models that also exploit information about the relation between context utterances, to approximate a typology of contexts ("interaction" models).

5.1 Main Effects Only Models

We enter each feature from Section 4.1 at a time in separate models (e.g., the *.semsim* models) and combine them (the *.combined* models). Moreover, we look at $Q_2.A_2$ features (*near.*, in the sense that we look at the nearest context element with respect to A_2), $A_1.A_2$ features (*far.*), and both (*complete.*). Table 1 summarizes our first set of experiments. All *near* and *far* single feature models, except *far.action*, perform significantly better than *baseline*, and in general the near context is more informative than the far one (group 1 vs. group 2). Combining different knowledge sources helps: *near.combined* is significantly better than the best non-combined model, *near.distsim*. Combining features only helps marginally when we look at the *far* setting (compare groups 2 and 4). The

Table 1. Mean ranks of correct A_2 out of 484 answer candidates in main effects only models

Group	Description	Model name	Mean rank	SD
0	A_2 picked at random	*baseline*	235.0	138.2
1	Single $Q_2.A_2$ feature	*near.lexsim*	80.4	106.0
		near.distsim	74.4	113.5
		near.semsim	101.2	115.2
		near.action	178.1	156.8
2	Single $A_1.A_2$ feature	*far.lexsim*	164.6	138.3
		far.distsim	157.3	145.3
		far.semsim	152.3	135.3
		far.action	231.5	153.5
3	Combined $Q_2.A_2$ features	*near.combined*	**57.6**	93.4
4	Combined $A_1.A_2$ features	*far.combined*	141.7	130.5
5	Single $Q_2.A_2$ and $A_1.A_2$ features	*complete.lexsim*	75.3	109.4
		complete.distsim	72.6	108.9
		complete.semsim	103.2	111.0
		complete.action	(= *near.action*)	
6	Combined $Q_2.A_2$ and $A_1.A_2$ features	*complete.combined*	**58.6**	97.4

complete.combined model (group 6) significantly outperforms the corresponding best single feature model of group 5 *complete.distsim*, but its performance is not distinguishable from the one of *near.combined*.

This first batch of analyses shows that the proposed features have a significant impact on correct A_2 prediction, that combining different knowledge sources considerably improves performance and that, if we do not consider an interaction with context type, the *far* features (comparing A_1 and A_2) are not helpful. In the next step, we will work with the two best main-effects-only models obtained so far, namely *near.combined* and *complete.combined* (groups 3 and 6 in Table 1), and we will investigate their behaviour when we add interaction terms that try to capture different context types.

5.2   Models with an Interaction

Table 2 reports the results obtained with models that add an interaction term that should capture contextual development patterns, and in particular the presence of a topic shift. As discussed in Section 4.2 above, depending on whether there is a topic shift, we should assign different weights to *near* and *far* features. Thus, we predict that the presence of an interaction term marking topic development should help *complete.** models (that encode both *near* and *far* features), but not *near.** models.

The prediction is fully confirmed by comparing the model in group 3 of Table 1 to the one in group 7 of Table 2, on the one hand, and the model in group 6 of Table 1 to the model in group 8 of Table 2 on the other. In both cases, we are adding interaction terms for each of the main effects (the A_2 prediction features) crossed with the binary feature recording manual annotation of the presence of a topic shift. For the *near* model (that only takes the $Q_2.A_2$ relations into account), there is *no* improvement whatsoever from adding the interaction. Indeed,

Table 2. Mean ranks of correct A_2 out of 484 answer candidates in interaction models

Group	Description	Model name	Mean rank	SD
7	*near.combined*: int. with manual *topshift*	*near.combined* × *topshift*	57.5 (= *near.combined*)	93.4
8	*complete.combined*: int. with manual *topshift*	*complete.combined* × *topshift*	**54.3**	93.2
9	*complete.combined*: interaction with approximation feature	*complete.combined* × $Q_1.Q_2.lexsim$	55.5	91.7
		complete.combined × $A_1.Q_2.lexsim$	56.7	93.8
		complete.combined × $Q_1.Q_2.distsim$	57.1	98.2
		complete.combined × $A_1.Q_2.distsim$	58.4	96.4
		complete.combined × $Q_1.Q_2.semsim$	60.0	100.8
		complete.combined × $A_1.Q_2.semsim$	**54.3**	90.9
		complete.combined × $Q_1.Q_2.action$	55.8	94.4
		complete.combined × $A_1.Q_2.action$	56.9	96.1

Table 3. Retained predictors, model *complete.combined* ∗ *topshift* (Table 2, group 8)

Predictor	β coef.	SE	z value	Pr(>|z|)	Predictor (cont'd)	β coef.	SE	z value	Pr(>|z|)
intercept	-11.6	0.7	-15.9	0.000	*far.distsim*	2.7	1.1	2.3	0.019
near.lexsim	6.5	1.0	6.4	0.000	*far.action*	0.0	0.3	-0.1	0.923
near.distsim	1.5	0.9	1.7	0.092	*topshift*	1.1	1.1	1.0	0.317
near.semsim	2.3	0.5	4.8	0.000	*topshift* × *near.distsim*	2.7	1.2	2.4	0.018
near.action	0.9	0.2	4.3	0.000	*topshift* × *far.distsim*	-3.2	1.6	-2.0	0.040
far.lexsim	3.2	0.6	5.1	0.000	*topshift* × *far.action*	-1.4	0.6	-2.5	0.014

the backward elimination procedure we use when estimating the model drops all interaction terms, leading to an estimated model that is identical to the one in group 3 (main effects only). Vice versa, some interaction terms are preserved in the model of group 8, that improves from the corresponding interaction-less model, from a mean rank of 57 to a mean rank of 54 (although the rank improvement itself is not statistically significant).

Table 3 reports the coefficients of the estimated group 8 model (trained, for these purposes only, on the complete corpus). The four *near* features have a major positive effect on the odds of an answer candidate being correct. Also, the first two *far* features listed in the table have positive effects. We interpret the retained interaction terms with *topshift* (the last three rows) as follows. If Q_2 is a topic shift, the weight given to *near.distsim* (a term measuring the similarity with Q_2) has an extra positive effect, while the weight given to semantic similarity and the repetition of the same action between A_1 and A_2 is significantly decreased, since in a topic shift the earlier context should be (nearly) irrelevant. These effects are in line with our hypothesis about topic shift FU Q processing. Thus, we confirm that knowing about topic shifts helps, and that it helps because we can assign different weights to *far* and *near* relations depending on whether the topic continues or changes.

Having established this, we now ask whether the manually coded *topshift* variable can be replaced by an automatically computed feature that captures topic shifting in terms of the similarity between context utterances (Q_1 vs. Q_2 or A_1 vs. Q_2). By looking at the results in group 9 of Table 2, we see that in fact one of these models (interaction with $A_1.Q_2.semsim$) performs as well as the model using the hand-annotated *topshift* feature, while outperforming the *complete.combined* model (barely missing full statistical significance, with $p = 0.05457$). Interpretation of the coefficients is harder in this case, since we deal with a continuous interaction term, but the main patterns are as for the model in group 8. Not only keeping track of topic development helps answer (re-)ranking, but a simple automatically assigned feature (WordNet-based similarity between A_1 and Q_2) is as good as manual annotation to cue basic topic development.

6   Conclusion

From our quantitative evaluation via LRM we can conclude that to answer FU Qs asked in a real help-desk setting, some form of shallow context detection and fusion should be considered. In particular, the system answer preceding the FU

Q seems to play an important role, especially because its similarity to the FU Q can cue a topic shift, that in turn requires a different context fusion strategy (more weight to FU Q, less to the preceding context).

Our shallow cues, though promising, need further refinement, in particular to deal with the following problems particular to real user interactions: (I) some Q_1 and Q_2 are quite similar, which could happen for example when users are not satisfied with the answer (even if it was correct for the question that was asked) and hence rephrase the question by, e.g., using a more specific entity, or they even repeat the same question in the hope to obtain a better answer; (II) Q_2 contains only *WH VERB ENTITY*, and the verb is a factotum verb. Both (I) and (II) require further investigation and seem to ask for more structured cues than those explored in this paper.

References

1. Yang, F., Feng, J., Di Fabbrizio, G.: A data driven approach to relevancy recognition for contextual question answering. In: Interactive Question Answering Workshop (2006)
2. Bertomeu, N.: A Memory and Attention-Bases Approach to Fragment Resolution and its Application in a Question Answering System. PhD thesis, Universität des Saarlandes (2007)
3. Van Schooten, B., Op den Akker, R., Rosset, S., Galibert, O., Max, A., Illouz, G.: Follow-up question handling in the imix and ritel systems: A comparative study. Nat. Lang. Eng. 15(1), 97–118 (2009)
4. Chai, J.Y., Jin, R.: Discourse structure for context question answering. In: Proceedings of the Workshop on Pragmatics of Question Answering at HLT-NAACL 2004 (2004)
5. Sun, M., Chai, J.: Discourse processing for context question answering based on linguistic knowledge. Know.-Based Syst. 20(6), 511–526 (2007)
6. Burek, G., De Roeck, A., Zdrahal, Z.: Hybrid mappings of complex questions over an integrated semantic space. In: Andersen, K.V., Debenham, J., Wagner, R. (eds.) DEXA 2005. LNCS, vol. 3588. Springer, Heidelberg (2005)
7. Tomás, D., Vicedo, J., Bisbal, E., Moreno, L.: Experiments with lsa for passage re-ranking in question answering. In: CLEF Proceedings (2006)
8. Moschitti, A., Quarteroni, S.: Kernels on linguistic structures for answer extraction. In: Proceedings of ACL 2008: HLT, Short Papers, pp. 113–116 (2008)
9. Kirschner, M., Bernardi, R.: An empirical view on iqa follow-up questions. In: Proc. of the 8th SIGdial Workshop on Discourse and Dialogue (2007)
10. Manning, C.D., Schütze, H.: Foundations of statistical natural language processing. MIT Press, Cambridge (1999)
11. Sahlgren, M.: The Word-Space Model. Dissertation, Stockholm University (2006)
12. Mihalcea, R., Corley, C., Strapparava, C.: Corpus-based and knowledge-based measures of text semantic similarity. In: Proceedings of AAAI (2006)
13. Fellbaum, C. (ed.): WordNet: An electronic lexical database. MIT Press, Cambrdige (1998)
14. Agresti, A.: Categorical data analysis. Wiley, New York (2002)

An Agent-Based Paradigm for Free-Hand Sketch Recognition

D.G. Fernández-Pacheco[1], J. Conesa[1], N. Aleixos[2], P. Company[3], and M. Contero[2]

[1] DEG. Universidad Politécnica de Cartagena, 30202 Cartagena, España
[2] Instituto en Bioingeniería y Tecnología Orientada al Ser Humano
(Universidad Politécnica de Valencia), España
[3] EMC. Universitat Jaume I, 12071 Castellón, España
{daniel.garcia,julian.conesa}@upct.es, naleixos@dig.upv.es,
pcompany@emc.uji.es, mcontero@dig.upv.es

Abstract. Natural interfaces for CAD applications based on sketching devices have been explored to some extent. Former approaches used techniques to perform the recognition process like invariant features extracted with image analysis techniques as neural networks, statistical learning or fuzzy logic. Currently, more flexible and robust techniques are being introduced, which consider other information as context data and other relationships. However, this kind of interfaces is still scarcely extended because they still lack scalability and reliability for interpreting user inputs.

Keywords: CAD, Natural interfaces, Agent-based systems, Gesture and symbol recognition.

1 Introduction

New devices created to substitute the traditional pen and paper support Calligraphic Interfaces [1], but automatic gesture recognition is a complex task since different users can draw the same symbols with a different sequence, shape, size or orientation, and still remains as a very active research field. Due to these inconveniences, many recogniser algorithms are not robust or present ambiguity in the results. To avoid this variability, techniques based on digital image processing can be applied. Moreover, the problems related to the sketch scale or orientation has to be solved by means of techniques that remain invariant in the face of these features [2]. Shape description from morphological features as length, width, perimeter, area, inertial moments, bounding box, etc., presents the drawback of dependency on the scale or orientation, which increases the percentage of mistakes in the classification stage of the algorithm [3], so invariant morphological features are more suitable for this purpose. For example, Fourier descriptors are an example of invariant features [4], and they have been largely used as a general technique for image information compression or classification of regular shapes [5] or handwriting characters [6]. There are also a set of regular invariant moments widely used as contour-based shape descriptors. These are the descriptors derived by Hu [7]. Along the time, some works are carried out using these

R. Serra and R. Cucchiara (Eds.): AI*IA 2009, LNAI 5883, pp. 345–354, 2009.

techniques in sketch recognition to detect symbols, diagrams, geometric shapes and other user command gestures. Some of them are scale and rotation dependent, others use invariant features related to shape factor ratios, Fourier descriptors or invariant moments, and others can accept multiple strokes but in a strict input order ([8-13]). As traditional techniques for classification, we can find the fuzzy logic, distances from ideal shape, linear and non-linear discriminant analysis, etc. ([14-17]). These traditional techniques have the drawback of being rigid and the more symbols are contained in the catalog the worse results are obtained.

As an alternative to traditional techniques, we can find a technology based on agents. The agent based systems have been widely used for process simulation, process control, traffic control and so on, but their use is being extended more and more to recognition processes for supporting natural interfaces. The main benefits we can take profit of are the flexibility of these systems and the autonomy of agents. This technology is being used even more for natural interfaces, as for instance Juchmes et al. [18] who base their freehand-sketch environment for architectural design on a multi-agent system. Also Achten and Jessurum [19] use agents for recognition tasks in technical drawings. So do Mackenzie et al. [20] in order to classify sketches of animals. Other examples can be found in Azar et al. [21], who base their system for sketch interpretation on agents, or in Casella et al. [22], who interpret sketched symbols using an agent-based system.

Our current goal is defining a new paradigm which is a) aimed at interpreting hand-drawn sketches used by product designers during the creative stages of design, and b) able to design a scalable solution. Hence, our aim is defining an agent-based structural approach, so as to take profit of the flexibility of agents systems and the autonomy of individual agents. Our structural approach works in three levels: a) *basic* agents in the low level act to partially recognise/classify simple strokes and distribute some recognition tasks, b) *primitive* agents in the intermediate level use these results for finding the syntactic meaning of strokes, and c) the upper level contains the *combined* agents aimed at obtaining the semantic meaning of groups of strokes.

This paper is organised as follows: first the recognition process is explained, second the structure of agent levels is shown and, finally, some reasoning about this methodology is given.

2   The Recognition Process Paradigm

To take advantage of the flexibility and autonomy of hierarchical multi-agent architecture for recognition process, a hierarchical decomposition for symbols is defined. So, symbols are made out of lines, where lines are like "phonemes" of an alphabet. One or several phonemes make a symbol (i.e. a "word"), that belongs to a set of symbols (a "dictionary"). In this way, our structural approach uses *basic* agents to recognise/classify lines, *primitive* agents for finding their syntactic meaning, and *combined* agents to obtaining the symbols they belong to.

What we call lines or "primitive symbols" are simple geometric shapes that appear repeatedly in the symbols. On the other hand, strokes are sets of points that are digitised by the input device between a consecutive pen-down and pen-up event. For the

sake of simplicity, along the rest of this paper, we have assumed the equivalence between lines and strokes. This simplistic assumption is valid as far as the current goal
is just testing whether the design and implementation of the agents-architecture is feasible for interpreting hand-drawn symbols. Figure 1 shows the set of lines/strokes
used as phonemes.

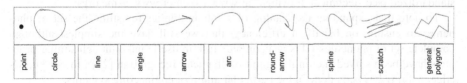

Fig. 1. Phonemes

However, it is clear that the final system will not continue assuming that each pen
stroke represents a single line. Overtracing and segmentation must be considered.
Overtracing is the use of multiple strokes to represent a single line. Readers interested
in this topic can find a recent contribution by Ku et al. [23]. Segmentation is the process of dividing a complex stroke into its geometrical primitives. Segmentation of
sketches is an open problem in the process of sketch recognition. Recent contributions
can be found in [24-25].

To compose phonemes into words, i.e. to identify symbols, we must find the semantic meaning of every group of phonemes. This means that the last stroke for a
symbol occurs when the recognition process finishes with a valid result. An improvement we accomplish is allowing *interspersing* of strokes from different objects.
Readers interested in this topic can find recent advances in [26].

Finally, we have defined a catalog or "dictionary", as shown in figure 2. Again, we
have just defined just a small catalog of symbols, although representative enough to
test the implementation of our agents-architecture. The symbols considered are representative of the modeling tasks that allow the users to construct parametric geometry
from sketches and to create basic solid models. They include command gestures (as
extrusion, scratch-out, etc.), symbols conveying information about dimensions and
geometric constraints (collinear, parallel, etc.), and symbols conveying other geometrical information (as axis, etc.).

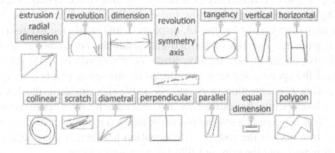

Fig. 2. Dictionary of gestures/symbols

Defining a full catalog is beyond the scope of the present work, but will be mandatory for the system to cope with actual design scenarios. At this end, we should perhaps consider the convenience of following an approach similar to the grammar of images described by Zhu and Mumford [27] or Lin et al. [28], which solve a hierarchical decomposition from scenes to objects. However, we should remember that most engineering symbols are already perfectly defined and catalogued, as they are standardized (see "technical drawings" at www.iso.ch or www.asme.org).

To note that expanding the catalog of symbols implies revisiting the set of phonemes to ensure, on behalf of efficiency, that we still have the simplest although complete set of phonemes. This is a complex task in itself, and although similar problems have been solved (i.e. in the ambit of character recognition [29]), to our knowledge, nothing has been done in the particular ambit of engineering symbols.

3 Agent Structure

As our goal is defining an agent-based paradigm proficient to design scalable solutions, we have defined an agents-architecture that works in three levels: low level for the *Basic Agents* aimed at interface, pre-processing and feature extraction; intermediate level for the *Primitive Agents* in charge of syntactic meaning of every single stroke, and the upper level that holds the *Combined Agents* in charge of semantic meaning of several strokes forming a symbol. All agents in the system run in parallel.

3.1 Basic Agent Level

Four agents work at this level:

- The **Broker Agent** (BA) that distributes the tasks between all agents living in the different levels of the system.
- The **Interface Agent** (IA) that supports the interaction with the user and provides the stroke points to the remaining basic agents.
- The **Pre-processing Agent** (PA) that beautifies every stroke in order to remove the noise and prepares them for further operations.
- The **Feature Agent** (FA) that extracts relevant features by means of image analysis techniques.

Regarding to the BA, it manages the delivering task in the system. Figure 3 shows the operating diagram of the BA for the Basic and Primitive agent levels.

When a user draws a stroke, the IA displays it and sends the digitised points to a global data structure. The BA is notified and sends the stroke points to the PA that produces the beautified stroke. Ideally, stroke points should be uniformly distributed. But the faster the speed a stroke is introduced at, the fewer points are digitised. That causes a varying concentration of points. To solve this problem, the PA detects and removes isolated points by estimating the distance between each point and their neighbors. Then PA passes a smoothing mask along the points of each stroke to avoid the effect of tremors. At the end, the line equation between two consecutive digitised points is calculated, and the gaps are filled in with new points.

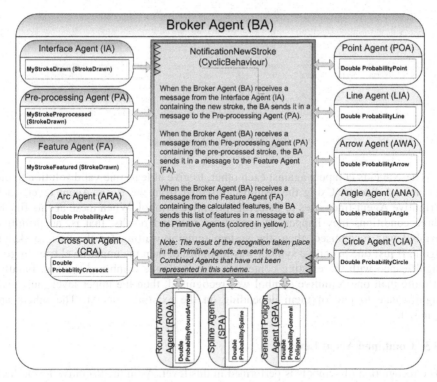

Fig. 3. Broker Agent operating diagram for Basic and Primitive agent levels

When beautification is done, BA asks for feature extraction to the FA. Once the FA has finished, the BA is informed and it passes the features to the Primitive Agents. When finished, the Primitive Agents pass their results to the BA, which will let them available for the Combined Agents to perform the second recognition step.

The feature extraction is sequential. The FA normalises the stroke points to a fixed number of points (FFT algorithm requires an input with 2n points) and then performs its analysis to calculate the FFT [30], and other features invariant to position, scale and orientation as Hu moments [7], perimeter, circularity, and so on [31]. Centroid, radius signature and the arc length versus cumulative turning angle signature [32] are also calculated. To note that not all the features will be used by all the Primitive Agents. They will use the significant ones to find their relevant cues, in order to recognise the primitive symbol they are in charge of.

3.2 Primitive Agent Level

Recognition of simple strokes is performed in this level. At least one Primitive Agent has been implemented for each primitive symbol (figure 4). The names of these agents have been chosen in the following way: the two first digits for the phoneme to recognise and the third digit is an *A* for Agent. So, the Primitive Agent that finds a point will be called POA (*PO* for point and *A* for Agent).

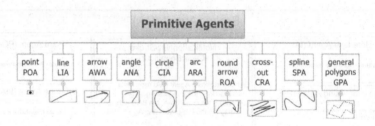

Fig. 4. Primitive Agents level

These agents compete against each other, to give at least one valid solution to the Combined Agents. Using the features provided from the Basic Agent level, each Primitive Agent finds cues to detect its primitive symbol. When these agents finish, three conditions may happen. If agents did not find results, that is, no primitive symbol was recognised, then, the system remains waiting for the user to sketch again. If just one primitive symbol was recognised, then the upper level tries to give significance with the existing context to recognise one combined symbol. Finally, if more than one primitive symbol was recognised, then the upper level must give significance to one of them depending on the existing context. The others are ignored.

3.3   Combined Agent Level

The recognition of symbols is performed in this level. As in the Primitive level, there is at least one Combined Agent for each symbol to recognise (figure 5).

Fig. 5. Combined Agents level

One iteration of the symbol recognition process starts when a stroke is drawn. It is first beautified and normalised, and, after that, their features are extracted by means of image analysis techniques. Then, it is recognised in two steps: as a simple stroke, and as a component of a symbol. In the first step, every stroke is matched to a specific phoneme. In the second step, consecutive phonemes are collected until they form a full symbol, i.e. until they reach a full semantic meaning. For the second step, the system checks out the following cases:

- If the phoneme is a single-stroke symbol (i.e. scratch or polygon) it is assigned and the recognition of the current symbol finishes.
- If it is suitable to pertain to some multiple-stroke symbol, then, it is added, and the system must wait for the next stroke to be introduced:

o If the next stroke is suitable to form a combined symbol altogether, then, it is added and the current symbol finishes.

o If the next stroke is suitable to pertain to, but there is not semantic meaning enough to form a combined symbol altogether, then,
it is added and the search continues.

o If the next stroke is not suitable to pertain to, it is ignored, and
the search continues.

Note that every stroke may be suitable to pertain to more than one multiple-stroke
symbol, and the same symbol may be drawn by way of different strokes (figure 6).
Hence, all of the valid candidates are searched *in parallel*, until one of them reaches a
full semantic meaning. The recognition system implements the blackboard paradigm,
where the decision is taken depending on the set of cues that have the maximum significance. The user can draw the strokes in any order, so the recogniser does not depend on the sequence.

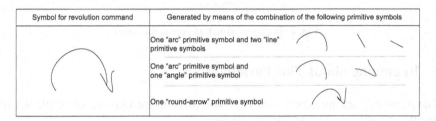

Fig. 6. Possibilities of sketching the revolve command gesture

4 Agent Architecture Implementation

The agents-architecture is described in Figure 7. The Basic Agents (already detailed
in figure 3), are colored in yellow and delimited by dotted lines. The intermediate
level manages the Primitive Agents (PA), and the upper level manages the
Combined Agents (CA). Their output is given to the BA, which passes the final
result to the external CAD editor or Sketched-Based Modeling system though the
IA agent.

The platform is implemented on top of the Jade agent-based platform [33] using
Java 1.6. Communication is in charge of messages that are encoded in FIPA-ACL
messages [34], which is a communication language natively supported by Jade. FIPA-
ACL specifies both the message fields and the message performatives, which consist
of communicative acts such as requesting, informing, making a proposal, accepting or
refusing a proposal, and so on. Jade offers the means for sending and receiving messages (both internal and to and from remote computers), in a way that is transparent to
the agent developer. It maintains a private queue of incoming ACL messages
per agent. Jade also allows the MAS developer to monitor the exchange of messages
using the *sniffer* built-in agent.

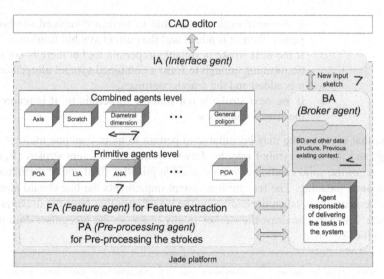

Fig. 7. The agent sketch system architecture

5   Reasoning about This Paradigm

The proposed paradigm consists of two main parts: the recognition of simple strokes, and the recognition of (multi-stroke) symbols.

To recognise simple strokes, several Primitive Agents compete to each other. Every agent collects cues related to its own stroke type and scores the probability of the stroke to belong to such a type. The phoneme reported to be the most probable is selected.

At this stage, the main difference from traditional recognition techniques as fuzzy logic, neural networks or other statistical learning is that specific algorithms are implemented to detect different phonemes. Traditional techniques are more rigid as they use the same classification method to recognise different symbols, that is, the same neural network or the same discriminant analysis is executed to classify different things. Whereas with this paradigm the classification method may be customized for each type of symbol to recognise, what can foresee a higher success ratio.

In the recognition of (multi-stroke) symbols, the main difference from traditional techniques is that in order to recognise a symbol, a specific code is executed, since there is one combined agent for each symbol. The agents are autonomous and adaptive too, and therefore capable of changing their decision depending on the context.

Another benefit of programming with agents is the flexibility of the system. We have tested that we are able to add new symbols to our catalog, just adding the corresponding Combined Agent for its recognition. Adding a new symbol to the recognition system (as for example a ♀ symbol), is as easy as adding its corresponding Combined Agent to the upper level, as it may use the implemented alphabet to classify the symbol.

If new strokes (phonemes) are also required, they are just added to the alphabet implementing its corresponding Primitive Agent. Moreover, in a traditional environment, the recognition process is carried out sequentially, but in here we have determined the specific tasks of the recognition process and have assigned them to different agents that can act independently, avoiding the sequential process and arranging the recognition process to execute in a more logical way. Hence, this method is not dependent on the number of strokes neither on the sketching sequence order.

6 Conclusions

Our current goal is defining a scalable solution aimed at interpreting hand-drawn symbols used by product designers during the creative stages of design.

At this end, we have described a hierarchical multi-agent architecture organised in three levels: the Basic Agents, the Primitive Agents and the Combined Agents. An experimental version was successfully implemented to demonstrate the validity of the architecture. Besides, the goal of flexibility was successfully checked too.

The next step will be benchmarking the effectiveness of this strategy compared to other current successful approaches. Finally, although the informal tests seem quite promising, the scalability of the approach should be properly checked.

Acknowledgments. The Spanish Ministry of Science and Education and the FEDER Funds, through the CUESKETCH project (Ref. DPI2007-66755-C02-01) partially supported this work.

References

1. Contero, M., Naya, F., Jorge, J., Conesa, J.: CIGRO: A minimal instruction set calligraphic interface for sketch-based modeling. In: Kumar, V., Gavrilova, M.L., Tan, C.J.K., L'Ecuyer, P. (eds.) ICCSA 2003. LNCS, vol. 2669, pp. 549–558. Springer, Heidelberg (2003)
2. Kan, C., Srinath, M.D.: Invariant character recognition with Zerkine and orthogonal Fourier-Mellin moments. Pattern Recognition 35, 143–154 (2002)
3. Blasco, J., Aleixos, N., Moltó, E.: Machine vision system for automatic quality grading of fruit. Biosystems Engineering 85(4), 415–423 (2003)
4. Gonzalez, R.C., Woods, R.E.: Digital Image Processing, 2nd edn. Prentice Hall, Upper-Saddle River (2002)
5. Mokhtarian, F., Abbasi, S.: Robust automatic selection of optimal views in multi-view free-form object recognition. Pattern Recognition 38, 1021–1031 (2005)
6. Chen, G.Y., Bui, T.D., Krzyzak, A.: Rotation invariant pattern recognition using ridgelets, wavelet cycle-spinning and Fourier features. Pattern Recognition 38, 2314–2322 (2005)
7. Hu, M.: Visual pattern recognition by moment invariants. IRE Trans. Inf. Theor. IT-8, 179–187 (1962)
8. Rubine, D.H.: Specifying Gestures by Example. Computer Graphics. In: Proceedings of the SIGGRAPH 1991, vol. 25(4), pp. 329–337 (1991)
9. Ajay, A., Vo, V., Kimura, T.D.: Recognising Multistroke Shapes: An Experimental Evaluation. In: Proceedings of the ACM (UIST 1993), Atlanta, Georgia, pp. 121–128 (1993)
10. Gross, M.D.: Recognising and Interpreting Diagrams in Design. In: Proceedings of ACM (AVI 1994), Bari, Italy, pp. 88–94 (1994)
11. Zang, D., Lu, G.: A Comparative Study of Fourier Descriptors for Shape Representation and Retrieval. In: The 5th Asian Conference on Computer Vision, ACCV 2002, Melbourne, Australia, pp. 1–6 (2002)
12. Harding, P.R.G., Ellis, T.J.: Recognising Hand Gesture Using Fourier Descriptors. In: Proceedings of the 17th International Conference on Pattern Recognition, vol. 3, pp. 286–289 (2004)
13. Zion, B., Shklyar, A., Karplus, I.: Sorting fish by computer vision. Comput. Electron. Agric. 23, 175–187 (1999)

14. Fonseca, M.J., Jorge, J.: Using Fuzzy Logic to Recognise Geometric Shapes Interactively. In: Proceedings of 9th IEEE Conference on Fuzzy Systems, vol. 1, pp. 291–296 (2000)
15. Xiangyu, J., Wenyin, L., Jianyong, S., Sun, Z.: On-Line Graphics Recognition. In: Pacific Conference on Computer Graphics and Applications, pp. 256–264 (2002)
16. Zhengxing, S., Liu, W., Binbin, P., Bin, Z., Jianyong, S.: User Adaptation for Online Sketchy Shape Recognition. In: Lladós, J., Kwon, Y.-B. (eds.) GREC 2003. LNCS, vol. 3088, pp. 305–316. Springer, Heidelberg (2004)
17. Park, C.H., Park, H.: Fingerprint classification using fast Fourier transform and non-linear discriminant analysis. Pattern Recognition 38, 495–503 (2005)
18. Juchmes, R., Leclercq, P., Azar, S.: A freehand-sketch environment for architectural design supported by a multi-agent system. Computers & Graphics 29(6), 905–915 (2005)
19. Achten, H.H., Jessurun, A.J.: An agent framework for recognition of graphic units in drawings. In: Proceedings of 20th International Conference on Education and Research in Computer Aided Architectural Design in Europe (eCAADe 2002), Warsaw, pp. 246–253 (2002)
20. Mackenzie, G., Alechina, N.: Classifying sketches of animals using an agent-based system. In: Petkov, N., Westenberg, M.A. (eds.) CAIP 2003. LNCS, vol. 2756, pp. 521–529. Springer, Heidelberg (2003)
21. Azar, S., Couvreury, L., Delfosse, V., Jaspartz, B., Boulanger, C.: An agent-based multimodal interface for sketch interpretation. In: Proceedings of International Workshop on Multimedia Signal Processing (MMSP 2006), British Columbia, Canada (2006)
22. Casella, G., Deufemia, V., Mascardi, V., Costagliola, G., Martelli, M.: An agent-based framework for sketched symbol interpretation. Journal of Visual Languages and Computing 19, 225–257 (2008)
23. Ku, D.C., Qin, S.F., Wright, D.K.: Interpretation of Overtracing Freehand Sketching for Geometric Shapes. In: WSCG 2006. Full papers proceedings, vol. G41, pp. 263–270 (2006)
24. Pu, J., Gur, D.: Automated Freehand Sketch Segmentation Using Radial Basis Functions. Computer-Aided Design (2009), doi:10.1016/j.cad.2009.05.005
25. Company, P., Varley, P.A.C., Piquer, A., Vergara, M., Sanchez-Rubio, J.: Benchmarks for Computer-based Segmentation of Sketches. In: Pre-Proceedings of the Eighth IAPR International Workshop on Graphics Recognition GREC 2009, pp. 103–114 (2009)
26. Sezgin, T.M., Davis, R.: Sketch recognition in interspersed drawings using time-based graphical models. Computers and Graphics 32(5), 500–510 (2008)
27. Zhu, S.C., Mumford, D.: A stochastic grammar of images. Foundations and Trends in Computer Graphics and Vision 2(4), 259–362 (2006)
28. Lin, L., Wu, T., Porway, J., Xu, Z.: A stochastic graph grammar for compositional object representation and recognition. Pattern Recognition 42(7), 1297–1307 (2009)
29. Ku, D.C., Qin, S.F., Wright, D.K.: Interpretation of Overtracing Freehand Sketching for Geometric Shapes. In: WSCG 2006. Full papers proceedings, vol. G41, pp. 263–270 (2006)
30. Tao, Y., Morrow, C.T., Heinemann, P.H., Sommer, H.J.: Fourier-Based Separation Technique for Shape Grading of Potatoes Using Machine Vision. Transactions of the ASAE 38(3), 949–957
31. Wojnar, L., Kurzydłowski, K.J.: Practical Guide to Image Analysis. ASM International, pp. 157–160 (2000) ISBN 0-87170-688-1
32. Yu, B.: Recognition of freehand sketches using mean shift. In: Proc. of IUI 2003, pp. 204–210 (2003)
33. Bellifemine, F., Poggi, A., Rimassa, G.: Developing multi-agent systems with JADE. In: Castelfranchi, C., Lespérance, Y. (eds.) ATAL 2000. LNCS (LNAI), vol. 1986, pp. 89–103. Springer, Heidelberg (2001)
34. FIPA ORG, FIPA ACL Message Structure Specification, Document no. SC00061G (2002)

Solving Integer Programming Problems by Using Artificial Bee Colony Algorithm

Bahriye Akay and Dervis Karaboga

Erciyes University, The Dept. of Computer Engineering,
38039, Melikgazi, Kayseri, Turkiye
bahriye@erciyes.edu.tr, karaboga@erciyes.edu.tr

Abstract. This paper presents a study that applies the Artificial Bee Colony algorithm to integer programming problems and compares its performance with those of Particle Swarm Optimization algorithm variants and Branch and Bound technique presented to the literature. In order to cope with integer programming problems, in neighbour solution production unit, solutions are truncated to the nearest integer values. The experimental results show that Artificial Bee Colony algorithm can handle integer programming problems efficiently and Artificial Bee Colony algorithm can be considered to be very robust by the statistics calculated such as mean, median, standard deviation.

1 Introduction

Integer programming is a form of linear programming in which the variables are required to have integer values. Capital budgeting, portfolio analysis, network and VLSI circuit design as well, as automated production systems are some applications in which integer programming problems are met [1, 2]. An integer programming problem in which all variables are integer is called a pure integer programming problem as stated in (1).

$$min \ f(x), \ x \in \mathcal{S} \subseteq Z^n \qquad (1)$$

where Z is the set of integers, and $\mathcal{S}$ is the feasible region. If some variables are integer and some are not then the problem is a mixed integer programming problem. The case where the integer variables are 0 or 1 is called pure (mixed) 0-1 programming problems or pure (mixed) binary integer programming problems. The integer programming techniques can be classified into three broad categories: approximate, exact, and heuristic [3]. In approximate techniques, variables to be optimized take on non-integer values and they are rounded to integers. Exact techniques such as Branch and Bound, dynamic programming divide the feasible region into smaller sub-regions or problem into sub-problems. However, they have high computational cost since they explore a search tree containing hundreds or more nodes on large-scale real-life problems [3, 4]. Branch and bound algorithms are also subjected to linear constraints with an objective function that need not be linear [3]. Main drawback of the Branch and Bound technique is that it

R. Serra and R. Cucchiara (Eds.): AI*IA 2009, LNAI 5883, pp. 355–364, 2009.

requires partitioning the feasible space accurately. Heuristic techniques give near optimal solutions in a reasonable time.

Heuristic optimization methods can also handle with integer programming problems. Most of them can cope with integer variables efficiently by truncating or rounding the real valued solutions. Particle Swarm Optimization (PSO) algorithm, introduced by Eberhart and Kennedy in 1995 [5], simulating the social behaviour of fish schools and bird flocks; and Artificial Bee Colony (ABC) algorithm, introduced by Karaboga [6], simulating the foraging behaviour of honey bee swarm are both heuristic optimization methods that can handle with integer programming problems. In this paper, results of the ABC algorithm was compared to those of PSO variants, Quantum-Behaved PSO (QPSO); and Branch and Bound tecniques on integer problems.

Rest of the paper is organized as follows: In the second section, Artificial Bee Colony Algorithm, and in the third section, other algorithms regarded in this study which are Particle Swarm Optimization, Quantum-behaved PSO, Branch and Bound technique, are summarized. In Section 4, experimental results are presented and finally it is concluded.

2   Artificial Bee Colony Algorithm

The minimal model of forage selection that leads to the emergence of collective intelligence of honey bee swarms consists of three essential components: food sources, employed foragers, and unemployed foragers, and defines two leading modes of the behaviour: recruitment to a nectar source and abandonment of a source [7]. The quality of a food source depends on many factors such as its proximity to the nest, richness or concentration of energy, and the ease of extracting this energy [7]. Employed foragers are associated with a particular food source which they are currently exploiting or are employed at. They carry information about this particular source, its distance and direction from the nest, and the profitability of the source and share this information with a certain probability by dancing on the dance area. Unemployed foragers look for a food source to exploit. There are two types of unemployed foragers: scouts searching the environment surrounding the nest for new food sources and onlookers waiting in the nest and finding a food source through the information shared by employed foragers.

Artificial Bee Colony algorithm is a recently proposed optimization algorithm that simulates the foraging behaviour of a bee colony [6]. The main steps of the algorithm are as below:

1: Initialize the population of solutions $x_i, i = 1 \ldots SN$
2: Evaluate the population
3: cycle=1
4: **repeat**
5:    **{Employed Bees' Phase}**

6:   Produce new solutions v_i for the employed bees by using (2) and evaluate them

$$v_{ij} = \begin{cases} x_{ij} + \phi_{ij}(x_{ij} - x_{kj}) \, , \, if \;\; R_j < MR \\ x_{ij} \qquad\qquad\qquad , \, \text{otherwise} \end{cases} \tag{2}$$

where $k \in \{1, 2,..., SN\}$ and $j \in \{1, 2,..., D\}$ are randomly chosen indexes. Although k is determined randomly, it has to be different from i. $\phi_{i,j}$ is a random number between [-1, 1]. It controls the production of neighbour food sources around $x_{i,j}$ and represents the comparison of two food positions visually by a bee. MR is a control parameter of ABC algorithm in the range of [0,1] which controls the number of parameters to be modified.

7:   Apply the greedy selection process between v_i and x_i

8:   **{Onlooker Bees' Phase}**

9:   Calculate the probability values p_i for the solutions x_i by (3)

$$p_i = \frac{fit_i}{\sum\limits_{n=1}^{SN} fit_n} \tag{3}$$

where fit_i is the fitness value of the solution i which is proportional to the nectar amount of the food source in the position i and SN is the number of food sources which is equal to the number of employed bees or onlooker bees.

10:   Produce the new solutions v_i for the onlookers from the solutions x_i selected depending on p_i and evaluate them

11:   Apply the greedy selection process between v_i and x_i

12:   **{Scout Bees' Phase}**

13:   Determine the abandoned solution for the scout, if exists, and replace it with a new randomly produced solution x_i by (4)

$$x_i^j = x_{\min}^j + \text{rand}[0, 1](x_{\max}^j - x_{\min}^j) \tag{4}$$

14:   Memorize the best solution achieved so far

15:   cycle=cycle+1

16:   **until** cycle=MCN

In ABC algorithm employed for integer programming problems, the position of a food source represents a possible solution consisting of integer numbers to the optimization problem and the nectar amount of a food source corresponds to the quality (fitness) of the associated solution. The number of the employed bees or the onlooker bees is equal to the number of solutions in the population. At the first step, the ABC generates a randomly distributed initial population $P(Cycle = 0)$ of SN solutions (food source positions), where SN denotes the size of employed bees or onlooker bees. Each solution x_i ($i = 1, 2, ..., SN$) is a D-dimensional vector. Here, D is the number of optimization parameters. After initialization, the population of the positions (solutions) is subject to repeated cycles, $Cycle = 1, 2, ..., MCN$, of the search processes of the employed bees, the onlooker bees and scout bees. An employed bee produces a modification on the position (solution) in her memory depending on the local information (visual information) and tests the nectar amount (fitness value) of the new source (new solution). Provided that the nectar amount of the new one is higher than that of the previous one, the bee memorizes the new position and forgets the old one. Otherwise she keeps the position of the previous one in her memory. After all employed bees complete the search process, they share the nectar information of the food sources and their position information with the onlooker bees. An onlooker bee evaluates the nectar information taken from all employed bees and chooses a food source with a probability related to its nectar amount. As in the case of the employed bee, she produces a modification on the position in her memory and checks the nectar amount of the candidate source. Providing that its nectar is higher than that of the previous one, the bee memorizes the new position and forgets the old one. These steps are repeated through a predetermined number of cycles called Maximum Number of Cycle (MCN) or until a termination criteria is satisfied. It is clear from the explanation that there are four control parameters in the ABC: The number of food sources which is equal to the number of employed or onlooker bees (SN), the modification rate (MR), the value of *limit* and the maximum cycle number (MCN).

3   Other Algorithms Considered in the Experiments

3.1   Particle Swarm Optimization Algorithm

In Particle Swarm Optimization algorithm [5], a population of particles starts to move in search space by following the current optimum particles and changes the positions in order to find out the optima. Main steps of the procedure are:

1: Initialize Population
2: **repeat**
3:     Calculate fitness values of particles
4:     Choose the best particle
5:     Calculate the velocities of particles by (5)

$$v(t + 1) = \begin{matrix} \chi(\omega v(t)+ \\ \phi_1 \text{rand}(0, 1)(p(t) - x(t))+ \\ \phi_2 \text{rand}(0, 1)(g(t) - x(t))) \end{matrix} \quad (5)$$

6:     Update the particle positions by (6)

$$x(t + 1) = x(t) + v(t + 1) \quad (6)$$

The velocity update is performed as indicated in Equation 6:
7: **until** requirements are met

The parameter ω is called the inertia weight and controls the magnitude of the old velocity $v(t)$ in the calculation of the new velocity, whereas ϕ_1 and ϕ_2 determine the significance of $p(t)$ and $g(t)$, respectively. χ parameter is constriction factor that insures the convergence of particle swarm optimization algorithm. Furthermore, v_i at any time step of the algorithm is constrained by the parameter v_{max}. The swarm in PSO is initialized by assigning each particle to a uniformly and randomly chosen position in the search space. Velocities are initialized randomly in the range $[v_{min}, v_{max}]$.

3.2   Quantum-Behaved Particle Swarm Optimization (QPSO)

In Quantum-Behaved Particle Swarm Optimization (QPSO), proposed by Sun et al. [8], quantum theory, following the principle of state superposition and uncertainty, was integrated with PSO. In QPSO, next position is calculated as:

$$x(t+1) = p \pm \beta * |mbest - x(t)| * \ln(1/u) \tag{7}$$

$mbest$, center-of-gravity position of all the particles, is defined by (8):

$$mbest = \sum_{i=1}^{M} p_i/M$$
$$mbest = \left(\sum_{i=1}^{M} p_{i1}/M, \sum_{i=1}^{M} p_{i2}/M, \ldots, \sum_{i=1}^{M} p_{id}/M \right) \tag{8}$$

where M is the population size and p_i is the pbest of particle i.

Parameter β is called Creativity Coefficient that affects the individual particles convergence speed and performance of the algorithm.

3.3   Branch and Bound Technique

The branch and bound method is based on tree structures of integer solutions. Keeping the tree as much as possible is the main purpose of the technique. Most promising nodes are mostly allowed to grow and these nodes are branched. When sub-nodes are determined to be on lower and upper bounds, these nodes are bounded.

4   Experimental Results and Discussion

In the experiments, nine problems widely used in the literature are employed in order to investigate the performance of the ABC algorithm on integer programming problems. These problems are listed below:

Test Problem 1 [9]:
$F_1(x) = \|x\|_1 = |x_1| + \ldots + |x_D|$.
with $x = (x_1, \ldots, x_D) \in [-100, 100]^D$ where D is the dimension. The solution is $x_i^* = 0$, $i = 1, \ldots, D$ and $F_1(x^*) = 0$.

Test Problem 2 [9]:

$$F_2(x) = x^\top x = (x_1 \quad \ldots \quad x_D) \begin{pmatrix} x_1 \\ \vdots \\ x_D \end{pmatrix}.$$

with $x = (x_1, \ldots, x_D)^\top \in [-100, 100]^D$ where D is the dimension. The solution is $x_i^* = 0$, $i = 1, \ldots, D$ and $F_2(x^*) = 0$.

Test Problem 3 [10]:
$$F_3(x) = -(\begin{matrix} 15 & 27 & 36 & 18 & 12 \end{matrix})x +$$
$$x^\top \begin{pmatrix} 35 & -20 & -10 & 32 & -10 \\ -20 & 40 & -6 & -31 & 32 \\ -10 & -6 & 11 & -6 & -10 \\ 32 & -31 & -6 & 38 & -10 \\ -10 & 32 & -10 & -20 & 31 \end{pmatrix} x.$$
The best known solutions are $x^* = (0, 11, 22, 16, 6)^\top$ and $x^* = (0, 12, 23, 17, 6)^\top$ and $F_3(x) = -737$.

Test Problem 4 [10]:
$F_4(x) = (9x_1^2 + 2x_2^2 - 11)^2 + (3x_1 + 4x_2^2 - 7)^2$. The solution is $x^* = (1, 1)^\top$ and $F_4(x^*) = 0$.

Test Problem 5 [10]:
$F_5(x) = (x_1 + 10x_2)^2 + 5(x_3 - x_4)^2 + (x_2 - 2x_3)^4 + 10(x_1 - x_4)^4$. The solution is $x^* = (0, 0, 0, 0)^\top$ and $F_5(x^*) = 0$.

Test Problem 6 [11]:
$F_6(x) = 2x_1^2 + 3x_2^2 + 4x_1x_2 - 6x_1 - 3x_2$. The solution is $x^* = (2, -1)^\top$ and $F_6(x^*) = -6$.

Test Problem 7 [10]:
$F_7(x) = -3803.84 - 138.08x_1 - 232.92x_2 + 123.08x_1^2 + 203.64x_2^2 + 182.25x_1x_2$. The solution is $x^* = (0, 1)^\top$ and $F_7(x^*) = -3833.12$.

Test Problem 8 [10]:
$F_8(x) = (x_1^2 + x_2 - 11)^2 + (x_1 + x_2^2 - 7)^2$. The solution is $x^* = (3, 2)^\top$ and $F_8(x^*) = 0$.

Test Problem 9 [10]:
$F_9(x) = 100(x_2 - x_1^2)^2 + (1 - x_1)^2$. The solution is $x^* = (1, 1)^\top$ and $F_9(x^*) = 0$.

In order to cope with integer programming problems, Particle Swarm Optimization and Artificial Bee Colony algorithms truncate the parameter values to the closest integer after producing new solutions.

In the experiments, the results of the ABC algorithm were compared to those of Quantum-Behaved Particle Swarm Optimization (QPSO) algorithm taken from [12] and those of Particle Swarm Optimization variants and Branch and Bound algorithm taken from [1]. In Table 1, the comparison of ABC and QPSO is presented. Results in Table 1 are success rates of 50 independent runs with different random seeds and mean number of iterations required for obtaining desired accuracy. In [12], desired accuracy is not given, we assumed that to be 10^{-6} which is precise enough. PSO-In indicates a variant of PSO algorithm with inertia weight and without constriction factor. Inertia weight, ω, is gradually decreased from 1.0 to 0.4. β parameter of QPSO algorithm is changed from 1.2 to

Table 1. Success Rates, Mean Iteration numbers, Swarm Size and Maximum number of Iterations of PSO-In, QPSO [12] and ABC algorithms for of test problems $F_1 - F_5$, F_8, F_9. SS:Swarm Size, MNI:Maximum Number of Iterations, D:Dimension of the problem, SR:Success Rate, MI: Mean of iteration numbers, Bold face indicates the best result.

				PSO-In		QPSO		ABC	
SS	MNI	F	D	SR	MI	SR	MI	SR	MI
20	1000	F_1	5	100	72.8	100	27.2	100	**20.03**
50	1000	F_1	10	100	90.26	100	48.26	100	**34.63**
100	1000	F_1	15	100	94.36	100	64.46	100	**48.93**
200	1000	F_1	20	100	96.3	100	72.24	100	**64.43**
250	1500	F_1	25	100	99.44	100	83.86	100	**81.33**
300	2000	F_1	30	100	103.14	100	**92.56**	100	99.76
20	1000	F_2	5	100	77.82	100	**21.2**	100	21.93
150	1000	F_3	5	100	**125.03**	100	166.9	100	165.43
20	1000	F_4	2	100	32.9	100	**8.95**	100	11.5
40	1000	F_5	4	100	79.6	100	65.7	100	**60.1**
20	1000	F_8	2	100	81.24	100	**19.35**	100	19.4
50	1000	F_9	2	100	41.4	100	14.9	100	**11.93**

Table 2. Dimension of the functions and Swarm Size used in the experiments for each function. D: Dimension of the problem, SS: Swarm Size.

F	F1	F1	F1	F1	F1	F1	F2	F3	F4	F5	F6	F7
D	5	10	15	20	25	30	5	5	2	4	2	2
SS	20	20	50	50	100	100	10	70	20	20	10	20

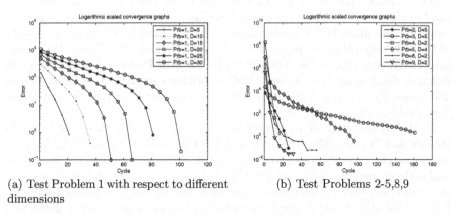

(a) Test Problem 1 with respect to different dimensions

(b) Test Problems 2-5,8,9

Fig. 1. Logarithmic scaled convergence graphs of the ABC algorithm

0.4 [12]. While obtaining the results in Table 1, the values of control parameters of ABC algorithm were $SN * D * 5$ and 0.8 for *limit* and MR, respectively. Maximum number of iterations and swarm sizes for each problem are also reported on Table 1. From the results in Table 1, for all functions, PSO, QPSO and ABC algorithms were able to find the solutions with desired accuracy in allowed maximum number of function evaluations. In terms of mean number of iterations to reach the desired accuracy, algorithms produce similar results for lower dimensional functions ($F_2, F_3, F_4, F_5, F_8, F_9$), while ABC algorithm is much better than

Table 3. Success Rates, Mean, Standard Deviation and Medians of evaluation numbers of PSO variants, BB Technique [1] and ABC Algorithm for Test Problems F_1-F_7, SR:Success Rate, StD: Standard Deviation, Bold face indicates the best result

Function	Method	SR	Mean	StD	Median	Function	Method	SR	Mean	StD	Median
	PSO-In	30/30	1646.0	661.5	1420		PSO-In	30/30	1655.6	618.4	1650
	PSO-Co	30/30	744.0	89.8	730		PSO-Co	30/30	428.0	57.9	430
F_1, $D=5$	PSO-Bo	30/30	692.6	97.2	680	F_2 $D=5$	PSO-Bo	30/30	418.3	83.9	395
	BB	30/30	1167.83	659.8	1166		BB	30/30	**139.7**	102.6	93
	ABC	30/30	**376**	64.6	380		ABC	30/30	449.3	56.7	440
	PSO-In	30/30	4652.0	483.2	4610		PSO-In	30/30	4111.3	1186.7	3850
	PSO-Co	30/30	1362.6	254.7	1360		PSO-Co	30/30	**2072.6**	536.4	2940
F_1, $D=10$	PSO-Bo	30/30	1208.6	162.7	1230	F_3 $D=5$	PSO-Bo	30/30	3171.0	493.6	3080
	BB	30/30	5495.8	1676.3	5154		BB	30/30	4185.5	32.8	4191
	ABC	30/30	**727.3**	64.4	740		ABC	24/30	13850	6711.3	11550
	PSO-In	30/30	7916.6	624.1	7950		PSO-In	30/30	304.0	101.6	320
	PSO-Co	30/30	3538.3	526.6	3500		PSO-Co	30/30	297.3	50.8	290
F_1, $D=15$	PSO-Bo	30/30	2860.0	220.2	2850	F_4 $D=2$	PSO-Bo	30/30	302.0	80.5	320
	BB	30/30	10177.1	2393.4	10011		BB	30/30	316.9	125.4	386
	ABC	30/30	**974**	60.5	960		ABC	30/30	**240.7**	79.4	240
	PSO-In	30/30	8991.6	673.3	9050		PSO-In	30/30	1728.6	518.9	1760
	PSO-Co	30/30	4871.6	743.3	4700		PSO-Co	30/30	1100.6	229.2	1090
F_1, $D=20$	PSO-Bo	29/30	4408.3	3919.4	3650	F_5 $D=4$	PSO-Bo	30/30	1082.0	295.6	1090
	BB	30/30	16291.3	3797.9	14550		BB	30/30	2754.0	1030.1	2714
	ABC	30/30	**1275.3**	97.7	1260		ABC	30/30	**193.3**	53.5	180
	PSO-In	30/30	11886.6	543.7	11900		PSO-In	30/30	**178.0**	41.9	180
	PSO-Co	30/30	9686.6	960.1	9450		PSO-Co	30/30	198.6	59.2	195
F_1, $D=25$	PSO-Bo	25/30	9553.3	7098.6	6500	F_6 $D=2$	PSO-Bo	30/30	191.0	65.9	190
	BB	20/30	23689.7	2574.2	25043		BB	30/30	211.1	15.0	209
	ABC	30/30	**1554.7**	108.6	1540		ABC	30/30	258.7	113.6	240
	PSO-In	30/30	13186.6	667.8	13050		PSO-In	30/30	334.6	95.5	340
	PSO-Co	30/30	12586.6	1734.9	12500		PSO-Co	30/30	324.0	78.5	320
F_1, $D=30$	PSO-Bo	19/30	13660.0	8863.9	7500	F_7 $D=2$	PSO-Bo	30/30	306.6	96.7	300
	BB	14/30	25908.6	755.5	26078		BB	30/30	358.6	14.7	355
	ABC	30/30	**1906**	129.9	1900		ABC	30/30	**106.7**	44.8	100

other two algorithms for higher dimensions of F1 function. From these results, it can be said that ABC algorithm shows a similar or better performance compared to QPSO and much better performance respect to PSO-In. Convergence graphs of the ABC algorithm are presented on Figure 1(a) and 1(b). Figure 1(a) shows the convergence rate of the ABC algorithm for Test Problem 1 with respect to different dimensions. For all cases, ABC algorithm converges the optima around 100 cycles. Figure 1(b) demonstrates the convergence of the ABC algorithm for Test Problems 2, 3, 4, 5, 8, 9.

In the second part of the experiments, ABC algorithm was compared to Branch and Bound (BB) technique and PSO variants which are PSO-In, with inertia weight and without constriction factor, PSO-Co, with constriction factor and without inertia weight, and PSO-Bo, with both constriction factor and inertia weight. For these comparisons, dimension and swarm sizes for all test problems are listed on Table 2, maximum number of evaluations are set to 25000 and algorithms are terminated when the accuracy is reached 10^{-6}. In these comparisons, algorithms were repeated through 30 independent runs with the control parameter values as follows: the constriction factor χ was set equal to 0.729; the inertia weight ω was gradually decreased from 1 towards 0.1; $\phi_1 = \phi_2 = 2$; and $v_{max} = 4$ [1]. Parameters of the ABC algorithm were set to the values given in previous experiments. Success Rates, Mean, Standard Deviation and Medians of evaluation numbers of PSO variants, BB Technique and ABC Algorithm for Test Problems F_1-F_7 are given in Table 3. Results in Table 3 indicate that ABC algorithm shows superior performance respect to other algorithms on all dimensions of the function F_1 (on all of six cases) in terms of mean, standard deviation and median values. It is also clear that BB technique produces the best result on one function, F_2. PSO-Co is the winner on one function, F_3, while ABC algorithm produces the worst result in terms of success rate in allowed maximum number of function evaluations on this function. On three functions, F_4, F_5 and F_7, ABC has the best performance. For F_6 function, PSO-In shows superior performance.

In overall evaluation, ABC algorithm can produce comparable or better performance respect to PSO variants and BB technique for integer programming problems considered in this study. Exact techniques such as BB produce solutions in computationally expensive period. ABC algorithm produces high quality solutions more quickly and is not subjected to linear constraints on the objective function that need not be linear. There is no limitation on the applicability of ABC algorithm. Control parameters certainly have effect on the performance of the algorithms. Lower values of MR has negative effect on the convergence rate of the algorithm. From the experiences gained from the previous studies, 0.8 is an appropriate value for MR. Lower values of $limit$ increases the exploration capability while decreasing the exploitation capability. Value of the control parameter limit should balance the exploration and exploitation processes.

5   Conclusion

Since most of combinatory problems can be converted to Integer Programming problems, integer programming problems are important in the subject of optimization field. In this paper, performance of the Artificial Bee Colony Algorithm was investigated on integer problems and compared to the results of Particle Swarm Optimization algorithm variants, and Branch and Bound method. From the results, it can be concluded that Artificial Bee Colony algorithm can cope with integer problems efficiently in a robust manner. This work is supported by Erciyes University, the Department of Research Projects under contract FBA–06–22.

References

[1] Laskari, E.C., Parsopoulos, K.E., Vrahatis, M.N.: Particle swarm optimization for integer programming. In: CEC 2002: Proceedings of the Evolutionary Computation on 2002. CEC 2002. Proceedings of the 2002 Congress, pp. 1582–1587. IEEE Computer Society, Los Alamitos (2002)
[2] Nemhauser, G.L., Wolsey, L.A.: Integer Progamming. In: Handbooks in Operations Research and Management Science, vol. 1. Elsevier Science and Technology, Amsterdam (1989)
[3] Misra, K.B., Sharma, U.: An efficient algorithm to solve integer-programming problems arising in system-reliability design. IEEE Transactions On Reliability 40(1), 81–91 (1991)
[4] Rouillon, S., Desaulniers, G., Soumis, F.: An extended branch-and-bound method for locomotive assignment. Transportation Research Part B: Methodological 40(5), 404–423 (2006)

[5] Kennedy, J., Eberhart, R.C.: Particle swarm optimization. In: IEEE International Conference on Neural Networks 1995, vol. 4, pp. 1942–1948 (1995)

[6] Karaboga, D.: An idea based on honey bee swarm for numerical optimization. Technical Report TR06, Erciyes University, Engineering Faculty, Computer Engineering Department (2005)

[7] Tereshko, V.: Reaction–diffusion model of a honeybee colony's foraging behaviour. In: Deb, K., Rudolph, G., Lutton, E., Merelo, J.J., Schoenauer, M., Schwefel, H.-P., Yao, X. (eds.) PPSN 2000. LNCS, vol. 1917. Springer, Heidelberg (2000)

[8] Sun, J., Feng, B., Xu, W.: Particle swarm optimization with particles having quantum behavior. In: Evolutionary Computation, CEC 2004, vol. 1, June 2004, pp. 325–331 (2004)

[9] Rudolph, G.: An evolutionary algorithm for integer programming. In: PPSN III: Proceedings of the International Conference on Evolutionary Computation. The Third Conference on Parallel Problem Solving from Nature, pp. 139–148. Springer, Heidelberg (1994)

[10] Glankwahmdee, A., Liebman, J.S., Hogg, G.L.: Unconstrained discrete nonlinear programming. Engineering Optimization 4, 95–107 (1979)

[11] Rao, S.S.: Engineering Optimization- Theory and Practice. Wiley Eastern, New Delhi (1996)

[12] Liu, J., Sun, J., Xu, W.: Quantum-Behaved Particle Swarm Optimization for Integer Programming. In: King, I., Wang, J., Chan, L.-W., Wang, D. (eds.) ICONIP 2006. LNCS, vol. 4233, pp. 1042–1050. Springer, Heidelberg (2006)

Representing Excuses in Social Dependence Networks

Guido Boella[1], Jan Broersen[2], Leendert van der Torre[3], and Serena Villata[1]

[1] Dipartimento di Informatica, University of Turin, Italy
{guido,villata}@di.unito.it
[2] Faculty of science, Universiteit Utrecht, The Netherlands
broersen@cs.uu.nl
[3] CSC, University of Luxembourg, Luxembourg
leendert@vandertorre.com

Abstract. In this paper, we propose a representation of excuses in the context of multiagent systems. We distinguish five classes of excuses, taking as starting point both jurisprudential and philosophical studies about this topic, and we discuss their acceptance criteria. We highlight the following classes of excuses: epistemic excuses, power-based excuses, norm-based excuses, counts as-based excuses and social-based excuses and we represent them using social dependence networks. The acceptance criteria individuate those excuses which success in maintaining the trust of the other agents, e.g. in the context of social networks, excuses based on norms seem better than counts as-based ones in achieving this aim.

1 Introduction

One of the aims of multiagent systems is the representation of agents' social behaviour using notions as, for instance, trust, social laws, cooperation and so on. A notion which seems to have received less attention in this context is the concept of excuse. In jurisprudence, an excuse means to grant or obtain an exemption for a group of persons sharing a common characteristic from a potential liability. This means that the shown behavior cannot be approved but some excuse may be found, e.g. that the accused was a serving police officer or suffering from a mental illness. The notion of excuses is also treated in the context of philosophy, for example in Austin's "A Plea for Excuses" [1]. *The New York Times* of March 6, 1984 presents a new social theory about excuses. Particularly, they reported *The garden variety - white lies that prevent hurt feelings ("Sorry, I can't make the party, I'm all tied up") - are so common, the research team says, because they are a social lubricant vital to the smooth operation of daily life*. The presence of excuses in real life is everywhere thus the use of excuses as social lubricant should be extended from humans to intelligent agents of multiagent systems, especially where they are governed by social laws.

In this paper, we are interested in the distinction of the possible classes of excuses, representing them using social dependence networks in different ways. Moreover, using social dependence networks, we discuss what kinds of excuses seem more useful in convincing the other agents to trust you again. We address the following research questions: How to distinguish classes of excuses?, What are their acceptance criteria? and How to represent excuses in multiagent systems?.

R. Serra and R. Cucchiara (Eds.): AI*IA 2009, LNAI 5883, pp. 365–374, 2009.

Norms and social laws can be represented as soft constraints, allowing for violations to be detected, instead of hard constraints that cannot be violated. Normally an agent fulfills its obligations, because otherwise its behavior counts as a violation that is sanctioned, and the agent dislikes sanctions. There are five categories of exceptions to this normal behavior which we call excuses. We distinguish these five classes of excuses considering the common human behaviours and the sources of the excuses. This classification seems the simplest one allowing anyway a certain variety of representation. The discussion of the acceptance criteria of excuses is strictly related to the way of representing them, by means of social dependence networks. First, epistemic excuses such as *I did not know, I was not aware, I forgot.* Second, excuses based on power such as *I did not have the ability, opportunity or power to do it, others prevented me from doing it.* Third, excuses based on normative conflicts such as *there was another more important norm.* Fourth, excuses based on difference of opinion on counts-as norms, such as *I did comply, my act did not count as a violation.* Finally, the common excuse *Everybody does it.*

We propose to use the methodology of dependence networks to represent these classes of excuses in multiagent systems' theory. The dependency modeling, used in software design by the TROPOS methodology [4], represents the multiagent system by means of goal-based dependencies and norm-based dependencies [10]. The advantage in using this methodology consists particularly in the representation of excuses like *Everybody does it* which are really difficult to be represented, for example in deontic logic. As in TROPOS, each agent has a dependence network representing its world knowledge and the network explicitly represents agents' excuses.

The reminder of the paper is as follows. In Section 2, we present the notion of excuses in philosophy and jurisprudence while section 3 describes the dependency modeling. In Section 4, we propose the five classes of excuses and their representation using social dependence networks. Related work and conclusions end the paper.

2   Excuses in Social Sciences and Jurisprudence

The word 'excuses' is strictly connected to terms such as 'plea', 'defence' and 'justification'. The main question sociologists and psychologists try to answer is *When are 'excuses' proffered?*. As described by Austin [1], the situation is one where someone is accused of having done something, or where someone is said to have done something which is bad, wrong, inept, unwelcome, or in some other of the numerous possible ways untoward. Thereupon he, or someone on his behalf, will try to defend his conduct or to get him out of it. One way of going about this is to admit flatly that he, X, did do that very thing, A, but to argue that it was a good thing, or the right or sensible thing, or a permissible thing to do, either in general or at least in the special circumstances of the occasion. To take this line is to justify the action, to give reason for doing it: not to say, to brazen it out, to glory in it, or the like. A different way of going about it is to admit that it was not a good thing to have done, but to argue that it is not quite fair or correct to say baldly 'X did A'. We may say it is not fair just to say X did it; perhaps he was under somebody's influence, or was nudged. A usual problem with excuses consists in their distinction from justifications. The two certainly can be confused. Let us consider the following example by Austin [1]. You dropped the tea-tray: Certainly, but an emotional storm was about to break out: or, Yes, but there was a wasp. In each case the defence insists on a fuller description of the event in its context but the first is a justification while the second is an excuse.

In jurisprudence [7], an excuse or justification is a form of immunity that must be distinguished from an exculpation. On the one hand, an exculpation is a defense in which the defendant argues that despite the fact that he has done everything to constitute the crime and so, in principle already has guilt for those actions or a liability to compensate the victim, he should be exculpated because of the special circumstances said to operate in favor of the defendant at the time the law was broken. On the other hand, a justification describes the quality of the act, whereas an excuse relates to the status or capacity (or lack of it) in the accused. To be excused from liability means that although the defendant may have been a participant in the sequence of events leading to the prohibited outcome, no liability will attach to the particular defendant because he belongs to a class of person exempted from liability. For instance, members of the armed forces may be granted a degree of immunity for causing prohibited outcomes while acting in the course of their official duties, e.g. for an assault or trespass to the person caused during a lawful arrest. Other excuses may due to a particular status and capacity. Finally, others may escape liability because the quality of their actions satisfied a general public good.

3   Dependency Modeling

Our model is a directed labeled graph whose nodes are instances of the metaclasses of the metamodel, e.g., agents, goals, and whose arcs are instances of the metaclasses representing relationships between them such as dependency and conditional dependency. In a multiagent system, since an agent is put into a system that involves also other agents, he can be supported by the others to achieve his own goals if he is not able to do them alone. In order to define these relations in terms of goals and powers, we adopt, as said, the methodology of dependence networks [8]. In this model, an agent is described by a set of prioritized goals, and there is a global dependence relation that explicates how an agent depends on other agents for fulfilling its goals. For example,

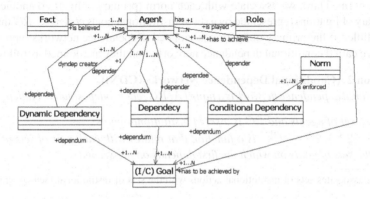

Fig. 1. The UML class diagram specifying the main concepts of the metamodel

$dep(\{a,b\},\{c,d\}) = \{\{g_1,g_2\},\{g_3\}\}$ expresses that the set of agents $\{a,b\}$ depends on the set of agents $\{c,d\}$ to see to their goals $\{g_1,g_2\}$ or $\{g_3\}$. A definition of dependence networks is provided in [5].

The introduction of norms in dependence networks is based on the necessity to adapt the requirements analysis phases to model norm-based systems. An example of this kind of application consists in the introduction of obligations in virtual Grid-based organizations where obligations are used to enforce the authorization decisions. The UML diagram of Figure 1 provides a unified vision of the concepts of the ontology representing our conceptual metamodel. We define a normative view to describe the normative system, as follows:

Definition 1 (Normative View). *Let the institutional view $\langle RL, IF, RG, X, igoals : RL \rightarrow 2^{RG}, iskills: RL \rightarrow 2^X, irules^1 : 2^X \rightarrow 2^{IF}\rangle$, the normative view is a tuple $\langle A, RG, N, oblig, sanct, ctd\rangle$ where:*

- *A is a set of agents, RG is a set of institutional goals, N is a set of norms;*
- *the function oblig : $N \times A \rightarrow 2^{RG}$ is a function that associates with each norm and agent, the institutional goals the agent must achieve to fulfill the norm. Assumption: $\forall n \in N$ and $a \in A$, $oblig(n,a) \in power(\{a\})^2$.*
- *the function sanct : $N \times A \rightarrow 2^{RG}$ is a function that associates with each norm and agent, the institutional goals that will not be achieved if the norm is violated by agent a. Assumption: for each $B \subseteq A$ and $H \in power(B)$ that $(\cup_{a \in A} sanct(n,a)) \cap H = \emptyset$.*
- *the function ctd : $N \times A \rightarrow 2^{RG}$ is a function that associates with each norm and agent, the institutional goals that will become the new goals the agent a has to achieve if the norm is violated by a. Assumption:$\forall n \in N$ and $a \in A$, $ctd(n,a) \in power(\{a\})$.*

We relate norms to goals following two directions. First, we associate with each norm n a set of institutional goals $oblig(n) \subseteq RG$. Achieving these normative goals means that the norm n has been fulfilled; not achieving these goals means that the norm is violated. Second, we associate with each norm a set of institutional goals $sanct(n) \subseteq RG$ which will not be achieved if the norm is violated and it represents the sanction associated with the norm. Third, we associate with each norm (primary obligation) another norm (secondary obligation) represented by a set of institutional goals $ctd(n) \subseteq RG$ that have to be fulfilled if the primary obligation is violated. Dependence networks representing norms are called conditional dependence networks and they are defined as follows:

Definition 2 (Conditional Dependence Networks (CDN)).
A conditional dependence network is a tuple $\langle A, RG, cdep, odep, sandep, ctddep\rangle$ where:

- *A is a set of agents and RG is a set of institutional goals;*
- *cdep : $2^A \times 2^A \rightarrow 2^{2^{RG}}$ is a function that relates with each pair of sets of agents all the sets of goals on which the first depends on the second.*

---

[1] *irules* associates sets of institutional actions with the sets of institutional facts to which they lead.

[2] Power relates each agent with the goals it can achieve.

- $odep : 2^A \times 2^A \to 2^{2^{RG}}$ is a function representing a obligation-based dependency that relates with each pair of sets of agents all the sets of goals on which the first depends on the second.
- $sandep \subseteq (OBL \subseteq (2^A \times 2^A \times 2^{2^{RG}})) \times (SANCT \subseteq (2^A \times 2^A \times 2^{2^{RG}}))$ is a function relating obligations to the dependencies which represent their sanctions. Assumption: $SANCT \in cdep$ and $OBL \in odep$.
- $ctddep \subseteq (OBL_1 \subseteq (2^A \times 2^A \times 2^{2^{RG}})) \times (OBL_2 \subseteq (2^A \times 2^A \times 2^{2^{RG}}))$ is a function relating obligations to the dependencies which represent their secondary obligations. Assumption: $OBL_1, OBL_2 \in odep$ and $OBL_1 \cap OBL_2 = \emptyset$.

4 Representing Excuses in Dependence Networks

4.1 Classes of Excuses

In this section we answer the research question: *How to distinguish classes of excuses?*. We highlight five classes of excuses: epistemic excuses, power-based excuses, norm-based excuses, counts as based excuses, social based excuses. These classes take into account different aspects of the social system from knowledge representation to norms and social laws. Particularly, the description of the classes of excuses is as follows:

1. *Epistemic excuses*: this class of excuses is related to the agents' knowledge. The distinction between epistemic excuses and moral ones consists in its relation with knowledge: an epistemic excuse transforms a true belief into knowledge. An example of this class of excuses consists in the following sentences: "I did not know, I was not aware, I forgot". Their peculiarity is the absence of knowledge which would make you able to see to your obligations or goals.
2. *Power-based excuses*: this class of excuses is based on an absence of the power to see to the obligation/goal the agent has to see to. An example of this class of excuses consists in the following sentences: "I did not have the ability, opportunity or power to do it, others prevented me from doing it". Their peculiarity is the absence of skills of the agent to perform the goal.
3. *Norm-based excuses*: this class of excuses is based on the presence of a norm with an higher priority which prevents the agent to see to its goals/obligations. An example of this class of excuses consists in the following sentence: "There was another more important norm I have to follow first". The peculiarity of this class of excuses consists in the presence of a preference order among the obligations of the agent.
4. *Counts as-based excuses*: this class of excuses is based on the idea that the excuse is about the fact that a violation did not occur because it did not count as one. An example of this class of excuses consists in the following sentence: "I did comply, my act did not count as a violation". This class of excuses could be seen as a sort of subclass of the epistemic class of excuses.
5. *Social-based excuses*: this class of excuses is based on society's common behaviours. The usual excuse of this kind is "Everybody does it".

4.2   Excuses in Dependence Networks

In this section we answer the research question: *How to represent excuses in multiagent systems?*. The representation of excuses using the dependency modeling is presented by means of a semiformal language of visual modeling where the goal-based dependency is depicted as a plain arrowed line and the obligation-based dependency is depicted as a dashed arrowed line. The label of the arrow is the goal on which the dependency is based (normative goals for obligation-based dependencies and material goals for goal-based dependencies).

As in TROPOS [4], each agent has its own dependence network which represents his view of the systems' stakeholders. Epistemic excuses are represented by means of these multiple dependence networks as depicted in Figure 2. In this figure, we show two kinds of epistemic excuses, the first one related to a goal-based dependency and the second one related to an obligation-based dependency. Agent a has his dependence network (the first square titled a) and it is involved as depender (see Figure 1 for the notion of depender) only in one norm-based dependency with agent d but, as can be noted observing the other dependence networks, a is also involved in a goal-based dependency with agent b. In this example, the excuse comes from the depender to convince the dependee, for instance if we see agent a as a node of a Grid network having to store a file too big for its memory he can say that he has not this goal in order to not be dependent on agent b. In the second example, agent a depends for a normative goal on agent d but d does not have in its dependence network this obligation-based dependency. In this case, the excuse comes from the dependee to convince the depender since the dependee does not want to support the depender in achieving its goal.

The representation of power-based excuses is achieved by means of absence of goal-based and obligation-based dependencies and it is connotated by grey arrows which represent a lack of power in terms of lacking skills. In Figure 3, agent a depends on agent b for achieving its goal but b has the excuse that it has not the power to achieve this goal for agent a. A second case of this kind of excuses' representation is provided by the dependence network of agent d which depends on itself for achieving its goal, for instance node d has to run a computation and it is able to do it alone, but it can say that it does not have the power to do it, in order to have an help by the other agents. Note that

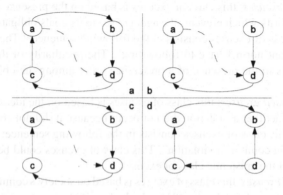

Fig. 2. Epistemic excuses

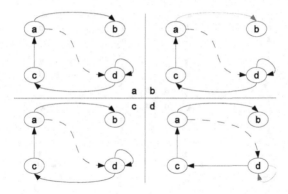

Fig. 3. Power-based excuses

the other agents of the system have in their dependence networks this self-dependence as an "active" one, since they "know" that agent d is able to achieve its goal, thus this is an excuse.

Norm-based excuses are represented by means of the addition to the label not only of the goal on which the dependency is based, but also the priority on agents' goals. The formal addition of goal-based preferences is presented in [10]. In Figure 4, we have two possible cases of norm-based excuses. The first one involves agents a, b and d. Agent a depends on agent b for normative goal $g1$ and on agent d for normative goal $g2$. Agent a prefers the dependency on b than the one on d so that in its dependence network $g1$ is preferred over $g2$. Agent b agrees with a while agent d "thoughts" that for a, $g2$ is preferred over $g1$. Agent a has an excuse for agent d regarding the priority of norm-based dependencies. The second case we highlight involves agents b, c and d. Agent d depends both on b and c respectively for normative goals $g3$ and $g4$. The difference consists in the preferences of agent b which is the dependee of two norm based dependencies. Node b can say to node d that $g1$ has the priority on $g3$, so that it cannot achieve the normative goal $g3$.

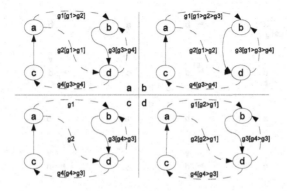

Fig. 4. Norm-based excuses

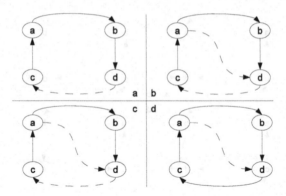

Fig. 5. Counts as-based excuses

Counts as based excuses are represented similarly to epistemic excuses. In Figure 5, counts as excuses are represented by the dependency between agent d and agent c. Concerning agent d, its dependency on agent c is a common one so that not achieving it does not mean a violation while for all the other agents this is an obligation-based dependency. The excuse of agent d, in the case this goal is not achieved, is that this does not count as a violation since it is not a normative goal.

Social-based excuses are the most difficult to be represented since they are summarized by sentences like "Everybody does it!". We represent this class of excuses by self-dependencies in the mind of the agent which uses this excuse since it sees that every agent is able in doing, for instance $g1$, by itself. In Figure 6, agent a is the agent using the excuse while, as shown by the dependence networks of the other agents of the system, it is not true that the other agents achieve $g1$ alone since all of them depend on d for $g1$. This class of excuses cannot be directly represented using logical formalisms such as deontic logic while, using the dependency modeling technique, it is represented using self-dependencies on norms.

Concerning the acceptance criteria, we can fix the following points. Dependence networks represent how agents need each other to reach their goals. Agents need each

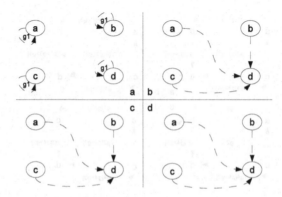

Fig. 6. Social-based excuses

other and make promises, which can be seen as a kind of contract, to each other. In some cases, something goes wrong and some agent does not deliver. He is not immediately expelled if he has a good excuse. We can note that not knowing seems a bad excuse in these contexts, especially if there are explicit contracts. Higher authority, such as the preference order between norms, might be a good excuse, although agents should not make to much promises. Counts-as excuses seem a little far-fetched here. 'Everybody does it' is an excuse, but not a very constructive one in this context. One agent can say to the others: "I break my promise because you broke yours", and, besides, nobody holds his promises around here but then the dependency causes the system to halt.

5   Related Work

As far as we know, the only approach in the field of multiagent systems related to the topic of excuses consists in the definition of the notion of forgiveness by Vasalou et al. [9]. Online offences are generally considered as frequent and intentional acts performed by a member with the aim to deceive others. However, an offence may also be unintentional or exceptional, performed by a benevolent member of the community. Vasalou et al. [9] examine whether a victim's decrease in trust towards an unintentional or occasional offender can be repaired in an online setting, by designing and evaluating systems to support forgiveness. In their experiments, an apology from the offender restores the victim's trust only if the offender cooperates in a future interaction; it does not alleviate the trust breakdown immediately after it occurs. By contrast, the forgiveness component restores the victims trust directly after the offence and in a subsequent interaction. Forgiveness is a prosocial process during which negative motivations towards the offender are reduced and replaced with positive motivations [6]. It should be noted that ,while a certain violation may be forgiven, other past behaviours may still impede one's trust towards another. Forgiveness mediates and resolves conflicts to sustain healthy long-term relationships [6]. A forgiving attitude promotes cooperation when compared to a vengeful outlook [2].

6   Summary

In this paper, we highlight five classes of excuses: epistemic excuses, power-based excuses, norm-based excuses, counts as-based excuses and social-based excuse. This classification seems to cover the main kinds of excuses, anyway it is possible that an excuse falls into two classes at the same time, such as for example epistemic excuses and counts as ones since the last one could be seen as a subclass of the former. The classification can be used to define different representations of the classes of excuses and consequently, to define different acceptance criteria due to the chosen representation. We represent excuses in multiagent systems using the methodology of social dependence networks. This methodology has the advantage, particularly in contrast with logical approaches, due to the representation of the last kind of excuses such as the common excuses "Everybody does it!". While, on the one side, these excuses are represented using dependence networks, as depicted in Figure 6, on the other side, they cannot be represented in deontic logic in a direct way. Moreover, dependence networks are used in design methodologies like the TROPOS one [4] and they are abstract, so they can be used, for example, for conceptual modeling, simulation, design and formal analysis.

Concerning future works, we are studying how excuses could help the formation of coalitions [3] acting as social lubricant. The idea is that each coalition has to maintain its stability but certain agents' behaviours may destroy this stability, such as laying to the other members of the coalition, decreasing in this way the level of trust inside it. Excuses can be seen as a social lubricant able to maintain, always due to time limits, coalition's stability despite negative behaviours of one or more of its members. Moreover, this study has to consider the addition of the notion of trust in dependence networks.

References

1. Austin, J.L.: A plea for excuses. Philosophical Papers (1961)
2. Axelrod, R.: The Evolution of Cooperation. Basic Books, New York (1984)
3. Boella, G., van der Torre, L., Villata, S.: Social viewpoints for arguing about coalitions. In: Bui, T.D., Ho, T.V., Ha, Q.-T. (eds.) PRIMA 2008. LNCS (LNAI), vol. 5357, pp. 66–77. Springer, Heidelberg (2008)
4. Bresciani, P., Perini, A., Giorgini, P., Giunchiglia, F., Mylopoulos, J.: Tropos: An agent-oriented software development methodology. Autonomous Agents and Multi-Agent Systems 8(3), 203–236 (2004)
5. Caire, P., Villata, S., Boella, G., van der Torre, L.: Conviviality masks in multiagent systems. In: Padgham, L., Parkes, D.C., Müller, J., Parsons, S. (eds.) AAMAS (3), pp. 1265–1268. IFAAMAS (2008)
6. McCullough, M., Worthington, E.L., Rachal, K.C.: Interpersonal forgiving in close relationships. Journal of Personality and Social Psychology 73, 321–336 (1997)
7. Robinson, P.H.: Criminal law defenses: A systematic analysis. Columbia Law Review (199) (1982)
8. Sichman, J.S., Conte, R.: Multi-agent dependence by dependence graphs. In: AAMAS, pp. 483–490. ACM, New York (2002)
9. Vasalou, A., Hopfensitz, A., Pitt, J.V.: In praise of forgiveness: Ways for repairing trust breakdowns in one-off online interactions. Int. J. Human-Computer Studies 66, 466–480 (2008)
10. Villata, S.: Institutional social dynamic dependence networks. In: Boella, G., Pigozzi, G., Singh, M.P., Verhagen, H. (eds.) NORMAS, pp. 201–215 (2008)

Statistical and Fuzzy Approaches for Atmospheric Boundary Layer Classification

Angelo Ciaramella[1,*], Angelo Riccio[1], Federico Angelini[2], Gian Paolo Gobbi[2], and Tony Christian Landi[2]

[1] Dept. of Applied Science, University of Naples "Parthenope",
Centro Direzionale, Isola C4, I-80143, Napoli, Italy
Tel.: +390815476674; Fax: +390815476514
{ciaramella,riccio}@uniparthenope.it
[2] Institute of Atmospheric Sciences and Climate, ISAC-CNR,
Via Fosso del Cavaliere, 100 I-00133, Roma, Italy
{f.angelini,t.landi,g.gobbi}@isac.cnr.it

Abstract. In this work we address the problem of inferring the height of atmospheric boundary layer from lidar data.

From one hand the problem to reconstruct the boundary layer dynamics is addressed using a Bayesian statistical inference method. Both parameter estimation and classification to mixed/residual layer, are studied. Probabilistic specification of the unknown variables is deduced from measurements. Hierarchical Bayesian models are adopted to relax the prior assumptions on the unknowns. Markov chain Monte Carlo (MCMC) simulations are conducted to explore the high dimensional posterior state space.

On the other hand a novel neuro-fuzzy model (Fuzzy Relational Neural Network) is used to obtain an "IF-THEN" reasoning scheme able to classify future observations. Experiments on real data are introduced.

Keywords: Bayesian statistical inference, Neuro-Fuzzy Model, Fuzzy Relations.

1 Introduction

The planetary boundary layer (PBL) is the region where air masses are strongly influenced by the surface processes and interact with air from the free troposphere [1].

Substances emitted or originated near the surface are gradually dispersed horizontally and vertically through the action of turbulence, and finally become completely mixed throughout the PBL, if sufficient time is given and if there are no significant sinks. So, the turbulent properties of this layer (diffusivity, mixing and transport) must be carefully assessed, since they modulate the exposure to particles and gaseous pollutants, and rule whether they are dispersed and diluted or whether they build up and lead to severe pollution episodes.

During the last decades, despite progress in experimental and theoretical activities for the characterization of atmospheric turbulent properties, the determination of PBL height is still one of the most uncertain parameter. Due to its ill-definition, it is not

---

* Corresponding author.

R. Serra and R. Cucchiara (Eds.): AI*IA 2009, LNAI 5883, pp. 375–384, 2009.
© Springer-Verlag Berlin Heidelberg 2009

surprising that a lot of methods exist for the experimental and modeling estimation of PBL height for the convective (CBL) and stable boundary layer (SBL); see, for example, [2] and references therein.

In this work we infer some atmospheric properties from lidar profiles. The lidar (light detection and ranging) is a remote sensing technique that allows to determine the concentration of particulate matter in the atmosphere. Information is obtained by sending pulses of laser light into the atmosphere and measuring the properties of the light after it is scattered by aerosols and gases. The continuos scanning of the atmosphere, at a high temporal resolution, can be used to work out a *scene process*. Scene processes are essentially sophisticated averaging schemes whose function is to identify and extract the base and top altitudes of clouds and aerosol layers from contiguous sequences of profile measurements, that is, from a lidar 'scene'. The lidar data analyzed in this paper were obtained by the Aerosol Remote Sensing Group, working at CNR-ISAC in Rome-Tor Vergata. A typical example of a scene process is shown in Figure 1.

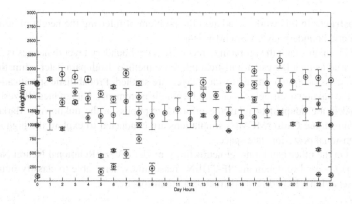

Fig. 1. An example of results obtained from lidar data. Blue dots show the heights of most intense backscatter signal gradients with corresponding gradients. Data concern the scene process derived from lidar data detected at Milano-Bicocca (45.45 degrees north, 9.16 degrees west) on April 13th, 2007.

The final aim is to recognize the top of the boundary layer. In order to do that, the labeling approach was guided by knowledge about the physics of the PBL evolution process. A novel Fuzzy Relational Neural Network model was applied to estimate the fuzzy rules to predict and classify new observations.

In section 2, we briefly describe the basic equations governing the physics of the process; next, the statistical procedure exploited to detect the PBL height (section 3). In section 4 we illustrate the Fuzzy Relation Neural Network model and in section 5 some experiments, obtained by using the approach on real lidar measurements, are described.

2   Basic Governing Equations

Similarly to previous studies on PBL modeling [3,4,5], we model the PBL evolution by means of two basic equations, one for the mean value of potential temperature and the other for the jump at the inversion height. The equation for the PBL height tendency comes from combination of these two equations and the continuity equation.

Under several assumptions (omitted for brevity), it is possible to derive the following governing equation for the potential temperature tendency

$$\frac{\partial \bar{\theta}}{\partial t} + \bar{u}_j \frac{\partial \bar{\theta}}{\partial x_j} = -\frac{1}{\rho c_p} \left[L_v E + \frac{\partial Q^*}{\partial x_j} \right] - \frac{\partial \overline{u'_j \theta'}}{\partial x_j} \tag{1}$$

The first term on the right hand side represents the net body source due to latent heat release ($L_v E$) and radiation divergence ($\partial Q^* / \partial x_j$); the last term represents the contribution from the turbulent heat flux divergence [1,6].

We can drastically simplify equation (1) if we:

1. assume horizontal homogeneity;
2. ignore source terms;
3. assume that a well-mixed boundary exists, capped by a stable boundary layer above;
4. fluxes vary linearly with height (see Figure 2).

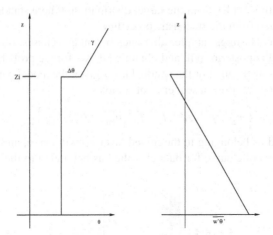

Fig. 2. Idealized slab mixed layer, with discontinuous jump at the mixed layer top

This idealization is sometimes called a *bulk* or *slab model* because the mixed layer is represented by a uniform slab of air, or a *jump model* because of the discontinuity across the top. The top of the boundary layer is defined as the height at which the temperature discontinuity occurs. A differential equation for the temperature discontinuity at the boundary layer top can be derived from (1).

3 The Estimation Procedure

Since the most part of aerosol is trapped within the boundary layer, a large discontinuity in aerosol concentration is usually detected at the top of the boundary layer and every layer where aerosol have been transported during previous convective processes; the last layer is usually named "residual layer".

These layers generate a more intense backscatter signal, so that lidar data can be used to detect the PBL height and the heights of all residual layers above it.

In order to distinguish between the PBL height and the residual layer, we exploit the previous physical model. This model contains several parameters, which were optimized to adapt to lidar data. In order to do that, we adopted a Bayesian estimation approach, that is, we selected an *a-priori* probability distribution function (pdf) for each parameter, and set up a Markov Chain Monte Carlo (MCMC) approach to draw samples from the *a-posteriori* pdf. A random walk Metropolis-Hastings procedure [7], was exploited to explore the *a-posteriori* pdf.

The basic procedure of Monte Carlo simulation is to draw a large set of samples $\left\{\theta_k^{(l)}\right\}_{l=1}^{L}$, from the target distribution (the posterior pdf in this work). One can then approximate the expectation of any function $f(\theta)$ by the sample mean as follows:

$$\mathrm{E}(f) = \int p(\theta|\cdot)f(\theta)d\theta \approx \frac{1}{L}\sum_{l=1}^{L}f(\theta^{(l)}), \qquad (2)$$

L is the number of samples from the target distribution. θ here stands for the array of parameters sampled from the stochastic procedure.

The Metropolis-Hastings sampler alternates two major phases: obtaining draws for parameters from the posterior pdf, and obtaining draws for the probability to belong to the mixed layer growth curve or the residual layer given the model parameters.

In the first phase, we drew a sequence of samples

$$\left\{w_L^{(l)}, \gamma^{(l)}, A^{(l)}, t_i^{(l)}, t_d^{(l)}, \bar{\theta}_{\mathrm{ini}}^{(l)}, \Delta\bar{\theta}_{\mathrm{ini}}^{(l)}, h_{\mathrm{ini}}^{(l)}\right\}_{l=1}^{L}$$

for data classified as belonging to the mixed layer growth curve, and for the three coefficients of a parabolic curve for data classified as belonging to the residual layer. In

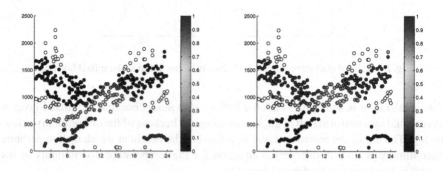

Fig. 3. An example of average results from the stochastic estimation procedure. Probability to belong to the mixed layer curve (left). The specular plot, i.e. the probability to belong to the residual layer curve, is shown on the right. Data plotted as white points have been classified as 'noise', since, on average, they have a small probability to belong to any class. Data concern the scene process derived from lidar data detected at Milano-Bicocca (45.45 degrees north, 9.16 degrees west) on April 9th, 2007.

the second phase, given the sampled parameters, each data is classified in two classes, (*mixed layer*, *residual layer*), depending on the ratio

$$\alpha = \frac{\text{pdf}_{ML}}{\text{pdf}_{RL}}$$

where pdf_{ML} is the value of a normal pdf evaluated at the given data, centered at the expected value of the mixed layer growth curve, and pdf_{ML} is the value of a normal pdf evaluated at the given data, centered at the expected value of the parabola fitted to the residual layer data. The variances of these pdfs were estimated from data as well at each iteration. Each data is assigned to the ML class with a probability α; otherwise to the RL class.

Figure 3 shows an example of results from this estimation procedure. The average probability to belong to the top of boundary layer or the residual layer is shown.

4 Fuzzy Relational Neural Network

Since a fuzzy relation expresses the dynamic features of a system described by a fuzzy model, it is natural to design a Fuzzy Neural Network based on the Fuzzy Relations (FRNN in the following) [8,9,11]. In this section we begin to show how we can design a FRNN for a complex fuzzy system. Let us assume that a fuzzy system with multi-input and one-output (it is simple the generalization to multi-output model) consists of the following fuzzy rules:

$$R_1: \text{If } x_1 \text{ is } A_{11} \text{ and } x_2 \text{ is } A_{12} \text{ and } \ldots \text{ and } x_m \text{ is } A_{1m} \text{ then } y \text{ is } B_1$$
$$\text{else}$$

$$\ldots \tag{3}$$

$$\text{else}$$
$$R_n: \text{If } x_1 \text{ is } A_{n1} \text{ and } x_2 \text{ is } A_{n2} \text{ and } \ldots \text{ and } x_m \text{ is } A_{nm} \text{ then } y \text{ is } B_n$$

where A_{ij} and B_i, $i = 1, \ldots, n$ and $j = 1, \ldots, m$, are fuzzy sets in $U \subset \Re$ and $V \subset \Re$, respectively, where U and V are the universes of discourse, and x_j and y are linguistic variables.

The connectionist model is presented in Figure 4. It designs the previous fuzzy system by using fuzzy relations (R_{rj}^i). The model is composed by 3 different hidden layers and it also contains the fuzzification (i.e. fuzzy sets A_r^i) and defuzzification phases. If we consider n inputs which are discretized into m_i input levels by a fuzzifier and k outputs which are obtained by the defuzzification of M discretized levels then, by using different t-norms and t-conorms, the output of the model is [10]

$$f_k(\mathbf{x}) = \frac{\sum_{j=1}^{M} \overline{y}_k^j \left[\mathbf{T}_{i=1}^n \left[\mathbf{S}_{r=1}^{m_i} (\mu_{A_r^i}(x_i) \mathbf{t} \mu_{R_{rj}^i}) \mathbf{s} \theta_{R^j}^i \right] \right]}{\sum_{j=1}^{M} \left[\mathbf{T}_{i=1}^n \left[\mathbf{S}_{r=1}^{m_i} (\mu_{A_r^i}(x_i) \mathbf{t} \mu_{R_{rj}^i}) \mathbf{s} \theta_{R^j}^i \right] \right]} \tag{4}$$

where $f : U \subset \mathcal{R}^n \to \mathcal{R}$, $\mathbf{x} = (x_1, x_2, \ldots, x_n) \in U$, s is the number of fuzzy rules, n the number of relation matrices, m_i is the number of input membership functions of

input hidden   first hidden   second hidden   third hidden
   layer   layer   layer   layer

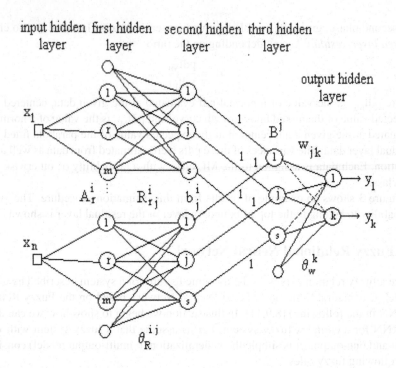

Fig. 4. Fuzzy Relational Neural Network

the i-th relation, $\mu_{A_r^i}$ is the membership function on the input space, $\overline{y}_k^j$ is the apex on the output space and $\mu_{R_{rj}^i}$ is the weight from the r-th input to j-th output of the j-th relation matrix [8,9].

If we use a sum of product for the relational model, the product for the inference, a simplified centroid defuzzification, and Gaussian input membership functions then the output of the model (with biases equal to zero) becomes (multi-input-one-output)

$$f(\mathbf{x}) = \frac{\sum_{j=1}^{M} \overline{y}^j \left[\prod_{i=1}^{n} \left[\sum_{r=1}^{m_i} (\mu_{A_r^i} \cdot \mu_{R_{rj}^i}) \right] \right]}{\sum_{j=1}^{M} \left[\prod_{i=1}^{n} \left[\sum_{r=1}^{m_i} (\mu_{A_r^i} \cdot \mu_{R_{rj}^i}) \right] \right]} \tag{5}$$

In [8,9] it is demonstrated that such as a model can approximate any function on a compact set (universal approximation properties). The learning algorithm that we use to tune the weights of the model is based on both Back-Propagation (BP) and on Pseudoinverse matrix strategies [8,9].

5   Boundary Classification

In this section we present some results obtained using the FRNN model to classify data previously labeled by the introduced estimation procedure. Data concern several scene processes obtained during an experimental campaign in Spring 2007.

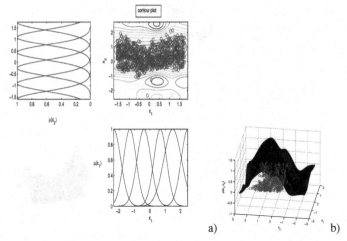

Fig. 5. FRNN contour of the first layer: a) 3 membership functions and contour plot; b) FRNN 3-dimensional mapping

In a first experiment we use a single day observation to verify the performance of the overall approach. From the labeling phase we obtain three partially overlapped classes, **mixed layer**, **residuum** and **outliers**. Applying the FRNN model to these data we obtain an "IF-THEN" reasoning scheme. Here we use a sum of product relational model, product inference, simplified centroid defuzzification, and Gaussian input membership functions. We use 3 membership functions for each input obtaining a classification performance of $\approx 92\%$. By using 5 membership functions we obtain a percentage of correct classification of $\approx 94\%$.

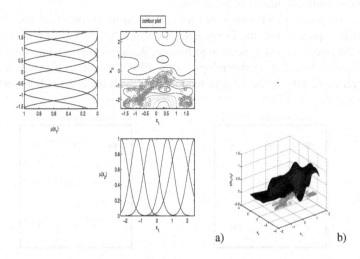

Fig. 6. FRNN contour of the second layer: a) 3 membership functions and contour plot; b) FRNN 3-dimensional mapping

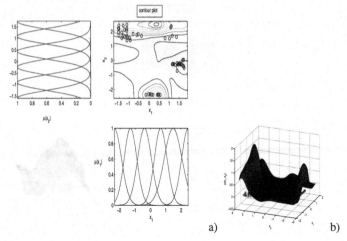

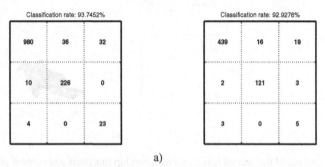

Fig. 7. FRNN contour of the outliers: a) 3 membership functions and contour plot; b) FRNN 3-dimensional mapping

A further experiment is conducted to test the FRNN capability to predict and classify new observations. First we label 7 observations by using the physically based estimation approach. Also in this case we identify three classes for each observation. We randomly extract 5 of these observations to obtain the training set. In Figure 5 we show the contour plot and the membership functions obtained classifying the residuum layer (3-dimensional contour plot in Figure 5b). In the Figures 6 and 7 we show the contours obtained classifying the mixed layer and the outliers, respectively. In this case we obtain a classification rate of $\approx 94\%$ as shown in Figure 8a. Moreover we use other 2 observations as test set. The classification obtained in this case is of $\approx 93\%$ as can be seen in the confusion matrix of Figure 8b.

Moreover we consider a new data set composed by 10 observations. We compare the FRNN approach with the Fuzzy Basis Function Network (FBFN) [12] model using a leave-one-out cross-validation technique. We use 6 membership functions for the FRNN approach and 10 memberships for the FBFN model. In Figure 9 we show the classification results on the 10 data sets obtained with the cross-validation technique.

Fig. 8. FRNN confusion matrices: a) training set; b) test set

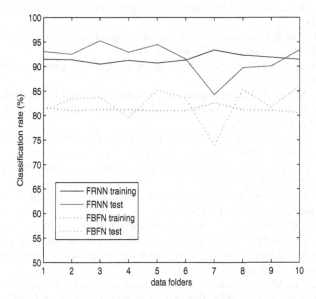

Fig. 9. Comparison between FRNN and FBFN by using a leave-one-out technique

Confirming the results presented in other works ([8,9,11]), the FRNN approach presents good performance and, how in this case, also better than the FBFN approach.

6   Conclusions

In this work we addressed the problem of inferring the height of atmospheric boundary layer from lidar data. The problem to reconstruct the boundary layer dynamics is addressed using a Bayesian statistical inference method. Both parameter estimation and classification to mixed/residual layer, are studied. Probabilistic specification of the unknown variables is deduced from measurements. Hierarchical Bayesian models are adopted to relax the prior assumptions on the unknowns. Markov chain Monte Carlo (MCMC) simulations are conducted to explore the high dimensional posterior state space. Moreover a novel neuro-fuzzy model (Fuzzy Relational Neural Network) has been used to obtain an "IF-THEN" reasoning scheme able to extract a boundary layer from noise observations. In the next future the authors will focus on the detection of the boundary layers also considering the temporal features of the observations and the application of the approach on new data observations (i.e. winter).

References

1. Stull, R.B.: An Introduction to Boundary Layer Meteorology. Kluwer Academic Publishers, Dordrecht (1998)
2. Seibert, P., Beyrich, F., Gryning, S.E., Joffre, S., Rasmussen, A., Tercier, P.: Review and inter-comparison of operational methods for the determination of the mixing height. Atmospheric Environment 34, 1001–1027 (2000)

3. Betts, A.K.: FIFE atmospheric boundary layer budget methods. J. Geophys. Res. 97, 18523–18531 (1972)
4. Carson, D.J.: The development of a dry inversion-capped convectively unstable boundary layer. Q. J. R. Meteorol. Soc. 99, 450–467 (1973)
5. Tennekes, H.: A model for the dynamics of the inversion above a convective boundary layer. J. Atmos. Sci. 30, 558–567 (1973)
6. Pielke, R.A.: Mesoscale Meteorological Model. Academic Press, London (2002)
7. Hastings, W.K.: Monte Carlo sampling methods using Markov chains and their applications. Biometrika 57, 97–109 (1970)
8. Ciaramella, A., Tagliaferri, R., Pedrycz, W., Di Nola, A.: Fuzzy Relational Neural Network. International Journal of Approximate Reasoning 41, 146–163 (2006)
9. Ciaramella, A.: Soft Computing Methodologies for Data Analysis, PhD Thesis, DMI-University of Salerno, Italy (2003)
10. Ciaramella, A., Pedrycz, W., Tagliaferri, R.: The Genetic Development of Ordinal Sums. Fuzzy Sets and Systems 151, 303–325 (2005)
11. Tagliaferri, R., Ciaramella, A., Di Nola, A., Bělohlávek, R.: Fuzzy Neural Networks Based on Fuzzy Logic Algebras Valued Relations. In: Nickravesh, M., Zadeh, L., Korotkikh, V. (eds.) Fuzzy Partial Differential Equations and Relational Equations: Reservoir Characterization of Modeling, Springer, Heidelberg (2004)
12. Wang, L.-X., Mendel, J.M.: Fuzzy Basis Functions, Universal Approximation, and Orthogonal Least-Squares Learning. IEEE Transactions on Neural Networks 3(5), 807–814

Learning Local Correspondences for Static Signature Verification

G. Pirlo[1,3], D. Impedovo[2,3], E. Stasolla[1,3], and C.A. Trullo[1,3]

[1] Dipartimento di Informatica – Università degli Studi di Bari – via Orabona 4, 70126 – Bari
[2] Dip. di Ing. Elettrotecnica ed Elettronica – Politecnico di Bari – via Orabona 4, 70126 – Bari
[3] Centro "Rete Puglia" – Università degli Studi di Bari – via G. Petroni 15/F.1 – 70100 Bari

Abstract. This paper presents a new approach for off-line signature verification. Signature verification is performed by matching only well-selected regions of the signature images. More precisely, from the analysis of lower and upper contours of a signature image, region stability is estimated and the most stable regions are selected for verification, during the enrollment phase. In the verification phase, an unknown specimen is verified through the analysis of the selected regions, on the basis of a well-defined similarity measure. The experimental results, carried out on signatures from the GPDS database, demonstrate the potential of the proposed approach.

1 Introduction

In the last years handwritten signature verification has been considered with renewed interest since it has been recognized that signature-based biometric has a very special place in the wide set of biometric techniques. Handwritten signatures are the most widespread means of personal verification and are well-recognized by administrative and financial institutions. In addition, verification by signature analysis requires no invasive measurements and people are familiar with the use of signatures in their daily life [1,2,3].

Unfortunately, signature is a complex biometric trait that depends strongly from the physical and psychological condition of the writer, as well as on the writing device. The net result is that automatic signature verification involves aspects from disciplines ranging from human anatomy to engineering, from neuroscience to computer science and system science [3]. Because of this fact, in recent years studies on signature verification have attracted researchers from different fields, working for universities and companies, which are interested in not only the scientific challenges but also the valuable applications this field offers [1, 3].

Of course, depending on the data acquisition method, two categories of signature verification systems can be identified [1]: static (off-line) systems and dynamic (on-line) systems. Static systems use off-line acquisition devices which perform data acquisition after the writing process has been completed. In this case, the signature is represented as a grey level image $\{I(u,v)\}_{0 \le u \le U, \ 0 \le v \le V}$, where $I(u,v)$ denotes the grey level at the position (u,v) of the image. Instead, dynamic systems use on-line acquisition devices that generate electronic signals representative of the signature during the

R. Serra and R. Cucchiara (Eds.): AI*IA 2009, LNAI 5883, pp. 385–394, 2009.

writing process. In this case, the signature is represented as a sequence $\{S(n)\}_{n=0,1...N}$, where $S(n)$ is the signal value sampled at time $n \cdot \Delta t$ of the signing process ($0 \leq n \leq N$), Δt being the sampling period. Thus, the off-line case involves the treatment of the spatio-luminance of a signature image, whereas the on-line case concerns the treatment of a spatio-temporal representation of the signature.

Although static signature verification is rightly considered as more difficult than dynamic signature verification, where dynamic information is available [4, 5], there are many important applications in which static signature verification is required, like for instance automatic bank-check processing and insurance form processing [6, 7, 8].

This paper addresses the problem of static signature verification. Precisely, starting from the consideration that handwritten signatures are highly variable patterns, signature verification is here performed by detecting and using only stable sub-regions of the signature image.

In the learning phase, regions from the upper and lower contours of the signature image are considered, since contour profiles have long been demonstrated to convey important information for signature verification purposes [9, 10, 11, 12]. Successively, stable sub-regions are selected according to a perturbation-based matching approach based on the Hamming distance. In the running phase, an unknown signature is verified through the analysis of the selected regions only.

The organization of the paper is the following. Section 2 presents the new technique for the analysis of relevant regions of signature images. In Section 3 the signature verification system is fully described. Section 4 reports the experimental results obtained on signatures from the GPDS database. The conclusion is reported Section 5.

2 Discriminant Regions in Off-Line Signatures

The selection of the discriminant regions of a signature is a relevant problem for signature verification. In fact, not all parts of a signature are equally distinctive nor useful for signature verification. This problem has been widely faced in dynamic signatures, in which the selection of stable parts in different representation domains has been used for selecting the most relevant strokes for verification aims [13, 14].

In order to select distinctive parts of a static signature, in this paper regions on the upper and lower contours of signature are considered, whose capability in conveying discriminant information is well known. More precisely, after the preprocessing phase, in which each signature is binarized and normalized to a fixed rectangular area (see Figure 1), the identification of the stable discriminant sub-regions starts. In order to describe the process, let us denote a signature image as

$$I = I(u, v) , \tag{1}$$

with $0 \leq u \leq U^{max}$, $0 \leq v \leq V^{max}$.

Furthermore, each sub-image of I of size $(2R+1, 2S+1)$ (R,S integers) is defined as $I_{x,y}$, i.e.

$$I_{x,y}(r,s) = I(x+r, y+s) , \tag{2}$$

with $-R \leq r \leq R$, $-S \leq s \leq S$.

(a)   *(b)*

Fig. 1. An input signature and its representation after preprocessing

Now, from a set of random sub-regions on the upper and lower contours of the signature image, a selection is done by estimating the regions that are most stable for the signer.

Let

$$I^1, I^2, \dots, I^k, \dots I^K \tag{3}$$

be a set of K genuine signature images and, for each k, let

$$I^k_{x^k_1, y^k_1}, \; I^k_{x^k_2, y^k_2}, \; \dots, \; I^k_{x^k_p, y^k_p}, \; \dots, \; I^k_{x^k_P, y^k_P} \tag{4}$$

be P sub-images on the upper (or lower) contour of I^k.

For each sub-image, an estimation of the stability of the signer in replicating the specific pattern in that sub-image is done. In particular, for the p-th sub-image $I^k_{x^k_p, y^k_p}$, the set

$$S(p) = \{\; D^*(I^q_{x^q_p, y^q_p}, I^t_{x^t_p, y^t_p}) \mid q, t = 1, 2, \dots K \;, q \neq t \;\} \tag{5}$$

is first defined, where

$$D^*(I^q_{x^q_p, y^q_p}, I^t_{x^t_p, y^t_p}) = \min \{\; D\,(I^q_{x^q_p, y^q_p}, I^t_{x^t_p, y^t_p}) \mid |x^q_p - x^t_p| < \theta, \; |y^q_p - y^t_p| < \theta \;\} \tag{6}$$

being θ a suitable threshold and $D(\cdot)$ the Hamming Distance sub-image matching:

$$D(I^q_{x^q_n, y^q_n}, I^t_{x^k_n, y^k_n}) = \sum_{r=-R}^{R} \sum_{s=-S}^{S} d(I^q_{x^q_n, y^q_n}(r, s), I^t_{x^t_n, y^t_n}(r, s)) \tag{7}$$

being $d(\alpha, \beta) = 0$, if $\alpha = \beta$; $d(\alpha, \beta) = 1$, if $\alpha \neq \beta$.

Successively for each set $S(p)$, the mean value μ_p and the variance σ_p are computed Of course the sub-images for which the variance is small are those replicated by the signer with the maximum similarity. In other words they are the most stable and hence they are considered to be the most useful for the verification purposes.

According to this consideration, the N sub-images to be used for the matching are determined as the N sub-images for which the variance is minimum. Figure 2 shows some regions selected on the upper contour of a specimen.

Fig. 2. Example of Stable Regions

3   The Verification Process

In this paper a test signature is verified according to the results of a local analysis. Therefore, the verification process follows a two-stages strategy:

- in the first stage each selected region on the test signature is verified individually. For this purpose, it is matched against the corresponding regions on the genuine samples;
- in the second stage the local verification results at the regional level are used to determine the verification decision for the test signature.

In the following the two stages are described.

3.1   The Regional Matching

In the regional matching stage a set of N sub-images are extracted from the test signature image and compared against N corresponding sub-images on the reference signature images. Let I^t be the image of a test signature and

$$I^1, I^2, \dots, I^k, \dots I^K \tag{8}$$

be a set of K reference signature images. Furthermore, let

$$I^t_{x^t_1, y^t_1}, \ I^t_{x^t_2, y^t_2}, \ \dots, \ I^t_{x^t_n, y^t_n}, \ \dots, \ I^t_{x^t_N, y^t_N} \tag{9}$$

be N sub-images of I^t selected for signature matching.

A sub-image $I^t_{x^t_n, y^t_n}$ of the test signature is considered as belonging to a genuine signature if and only if

$$\left| \Delta(I^t_{x^t_n, y^t_n}) - \mu_n \right| \leq \sigma_n \tag{10}$$

where:

$$\Delta(I^t_{x^t_n, y^t_n}) = \min_k \ \min_{(x^k_n, y^k_n)} D(I^t_{x^t_n, y^t_n}, I^k_{x^k_n, y^k_n}) \tag{11}$$

with $|_xk_p-x^v_p|<\theta$ and $|_yk_p-y^v_p|<\theta$, and where the Hamming Distance is considered for sub-image matching.

Figure 3 shows an example of corresponding regions, on the test (Fig. 3a) and a genuine reference signature (Fig. 3b), as determined by the regional matching approach.

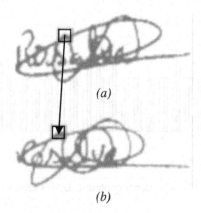

(a)

(b)

Fig. 3. Regional Matching: Identification of Corresponding Regions

3.2 The Final Verification Decision

Depending on the results obtained at the regional level, the signature I^t is considered to be genuine if and only if

$$\frac{R_g}{R_{tot}} \geq \lambda_s \tag{12}$$

where:

- R_g is the number of genuine sub-regions;
- R_{tot} is the total number of discriminant sub-regions considered for the verification of the signatures of the signer s;
- λ_s is a personal threshold value of the signer s, that estimates his/her personal variability in signing.

4   Experimental Results

In this paper the GPDS database has been considered: 2106 signatures from 80 individuals (40 signers and 40 forgers) have been used [15]:

- 880 genuine signatures have been affixed by 40 signers (24 signatures from each)
- 1200 counterfeit signatures have been affixed by the 40 forgers (30 skilled signatures from each forgers).

For each signer the most discriminant regions on the upper and lower contours of his/her signatures are identified according to the procedure described in Section 2.

Specifically, for each signer, K=23 signatures have been used for reference and diverse squared regions have been considered having different size *lxl*, where *l* is equal to 11, 15, 19 pixels. The number of selected regions for the signers is reported in Figure 4, in the case in which the upper contour or the lower contour is considered (for the case in which *l*=11). The result shows that from 3 to 6 regions are selected for each contour of the signature image.

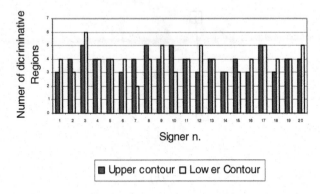

Fig. 4. Number of discriminant regions selected for each signer

Tables 1 report the performance of the system in terms of FRR (False Rejection Rate) and FAR (False Acceptance Rate), for different sizes of discriminant regions. More precisely, Table 1a concerns the results provided by the analysis of the upper contour. Table 1a shows that the best result is obtained in the case in which the region size is equal to 11x11 pixels. In this case FRR=23% and FAR=25%, on average. On the contrary, Table 1b concerns the results provided by the analysis of the lower contour. Also in this case the best result is obtained for a region size of 11x11 pixels. In this case FRR=29% and FAR=29%, on average. This result shows that the region size of 11x11 pixels always allows to achieve the best performances.

Table 1a. Upper Contour: Performances

Region Size (*lxl*)	FRR	FAR
11x11	23%	25%
15x15	32%	30%
19x19	33%	38%

Table 1b. Lower Contour: Performances

Region Size (*lxl*)	FRR	FAR
11x11	29%	29%
15x15	35%	37%
19x19	45%	42%

Figures 5 and 6 show the results in terms of FRR and FAR, for the case in which a region size of 11x11 pixels is used. More precisely, Figure 5 shows system performances when the upper contour is considered. Figure 6 shows system performances when the lower contour is considered. From this result it follows that the upper contour is generally better than lower contour for verification tasks.

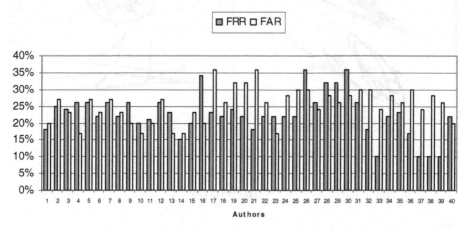

Fig. 5. Performance from the analysis of the upper contour (*lxl*=11x11)

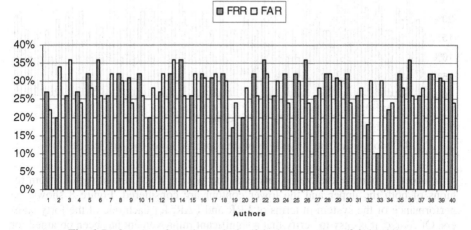

Fig. 6. Performance from the analysis of the lower contour (*lxl*=11x11)

Indeed, it is quite interesting to note that from the analysis of these results it follows that the best outcomes are obtained from signer in which characters are clearly written or also when the upper profile is very distinctive. Figure 7a shows some examples of "good" signatures producing satisfactory results in terms of FRR and FAR. Conversely, Figure 7b shows three examples of "poor" signatures for which the results are not good. This is the case of signatures in which there are no distinctive patterns in the upper contour, like for instance those included in a circular stroke.

(a) "good" signatures   (b) "poor" signatures

Fig. 7. Examples of "good" and "poor" signatures

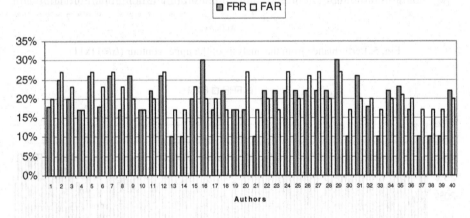

Fig. 8. Performance from the analysis of the upper and lower contour (*lxl*=11x11)

In order to increase the performances of the system, discriminant regions from regions of both the upper and the lower contour are considered. Figure 8 shows the performance of the system in terms of FRR and FAR, for each one of the forty signers. Of course, it is easy to verify that a significant improvement has been obtained for all signers. The result is FRR=19% and FAR=21%, on average.

5 Conclusion

In this paper a system for off-line signature verification is presented. The system uses a regional matching strategy. In particular, from the analysis of upper and lower contours some high-discriminant sub-regions are selected for each signer, based on Hamming Distance, and their statistical variability is estimated. An unknown signature is verified by evaluating its similarity with respect to the reference specimens only in the selected sub-regions.

The experimental results have been carried out on signatures from the GPDS database. The results shows interesting evidences, such as the fact that the upper contour conveys much more information than lower contour. As matter of the fact, performances achieved by the analysis of the upper contour are generally better that those derived from the lower contour. In addition, experimental tests have demonstrated that a significant improvement can be obtained by using information provided both from upper and lower contour.

In conclusion, the results show that the approach here presented can lead to fast verification process and reduced data storage requirements, since only well-suited parts of a signature are considered. In other word, this kind of approach seems to be flexible enough to support diverse application requirements by determining the most profitable tradeoff between high performance (for which a considerable number of sub-regions can be considered) and reduced waste of resources (for which a relatively small number of sub-regions should be used).

References

[1] Plamondon, R., Lorette, G.: Automatic Signature Verification and Writer Identification – The State of the Art. Pattern Recognition 22(2), 107–131 (1989)

[2] Leclerc, F., Plamondon, R.: Automatic Signature Verification: The State of the Art – 1989 1993. International Journal of Pattern Recognition and Artificial Intelligence (IJPRAI) 8(3), 643–660 (1994); In: Plamondon, R. (ed.): Special Issue on Progress in Automatic Signature Verification. MPAI. World Scientific, Singapore (1994)

[3] Impedovo, D., Pirlo, G.: Automatic Signature Verification – The State of the Art. IEEE Transactions on Systems, Man and Cybernetics - Part C: Applications and Review 38(5), 609–635 (2008)

[4] Plamondon, R., Srihari, S.N.: On line and Off line Handwriting Recognition: A Comprehensive Survey. IEEE Transactions on Pattern Analysis and Machine Intelligence (T-PAMI) 22(1), 63–84 (2000)

[5] Dimauro, G., Impedovo, S., Lucchese, M.G., Modugno, R., Pirlo, G.: Recent Advancements in Automatic Signature Verification. In: 9th International Workshop on Frontiers in Handwriting Recognition (IWFHR-9), Kichijoji, pp. 179–184, October 25-29 (2004)

[6] Dimauro, G., Impedovo, S., Pirlo, G., Salzo, A.: A multi-expert signature verification system for bankcheck processing. International Journal of Pattern Recognition and Artificial Intelligence (IJPRAI) 11(5), 827–844 (1997)

[7] Yoshimura, M., Yoshimura, I.: Investigation of a verification system for Japanese countersignatures on traveler's checks. Transactions of the IEICE J80 D II(7), 1764–1773 (1997)

[8] Lee, L.L., Lizarraga, M.G., Gomes, N.R., Koerich, A.L.: A prototype for Brazilian bankcheck recognition. International Journal of Pattern Recognition and Artificial Intelligence (IJPRAI) 11(4), 549–569 (1997)

[9] Ferrer, M.A., Alonso, J.B., Travieso, C.M.: Offline Geometric Parameters for Automatic Signature Verification Using Fixed-Point Arithmetic. IEEE Trans. on Pattern Analysis Machine Intelligence 27(6), 993–997 (2005)

[10] Nouboud, F.: Handwritten signature verification: a global approach. In: Impedovo, S. (ed.) Fundamentals in Handwriting Recognition, pp. 455–459. Springer, Berlin (1994)

[11] Ramesh, V.E., Narasimha Murty, M.: Off-line signature verification using genetically optimized weighted features. Pattern Recognition 32(2), 217–233 (1999)

[12] Bajaj, R., Chaudhury, S.: Signature Verification Using Multiple Neural Classifiers. Pattern Recognition 30(1), 1–7 (1997)

[13] Congedo, G., Dimauro, G., Forte, A.M., Impedovo, S., Pirlo, G.: Selecting Reference Signatures for On Line Signature Verification. In: Braccini, C., Vernazza, G., DeFloriani, L. (eds.) ICIAP 1995. LNCS, vol. 974, pp. 521–526. Springer, Heidelberg (1995)

[14] Di Lecce, V., Dimauro, G., Guerriero, A., Impedovo, S., Pirlo, G., Salzo, A., Sarcinella, L.: Selection of Reference Signatures for Automatic Signature Verification. In: 5th International Conference on Document Analysis and Recognition (ICDAR-5), Bangalore, India, pp. 597–600, September 20-22 (1999)

[15] Vargas, J.F., Ferrer, M.A., Travieso, C.M., Alonso, J.B.: Off-line Handwritten Signature GPDS-960 Corpus. In: Proc. 9th Int. Conf. on Document Analysis and Recognition (ICDAR), vol. 2, pp. 764–768, September 23-26 (2007)

ML Techniques for the Classification of Car-Following Maneuver

Fabio Tango and Marco Botta

University of Torino, Computer Science Department, corso Svizzera, 185
10149 Torino, Italy
{tango,botta}@di.unito.it

Abstract. The goal of this paper is to apply some of the best-known machine learning techniques to a practical problem in the automotive field: the identification and classification of the user's intentions in performing specific driving maneuvers. Data have been collected by a static driving simulator. These models are then analyzed and compared, in order to select the best car-following maneuver classifier.

Keywords: Machine Learning, Support Vector Machine, Neural Networks, Hidden Markov Models, classification and multi-classification problems, cognitive architecture.

1 Introduction

Since its introduction over a century ago, automobiles have enabled individual mobility for an ever growing part of human population. Unfortunately, motorised traffic has also several drawbacks; in particular, *traffic safety* is one of the most urgent and big problem that European society has to deal with: every year, about 45,000 people die and 1.5 millions people are injured in traffic accidents in Europe (www.aide-eu.org; www.prevent-ip.org; www.istat.it, http://www.irtad.com). Several countermeasures have been adopted or are under evaluation, to cope with this big challenge: systems, able to support the drivers, have been developed in these last years and nowadays some of them are already available on the market. They are called *Advanced Driver Assistance Systems* (ADAS) and include anti-collision, lane change support and blind spot detection, pedestrian detection, etc. A fundamental aspect in their design process is to guarantee high performances and dependability under all possible combinations of traffic scenarios, operating conditions and failure modes. In fact, the possible side-effects are the addition of information towards drivers that can increase – instead of reducing – the attention load on drivers themselves and potentially make the driving task even more difficult. Without a proper understanding of driver's mental state it is conceivable that the concurrent activation of many such systems, or also their unintuitive intervention can cause confusion in the user, so being distracting or even overwhelming. Hence, a method for detecting drivers' intentions is essential to facilitate operating mode transitions between driver and ADAS applications.

In this context, this paper aims at illustrating the applications of some of the best-known Machine Learning (ML) techniques to a practical problem: the identification

R. Serra and R. Cucchiara (Eds.): AI*IA 2009, LNAI 5883, pp. 395–404, 2009.

and the classification of a specific driving maneuver: the Car-Following (CF). In details, we have investigated three methods: Support Vector Machines (SVMs), Neural Networks (NNs) and Hidden Markov Models (HMMs). These models have been trained and tested using data collected by a dedicated experimental phase at the static driving simulator. This paper is so structured. The next section gives a brief description of the experimental design. Section 3 deals with the methodology followed by our research. Then, Section 4 presents the main classifiers we used, as well as it illustrates the data analysis and the main results achieved. Finally, Section 5 provides some conclusions, including a brief comparison with the current state of the art and some possible further researches in this direction.

2   Experimental Design and Set-Up

The experiments have been carried out with "real subjects" using the static driving simulator in the *Human Machine Interaction Laboratory of University of "Modena e Reggio Emilia"* (Italy). The simulator is the OKTAL Driving Simulator, whose main features can be found at the following links: www.scaner2.com, www.oktal.fr. The experiment was designed to reproduce different driving conditions, as requested by the development of manoeuvre classifier.

Six participants between 22 and 35 years old were involved in the experiment, with at least 3 years of driving license and more than 16000 km/year of experience. Participants were asked to: drive along a simulated track, respecting road rules and safety; perform the action indicated in a dedicated landmark: lane-change and car-following.

The experimental track was 24000 m (= 24 km) long and articulated into nine segments, each of them corresponding to different driving conditions. These ones were defined by different combinations of environmental and road traffic situations. The objective of the tests was to collect data on vehicle dynamics and driver's behavior during the execution of these tasks:

- Lane-Change (LC) ⇒ Two double lane-changes maneuver
- Car-Following (CF) ⇒ Driver has to complete two maneuvers

These maneuvers have been performed at two different leader-vehicle velocities: 20 km/h and 30 km/h. Such a difference was tested in [1] where it showed a significant effect on driver's approach to car following. CF starts when the leader vehicle moves from the parking position (in the simulator) and it reaches the target speed in the same lane of the follower vehicle (the vehicle driven by subject). At the end of the task, the leader comes back to the parking position on the right side of the road. Between the two maneuvers of LC or CF, a "No Event" (NE) situation was present, that is a lane-keeping action in free ride mode.

Since manoeuvres are classified using a ML approach, data has been collected for the training and testing of the models. The input variables, recorded in the experiment, concern the vehicle dynamics and the drivers' behavior and are listed as following (with the related unit of measures):

- Steering Angle (SA) ⇒ [SA] = deg
- Lateral Position (LP) ⇒ [LP] = m

- Speed of the host-vehicle (S) $\Rightarrow$ [S] = m/s
- Deceleration Jerk (DJ) $\Rightarrow$ [DJ] = m/s^3
- Time-To-Line-Crossing (TTLC) $\Rightarrow$ [TTLC] = s
- Time-To-Collision (TTC) $\Rightarrow$ [TTC] = s

Here, TTC is defined as the rate between the distance from an obstacle ahead and their relative speed; it characterizes the CF maneuver.

In addition, a reference indicator (e.g. a target output) was needed for training the system. As mentioned above, this is represented by the labels 0 and 1, which was marked manually by the experimenter:

- FLAG = 1 $\Rightarrow$ Presence of Maneuver (namely, maneuver in action)
- FLAG = 0 $\Rightarrow$ Absence of Maneuver

For each parameter in the list, the mean and the standard deviation have been computed, as method to group the data. We have avoided to use only the punctual measures (speed, steering angle, etc.) in order to reduce the amount of data; furthermore, each minimum variation of steering and speed would be used as valid data, while they represent rather a noise and a disturbance to the measure. Using the average or the standard deviation, such a noise can be minimized.

Following the ordinary procedure for learning, each dataset has been split in three different subsets:

- *Training data* (about 60% of the whole dataset) $\Rightarrow$ These are presented to the learner during training.
- *Checking data* (about 15% of the whole dataset) $\Rightarrow$ These are used to measure learner generalization and to halt training when generalization stops improving
- *Testing data* (about 25% of the whole dataset) $\Rightarrow$ These have no effect on training and so provide an independent measure of learner performance during and after training

Finally, we discuss here the sampling method of the data. This type of simulator can output data at 20 Hz (1 data-point each 0.05s).

3 Formulation of Maneuver Classifiers

In our research we have focused on Lane-Change (LC) and Car-Following (CF), since they are considered as the fundamental maneuvers of a driving task. In particular, in this paper, we present the results regarding the CF manoeuvre, being the most important one in longitudinal control: accident analysis data shown that rear-crashes are responsible for 73.3% of all accidents   (see the European co-funded project ISI-PADAS for more details: www.isi-padas.eu).

After data collection, an analysis of the related correlation matrix have been carried out, in order to obtain the set of variables to be used for the model development; the following combinations of variables have been obtained for CF manoeuvre:

- Standard Deviation of Steering Angle (SDSA)
- Standard Deviation of Speed (SDS)

- Mean of Time-To-Collision (MTTC)
- Standard Deviation of Deceleration Jerk (SDDJ)
- Standard Deviation of Lateral Position (SDLP)

Since we focus on car following, all input variables include only the longitudinal aspect. In details, the following combinations have been considered to train and test all the different models:

- **Combination 1** (C1) ⇒ SDS, SDSA, MTTC
- **Combination 2** (C2) ⇒ MS, MSA MTTC
- **Combination 3** (C3) ⇒ MS, MSA, DJ, MTTC

In order to develop the classifiers of CF manoeuvre, we have considered the following machine learning techniques: Support Vector Machines; Multi-layer Perceptron; Recurrent Neural Network; Hidden Markov Models. These are regarded from the literature as the most promising and effective in classification problem. The rationale to follow a so-called ML approach, is based on a twofold reason. From a "philosophical" point of view, one of the most ambitious goals of automatic learning systems is to mimic the learning capability of humans; this concept seems to fit absolutely well all the needs and constraints typical of driver's model (humans capability of driving is widely based on experience and possibility to learn from experience). Moreover, from a more technical viewpoint, data that are collected about vehicle dynamics and external environment, characterising the driver, are definitely non-linear. From literature, several studies have demonstrated that in such situations machine learning approaches can outperform the traditional analytical methods. Moreover, human's driver mental and physical behaviour is non-deterministic and it affects the user's intentions we want to predict.

4   Data Analysis and Results for the Different CF Classifiers

In order to measure the performances of the CF classifiers, we have used the following indexes: **Correct Rate** (CR), defined as the ratio between the number of instances correctly classified and the number of instances presented in the test dataset; **Mean Squared Error** (MSE) defined as the *average squared difference between outputs and targets*.

Finally, in the post-processing phase we have clustered data using three kinds of mobile-window (Win): Win = 0 (data are left as they are); Win = 2 (two rows are grouped and then the window is shifte); Win = 3 (three rows are grouped and then the window is shifted). We have analysed and compared the three combinations of data named as C1, C2 and C3, for the different values of "Win", for each type of classifier.

4.1   Car-Following Classifier Using Support Vector Machine

Support Vector Machines (SVMs) [2] are arguably one of the most important development in supervised classification of recent years. Their main idea is to implicitly map data to a higher dimensional space via a kernel function and then solve an optimization problem to identify the maximum-margin hyperplane that separates training instances. The hyperplane is based on a set of boundary training instances, called *support vectors*. New instances are classified according to the side of the hyperplane they fall into.

We have used SVMs in their "classical" application of binary classification task, with the following characteristics. Firstly, since several functions can be a kernel, provided that they satisfy the Mercer's Theorem, we had to choose the most appropriate to our case. Therefore, different types of Kernel functions have been tested. Secondly, different methods to find the separating hyper-plane have been also analysed; depending on the samples, we used the **Quadratic Programming** (QP) and the **Sequential Minimal Optimisation** (SMO). SMO is a fast method to solve huge QP problems that take a lot of time, widely used to speed up the training of SVMs. Hereafter, the main results achieved by SVM classifier for the CF maneuver are reported:

Table 1. Performances of SVM classifiers for different Kernels, *Win* values and combinations. α is the number of support vectors; n is the polynomial order; σ is the scaling factor.

Combinations	Win Value	Kernel	Parameters	CR Value	α
C2	2	RBF	$\sigma = 0.1$	0.99	415
	3	POLY	$n = 5$	0.99	14
C3	2	POLY	$n = 3$	0.99	23
	3	RBF	$\sigma = 0.5$	0.99	282

Some general considerations can be made. First, post-processing of data improves the classification rate of correct instances (so with Win = 2 and Win = 3) for which we obtained best results (CR index higher). Then, combinations C2 and C3 outperform C1. In addition, different Kernels show very good performances: for all the combinations, the CR of the classified instances is equal to around 99%. Also the computational time is very similar: (2÷3)s. What is different between the different Kernel functions, is the number of support vectors: *Polynomial* Kernel results in a value of α less than the *RBF* Kernel (the number of support vectors gives an idea of how complex is the function: the less the number, the simpler the function; in a real-time implementation of the classifier, a smaller number can save memory resources).

4.2 Car-Following Classifier Using Neural Networks

Neural Networks (NNs) are an information processing system, that are inspired by biological nervous system (the brain) and that consist in a large number of highly interconnected processing elements, working together to solve specific problems [3]. NNs can be classified into two main categories: *dynamic* and *static*. Static networks, the *Feed-forward Neural Networks* (FFNN), have a layered structure, with no feedback elements and contain no delays; the output is calculated directly from the input through feed-forward connections. For our goals, we have considered the *Multi-Layers Perceptron* (MLP). In dynamic networks, the output depends not only on the current input to the network, but also on the previous inputs, outputs, or states of the network. Despite the fact that dynamic networks are in some ways more difficult to train, nonetheless they are more powerful than static networks. We considered the *(Layer) Recurrent Neural Networks* (RNN), introduced by Elman [4].

For what concerning the training mode, we have adopted the so-called *batch training*, where the weights and biases of the network are updated only after the entire training set has been applied to the network. Another fundamental issue when using neural networks, is the choice of the network topology; in our case, the *Number of Hidden Layer* is 1 (the cases in which is needed to have more than 1 HL are very particular and rare those with more than 2); the *Transfer Function for Hidden Layer* is Tangent-Sigmoid; the *Transfer Function for Output Layer* is Linear; the Optimisation method for training is Levenberg-Marquardt; the training parameter values (most important) are: $goal = 0$; *Min Gradient* = 1e-10; $\mu = 0.001$;

Of course, the number of Hidden Neurons (HN) in the HL and the number of epochs depends on the specific run and applications. So, the following results have been obtained:

Table 2. Performances of NN classifiers for types of network, different mobile-windows (*Win*) and different number of hidden neurons (*HN*) in the hidden layer

Combinations	Win value	Kernel	CR Value	HN
C2	0	MLP	0.97	50
C3	3	RNN	0.99	30

It is worth noting that MLP reaches the best performances with no mobile window and with 50 HN in the HL; while for RNN, we have Win = 3 and 30 HN in the HL respectively. Comparing the best performances of MLP and RNN models, the latter outperforms the former. In fact, considering the CR value, we get around 97% of instances correctly classified for MLP and around 99% (like SVM) for RNN. The following figure shows the RNN output data (for sake of figure comprehensibility, we show an example with Win = 0):

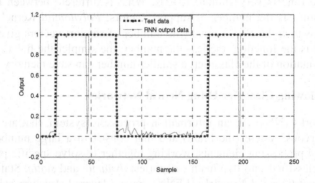

Fig. 1. Example of test-data vs. output data for RNN classifier with *Win = 0* and combination C3

From the figure, it is possible to see the very good agreement between the data used for the test and the output provided by the model. In this example the cases of misclassification are also shown (the two spikes near to samples 50 and 200). Actually, the choice of a model rather than the other is depending not only on the performance level, but also on the specific needs and application to be developed; in fact, if we consider also the computational time needed for the training phase, the situation is really different: for the MLP model, the necessary training time was 135.33s on average, while it was around 3600s for RNN classifier. For this reason, we can conclude that the MLP model for CF classification is enough for most applications of CF classification.

4.3 Car-Following Classifier Using Hidden Markov Model

Up to now, we have considered a single-classification problem (given a class, an instance may belong or not belong to it), but most real-life diagnostic tasks are not binary and we often have to tackle problems involving $k > 2$ classes. In our case, when driving, more maneuvers can occur and so it is necessary to distinguish, for example, among lane-change, car-following and lane keeping in a free lane. There are several algorithms that allow **multi-category classification**, including those which have been developed to allow multi-classification using SVM. The dominating approach is to reduce the single multiclass problem into multiple binary problems. Two common methods to build such binary classifiers consist in the fact that each classifier distinguishes between one of the labels to the rest (*one-versus-all*, or OVA) or between every pair of classes (*one-versus-one*, or OVO). As pointed out in [5], both these approaches present some limits and drawbacks about inconsistent results or too high computational costs. Despite some temtatives to find new methods, we have preferred a "traditional" method for this problem: the use of Hidden Markov Models where such a classification of more-than-2 classes is solved quite naturally. In mathematics, a **Markov chain** is a discrete-time *stochastic process* with the *Markov property*, meaning that the knowledge of the previous states is irrelevant for predicting the probability of subsequent states (thus a Markov chain is "memory less"). Hidden Markov Models (HMMs) are finite models, describing a probability distribution over an infinite number of possible sequences, which is composed of a number of states and each state 'emits' symbols (residues) according to symbol emission probabilities; the states are interconnected by state-transition probabilities. Starting from some initial state, a sequence of states is generated by moving from state to state according to the state-transition probabilities until an end state is reached. Each state then emits symbols according to that state emission probability distribution, creating an observable sequence of symbols. The sequence of states is a Markov chain, but it is hidden; only the symbol sequence that these hidden states generate is observed [6].

So, the main challenges with HMM is to determine the hidden parameters (the matrices of transitions and emissions probabilities between states and emitted symbols) from the observable parameters. The *internal states* are the driver's intentions (CF manoeuvre) and, of course, they are hidden. The *emitted variables*, that is the *observations*, are:

- Steering angle (SA) and Speed (S) of the vehicle
- Lateral Position (LP)
- Jerk (first derivative of acceleration) (J)
- Time To Collision (TTC)

We have proposed a structure of HMM for CF manoeuvre recognition with 5 states and a *left-right configuration*, which did not allow for skipping of states or backward states transition [7]. We used the following grammar for the HMM classifier of CF:

Fig. 2. HMM grammar for CF maneuver

For CF, the 5-states highlighted in the figure are:

- Lane Keeping (state "LK")
- Detection of an object ahead, that is, approaching of the obstacle in front of the host-vehicle, with variation of the TTC parameter (state "*Detect*")
- Slow down or braking action (state "*Slow / Brake*")
- Stabilization of speed (state "*Stabilization*")
- Regularization and adaptation of distance and speed respect to the vehicle ahead (state "*Follow-up*")

For the CF recognition, we have solved the three basic problems of HMM, considering an alphabet of 32 symbols [6]. In order to deal with multi-classification problems, since driving task is characterised by the contemporary presence of more than one manoeuvre, or anyway, there is a continuous transfer from one state to another, an "ad-hoc" architecture has been developed, as shown in Figure 3.

The model associated with the highest likelihood is used as the "winner" for the manoeuvre recognition and – in a broad sense – this automatically solves the problem of multi-classification for different manoeuvres. So, given an observation sequence and a model, the probability that the observed sequence would be generated by the model is needed. The *forward-backward algorithm* is often used in practice to compute this probability. Once it has been evaluated for all competing models with respect to an observation data sequence (that is, a set of driving action data) then the model with the highest probability supports recognition of the manoeuvre it characterizes. In our case, we have compared the sequence of states emitted by the HMM classifier, with the test dataset collected during the experiments.

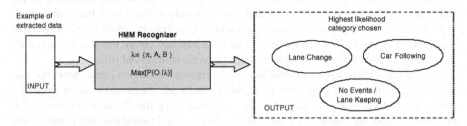

Fig. 3. System Architecture for HMM classifier of CF maneuver

The recognition rate by category for different observation data is shown as following:

- CR for CF manoeuvre when present = 0.9666 (97%)
- CR for CF manoeuvre when absent = 2.6603e-4

The model for CF classification achieves good performances, with around the 97% of instances correctly classified. Moreover, in the situation where the manoeuvres are mixed, the CR index is very good when CF was not present, meaning that the system, represented in Figure 3, is able to distinguish if a specific manoeuvre is present or not, even in case of multi-classification situation.

5   Conclusions and Future Research

In this paper we have shown how to use a ML approach for the CF manoeuvre classification, proving once more it is an excellent solution to deal with problems characterised by highly non-linear and non-deterministic data. In an initial stage of our research, we have considered a binary classification problem. We have selected the most appropriated Kernel type and parameters for SVM (*Gaussian Radial Basis* and *Polynomial* the most satisfying) and the best combination of NN topology (Multi Layer Perceptron). We have achieved a rate of instances correctly classified (CR index) equal to 99%, which is excellent. For this kind of application, NN gave out comparable performances with respect SVM. Yet, the main difference is about the computational time taken for learning; in fact, in the case of SVM, this process took typically $(3 \div 5)$s, while for Feed-forward neural networks we had a training time of $(90 \div 93)$s. Despite this fact, it is also worth to saying here that once the network is trained, its response time is very short. For SVM, things are different: they are faster to be trained, but then they take more time to classify instances, since they have more complex calculations to do and so higher computational load (in fact, they deal with a Kernel function, to map data in the appropriate hyper-space). At the very end, the choice of one technique rather than another, is strongly depending on the specific application to implement, stated that both can assure a very good performance level: around 99% of test instances have been correctly classified by the models.

At a second stage of this research, we have considered the multi-classification problem, selecting a classifier based on HMM. In particular, we have obtained an architecture for the recognition system that is able to classify between different classes. In this context of more manoeuvres mixed together, for CF, the model has achieved a very good result, around 97% of test instances correctly classified (while, the CF manoeuvre was not classified when not present). Although the performance value is a bit lower than the one achieved in binary classification problem, however, HMM allows to solve the multi-classification problem in a quite natural way, providing very good results in the distinction between the case when manoeuvre is present and when not, as well as providing the basis for an architecture that allows the implementation of a classification system of different manoeuvre in real-time.

With respect to current state of the art, we have obtained similar results (actually, even little better). All in all, many studies have found that driver's behaviour can be regarded as a sequence of basis actions, each associated with a particular state and characterised by a set of observable features. Therefore, most of the approaches used a method based on HMM, like [7] and [8]. Only [9] presents a system for detecting driver's lane changes based on SVM, but achieving different results about the best Kernel: linear function was regarded as good as other functions (anyway, with no significant improvements), which is in contrast with our findings. All these works use two types of experimental data (concerning the vehicle dynamic, as ours), coming from driving simulator or from real vehicle. Using data from simulator, the best CR were around 95% (and around 80% using real data) which is in line with our results.

Although these encouraging results, nonetheless this does not mean we have built the "perfect classifier", since some limitations in our work exist (limited range of scenarios, necessity of more extended dataset for training, exploration of alternative approaches for multi-classification, etc.). Therefore, some future researches in this direction are possible. First, a deeper investigation of the multi-classification problem is needed, including other techniques like the Augmented Binary SVM or Relevance Vector Machine. In the meantime, the refinement and the extension of the HMM classifier to other maneuvers (i.e. overtaking) has to be carried out. Then, an important aspect concerns the use of more extended scenarios, both in terms of data points and of variety of scenarios considered. In addition, these experiments should be done also using a real-prototype car (to get real-data for the models). In fact, the ultimate goal is to produce an "executive" programme for real-time implementation on a car; then, tests will be carried out, in order to measure and evaluate the performances during travelling on ordinary roads.

References

1. Fiorani, M.: Monitoring time-headway in car-following task. In: CHI Conference Italy (2008)
2. Vapnik, V.: Statistical Learning Theory. John Wiley and Sons, Chichester (1998)
3. Haykin, S.: Neural Networks: a comprehensive foundation. Prentice-Hall, Englewood Cliffs (1999)
4. Elman, J.L.: Finding structure in time. Cognitive Science 14, 179–211 (1990)
5. Bishop, C.M.: Pattern Recognition and Machine Learning, new edn. Springer-Verlag New York Inc., Heidelberg (2008)
6. Rabiner, L.R.: A Tutorial on Hidden Markov Models and selected Applications in Speech Recognition. Proceeding of the IEEE 77(2) (1989)
7. Kuge, N.: A Driver Behaviour Recognition Method based on a Driver Model Framework – Paper SAE 2000-01-0349 (2000)
8. Pentland, A., Liu, A.: Modelling and Prediction of human Behaviour; Nerual Computation (1999)
9. Salvucci, D.D.: Lane Change Detection using a computational Driver Model. Human Factors 49(3), 532–542 (2007)

An Asynchronous Cellular Automata-Based Adaptive Illumination Facility

Stefania Bandini[1], Andrea Bonomi[1], Giuseppe Vizzari[1], and Vito Acconci[2]

[1] Complex Systems and Artificial Intelligence (CSAI) research center
Department of Computer Science, Systems and Communication (DISCo)
University of Milan - Bicocca
Viale Sarca 336/14, 20126 Milano, Italy
{bandini,bonomi,vizzari}@disco.unimib.it
[2] Acconci Studio
20 Jay St., Suite #215, Brooklyn, NY 11201, USA
studio@acconci.com

Abstract. The term Ambient Intelligence refers to electronic environments that are sensitive and responsive to the presence of people; in the described scenario the environment itself is endowed with a set of sensors (to perceive humans or other physical entities such as dogs, bicycles, etc.), interacting with a set of actuators (lights) that choose their actions (i.e. state of illumination) in an attempt improve the overall experience of these users. The model for the interaction and action of sensors and actuators is an asynchronous Cellular Automata (CA) with memory, supporting a self-organization of the system as a response to the presence and movements of people inside it. The paper will introduce the model, as well as an ad hoc user interface for the specification of the relevant parameters of the CA transition rule that determines the overall system behaviour.

1 Introduction

The main aim of research on Ambient Intelligence [1] is the definition of models and tools for the realization of environments endowed with a large number of electronic devices, interconnected by means of wireless communication facilities, able to perceive and react to the presence of people. These facilities can have different goals, ranging from explicitly providing electronic services to humans accessing the environment by means of computational devices (e.g. personal computers or PDAs), to simply providing some form of ambient adaptation to the users' presence (or voice, or gestures), without requiring an explicit interaction though a traditional computational device. An Ambient Intelligence system can be viewed in terms of autonomous entities, managing internal resources and interacting with surrounding ones in order to obtain the desired overall system behaviour as a result of local actions and interactions among system components. Approaches that take this perspective share a growing interest on models and mechanisms supporting forms of self-organization and management of the components (both hardware and software) of such systems.

This paper describes an asynchronous Cellular Automata (CA) based approach to the modeling and realization of a self-organizing ambient intelligence system; the latter

R. Serra and R. Cucchiara (Eds.): AI*IA 2009, LNAI 5883, pp. 405–415, 2009.

is viewed in terms of cells comprising sensors and actuators. The former can trigger the behaviours of the latter, both through the interaction of elements enclosed in the same cell and by means of the local interaction among adjacent cells. Since the modeled system is not necessarily characterized by the presence of a global clock synchronizing system operations we adopted an asynchronous approach to the activation of cell transition rules.

The transition rule adopted for the CA was derived by previous applications to reproduce natural phenomena such as percolation processes of pesticides in the soil, in specific percolation beds for the coffee industry and for the experimentation of elasticity properties of batches for tires [2,3], by modeling mechanisms of reaction and diffusion. In this specific application this rule is used to manage the interactions of cells arranged through a multilayered architecture [4], better suited to represent an artificial environment comprising a set of sensors that perceive the presence of humans (or other physical entities such as dogs, bicycles, cars), and actuators that choose their actions in an attempt improve the overall experience of these users.

The developed model is the core component of an overall system supporting the design and definition of the relevant parameters of the transition rule, through the simulation and envisioning of its dynamic behaviour. Due to the model nature it is not trivial to foresee the effect of altering the value of parameters on the overall system behaviour. Therefore we developed an ad hoc user interface to let the user specify the value for the relevant parameter and see how the system reacts to external stimuli, generated through the mouse, simulating the activation of sensors in the environment.

The following section will introduce the specific scenario in which this research effort is set, describing the requirements for the adaptive illumination system and the environment adaptation model. Section 3 introduces the modeling approach, setting it in the relevant literature, while section 4 describes the developed model in details. A description of the developed environment supporting designers will follow, then conclusions and future works will end the paper.

2   The Application Scenario

The Acconci Studio was founded in 1988 to help realize public-space projects through experimental architecture and public art afforts. The method of Acconci Studio is on the one hand to make a new space by turning an old one inside-out and upside-down; and on the other hand to insert within a site a capsule that grows out of itself and spreads into a landscape.

The Studio has recently been involved in a project for the renovation of a tunnel in the Virginia Avenue Garage in Indianapolis. The tunnel is currently mostly devoted to cars, with relatively limited space on the sidewalks and its illumination is strictly functional. The planned renovation for the tunnel comprises a set of interventions along the direction defined by the following narrative description of the project:

> The passage through the building should be a volume of color, a solid of color.
> It's a world of its own, a world in itself, separate from the streets outside at either end. Walking, cycling, through the building should be like walking through a solid, it should be like being fixed in color.

The color might change during the day, according to the time of day: pink in the morning, for example, becomes purple at noon becomes blue, or blue-green, at night. This world-in-itself keeps its own time, shows its own time in its own way.

The color is there to make a heaviness, a thickness, only so that the thickness can be broken. The thickness is pierced through with something, there's a sparkle, it's you that sparkles, walking or cycling though the passage, this tunnel of color. Well no, not really, it's not you: but it's you that sets off the sparkle – a sparkle here, sparkle there, then another sparkle in-between – one sparkle affects the other, pulls the other, like a magnet – a point of sparkle is stretched out into a line of sparkles is stretched out into a network of sparkles.

These sparkles are above you, below you, they spread out in front of you, they light your way through the tunnel. The sparkles multiply: it's you who sets them off, only you, but – when another person comes toward you in the opposite direction, when another person passes you, when a car passes by – some of these sparkles, some of these fire-flies, have found a new attractor, they go off in a different direction.

The above narrative description of the desired adaptive environment comprises two main effects of illumination, also depicted in a graphical elaboration of the desired visual effect shown in Figure 1:

- an overall effect of uniformly coloring the environment through a background, ambient light that can change through time, but slowly with respect to the movements and immediate perceptions of people passing in the tunnel;
- a local effect of illumination reacting to the presence of pedestrians, bicycles, cars and other physical entities.

The first type of effect can be achieved in a relatively simple and centralized way, requiring in fact a uniform type of illumination that has a slow dynamic. The second point requires instead a different view on the illumination facility. In particular, it must be able to perceive the presence of pedestrians and other physical entities passing in it, in other words it must be endowed with sensors. Moreover, it must be able to exhibit local changes as a reaction to the outputs of the aforementioned sensors, providing thus

Fig. 1. A visual elaboration of the desired adaptive illumination facility (the image appears courtesy of the Acconci Studio)

for a non uniform component to the overall illumination. The overall environment must be thus split into parts, proper subsystems.

However, these subsystems cannot operate in isolation, since one of the requirements is to achieve patterns of illumination that are local and small, when compared to the size of the tunnel, but that can have a larger extent than the space occupied by a single physical entity ("sparkles are above you, below you, they spread out in front of you, they light your way through the tunnel"). The subsystems must thus be able to interact, to influence one another to achieve more complex illumination effects than just providing a spotlight on the occupied positions.

3   Related Works

Cellular Automata (CA), introduced by John von Neumann as an environment for studying self-replicating systems [5], have been primary investigated as theoretical concept and as a method for simulation and modeling [6]. They have also been used as computational framework for specific kind of applications (e.g. image processing [7], robot path planning [8]) and they have also inspired several parallel computer architectures, such as the Connection Machine [9] and the Cellular Automata Machine [10].

3.1   Asynchronous Cellular Automata

Cellular Automata have traditionally treated time as discrete and state updates as occurring synchronously and in parallel. The state of every cell of the automaton is updated together, before any of the new states influence other cells. The synchonous approach assumes the presence of a global clock to ensure all cells are updated together.

Several authors (e.g. [11,12]) have argued that asynchronous models are viable alternatives to synchronous models and suggest that asynchronous models should be preferred where there is no evidence of a global clock. Nehaniv [13] has demonstrated an asynchronous CA model that can behave as a synchronous CA, due to the addition of extra constraints on the order of updating.

Cornforth, Green, and Newth argue that asynchronous updating is widespread and ubiquitous in both natural and artificial networks [14]. They identified two classes of asynchronous behavior: Random Asynchronous (RAS), and Ordered Asynchronous (OAS) updating. Random Asynchronous includes any process in which at any given time individuals to be updated are selected at random according to some probability distribution; Ordered Asynchronous includes any process in which the updating of individual states follows a systematic pattern.

3.2   Dissipative Cellular Automata

The Dissipative Cellular Automata (DCA) are a class of cellular automata that have been defined as dissipative, i.e., cellular automata that are open and makes it possible for the environment to influence their evolution [15].

The two main characteristics of the DCA are the asynchronous time-driven dynamics and openness. DCA are Asynchronous Cellular Automata: according to the asynchronous dynamics [16,17], at each time, one cell has a probability of rate λ_a to autonomously wake up and update its state.

The above characteristics of modern software systems are reflected in DCA. can be considered as a minimalist open agent system (or, more generally, as a minimialist open software system). As that, the dynamic behavior of DCA is likely to provide useful insight into the behavior of real-world open agent systems and, more generally, of open distributed software systems.

3.3 Cellular Automata with Memory

Standard CA are ahistoric (memoryless): the cells have no memory of previous states, except the last one in the case the central cell is included in the neighborhood. Historic memory can be embedded in CA increasing the number of states and modifying the transaction function. Alonso-Sanz proposed to maintain the transaction rule unaltered, but make them act not only to the current state but weighted mean value of their previous states [18]. According to the author, CA with memory can be considered as a promising extension of the basic CA paradigm.

4   The Proposed Approach

The proposed approach adopts an Asynchronous Cellular Automata with memory to realize a distributed control system able to face the challenges of the previously presented scenario. The control system is composed of a set of controllers distributed throughout the system; each of them has both the responsibility of controlling a part of the whole system as well as to collaborate with a subset of the other controllers (identified according to the architecture of the CA model) in order to achieve the desired overall system behavior. In the proposed architecture, every node is a cell of an automata that can communicate only with its neighbors, it processes signals from sensors and it controls a predefined set of lights associated to it. The approach is totally distributed: there is no centralized control and no hierarchical structuring of the controllers, not only from a logical point of view but also a physical one.

The designed system is an homogeneous peer system, as described in Figure 2: every controller has the responsibility of managing sensors and actuators belonging to a fixed area of space. All controllers are homogeneous, both in terms of hardware and software capabilities. Every controller is connected to a motion sensor, which roughly covers the controlled area, some lights (about 40 LED lights) and neighbouring controllers.

The state of the motion sensor influenced the internal state of the cell. The state of the sensor is represented by a single numerical value $v_s \in \mathbb{N}_{8bit}$, where

$$\mathbb{N}_{8bit} \subset \mathbb{N}_0, \forall x : x \in \mathbb{N}_{8bit} \Rightarrow x < 2^8$$

The limit value was chosen for performance reasons because 8-bit microcontrollers are widely diffused and they can be sufficiently powerful to manage this kind of situation. The value of v_s is computed as

$$v_s(t+1) = v_s(t) \cdot m + s(t+1) \cdot (1-m)$$

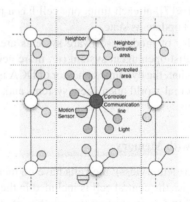

Fig. 2. The proposed architecture for the distributed control system to be managed through an Asynchronous Cellular Automata with memory

where $m \in \mathbb{R}, 0 \leq m \leq 1$ is the *memory coefficient* that indicates the degree of correlation between the previous value of v_s and the new value, while $s(t) \in N_{8bit}$ is the reading of the sensor at the time *s(t)*. If the sensor is capable of distance measuring, *s(t)* is inverse proportional to the measured distance (so, if the distance is 0, the value is 255, if the distance is ∞ the value is 0). If the sensor is a motion detector sensor (it able to signal 1 if an object is present or 0 otherwise) *s(t)*, s(t) is equal to 0 if there is not detected motion, c in case of motion, where $c \in N_{8bit}$ is a constant (in our tests, 128 and 192 are good values for c).

The diffusion rule is used to propagate the sensors signals throughout the system. At a given time, every level 2 cell is characterized by an intensity of the signal, $v \in N_{8bit}$. Informally, the value of v at time $t+1$ depends of the value of v at time t and on the value of $v_s(t+1)$, to capture both the aspects of interaction with neighbouring cells and the memory of the previous external stimulus caused by the presence of a physical entity in the area associated to the cell.

The intensity of the signal decreases over time, in a process we call evaporation. In particular, let us define $\epsilon_{evp}(v)$ as the function that computes the quantity of signal to decrement from the signal and is defined as

$$\epsilon_{evp}(v) = v \cdot e_1 + e_0$$

where $e_0 \in \mathbb{R}^+$ is a constant evaporation quantity and $e_1 \in \mathbb{R}, 0 \leq e_1 \leq 1$ is the evaporation rate (e.g. a value of 0.1 means a 10% evaporation rate).

The evaporation function $evp(v)$, computing the intensity of signal v from time t to $t+1$, is thus defined as

$$evp(v) = \begin{cases} 0 & \text{if } \epsilon_{evp}(v) > v \\ v - \epsilon_{evp}(v) & \text{otherwise} \end{cases}$$

The evaporation function is used in combination with the neighbours' signal intensities to compute the new intensity of a given cell.

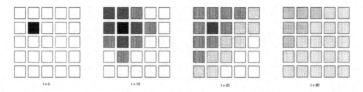

Fig. 3. An example of the dynamic behaviour of a diffusion operation. The signal intensity is spread throughout the lattice, leading to a uniform value; the total signal intensity remains stable through time, since evaporation was not considered.

The automaton is contained in the finite two-dimensional square grid $\mathbb{N}^2$. We suppose that the cell $C_{i,j}$ is located on the grid at the position i, j, where $i \in \mathbb{N}$ and $j \in \mathbb{N}$. According to the von Neumann neighbourood [19], a cell $C_{i,j}$ (unless it is placed on the border of the lattice) has 4 neighbours, denoted by $C_{i-1,j}, C_{i,j+1}, C_{i+1,j}, C_{i,j-1}$.

For simplicity, we numbered the neighbours of a cell from 1 to 4, so for the cell $C_{i,j}$, N_1 is $C_{i-1,j}$, N_2 is $C_{i,j+1}$, N_3 is $C_{i+1,j}$, and N_4 is $C_{i,j-1}$.

At a given time, every cell is characterized by an intensity of the sensor signal. Each cell is divided into four parts (as shown in Figure 4), each part can have a different signal intensity, and the overall intensity of the signal of the cell is the sum of the parts intensity values. The state of each cell $C_{i,j}$ of the automaton is defined by $C_{i,j} = \langle v_1, v_2, v_3, v_4 \rangle$ where $v_1, v_2, v_3, v_4 \in \mathbb{N}_{8bit}$ represent the intensity of the signal of the 4 subparts. $V_{i,j}(t)$ represents the total intensity of the signals (i.e. the sum of the subparts signal intensity) of the cell i, j at time t. The total intensity of the neighbours are denoted by $V_{N1}, V_{N2}, V_{N3},$ and V_{N4}. The signal intensity of the subparts and the total intensity are computed with the following formulas:

$$v_j(t+1) = \begin{cases} \frac{evp(V(t)) \cdot q + evp(V_{Nj}(t)) \cdot (1-q)}{4} & \text{if } \exists N_j \\ \frac{evp(V(t))}{4} & \text{otherwise} \end{cases}$$

$$V(t+1) = \sum_{i=1}^{4} v_i(t+1)$$

where $q \in \mathbb{R}, 0 \leq q \leq 1$ is the conservation coefficient (i.e. if q is equals to 0, the new state of a cell is not influenced by the neighbours values, if it is equals to 0.5 the new values is a mean among the previous value of the cell and the neighbours value, if it is equals to 1, the new value does not depend on the previous value of the cell but only from the neighbours). The effect of this modeling choice is that the parts of cells along the border of the lattice are only influenced through time by the contributions of other parts (that are adjacent to inner cells of the lattice) to the overall cell intensity.

In this project the actuators are LED lamps that are turned on and off according the the state of the cell. Instead of controlling a single LED from a cell, every cell is related to a group of LEDs disposed in the same (small) area.

There are different approaches, called "coloring strategies", to associate LED activity (i.e. being on or off, with which intensity) to the state of the related actuator cell. An

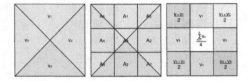

Fig. 4. Correlation between the upper layer cell subparts and the actuators layer cells

example of coloring strategy consists in directly connecting the lights' intensity to the signal level of the correspondent cell; more details on this will be given in the following Section.

5   The Design Environment

The design of a physical environment (e.g. building, store, square, road) is a composite activity, comprising several tasks that gradually define the initial idea into a detailed project, through the production of intermediate and increasingly detailed models. CAD softwares, and also 3D modelling applications are generally used to define the digital models for the project and to generate photo realistic renderings and animations. These applications are extremely useful to design a lights installation like the one related to this scenario, but mainly from the physical point of view.

In order to generate a dynamics in this kind of structure, to grant the lights the ability to change illumination intensity and possibly color, it is also possible to "script" these applications in order to characterize lights with a proper behaviour. Such scripts, created as text files or with graphical logic editors, define the evolution of the overall system over time. These scripts are however heavily dependent on the adopted software

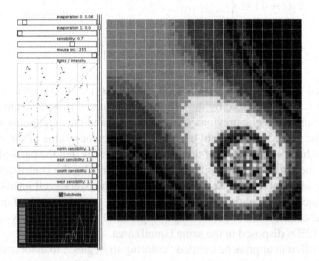

Fig. 5. A screenshot of the design environment. On the left, there is the system configurator and the global intensity graph, on the right the lights view.

and they are not suitable for controlling real installations, even though they can be used to achieve a graphical proof of concept. Another issue is that these tools are characterized by a "global" approach, whereas the system is actually made up of individual microcontrollers' programs acting and interacting to achieve the global desired effect.

In this experience, our aim was to facilitate the user in designing the dynamic behavior of a lights installation by supporting the envisioning of the effects of a given configuration for the transition rule guiding lights; therefore we created an ad-hoc tool, also shown in Figure 5, comprising both a simulation environment and a graphical parameters configurator. This tool support the specification of the values for some of the parameters of the transition rule, affecting the global behavior of the overall system. The integrated simulation helps understanding how the changes of the single parameters influence the overall behavior of the illumination facility: every changed parameter is immediately used in the transition rule of every cell.

In the following paragraphs, the tool's main components are described. Ad the end of this section, some experimental configurations and the related dynamic evolution are presented.

5.1   The Cells Simulator

The main component of the design environment is the simulator. This component simulates the dynamic evolution of the cell over the time, according to the transition rule. The simulated cells are disposed over a regular grid and each cell is connected to its neighbors according to the von Neumann neighbourood. By default, the tools is configured to simulate 400 cells, organized in a 20x20 grid. The grid is not toroidal, to better simulate a (portion of) the real installation space. Each cell has an internal state represented as an 8 bits unsigned number. In order to better simulate the real asynchronous system, an independent thread of control, that re-evaluates the internal state of the cell every 200 ms is associated to each cell. At the simulation startup, each thread starts after a small (< 1 s) random delay, in order to avoid a sequential activation of the threads, that is not realized in the real system. The operating system scheduler introduces additional random delays during both the activation and the execution cycle of the threads.

5.2   The Lights View

The aim of the Lights View is to realize an interactive visualization of the dynamic evolution of the system. In particular, the user can simulate the presence of people in the simulated environment by clicking on the cells and moving the mouse cursor. Each cell of the simulated system is associated an area of the screen representing a group of lights controlled by the cells. More precisely, it is possible to define at runtime if the area controlled by each cell is subdivided in 9 sub-areas (9 different lights groups) or if it is a single homogeneous light group. Each simulated group of lights is characterized by 256 different light intensity levels.

On the left of lights view, there is a graph showing the evolution over the time of the sum of all the cells intensity levels. This graph is particularly useful to set the coefficients of the evaporation function.

5.3   The System Configurator

Through this component, the user can define most of the parameters related to the transition rule of the simulated system.

The first two sliders control the evaporation coefficients e_0 and e_1, the next one controls the sensibility parameters q (see Section 4 for the parameters' semantics). The "mouse increment" slider defines the amount of the increment in the cell intensity when a user clicks on the cell: it represents the sensitiveness of the cell to sensor stimulus in the real system.

Under the four sliders there is a small panel that supports drawing the function that correlates the internal cell intensity value and the correspondent light group intensity value. The default function, represented by a diagonal segment between the (0,0) position and the (255,255) position, is the "equal" function (i.e. if the cell intensity has value x, the lights intensity has value x). It is possible to draw an arbitrary function, setting for each cell intensity value a correspondent light intensity value simply drawing the function over the graph.

The last four sliders control the sensitivity of each cell to the neighbors in the four directions (q_N, q_E, q_S, q_W); by keeping these values separated it is possible to configure the cell to be more sensitive to the cells in a specific direction (e.g. left or right).

Finally, there is a check-box to switch between 1 and 9 lights groups per cell.

6   Future Development

The paper introduced an ambient intelligence scenario aimed at improving the everyday experience of pedestrians and people passing through the related environment. A specific scenario related to the definition and development of an adaptive illumination facility was introduced, and a CA-based model supporting the specified behaviour for the illumination facility was defined. A prototype of a system supporting designers in the definition of the relevant parameters for this model and for the overall illumination facility was also introduced.

The realized prototype explored the possibility of realizing an ad hoc tool that can integrate the traditional CAD systems for supporting designers in simulating and envisioning the dynamic behaviour of complex, self-organizing installations. It has been used to understand the adequacy of the modeling approach in reproducing the desired self-organized adaptive behaviour of the environment to the presence of pedestrians. We are currently improving the prototype to realize a more general framework for supporting designers of dynamic self-organizing environments.

References

1. Shadbolt, N.: Ambient Intelligence. IEEE Intelligent Systems 18(4), 2–3 (2003)
2. Bandini, S., Erbacci, G., Mauri, G.: Implementing cellular automata based models on parallel architectures: The capp project. In: Malyshkin, V.E. (ed.) PaCT 1999. LNCS, vol. 1662, pp. 167–179. Springer, Heidelberg (1999)

3. Bandini, S., Mauri, G., Pavesi, G., Simone, C.: Parallel simulation of reaction-diffusion phenomena in percolation processes: A model based on cellular automata. Future Generation Comp. Syst. 17(6), 679–688 (2001)
4. Bandini, S., Mauri, G.: Multilayered cellular automata. Theor. Comput. Sci. 217(1), 99–113 (1999)
5. von Neumann, J.: Theory of Self-Reproducting Automata. University of Illinois Press, Urbana and London (1966)
6. Weimar, J.R.: Simulation with Cellular Automata. Logos Verlag Berlin (1997) ISBN 3-89722-026-1
7. Rosin, P.L.: Training cellular automata for image processing. IEEE Transactions on Image Processing 15(7), 2076–2087 (2006)
8. Behring, C., Bracho, M., Castro, M., Moreno, J.A.: An algorithm for robot path planning with cellular automata. In: Bandini, S., Worsch, T. (eds.) ACRI, pp. 11–19. Springer, Heidelberg (2000)
9. Hillis, W.D.: The Connection Machine. MIT Press, Cambridge (1985)
10. Margolus, N., Toffoli, T.: Cellular Automata Machines. A new environment for modelling. MIT Press, Cambridge (1987)
11. Paolo, E.A.D.: Searching for rhythms in asynchronous random boolean networks. In: Bedau, M. (ed.) Alife VII: Proceedings of the Seventh International Conference, pp. 73–80. MIT Press, Cambridge (2000)
12. Thomas, R., Organization, E.M.B.: Kinetic logic: a Boolean approach to the analysis of complex regulatory systems. In: Thomas, R. (ed.) Proceedings of the EMBO course "Formal analysis of genetic regulation," held in Brussels, September 6-16, 1977. Springer, Berlin (1979)
13. Nehaniv, C.L.: Evolution in asynchronous cellular automata. In: ICAL 2003: Proceedings of the eighth international conference on Artificial life, pp. 65–73. MIT Press, Cambridge (2003)
14. Cornforth, D., Green, D.G., Newth, D.: Ordered asynchronous processes in multi-agent systems. Physica D: Nonlinear Phenomena 204(1-2), 70–82 (2005)
15. Zambonelli, F., Mamei, M., Roli, A.: What can cellular automata tell us about the behavior of large multi-agent systems. In: Garcia, A.F., de Lucena, C.J.P., Zambonelli, F., Omicini, A., Castro, J. (eds.) Software Engineering for Large-Scale Multi-Agent Systems. LNCS, vol. 2603, pp. 216–231. Springer, Heidelberg (2003)
16. Buvel, R.L., Ingerson, T.E.: Structure in asynchronous cellular automata. Physica D 1, 59–68 (1984)
17. Lumer, E.D., Nicolis, G.: Synchronous versus asynchronous dynamics in spatially distributed systems. Phys. D 71(4), 440–452 (1994)
18. Alonso-Sanz, R.: The beehive cellular automaton with memory. Journal of Cellular Automata 1(3), 195–211 (2006)
19. Gutowitz, H.: Cellular Automata: Theory and Experiment. MIT Press/Bradford Books, Cambridge Mass. (1991), ISBN 0-262-57086-6

Relational Temporal Data Mining for Wireless Sensor Networks

Teresa M.A. Basile, Nicola Di Mauro, Stefano Ferilli, and Floriana Esposito

Università degli Studi di Bari, Dipartimento di Informatica, 70125 Bari, Italy
{basile,ndm,ferilli,esposito}@di.uniba.it

Abstract. Wireless sensor networks (WSNs) represent a typical domain where there are complex temporal sequences of events. In this paper we propose a relational framework to model and analyse the data observed by sensor nodes of a wireless sensor network. In particular, we extend a general purpose relational sequence mining algorithm to take into account temporal interval-based relations. Real-valued time series are discretized into similar subsequences and described by using a relational language. Preliminary experimental results prove the applicability of the relational learning framework to complex real world temporal data.

Keywords: Wireless Sensor Networks, Relational Sequence Mining, Temporal Patterns.

1 Introduction

Wireless sensor networks (WSNs) represent a recent technology able to monitor the physical world such as health, micro-climate and habitat, or earthquake and building health [1,2,3,4]. A WSN represents a typical domain where there are complex temporal sequences of events, such as computer security and planning. In this paper we aim at applying a relational framework to model and analyse the data observed by nodes involved in a sensor network. The main contribution of this work regards the proposal of a powerful and expressive description language able to represents the spatio-temporal relations appearing in a sensor network, and a general purpose system able to elicit hidden frequent temporal correlations between sensor nodes. In particular the objective of this paper is twofold: exploiting a relational language to describe the temporal evolution of a sensor network, and using relational learning techniques to discover interesting and more human readable patterns relating spatio-temporal correlations. Furthermore, we can combine temporal and content-based relations into a heterogeneous language providing a general framework applicable to other domains characterized by temporal and content relational features. Indeed, the data generated by sensor nodes involved in a sensor network are type-related (the humidity depends on the temperature), time-related (the temperature may change over time) and spatio-related. All these relations could be easily represented by means of a relational language such the one proposed in this paper, trying to

R. Serra and R. Cucchiara (Eds.): AI*IA 2009, LNAI 5883, pp. 416–425, 2009.

shift the basic time-series description language to a higher one. Algorithms proposed for sequential pattern mining generally consider events occurring in a time instant, while in some applications, like in sensor networks, events may occur in a time interval. The paper proposes an efficient algorithm able to mine frequent relational patterns representing arrangements of related intervals.

2   Background: Relational Pattern Mining

The algorithm we present in this paper is based on the work described in [5] where the authors proposed a framework for mining complex patterns, expressed in first-order language, in which events may occur along different dimensions. Specifically, multi-dimensional patterns were defined as a set of atomic first-order formulae in which events are explicitly represented by a variable and the relations between events were represented by a set of dimensional predicates. Here, that framework has been extended in order to take into account interval-based temporal data. We used Datalog [6] as representation language for the domain knowledge and patterns. Sequences and patterns are represented by a set of logical atoms. An atom $p(t_1, \ldots, t_n)$ is a predicate symbol p of arity n applied to n terms t_i (constants or variables).

Definition 1 (Subsumption). *A set of logical atoms c_1 θ-subsumes a set of logical atoms c_2 if and only if there exists a substitution θ such that $c_1\theta \subseteq c_2$. A substitution θ is defined as a set of bindings $\{X_1 \leftarrow a_1, \ldots, X_n \leftarrow a_n\}$ where $X_i, 1 \leq i \leq n$ is a variable and $a_i, 1 \leq i \leq n$ is a term. A substitution θ is applicable to an expression e, obtaining the expression $e\theta$, by replacing all variables X_i with their corresponding terms a_i.*

Definition 2 (Relational sequence). *A relational sequence may be defined as an ordered list of atoms separated by the operator $<: l_1 < l_2 < \cdots < l_n$.*

In order to make the framework more general, the concept of *fluents* has been considered. Let a sequence an ordered succession of events, a fluent is used to indicate that an atom holds for a given event. In this way we are able to distinguish *dimensional* and *non-dimensional* atoms. Specifically, the first ones refer to the dimensional relations between events involved in the sequence while the non-dimensional atoms introduce an event and the objects involved in it (fluent atoms) or the proprieties and the relations of the objects introduced by an event (non-fluent atoms). A further generalization consisted in introducing some dimensional operators able to represent general event relationships in the multi-dimensional relational patterns description: a) $<_i$, *next step on dimension* i; b) $\lhd_i$, *after some steps on dimension* i; and c) $\bigcirc_i^n$, *exactly after n steps on dimension* i.

Definition 3 (Subsequence [7]). *Given a sequence $\sigma = (e_1 e_2 \cdots e_m)$ of m elements, a sequence $\sigma' = (e'_1 e'_2 \cdots e'_k)$ of length k ($1 \leq k \leq m$) is a subsequence (or pattern) of the sequence σ if i) $\forall i, 1 \leq i \leq k, \exists j, 1 \leq j \leq m : e'_i = e_j$; ii) $\forall i, j, 1 \leq i < j \leq k, \exists h, l, 1 \leq h < l \leq m : e'_i = e_h$ and $e'_j = e_l$.*

The frequency of a subsequence in a sequence is the number of different mappings from elements of σ' into the elements of σ such that the previous conditions hold.

Note that this is a general definition of subsequence, in our case the *gaps* represented by the ii) condition are modelled by the $\triangleleft_i$ and $\bigcirc_i^n$ operators as reported in the following definition.

Definition 4. *[Multi-dimensional relational pattern] A multi-dimensional relational pattern is a set of atoms, involving k events and regarding n dimensions, in which there are non-dimensional atoms and each event may be related to another event by means of the operators $<_i$, $\triangleleft_i$ and $\bigcirc_i^n$, $1 \leq i \leq n$.*

We are interested in mining maximal frequent patterns.

Definition 5 (Maximal pattern). *A pattern σ' of a sequence σ is maximal if there is no pattern σ'' of σ more frequent than σ' and such that σ' is a subsequence of σ''.*

In order to calculate the frequency of a pattern over a sequence it is important to define the concept of sequence subsumption.

Definition 6 (Pattern Subsumption). *Given P a multi-dimensional relational pattern and S a multi-dimensional relational sequence. P subsumes S, written as $P \subseteq S$, iff there exists an SLD_{OI}-deduction[1] of P from S.*

Indeed, as the multi-dimensional relational sequences and patterns are represented by means of a set of logical atoms and dimensional operators, that can be expressed as a set of logical atoms, the frequency of a pattern P over a sequence S can be calculated as the number of substitutions θ_i such that P subsumes S.

2.1   The Algorithm

The algorithm for frequent multi-dimensional relational pattern mining is based on the same idea of the generic level-wise search method, known in data mining from the APRIORI algorithm [8]. The generation of the frequent patterns is based on a top-down approach. Specifically, it starts with the most general patterns of length 1 generated by adding to the empty pattern a non-dimensional atom. Then, at each step it *specializes* all the frequent patterns, discarding the non-frequent patterns and storing the ones whose length is lesser than the parameter *maxsize*. Furthermore, for each new refined pattern, semantically equivalent patterns are detected, by using the θ_{OI}-subsumption relation, and discarded.

In the specialization phase, the refinement of patterns is obtained by using a refinement operator ρ that maps each pattern to a set of specializations of the pattern, i.e. $\rho(p) \subset \{p' | p \preceq p'\}$ where $p \preceq p'$ means that p subsumes p'.

The algorithm uses a background knowledge $\mathcal{B}$ (a set of Datalog clauses) containing the sequence and a set of constraints that must be satisfied by the generated patterns. In particular $\mathcal{B}$ contains the following predicates:

---

[1] An SLD$_{OI}$-deduction is an SLD-deduction under Object Identity. In the Object Identity framework, within a clause, terms that are denoted with different symbols must be distinct, i.e. they must represent different objects of the domain.

- *maxsize(M)*: the maximal pattern length (i.e., the maximum number of non-dimensional predicates that may appear in the pattern);
- *minfreq(m)*: the lower bound of pattern frequency;
- *dimension(next_i)*: the dimension involved in the sequence. Each of such atoms denotes a different dimension. In particular, the number of these atoms represents the number of the dimensions described in the sequence.

Constraints. Furthermore the background knowledge contains some constraints useful to avoid the generation of unwanted patterns and, hence, to prune the solution space. Specifically, they are:

- *negconstraint([$p_1, p_2, \ldots, p_n$])*: specifies a constraint that the patterns must not fulfill, i.e. if the clause $\{p_1, p_2, \ldots, p_n\}$ subsumes the pattern then it must be discarded.
- *posconstraint([$p_1, p_2, \ldots, p_n$])*: specifies a constraint that the patterns must fulfill by discarding all the patterns that are not subsumed by the clause $\{p_1, p_2, \ldots, p_n\}$;
- *atmostone([$p_1, p_2, \ldots, p_n$])*: this constraint discards all the patterns that make true more than one predicate among $p_1, p_2, \ldots, p_n$.

Efficiency Issues. In order to avoid the generation of patterns containing not linked variables (that would represent objects and events as isolated points in the sequence) we used the classical types and modes declaration. They specify a language bias that indicates the predicates to use in the patterns and formalizes constraints on the binding of the variables. The efficiency of the algorithm is improved as patterns containing unrelated atoms are not generated.

- *type(p)*: denotes the type of the predicate's arguments p;
- *mode(p)*: denotes the input output mode of the predicate's arguments p.

Finally, the background knowledge contains the predicate $key([p_1, p_2, \ldots, p_n])$ specifying the set of predicates $\{p_1, p_2, \ldots p_n\}$ as predefined keys. Each pattern must have as first literal one of these predicates.

Hence, since a) the first literal of each pattern must be a non-dimensional predicate (specifically a fluent atom introducing an event and the objects involved in it), or a predefined key, and b) the pattern frequency must be less than the sequence length, the frequency of a pattern can be defined as follows.

Definition 7 (Pattern Frequency and Support). *Given a relational pattern* $P = (p_1, p_2, \ldots, p_n)$ *and S a relational sequence, the frequency of the pattern P, denoted as f_P, is equal to the number of different ground literals used in all the possible SLD_{OI}-deductions of P from S that make true the first literal of P, i.e. p_1. The support of P on S is $s_P = \frac{f_{p_1}}{f_P}$, i.e. is equal to the frequency of the pattern $\{p_1\}$ over the frequency of the pattern P.*

In case of more than one sequence, the support is calculated as the number of covered sequences over the total number of sequences.

In order to improve the efficiency of the algorithm, for each pattern $P = (p_1, p_2, \ldots, p_n)$ the set Θ of the substitutions defined over the variables in p_1 that make true the pattern P are recorded. In this way, the support of a specialization P' of P is computed by first applying a $\theta \in \Theta$ to P'. It is like to remember all the keys of a table that make true a query.

3   Relational Interval-Based Temporal Sequences

In this section we present the extension of the framework to the case of interval-based sequences. Specifically, the extension will concern the segmentation and labelling of the real-valued time series into similar subsequences and the successive integration of such a new knowledge representation, describing interval-based sequences, along with the relative interval-based operators, able to deal with it, in the relation pattern mining framework presented in Section 2.

In a sensor network made up of n nodes, each node i is located in the environment at the position p_i and senses a set of properties $\mathcal{P}_i$ at every time instance t. In other words, each sensor produces a continuous time series describing its reading over time, hence we have an observation at every instant of time.

The high values variability, even in short time intervals, showed by the continuous time series makes quite difficult to model its general trend. Thus a segmentation step may be performed in order to reduce this complexity and to make the values more homogeneous. A strategy to segment this signal could be by looking for a sequence of measurements over which a property holds, such as below a given threshold, and by labelling it. A method to segment a sequence is to iteratively merge two similar segments based on the squared error minimization criteria. Another approach is using clustering, by firstly finding the set of subsequences with length w, by sliding a window of width w, and then clustering the set of all subsequences. A different symbol is associated with each cluster. Other approaches use self-organizing maps.

Here, we concentrate on the abstraction process that translates the initial sequence (with real-valued elements) to a discretized sequence made up of symbols taken from a given alphabet. In particular, our segmentation method is a supervised process that assigns labels to a portion of a time series by using a set of predefined attributes. Future extensions of proposed approach include the use of more powerful techniques to partition time series, like that proposed in [9].

Specifically, given a real-valued time series $(t_i, x_i)_{1 \leq i \leq n}$, $x_i \in \mathbb{R}$, the goal is to transform it into a discrete time series $(t_i, c_i)_{1 \leq i \leq n}$, $c_i \in \{1, \ldots, \mathcal{C}\}$ ($\mathcal{C}$ denoting the set of descriptive labels, such as "temperature is high") by exploiting some defined abstraction rules.

An abstraction rule on a time series $a = (t_i, x_i)_{1 \leq i \leq n}$, denoted by $(t, x)_{1\_n}$, is defined as a function $\phi_a((t, x)_{1\_n})$ returning a set of m intervals for the time series a. In particular,

$$\phi_a((t, x)_{1\_n}) = \{\delta_a(l, t_i, t_{i+h}, c_k) | t_j \in \mathcal{D}_a^k, i \leq j \leq i + h \wedge c_k \in \mathcal{C}\}_{1 \leq l \leq m}$$

where $\delta(l, t_i, t_{i+h}, c_k)$ denotes the l-nth interval of the time series a starting from t_i and ending to t_{i+h}, and $\mathcal{D}_a^k$ represents the domain of values for the function ϕ_a

associated to the label $c_k \in \mathcal{C}$. The definition of such functions is crucial for the effectiveness of the process as different functions could lead to different results.

For instance, for the temperature time series in the wireless sensor network domain, an analysis on data distribution led to define the abstraction function $\phi_t((t,x)_{1\_n}) = \{\delta_t(l, t_i, t_{i+h}, c_k)|t_j \in \mathcal{D}_t^k, c_k \in \mathcal{C}_t\}$ where

$$\mathcal{D}_t^{vl} = \{x|x < 13\}, \qquad \mathcal{D}_t^l = \{x|13 \le x < 22\}, \mathcal{D}_t^m = \{x|22 \le x < 31\}$$
$$\mathcal{D}_t^h = \{x|31 \le x < 40\}, \mathcal{D}_t^{vh} = \{x|x \ge 40\}$$

and $\mathcal{C}_t = \{$ very_low, low, medium, high, very_high $\}$.

Defined the discretization process of the time series into intervals, we can extend the definitions of sequence and pattern to the case of interval-based relational sequences and introduce the interval-based operators to be exploited.

Definition 8 (Relational Interval-based Sequence). *Given a set $\mathcal{T}$ of time series and the sets $\mathcal{C}_1, \ldots, \mathcal{C}_{|\mathcal{T}|}$ of descriptive labels, a relational interval sequence is a sequence of relational atoms*
$$\delta_{a_1}(id_1, b_1, e_1, v_1), \delta_{a_2}(id_2, b_2, e_2, v_2), \ldots, \delta_{a_n}(id_n, b_n, e_n, v_n)$$
where $v_j \in \mathcal{C}_i$ is a descriptive label, b_j and e_j represent, respectively, the starting and ending time, $id_j \in \mathbb{N}$ represents the interval identifier, and δ_{a_j} is the corresponding name of the time series $a_j \in \mathcal{T}$. (The interval $\delta(id, b, e, v)$ can be written also by means of three literals as $\delta(id, v)$, begin(id,b), end(id,e)).

In particular a relational interval sequence could describe several labeled interval sequences into a single one, enabling one to take into account the multivariate analysis in case of different time series. Thus interval-based operators able to deal with relations between time intervals are introduced by exploiting the Allen's temporal interval logic [10], as reported in Figure 1.

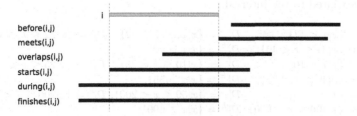

Fig. 1. Allen's temporal intervals [10]

Finally, a pattern on the interval-based sequence is:

Definition 9 (Relational Interval-based Pattern). *Given $\mathcal{S}$, the set of interval relation symbols, a relational temporal pattern is a set of relational atoms*

$$P = I \cup R = \{\delta_i(id_i, b_i, e_i, v_i)\}_{i=1\ldots n} \cup \{rel_j(id_j^1, id_j^2)\}_{j=1\ldots m}$$

where $rel_j \in \mathcal{S}$, and $\forall rel_j(id_j^1, id_j^2) \in R \; \exists \delta_h(id_h, b_h, e_h, v_h), \delta_k(id_k, b_k, e_k, v_k) \in I$ such that $id_j^1 = id_h$ and $id_j^2 = id_k$.

4 Experiments

The approach was evaluated on data, freely available from [11], collected from a wireless sensor network made up of 54 Mica2Dot sensors deployed in the Intel Berkeley Research Lab and arranged in the laboratory as shown in Figure 2.

A sensor network node is a small autonomous unit, often running on batteries, with hardware to sense environmental characteristics. Such nodes usually communicate using a wireless network. A sensor network is composed of a large number of sensors deployed in a natural environment. The sensors gather environmental data and transfer the information to the central base station with external power supply. The 54 sensors have been monitored from February 28th to April 5th 2004, and the data, about 2.3 million readings, were collected using the TinyDB in-network query processing system, built on the TinyOS platform. Each sensor collected topology information, along with humidity, temperature, light and voltage values once every 31 seconds.

We selected the measurements (temperature, humidity, light and voltage) from the sensors 41, 42, and 24, for the time period from 2004-03-18 to 2004-03-21 corresponding to 23178 log rows. The aim is to discover some correlations between sensors and/or measurements useful for anomaly detection. For instance, there is a strong correlation between the temperature and humidity, as we can see from Figure 3 that reports the corresponding graphs for the sensor 41. The first task is to discretize the time series corresponding to each information in order to obtain an interval-based temporal sequence like that reported in Figure 4 where each interval is labeled with a specific name. The discretization step has been carried out exploiting the functions ϕ_t, ϕ_h, ϕ_l, and ϕ_v with the corresponding domains $\mathcal{D}_i^j$ where i is the time series name (temperature, humidity, light and voltage) and j is the descriptive label (very low, low, medium, high and very high) associated to the interval:

$$\mathcal{D}_t^{vl} = \{x | x < 13\}, \qquad \mathcal{D}_t^l = \{x | 13 \leq x < 22\}, \quad \mathcal{D}_t^m = \{x | 22 \leq x < 31\},$$
$$\mathcal{D}_t^h = \{x | 31 \leq x < 40\}, \quad \mathcal{D}_t^{vh} = \{x | x \geq 40\},$$
$$\mathcal{D}_h^{vl} = \{x | x < 10\}, \qquad \mathcal{D}_h^l = \{x | 10 \leq x < 25\}, \quad \mathcal{D}_h^m = \{x | 25 \leq x < 40\},$$
$$\mathcal{D}_h^h = \{x | 40 \leq x < 55\}, \quad \mathcal{D}_h^{vh} = \{x | x \geq 55\},$$
$$\mathcal{D}_l^{vl} = \{x | x < 50\}, \qquad \mathcal{D}_l^l = \{x | 50 \leq x < 200\}, \mathcal{D}_l^m = \{x | 200 \leq x < 400\},$$
$$\mathcal{D}_l^h = \{x | 400 \leq x < 600\}, \mathcal{D}_l^{vh} = \{x | x \geq 600\},$$
$$\mathcal{D}_v^l = \{x | x < 2\}, \qquad \mathcal{D}_v^m = \{x | 2 \leq x < 2.4\}, \quad \mathcal{D}_v^h = \{x | 2.4 \leq x < 2.75\},$$
$$\mathcal{D}_v^{vh} = \{x | x \geq 2.75\}.$$

These intervals were built by considering the distribution of data on the time series. Obviously, different abstraction functions could lead to different results.

Adopting these functions we obtained a temporal sequence made up of 816 intervals (81 for temperature, 94 for humidity, 255 for light and 386 for voltage). Then we added all the Allen's temporal relations between the intervals (332402 before, 1052 meets, 8872 overlaps, 163 starts, 7438 during, 131 finishes and 42 matches atoms) obtaining a relational sequence of about 350000 literals. The following literals represent a fragment of a sequence describing the relational representation of some time series, where each interval is described by three

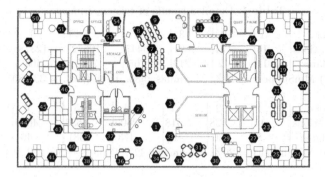

Fig. 2. Sensors in the Intel Berkeley Research lab

Fig. 3. Correlation between temperature (bottom) and humidity (top) time series

Fig. 4. Intervals of the temperature (bottom) and humidity (top) time series

predicates α*(sensor, interval, label), begin(interval, s), end(interval, e)* where $\alpha \in$ {temperature, humidity, light, voltage}.

```
near(41,42). far(41,24). far(42,24).
temperature(24,i1,high). begin(i1,0). end(i1,20).
light(41,i30,very_low). begin(i30,2). end(i30,16). ...
starts(i1,i2), before(i12,i34), ...
```

Table 1 reports the results of the algorithm reported in Section 2.1 when applied on the sequence previously described and using two different values for the minimum support (Table 1(a) and Table 1(b)). The third rows report the number of patterns belonging to all the possible specializations whose support is greater than MinSupport. The number of maximal patterns fulfilling all the constraints provided to the algorithm and the execution time are reported in sub-tables caption. Some interval-based patterns discovered by the algorithm and expressing the time correlation and the information correlation are:

temperature(A,B,low), before(B,D), temperature(A,D,medium) [*s* = 32.1%],
temperature(A,B,low), meets(B,D), humidity(A,D,high) [*s* = 13%],
temperature(A,B,medium),overlaps(B,D),humidity(24,D,high),far(24,A) [*s* = 24%].

Table 1. Detailed results of two experiments

Level	1	2	3	4	5	6	7
Specializations	16	174	691	3447	10672	12378	3992
Candidates	11	42	213	765	1060	408	0

(a) MinSupport = 10%, Maximals=585, Time (secs.)= 337.805.

Level	1	2	3	4	5	6	7
Specializations	16	128	518	2609	6636	4639	452
Candidates	8	30	160	473	381	46	0

(b) MinSupport = 20%, Maximals=327, Time (secs.)= 222.313.

5   Conclusion

Previous works on mining temporal patterns for interval-based sequences, as opposed to point-based events, include [12,13,14]. However, they do not use a logical description language. Other works in the research area of spatial data mining try to take into account complex representations with a logical language, however without considering temporal-based relations [15,16].

The work presented in this paper can be related to that proposed in [17,18] and [19], where the authors represent a single sequence as a set of predicates and temporal relations. Each predicate is assumed to be hold in a given temporal interval, while the temporal relations are predicates expressing the Allen's temporal correlation between two predicates. Furthermore, each predicate is associated to an unique symbolic identifier indicating a defined temporal interval and the temporal relations are expressed between such identifiers. Hence, their framework is purposely designed for the application domain as they cannot express any other structural relation between objects. Moreover, as reported in [17], the algorithm they presented is not applicable to real world problems due to its high complexity. Indeed, they specialize a pattern by adding a literal, or by variable unification, or by introducing k^n (where k is the number of different Allen's relations and n corresponds to the number of possible predicate pairs) temporal restrictions between predicate pairs leading to an exponential time complexity.

On the contrary, the framework we proposed is based on a powerful and general purpose multi-dimensional relational pattern mining system [5] and its applicability to a real world wireless sensor network data was showed. Specifically, it is an extension of the framework presented in [5] with new dimensional operators and preliminary experimental results prove that the framework can be applied to real world domains such as wireless sensor networks. Furthermore, It can be used to solve complex temporal data mining tasks by using a relational interval-based description.

Acknowledgment

This work is partially supported by the Italian project PS121 "Infrastrutture di Telecomunicazione e Reti Wireless di Sensori nella Gestione di Situazioni di Emergenza".

References

1. Estrin, D., Culler, D., Pister, K., Sukhatme, G.: Connecting the physical world with pervasive networks. IEEE Pervasive Computing 1(1), 59–69 (2002)
2. Akyildiz, I.F., Su, W., Sankarasubramanian, Y., Cayirci, E.: A survey on sensor networks. IEEE Communication Magazine 40(8), 102–114 (2002)
3. Akyildiz, I.F., Su, W., Sankarasubramaniam, Y., Cayirci, E.: Wireless sensor networks: a survey. Computer Networks 38, 393–422 (2002)
4. Mainwaring, A., Culler, D., Polastre, J., Szewczyk, R., Anderson, J.: Wireless sensor networks for habitat monitoring. In: Proceedings of the 1st International Workshop on Wireless sensor networks and applications, pp. 88–97. ACM, New York (2002)
5. Esposito, F., Di Mauro, N., Basile, T., Ferilli, S.: Multi-dimensional relational sequence mining. Fundamenta Informaticae 89(1), 23–43 (2008)
6. Ullman, J.D.: Principles of Database and Knowledge-Base Systems, vol. I. Computer Science Press (1988)
7. Jacobs, N., Blockeel, H.: From shell logs to shell scripts. In: Rouveirol, C., Sebag, M. (eds.) ILP 2001. LNCS (LNAI), vol. 2157, pp. 80–90. Springer, Heidelberg (2001)
8. Agrawal, R., Manilla, H., Srikant, R., Toivonen, H., Verkamo, A.: Fast discovery of association rules. In: Fayyad, U., Piatetsky-Shapiro, G., Smyth, P., Uthurusamy, R. (eds.) Advances in Knowledge Discovery and Data Mining, pp. 307–328. AAAI Press, Menlo Park (1996)
9. Hoppner, F.: Knowledge Discovery from Sequential Data. PhD thesis (2003)
10. Allen, J.: Maintaining knowledge about temporal intervals. Commun. ACM 26(11), 832–843 (1983)
11. Intel Berkeley Research Lab (2004),
    http://db.csail.mit.edu/labdata/labdata.html
12. Kam, P., Fu, A.W.: Discovering temporal patterns for interval-based events. In: Kambayashi, Y., Mohania, M., Tjoa, A.M. (eds.) DaWaK 2000. LNCS, vol. 1874, pp. 317–326. Springer, Heidelberg (2000)
13. Hoppner, F.: Learning dependencies in multivariate time series. In: Proc. of the ECAI Workshop on Knowledge Discovery in (Spatio-)Temporal Data, pp. 25–31 (2002)
14. Laxman, S., Unnikrishnan, K.P., Sastry, P.S.: Generalized frequent episodes in event sequences. In: 8th ACM SIGKDD Int. Conf. on Knowledge Discovery and Data Mining, Workshop on Temporal Data Mining (2002)
15. Koperski, K., Han, J.: Discovery of spatial association rules in geographic information databases. In: Egenhofer, M.J., Herring, J.R. (eds.) SSD 1995. LNCS, vol. 951, pp. 47–66. Springer, Heidelberg (1995)
16. Malerba, D., Lisi, F.: An ILP method for spatial association rule mining. In: Working notes of the First Workshop on Multi-Relational Data Mining, pp. 18–29 (2001)
17. Lattner, A.D., Herzog, O.: Unsupervised learning of sequential patterns. In: ICDM Workshop on Temporal Data Mining: Algorithms, Theory and Applications (2004)
18. Lattner, A.D., Herzog, O.: Mining temporal patterns from relational data. In: Lernen Wissensentdeckung Adaptivität (LWA), GI Workshops, pp. 184–189 (2005)
19. Papapetrou, P., Kollios, G., Sclaroff, S., Gunopulos, D.: Discovering frequent arrangements of temporal intervals. IEEE ICDM, 354–361 (2005)

Ontology-Driven Co-clustering of Gene Expression Data

Francesca Cordero[1,2,3], Ruggero G. Pensa[2], Alessia Visconti[2,3],
Dino Ienco[2,3], and Marco Botta[2,3]

[1] Department of Clinical and Biological Sciences, University of Torino
[2] Department of Computer Science, University of Torino
[3] Center for Complex Systems in Molecular Biology and Medicine - SysBioM,
University of Torino
{fcordero,pensa,visconti,ienco,botta}@di.unito.it

Abstract. The huge volume of gene expression data produced by microarrays and other high-throughput techniques has encouraged the development of new computational techniques to evaluate the data and to formulate new biological hypotheses. To this purpose, co-clustering techniques are widely used: these identify groups of genes that show similar activity patterns under a specific subset of the experimental conditions by measuring the similarity in expression within these groups. However, in many applications, distance metrics based only on expression levels fail in capturing biologically meaningful clusters.

We propose a methodology in which a standard expression-based co-clustering algorithm is enhanced by sets of constraints which take into account the similarity/dissimilarity (inferred by the Gene Ontology, GO) between pairs of genes. Our approach minimizes the intervention of the analyst within the co-clustering process. It provides meaningful co-clusters whose discovery and interpretation is increased by embedding GO annotations.

1   Introduction

Microarrays, and other high-throughput techniques, measure the expression level of thousands of genes in different samples captured at different time points or in different experimental conditions. The volume of data produced by these techniques is huge and grows up day by day. This requires the development of new computational techniques to store and evaluate the data and to formulate new biological hypotheses.

To this purpose, clustering techniques are widely used in microarrays data analysis that enable to discover homogeneous experimental clusters or genes clusters based on a distance measure quantifying the degree of correlation of expression profiles [1]. A limitation of traditional clustering techniques is that they are applied on gene sets or sample sets independently. To exceed this view, *co-clustering* algorithms [2] have been proposed: these identify groups of genes that show similar activity patterns under a specific subset of the experimental conditions.

R. Serra and R. Cucchiara (Eds.): AI*IA 2009, LNAI 5883, pp. 426–435, 2009.

The goal of co-clustering algorithms emphasizes one of the major target in computational biology: the discovery of regulatory modules that control gene transcription in biological model systems. Approaches based on co-clustering simultaneously cluster genes and conditions, and enable the discovery of more coherent and meaningful groups. The main practical reasons are that biological systems are inherently modular and that grouping genes into modules reduces the effective complexity of a given data set. For instance, the association of these modules with a specific histological cancer class may be exploited within an effective diagnostic tool.

In many applications, distance metrics based only on expression levels fail in capturing biologically meaningful clusters. Moreover, approaches based on clustering that identify gene signatures in specific conditions tend to base the analysis on their signal in the conditions under study. However, a simple list of genes associated with a certain tumor type is far from identifying the regulatory modules in which genes are involved. Several works proposed to define distance metrics based on different sources of information. As an advantage, additional information could help in resolving ambiguities or in avoiding erroneous linking based on spurious similarities.

The pioneer of this stream of works is Hanisch et al. [3]. They proposed a novel approach that allows for an entirely exploratory joint analysis of gene expression data and biological networks. The authors proposed a combined measure derived from gene expression data and metrics based on biological networks into a single distance function that they use as distance measure in a hierarchical average linkage clustering algorithm. Starting from Hanish's work, Steinhauser et al.[4] proposed a new measure that involves operon annotations, intergenic distance and transcriptional co-response data into a distance metric used in hierarchical clustering algorithms. More recently, Brameier et al. [5] presented a co-clustering approach based on self-organizing maps, where center-based clustering of standard SOMs have been combined with a representative-based clustering. The authors developed a two-level cluster selection where the nearest cluster according to GO distance is selected among the best matching clusters w.r.t. gene expression distance. In this work, co-clustering means that the GO-based clustering and expression-based clustering are performed in parallel. None of these methods perform co-clustering on both genes and samples at the same time.

Instead of combining ontology-based metrics and expression-based metrics within the same distance measure, we propose a methodology in which a standard expression-based co-clustering algorithm is enhanced by sets of constraints which take into account the similarity/dissimilarity (inferred by some background knowledge) between pairs of genes. Using constraints has been proved to be very effective in many applications, including gene expression analysis [6] and sequence analysis [7], since the user can decide which type of biological knowledge leads to the association among gene clusters and condition clusters. In this way the list of genes associated with a set of conditions may assume a specific meaning. Moreover, constraints can be generated by mixing different semantics, while combining different semantics in a single measure is not an easy task. Defining these constraints by hand is not that simple either. Since the advantage of modularity is crucial in learning biological meaningful clusters from data, we decided to use the expressive power provided by Gene Ontology [8] to construct a set of similarity (must-link) and dissimilarity (cannot-link) constraints automatically.

Furthermore, for a correct usage of the technique presented in [6], similarly to all co-clustering techniques, the user has to specify the desired number of clusters on rows and columns. Deciding an adequate number of clusters is not trivial, and a bad choice may influence negatively the quality of co-clustering results. Thus, we adopt a preprocessing method that automatically determines a congruent number of clusters per rows and columns.

In a nutshell, we propose a new methodology that minimizes the intervention of the analyst within the co-clustering process and that provides meaningful co-clusters whose discovery and interpretation is enhanced by embedding GO annotations. To show the effectiveness of our approach, we apply our methodology on a gene expression dataset consisting on different stress conditions on the *S. Cerevisiae* yeast.

2   Constrained Co-clustering

In this section we briefly describe the constrained co-clustering algorithm presented in [6], and which is central to our methodology.

Let $X \in \mathbb{R}^{m \times n}$ denote a data matrix. Let x_{ij} be the expression level corresponding to gene (row) i and condition (column) j.

A co-clustering $C^{k \times l}$ over X simultaneously produces a set of $k \times l$ co-clusters (a partition C^r into k groups of rows associated to a partition C^c into l groups of columns) which optimize a given objective function. In this work we use the Cheng and Church residue [9] as objective function. Given an element x_{ij} of X, the residue of x_{ij} in the co-cluster defined by the sets of indices I and J, and whose respective cardinalities are $|I|$ and $|J|$, is given by $h_{ij} = x_{ij} - x_{Ij} - x_{iJ} + x_{IJ}$, where $x_{IJ} = \frac{\sum_{i \in I, j \in J} x_{ij}}{|I| \cdot |J|}$, $x_{Ij} = \frac{\sum_{i \in I} x_{ij}}{|I|}$, $x_{iJ} = \frac{\sum_{j \in J} x_{ij}}{|J|}$.

Let $H = [h_{ij}] \in \mathbb{R}^{m \times n}$ denote the matrix of residues computed using the previous definition. The objective function to be minimized is the sum of squared residues [10] computed as follows:

$$||H||^2 = \sum_{I,J} ||h_{IJ}||^2 = \sum_{I,J} \sum_{i \in I, j \in J} h_{ij}^2 \qquad (1)$$

The kind of constraints we consider in this work are the two well-known **must-link** (similarity) and **cannot-link** (dissimilarity) constraints, also referred as pairwise constraints, since they involve pairs of objects.

If rows i_a and i_b (resp. columns j_a and j_b) are involved in a **must-link** constraint, denoted $c_=(i_a, i_b)$ (resp. $c_=(j_a, j_b)$), they must be in the same cluster of $C^r = r_1, \ldots, r_k$ (resp $C^c = c_1, \ldots, c_k$). If rows i_a, i_b (resp. columns j_a and j_b) are involved in a **cannot-link** constraint, denoted $c_{\neq}(i_a, i_b)$ (resp. $c_{\neq}(j_a, j_b)$), they cannot be in the same cluster of $C^r = r_1, \ldots, r_k$ (resp $C^c = c_1, \ldots, c_k$).

We can then transform a set of must-link constraints over rows into a collection $\mathcal{M}_r = M_1, \ldots, M_N$, where each M_i is a set of rows involved by the same transitive closure of must-link constraints. Let us denote $\mathcal{M}_c$ the same set built for columns and let $\mathcal{C}_r$ and $\mathcal{C}_c$ be the sets of cannot-link constraints for rows and columns respectively. The co-clustering algorithm builds a $k \times l$ co-clustering over X, trying to minimize the objective function (1), and satisfying constraints $\mathcal{M}_r$, $\mathcal{M}_c$, $\mathcal{C}_r$, and $\mathcal{C}_c$. We skip the algorithmic details of this approach (see [6] for the complete algorithm).

3   Methodology Overview

Our framework is motivated by the necessity of using the previously described co-clustering algorithm on gene expression data, by limiting the number of user-defined parameters. So far, the user has to provide the following parameters: (i) number of row clusters; (ii) number of column clusters; (iii) a set of pairwise constraints (optional); (iv) convergence criterion (optional).

Providing a correct number of clusters is crucial for every clustering algorithm, since a wrong number might considerably alter the quality of the results. Unfortunately this is not an easy task, and some heuristics should be used to determine a correct number of clusters.

Providing a coherent and useful set of constraints is also a hard task. In classic semi-supervised applications, constraints are automatically selected from labeled samples, by selecting random pairs of labeled objects and setting a must-link or a cannot-link constraint depending on their class label. We will show how to extend this setting to gene expression data analysis.

Figure 1 shows an overview of the crucial steps of our methodology. The first two steps are performed independently. From one side, microarray experiments (A) are preprocessed to build a gene expression matrix consisting of normalized expression values. Other preprocessing techniques, such as missing value

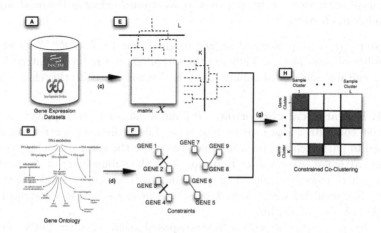

Fig. 1. Methodology overview

replacement, gene or sample filtering, are performed (c). The resulting matrix is then processed using the method described in [11], in order to determine a congruent number of row/column clusters (E). From the other side, the Gene Ontology graph (B) is processed in order to obtain nodes consisting of multiple regulated GO terms, linked by similarity relationships (d). The retained set of genes is mapped into the obtained GO graph to identify groups of similar and dissimilar genes, that contributes to the definition of must-link and cannot-link constraints (F). The central step is the constrained co-clustering algorithm (g) performed over the gene expression matrix obtained at step (c), using the number of row/column clusters discovered at step (E), and embedding the set of constraints built during step (F). The results (H) are a set of row clusters described by GO annotations, and a set of columns clusters described by experiment information.

4   Methodology Detailed Description

In this section we describe in full detail steps E and F of the methodology described beforehand.

4.1   Determining a Suitable Number of Co-clusters

To estimate parameters k and l (i.e., the number of row and column clusters) we adopt the method described in [11], namely *L-method*, which aims at selecting the number of clusters that provides the best result in a hierarchical clustering setting. It consists in the following steps:

1. Generation of a hierarchical clustering using a distance matrix over the set of objects to be clustered;
2. Analysis of the resulting dendrogram, and assessment of the goodness (homogeneity) of the obtained clusters for each dendrogram level;
3. Analysis of the previously performed evaluations to identify the most suitable number of clusters.

These steps are first performed on the original matrix $X \in \Re^{m \times n}$ to identify the number of row clusters. They are then applied on a reduced matrix X^R, to obtain a suitable number of column clusters (we will motivate this choice later).

Step 1. Hierarchical clustering. To build a cluster hierarchy over rows, we must first compute a distance matrix. The chosen distance metric is the one described in [9], which has been modified to enable the comparison between two rows. In particular, we consider Equation 1 for the submatrices of X consisting in each pair of rows (considered as singleton clusters), and n singleton column clusters. The resulting distance matrix is then processed using a standard hierarchical clustering algorithm.

For columns, we do not process the transposed matrix directly, but we first reduce its dimensions using PCA. This choice is motivated by the fact that in gene expression data analysis usually the number of conditions is much lesser than the number of genes. On the resulting matrix we apply a standard hierarchical clustering algorithm using Euclidean distance.

Step 2. Analysis of the dendrograms. Once the dendrogram has been computed, we analyze its levels. This analysis is performed in three steps:

1. starting from the bottom of the hierarchy, for each level we identify the two clusters that are joined at the next level;
2. we compute a representative for each of the two clusters, by averaging the features of the members belonging to each of them;
3. we compute a distance between the two representatives using the above mentioned metrics (Cheng and Church for rows, and Euclidean distance for columns).

This process results in pairs of values (number of clusters and distance) which identify a series of points.

Step 3. Determining the number of clusters. The crucial phase of the *L-method* presented in [11] consists in determining a suitable number of clusters. It is performed in four steps:

1. We consider the $n-1$ points (where n is the number of rows or columns) generated at the previous step. We choose a point c (which, in the first iteration, is equal to 2). This point divides the whole set of points in two subsets that, graphically represents two intervals: the left interval L_c, containing points $[1, c]$, the right interval R_c, containing points $[c + 1, n - 1]$.
2. For each of these subsets, we computed the line approximating them, using linear regression.
3. For each line, we compute the Root Mean Square Error (RMSE). We obtain a value for the line built on the left subset ($RMSE(L_c)$) and another one for the line built on the right subset ($RMSE(R_c)$).
4. Those values are then combined using the following formula [11]:

$$RMSE_c = \frac{c-1}{n-1} * RMSE(L_c) + \frac{n-c}{n-1} * RMSE(R_c)$$

We iterate these steps until $c = n - 1$. The estimated number of clusters is then given by $\hat{c} = \min_c RMSE_c$. Clearly, this step is performed for both rows and columns.

4.2 Definition of Constraints

To create sets of must-link and cannot-link constraints, we decided to use the information stored in Gene Ontology [8].

The Gene Ontology (GO) is a controlled vocabulary for the consistent description of attributes of genes and gene products maintained by the Gene Ontology

Consortium. The ontologies are in the form of direct acyclic graphs whose nodes represent GO terms and edges represent the relationships between them. The nodes can be associated by five types of relationships: *is_a*, *part_of*, *regulates*, *positively_regulates* and *negatively_regulates*. This ontology is organized in three key domains that are shared by all organisms: **molecular function, biological process** and **cellular component**. These domains are represented by separate disconnected sub-graphs of the root node.

The *is_a* relationship is a class-subclass relationship, where A is_a B means that A is a subclass of B. Instead, it is defined C *part_of* D if whenever C is present, it is always a part of D, but C does not always have to be present. In other words, a child class is either a *part_of* the parent class or *is_a* more specific variant. The *regulates*, *positively_regulates* and *negatively_regulates* relationships describe interactions between biological processes and other biological processes, molecular functions or biological qualities. When a biological process E regulates a function or a process F, it modulates the occurrence of F. If F is a biological quality, then E modulates the value of F. In this work, we do not consider the cellular component domain.

We reformat the regulative information contained in GO in order to obtain a more concise representation of regulative relationships between the GO classes.

We built a weighted graph, where each node is a set of GO term linked together by *regulative* relationships. If there is at least one *is_a* relationship among GO terms in two different nodes we put an edge among these two vertices. The resulting graph contains 1537 GO macro-nodes. The weight associated to each edge is given by the number of *is_a* relationships existing among two nodes. By extracting the cliques in that graph, we obtain 202 cliques that represent strongly connected *regulative modules*.

Then, each gene of X is mapped into GO cliques following its GO annotation. Clearly, genes that are involved in multiple biological process/molecular functions, are likely to belong to more than one clique. To construct the set of must-link constraints, we perform the following steps: first, since we perform hard (non overlapping) co-clustering, we do not consider genes that belong to more than one clique; then those genes (among the remaining ones) belonging to one clique are associated to a unique transitive closure of must-link constraints.

Finally, to provide a set of cannot-link constraints, we consider all pair of cliques associated to the transitive closures of must-link constraints generated before. If they do not share any gene, then we set a cannot-link constraint between an arbitrary pair of genes belonging to the associated transitive closures.

5 Application

To evaluate the performance of our method, we used gene expression experiments for the organism *S. Cerevisiae* (yeast). This data set consists in 5 different microarray experiments (GEO accession series: GSE1312, GSE5301, GSE4660, GSE2224, GSE1723), downloaded from Gene Expression Omnibus (GEO[1]). In

---

[1] http://www.ncbi.nlm.nih.gov/geo/

these experiments, yeast is treated with different types of stress. Gene expression levels are given as log (base 2) ratios of the measured level and a reference (control) level. All these experiments are hybridised on the same platform GPL90, Affymetrix Yeast Genome S98 Array. From GEO site we also extracted the gene ontology annotations files of each experiment.

5.1 Instantiation of the Methodology

Following the steps described in the work-flow reported in Figure 1, we describe in details the instantiation of our approach:

- *Construction of matrix X* From the gene expression dataset, we build a matrix X with 9335 rows and 29 columns.
- *Selection of a suitable number of row/column clusters* We process matrix X using the method described in Section 4.1. We obtained a suggested number of 1677 row clusters ($k = 1677$) and 9 column clusters ($l = 9$).
- *Generation of a collection of constraints* We generated a collection of must-link constraints $\mathcal{M}$ and a set of cannot-link constraints $\mathcal{C}$ as described in Section 4.2. A total number of 2151 genes were constrained in 52 transitive closures of must-link constraints, and 39 cannot-link constraints.
- *Constrained Co-Clustering* Using the previously discovered k and l parameters and the collections $\mathcal{M}$ and $\mathcal{C}$ of constraints, we performed 40 trials of the co-clustering algorithm. The co-clustering process stops when $||X||^2_{t-1} - ||X||^2_t < 10^{-5}$, where $||X||^2_t$ and $||X||^2_{t-1}$ are the values of the objective function at iteration t and iteration $t-1$.

5.2 Validation of the Results

To be able to assess the quality of the clustering results, we evaluated how accurate the partition over columns is w.r.t. the reference partition given by the GEO accession ID. To measure the accuracy we used the Normalized Mutual Information [12]. We denote by $\mathbf{C} = \{C_1 \ldots C_J\}$ the partition built by the clustering algorithm on objects, and by $\mathbf{P} = \{P_1 \ldots P_I\}$ the partition inferred by the original classification. J and I are respectively the number of clusters $|\mathbf{C}|$ and the number of classes $|\mathbf{P}|$. We denote by n the total number of objects. The Normalized Mutual Information (NMI) provides an information that is impartial with respect to the number of clusters [12]. It measures how clustering results share the information with the true class assignment. NMI is computed as the average mutual information between every pair of clusters and classes:

$$\mathbf{NMI} = \frac{\sum_{i=1}^{I} \sum_{j=1}^{J} p_{ij} \log \frac{n p_{ij}}{p_i p_j}}{\sqrt{\sum_{i=1}^{I} p_i \log \frac{p_i}{n} \sum_{j=1}^{J} p_j \log \frac{p_j}{n}}}$$

where p_{ij} is the cardinality of the set of objects that occur both in cluster C_j and in class P_i; p_j is the number of objects in cluster C_j; p_i is the number of objects in class P_i. Its values range between 0 and 1.

In our experiments, the column partitioning was quite stable: we achieved an average NMI of 0.8514 with a standard deviation of 0.0661. The best trial corresponds to a NMI value of 0.9199. We selected this trial for a detailed analysis on gene partition. The selected row partition contains 4 column clusters and 1310 row clusters. Among them, 4 clusters contain only genes involved by constraints, 1259 clusters contain only genes not involved by any constraints and 47 clusters contain a mix of constrained and unconstrained genes.

To assess the homogeneity of the discovered clusters, we use an *homogeneity* score, defined as the ratio of each involved GO term (for both biological process domain and molecular function domain) in each cluster. It takes values between 0 and 1, where a value of 1 means that all genes are involved in at least one common GO term. We consider that a cluster is consistent if the *homogeneity* of at least one domain is high, as a consequence we retained the maximum between the two domain values. The average score is 0.6105 (with a standard deviation of 0.3966), and in 70% of the cases it is greater than 0.50.

Low homogeneity clusters contain genes spread in multiple GO terms, but these terms might still be strongly connected each other. To verify this hypothesis, we performed a in-depth analysis on the 229 clusters which contain more than 5 and less than 50 genes. We found 4 clusters containing only genes involved by constraints, 191 clusters containing only genes not involved by any constraints and 34 clusters containing a mix of constrained and unconstrained genes. In most cases, the GO terms are involved in the same biological process: for instance in one cluster (containing 7 unconstrained genes; Gene IDs: UTP14, ERB1, RNT1, DBP6, RIX1, URB2, RPA12, UTP13), whose homogeneity value for the biological process domain is low, all GO terms are related to *ribosomal processing*. Moreover, another cluster (containing 6 unconstrained genes; Gene IDs: ALD6, DLD1, SNA2, MPM1, OPI3, MCR1 and 5 constrained genes; Gene IDs: SDH2, COQ10, SDH4, SDH3, SDH1) has a low homogeneity value and all its genes are related to *cellular respiration*.

Gene expression regulation is controlled by a complex network of interactions involving DNA cis-regulatory elements and transcription factors (TF). TFs control, by promoting or blocking, the transcription of genetic information from DNA to mRNA. Therefore, as we built (using our approach in Section 4.2) constraint sets from the GO regulation relationship, we expect the obtained clusters to contain genes belonging to specific transcriptional modules. Since in literature a strong dependency between the transcriptional units and the experimental conditions is proved, we checked how many clusters contain at least one TF. From the YTF website[2], we extracted a list of yeast TFs, and we found that 101 of the 229 analyzed clusters contain at least one TF.

6   Conclusion

In this paper we presented an ontology-driven co-clustering approach for the identification of gene clusters characterized by similar expression profiles and

---

[2] http://biochemie.web.med.uni-muenchen.de/YTFD/index.htm

involved in similar biological processes or functions. This leads to discovery more biological coherent and meaningful gene groups. Therefore, we provided a methodology to cluster genes following their expression signature in specific conditions with the help of constraints built over the Gene Ontology. The methodology is based on a constrained co-clustering algorithm, and automatically suggest a number of column/row clusters, as well as a congruent set of constraints.

Acknowledgments. Francesca Cordero and Ruggero G. Pensa are co-funded by Regione Piemonte.

References

1. Eisen, M., Spellman, P., Botstein, P.B.D.: Cluster analysis and display of genome-wide expression patterns. Proc. Natl. Acad. Sci. USA 95, 14863–14868 (1998)
2. Madeira, S., Oliveira, A.: Biclustering algorithms for biological data analysis: a survey. IEEE/ACM Trans Comput Biol Bioinform 1, 24–45 (2004)
3. Hanisch, D., Zien, A., Zimmer, R., Lengauer, T.: Co-clustering of biological networks and gene expression data. Bioinformatics 18, S145–S154 (2002)
4. Steinhauser, D., Junker, B., Luedemann, A., Selbig, J., Kopka, J.: Hypothesis-driven approach to predict transcriptional units from gene expression data. Bioinformatics 20, 1928–1939 (2004)
5. Brameier, M., Wiuf, C.: Co-clustering and visualization of gene expression data and gene ontology terms for saccharomyces cerevisiae using self-organizing maps. J. Biomed. Inform. 40, 160–173 (2007)
6. Pensa, R., Boulicaut, J.: Constrained co-clustering of gene expression data. In: Proceedings of SIAM SDM, pp. 25–36 (2008)
7. Cordero, F., Visconti, A., Botta, M.: A new protein motif extraction framework based on constrained co-clustering. In: Proceedings of the 24th Annual ACM Symposium on Applied Computing, pp. 776–781 (2009)
8. Ashburner, M., et al.: Gene ontology: tool for the unification of biology. the gene ontology consortium. Nat Genet. 25, 25–29 (2000)
9. Cheng, Y., Church, G.M.: Biclustering of expression data. In: Proceedings ISMB 2000, pp. 93–103 (2000)
10. Cho, H., Dhillon, I.S., Guan, Y., Sra, S.: Minimum sum-squared residue co-clustering of gene expression data. In: Proceedings of the Fourth SIAM International Conference on Data Mining, pp. 114–125 (2004)
11. Salvador, S., Chan, P.: Determining the number of clusters/segments in hierarchical clustering/segmentation algorithms. In: Proceedings of the 16th IEEE International Conference on Tools with AI, pp. 576–584 (2004)
12. Strehl, A., Ghosh, J.: Cluster ensembles - a knowledge reuse framework for combining multiple partitions. Journal of Machine Learning Research 3, 583–617 (2002)

Value-Driven Characters for Storytelling and Drama

Rossana Damiano and Vincenzo Lombardo

Dipartimento di Informatica
Centro Interdipartimentale per la Multimedialità e l'Audiovisivo
Università di Torino, Italy
{rossana,vincenzo}@di.unito.it

Abstract. Agent architectures have proven to be effective in the realization of believable characters, but they stay at odds with the notion of story direction, that is difficultly reconciled with the characters' autonomy.

In this paper we introduce the notion of character's values to mediate between agent architecture and story direction in storytelling systems. Modern theories of drama view story advancement as the result of the characters' attempt to maintain or restore their values, put at stake by unexpected events or antagonists. We relate characters' values with their goals; the activation and suspension of goals depend on the values that are put at stake by the progression of story incidents. Values and goals are integrated in a computational framework for the design of storytelling systems in which the direction is defined in terms of characters' values.

1 Introduction

In the last decade, storytelling has emerged as a powerful instrument to convey meaning, in a range of applications that span from education to entertainment [1,2,3]. The effectiveness of stories depends on their dramatic quality, that creates an emotional bond between the audience and the characters. This quality can be ascribed to two main properties: on the one side, the story must proceed from some initial conflict to its final solution according to a clear direction (the 'unity of action' dating back to Aristotle's poetics and clearly stated by [4,5]); on the other side, story characters must act according to recognizable aims and react to story events with appropriate feelings, in order to gain the naturalness that leads to Coleridge's 'suspension of disbelief'.

The twofold nature of dramatic quality corresponds to two parallel lines of research in computational storytelling. The attempt to formalize the structure of stories has lead to the development of story models, mainly inspired by the semiotic tradition in the analysis of narratives [6,7,8]; the research in character design has investigated the applicability of intelligent agent theories and technologies in storytelling systems [1,9]. However, these two dimensions are difficulty reconciled, since the autonomy that characterizes intelligent agents contrasts with the imposition of constraints on the story development [10]. This problem is exacerbated by the interactive storytelling paradigm, in which the user's actions interfere with the establishment of the story direction and must be addressed by the storytelling system in order to constrain their consequences to the admissible outcomes [11,12].

The notion of character's value has emerged in scriptwriting theory as a major propulsive force in story advancement. First stated in Egri's definition of drama premise

R. Serra and R. Cucchiara (Eds.): AI*IA 2009, LNAI 5883, pp. 436–445, 2009.

[22], the notion of value underpins most of the subsequent work conducted in scriptwriting [23], until the recent formulation by McKee [24] about cinematographic stories. The progression of the story follows a cyclic pattern: a character follows a line of action suitable to preserve its balance of values, when some event (typically, a twist of fate or an antagonist's action) occurs that invalidates its line of action and puts other (more important) values at stake, requiring the character to abandon or suspend its current line of action and to devise yet another line of action to restore them.

For example, in Bond movies, the hero must defeat an arch-vilain who threatens the human kind. As he devises a clever plan to neutralize his antagonist, the value at stake, initially limited to the 'security of the country', becomes increasingly higher as an effect of the counter attacks of the antagonist, with a climax that invariably ends with the removal of the threaten. In courtroom dramas, a solitary lawyer fights against injustice; the climax puts at stake the lawyer's self-achievement, then the fate of the victim of the injustice, until the lawyer becomes involved in some kind of direct opposition against the law institution. Abstracting from the actions actually carried out by the characters, the story can be conceptualized in terms of the values it puts at stake, which form the 'direction' of the story [5]. The direction conveys a type of universal meaning, independent from the actual the actions carried out by the characters in the story. For example, Bond movies tell the audience – through infinite variations – that evilness can be defeated by the courage of a solitary hero, courtroom dramas say that justice will infallibly triumph.

In this paper, we propose the notion of character values to reconcile story direction and characters. We define the story direction in terms of a set of values that are subject to change, and we propose a model of how characters react when their values are put at stake. The link between the characters' intentions and their values forms the meaning of the story. This framework can be employed to drive the generation of stories, by constraining the autonomy of the characters to the value dynamics prescribed by the author's direction. We believe that agent theories and architectures are effective to embed values in storytelling systems and that story direction results from the actions charaters undertake to reestablish their values put at stake. In particular, we address the following research questions:

- How can characters' values be formulated?
- How do values affect the behavior of characters, modeled as intelligent agents?

In the following, we first briefly examine the use of intelligent agents in the context of storytelling, and then introduce the notion of value as a propulsive force for the advancement of story direction.

2   Intelligent Agents and Values in Storytelling

The paradigm of intelligent agents offers an operational way to design and implement characters. Most agent architectures rely on the Belief-Desire-Intention (BDI) characterization of agents [13]. According to the BDI paradigm, agents are driven by their goals (or desires), and form plans to achieve them. Once an agent commits to a plan (i.e., the plan becomes the agent's current intention), it actively pursues the execution of

World War II. 1944, Italy. *Two partisans, Tenebra and Echo, are on a mission in the North of Italy (0). On the way to the meeting with an Allied officer, they decide to stop at a farm to get food and water, only to discover that the farmer, Agnese, is being tortured by a brigade of nazifascists (1). While Tenebra decides to stick with the mission, Echo decides to rescue Agnese. With a strategem, he kills most of the brigade, except the officer, who asks him to surrender in exchange for Agnese's life. While Echo hesitates, Tenebra, who's gone back as he cannot find his way to the meeting (2), gets close to the officer, unnoticed, and shoots him.*

Fig. 1. The synopsis of the example story, the short film "1944", directed by A. Scippa, Italy, 2007. The number in brackets mark the relevant time points for the value-based analysis illustrated in Table 1 (first column).

the actions that are prescribed by that plan. From time to time (typically, after the execution of an action), the agent monitors the world by performing sensing actions; sensing actions enable the agent to recognize other agents' plans and goals from the observation of their actions, so that the agent is likely to anticipate their consequences. BDI-based architectures have proven to provide a solid and effective basis for the implementation of character-based storytelling systems [14,15,1,9], thanks also to the availability of programmable agent frameworks [16,17]. Moreover, [9] argue that the BDI characterization of agency [13] provides the necessary basis to model emotions in characters.

From a cognitive perspective, it has been argued that characters are expected to be rational agents by the audience [18]: according to this claim, characters must manifest an intentional behavior to acquire believability. For example, the generation of interactive stories by the system described in [11] relies on the assumption that a rational model of the characters' behavior is a precondition for equipping characters with an 'expressive behavior' [19], i.e. descriptive elements that help the audience making sense of the motivations underlying their behavior.

Following the line of research summarized above, we assume that a character is modeled as a BDI agent, with beliefs, high-level goals (or desires) and intentions (or plans). For example, in the story summarized in Figure 1, the protagonists recognizably display an intentional behavior: they have the goal to accomplish the mission they have been assigned, and have a plan to accomplish it. There are obvious antagonists in the story, the Nazifascists, who threaten the protagonists' goal. However, some relevant aspects of the story cannot be grasped by pure rationality. In the example story, the protagonists exhibit two different 'scales of values' [20]: when they realize that Agnese's life is threatened by the Nazifascists, they make different choices, as a result of the different importance they attribute to the life of the individual, traded off against the interest of the mankind. The notion of value belongs to the realm of ethics and economics [21]; a value is an assignment of importance to some type of abstract or physical object. Recognizably subjective, values are related with the regulation of behavior, but they retain a more abstract, symbolic meaning and do not exhibit a direct correspondence within the theory of rational action; at the same time, they are very relevant for the establishment of the story direction.

In order to link the definition of characters, encoded by the BDI model, with the direction of the story, we propose to augment the character definition with the notion of values, and we describe how characters modify their goals (and plans) in response to values at stake. Incorporating explicit values in characters' deliberation is the precondition to designing a storytelling system that manipulates values (for example, by putting them at stake) according to a pattern established by the author in order to affect the behavior of characters.

We model the **character** as a 4-tuple $\{B, D, I, V\}$, where, beside beliefs B, Desires D and Intentions I, V represents the character's values. Note that, here, we consider only the subjective dimension of values, and do not consider their relation with an external social system. Each value is polarized, with a negative or positive polarity p, and is associated with a condition c. The negative polarity of a value means that, when the value condition holds in a certain state of the world, the value is *violated*; the positive polarity corresponds to the value being *in force*. In order to let characters arbitrate among their values, values are associated to a priority r that ranks them according to their importance.

So, a **value** v is defined by a set of constructs of the form (p, c, r) where p is a negative or positive polarity, c is a ground formula and r is a ranking (a real number). We pose the restriction that, for some value v, the conditions c cannot be inconsistent and the priority r must be the same for all constructs. If the condition c of a value holds in a state of the world, represented in the character's beliefs, that value is *at stake*.

A character's **record of the values at stake**, VaS, is a set of triples (c, p, r) for which c holds in the current state of the world or is expected to hold in the future, according to the character's beliefs. The character's record of the values at stake (VaS) is a dynamic structure: along the progression of the story, the character updates its VaS by matching the conditions of its values V with its current belief state B. From the point of view of agent architectures, the monitoring of values can be expensive. However, we assume that the set of character's values V has a limited size, since they are intended as general instruments for the regulation of behavior.

As observed by McKee, stories often put at stake a character's value by offering her/him the possibility of bringing about a state in which that value is true (e.g. self-realization for Nora in Ibsen's "A Doll's house), possibly in conflict with some other values (family values in Ibsen's drama). In agent systems, strong limitations are posed to opportunistic behaviors, since they conflict with the requirement of behavior stability, and pose complexity issues for the management of multi-agent systems. On the contrary, in fictional worlds, provided that the behavior of the character is consistently believable along the story, stability is not a desirable feature, since the story tends to prefer changes to stability. In order to make the character proactive with respect to the compliance with its values, the character's beliefs include not only the current state of the world, but also its expectations about how it may evolve. Expectations are computed by verifying if, from the current state of the world, any state can be derived in which any conditions of its values hold, through the character's own actions or other characters' actions.[1]

---

[1] Since expectations pose problems for the computational complexity of practical systems, the look–ahead process may be limited in practical applications, for example by constraining it to one or few steps.

Notice that this definition of values is intended as a subjective one, i.e., it concerns the beliefs of the character, not the beliefs established by the story in the audience. In fact, the gap between the character's belief and the beliefs held by the audience must be preserved in drama since it is the primary source of the effect known as dramatic irony, i.e. the audience knowing facts, relevant for a character, that are not known to the character itself.

3   Values at Work

When a character realizes that some value is at stake, it is expected to modify its commitment accordingly, by forming a goal (**value–dependent goal**) that contributes to re-establish the value at stake. The way goals arise and are affected by values can be grasped effectively by the framework by [25], which provides a unifying account of goal types and describes how goal state is transformed as a consequence of the modifications of an agent's beliefs according to the operational architecture in [25], *adopted* goals remain *suspended* until they are ready for execution (i.e., they are in *active state*), than possibly suspended again if a more important goal is adopted. Goals can eventually be *dropped* if certain conditions hold, namely, when the rationality constraints stated by [26] are met.[2] In the following, we assume this framework, and we only specify the role of values at stake in the state transitions of goals, according to simple automaton represented Figure 2.

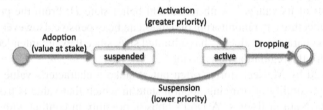

Fig. 2. A graphical representation of goal states and transitions including values (adapted from [25])

- Given a construct $v(p, r, c)$ in VaS (the record of the values at stake), the character formulates and *adopts* a set of goals that have the conditions of its values at stake as an object.
  1. If the condition c holds in the character's *expectations* and the associated polarity p is *positive* (i.e., it corresponds to the value being in force in that state), the character forms the goal to achieve that state of affairs (*achievement goal*). Expectations model the opportunity for the character to achieve the compliance with her/his own values.

---

[2] In order to separate commitment to goals from commitment to plans, the model in [25] acknowledges different conditions to manage the status change of a goal in the presence or in the absence of a plan to achieve the goal. Since this difference is irrelevant for our purposes, here we consider only the overall set of goal conditions.

2. If the condition holds in the *present* or in the character's *expectations* and the associated polarity is *negative*, the character forms the *reactive maintenance goal* to achieve any state of the world in which that condition does not hold (any state s such that $c \not\subseteq s$). In most cases, this goal involves the need to contrast the behavior of an antagonist who is about to bring about just a state of affairs in which the condition holds.[3]

– Goals become *suspended* immediately after their adoption. In our model, the activation of goals depends on the priority of the associated values, which reflects their importance for the character:

- If the priority r of the value at stake v is higher than the priority of the current values at stake ($\forall\ (p_i, c_i, r_i), (v \neq v_i) \rightarrow (r_i < r)$), the related goal becomes *active*. At this point, the character can *generate a plan* to achieve it and starts to *execute* it. The currently active goal may be have already been active in the past: if so, the character may resume the execution of a previously suspended plan, if still valid.

- Or, the new goal may not become active immediately, and be activated later when no other goals with higher priority are left (because they have been achieved or abandoned, two conditions that equally result in the dropping of the goal) or never move to the active state. If a new goal is formed in reaction to a value at stake that has a higher priority, the active goal is *suspended*.

– A goal is finally *dropped* if the character recognizes that it is impossible to find a plan to achieve it or if it has been achieved, or if it is not relevant anymore, independently of its priority.

In order to illustrate how the model works, we resort to the example story (see Fig. 1). Table 1 reports the advancement of the story.

At time point (0), the value at stake, for both characters, Tenebra and Echo, is the $freedom$ of mankind, put at stake by the occupation of the country by the Nazifascists. In order to remove this state of affairs, Echo and Tenebra have formed the goal to accomplish the mission of connecting the partisan army with the Allied forces (formalized in Table 1 as $accomplished(mission)$). When the story begins, the two characters are committed to a shared plan, that consists in going to a farm to get food and water, then walking to the meeting point with the Allied officer and meeting him.

At time point (1), when they finally arrive at the farm, they discover that Agnese is held by the Nazifascists, ($prisoner\_of(Agnese, Nazis)$ in the table) and they realize that the threat to Agnese's life puts the $human\_life$ value at stake. Since the associated polarity is negative, this state of affairs represents the violation of the value. For Echo, $human\_life$ has a higher priority than any other value; in his scale of values, the 'concrete particular' is more relevant than 'state of affairs', in line with the definition of 'moral ontology' provided by [21]. So, when he *adopts* the value–dependent goal of preserving Agnese's life ($freed(Agnese)$), it becomes the *active* goal, since the value

---

[3] The number of states that match the definition above can be effectively limited by considering the specific events or actions that have established or are going to establish the condition c, and focusing the character's deliberation on the possible ways to undo the effects of these actions or events.

Table 1. Overview of the values put at stake in the plot of "1944". Timepoints (first column) represents the temporal ordering according to which values v are put at stake (second column) for a character (third column), with a certain polarity (fourth column). The fifth column represents the the condition c by which the value is put at stake, and the associated polarity p (fifth column). The goal adopted by the character in response to the value at stake is represented in the sixth column.

time point	character	v	r	c	p	value-dependent goal
(0)	Tenebra	freedom	2	$is\_occupied\_by(Italy, Nazis)$	-	accomplished(mission)
	Echo	freedom	1	$is\_occupied\_by(Italy, Nazis)$	-	accomplished(mission)
(1)	Echo	human_life	2	$prisoner\_of(Agnese, Nazis)$	-	freed(Agnese)
	Tenebra	human_life	1	$prisoner\_of(Agnese, Nazis)$	-	accomplished(mission)
(2)	Tenebra	freedom	2	$unfeasible(mission)$	-	accomplished(mission)

at stake to which this goal is related has a priority of 2 (Table 1, row 3), higher than the priority of the previously active goal. Echo *generates a plan* to rescue Agnese and immediately starts to *execute* it: he kills one of the officers who are torturing Agnese, but the other catches Agnese and points the gun at her head.

For Tenebra, the situation is appraised differently: the value of freedom of mankind (*freedom*, priority 2) is more important than the life of the individual (*human_life*, priority 1); this different ranking of values solves the conflict between the continuation of the mission and the rescue of Agnese in favor of the former. For him, the goal of rescuing Agnese is *adopted* but it does not become *active*. However, as Tenebra realizes, at time point (2), that his plan to accomplish the mission is not valid anymore (*unfeasible(mission)*), he adopts the goal to help Echo only as an instrumental way to regain him to the mission. According to the mechanism described in [25], he *drops* the goal of accomplishing the mission only to *generate plan* a new modified plan to accomplish the mission that includes the cooperation with Echo as a necessary step. So, Tenebra goes back to the farm; unseen, he attacks the officer by surprise and kills him.

Finally, notice that in the story analyzed, the characters' goal are only reactive, i.e., they do not stem from the opportunity of achieving the compliance with a value. The reason lies in the fact that this type of story does not leave much room for the initiative of characters, who tend to be driven by the events. However, most stories contain also instances of opportunistic behavior in respose to values at stake: for a very well known example, consider the excerpt of Shakespeare's Hamlet, the so–called "nunnery scene", analyzed in a BDI perspective in [27]: Hamlet, by a fortuitous encounter, becomes aware of the possibility of saving Ophelia from the corruption of the court, forms the goal to convince her go to a nunnery and (unsuccessfully) tries to achieve it through a well-articulated rhetorical plan.

4   Discussion

The integration of values in the rational model of characters posits a computational system for storytelling in the position to account for the basic patterns of drama. In summary, when a character suspends the current goal (and the related plan) in favor of a new goal, associated with a value at stake of higher priority, the new goal becomes the active goal, and the patterns is repeated until the story reaches its climax. The possibility, for a character, to merge plans with the goal of maximizing the stability of its commitment is not prescribed by the model we propose. In order to implement such characters, the agent model must be equipped with meta-deliberation capabilities [28].

Since a story encompasses multiple characters[4], the dramatic direction of the story is given by the temporal pattern of all value changes. Although changes of polarity can happen fortuitously, they normally are intentionally caused by the characters: as a result, what the plot communicates with its up- or down-ending is the adequacy of a certain scale of values in a situation that the author assumes to be paradigmatic.

The important consequence of embedding an explicit model of values in characters is that the author (a *procedural author* in the sense of [29]), when establishing the direction of the story in a storytelling system, can deal only with values and manipulate them to obtain communicative effects, relying on the functioning of the characters to connect values with the character's behavior. Value–sensitive characters can be employed to design interactive storytelling systems in which a story engine 'directs' the characters, by putting their values at stake and relying on their individual deliberation to make the story progress. The subjective ranking of values in characters, established by the author, provides the system with the possibility to generate a climax in the values at stake, thus effectively contributing to the user engagement. Every time a character devises alternative ways to achieve a value–dependent goal, the system can use these alternatives to generate user options, letting the user choose among the possible behaviors of one or more characters. Or, else, the user can be asked to choose the value at stake among multiple options, thus playing the role of the author.

The story engine can derive the necessary knowledge to generate events that put values at stake from the characters' definition of values, and rely on interplay of characters' behavior to generate unexpected state of affairs that put values at stake. Differently from [8], in such system, the conflict among values is not planned but obtained by constraining the emergent behavior of a multi–agent system to follow a pattern of values at stake established by the author.

In practical applications, the way values at stake affect the characters' goals may be integrated by more sophisticated models. For example, the utilitarian character may compute the cost of the plans that achieve a value-related goal (with number of steps providing a gross-grained cost evaluation), and activate only the goals that can be achieve by plans whose cost is below a certain threshold. Or, for the conservative character, the goal activation condition may account for the compatibility of the new goals with the goals it is currently committed, so that its behavior acquires stability.

Storytelling theories also provide some useful insight into goal transitions. For example, since personal values change more slowly then extra-personal ones, a different policy, among the ones sketched above, may be associated with different value types, so that the character results more inclined to struggle for certain types of values. The system of genres is also relevant: a coming-of-age drama is likely to show the slow change of a character's inner values; in a sport drama, the protagonist is usually

---

[4] In some case, characters may non-human, e.g. the forces of the nature: in this cases, the metaphor of agency becomes relevant as well.

stubbornly obstinate to achieve its self-realization through victory, and more less inclined to consider other options, no matter the importance of the value at stake. Finally, in some media, like television, persistence of intentions is intrinsically lower, while it is higher in other ones, like books. This suggests that the selected policy to manage goal adoption and activation should take the medium into account.

5   Conclusions

In this paper, we have proposed a framework for the definition of characters' values and a specification of how values at stake can be accounted for by characters' deliberation, governing the formation of new goals that eventually affect the state of values. The behavior of the characters is not meaningful by itself, but only as it contributes to the pattern of value at stake that constitutes the expressive language through which the author conveys its message.

Despite the explanatory power of values in storytelling, the framework we propose may suffer from some relevant complexity problems when employed for story generation. In particular, the monitoring of values by characters and the formation of value–dependent goals may pose problems for practical applications. Although we believe that narrative–specific patterns can contribute to alleviate these issues, the ultimate answer can be provided only by the design and testing of prototypes.

In this paper, we have only tackled the problem of interactivity, that constitutes one of the main concerns of computational storytelling. Since values provide a high–level guidance to the story progression, they are likely to be an important tool for designing character–based interactive systems. Future work includes testing the validity of this hypothesis in the context of practical applications and assessing its compliance with authorial practices.

As a final remark, note that modeling the achievement of a story direction through the characters' value-related deliberation is not in conflict with the use of story models, which represent an additional source of knowledge about cultural patterns of storytelling.

References

1. Aylett, R., Vala, M., Sequeira, P., Paiva, A.: FearNot! – an emergent narrative approach to virtual dramas for anti-bullying education. In: Cavazza, M., Donikian, S. (eds.) ICVS-VirtStory 2007. LNCS, vol. 4871, pp. 202–205. Springer, Heidelberg (2007)
2. Mateas, M., Stern, A.: Structuring content in the façade interactive drama architecture. In: Proceedings of Artificial Intelligence and Interactive Digital Entertainment (2005)
3. Pizzi, D., Charles, F., Lugrin, J., Cavazza, M.: Interactive storytelling with literary feelings. In: Paiva, A.C.R., Prada, R., Picard, R.W. (eds.) ACII 2007. LNCS, vol. 4738, pp. 630–641. Springer, Heidelberg (2007)
4. Stanislawski, K.S.: An Actor Prepares. Eyre Methuen, London (1980/1936)
5. Styan, J.L.: The Elements of Drama. University Press, Cambridge (1963)
6. Sgouros, N.: Dynamic generation, management and resolution of interactive plots. Artificial Intelligence 107(1), 29–62 (1999)

7. Szilas, N.: Idtension: a narrative engine for interactive drama. In: Proc. 1st International Conference on Technologies for Interactive Digital Storytelling and Entertainment (TIDSE 2003), Darmstadt, Germany (2003)
8. Barber, H., Kudenko, D.: Generation of dilemma-based interactive narratives with a changeable story goal. In: Proc. of 2nd INTETAIN, Brussels (2008)
9. Peinado, F., Cavazza, M., Pizzi, D.: Revisiting Character-based Affective Storytelling under a Narrative BDI Framework. In: Proc. of ICIDIS 2008, Erfurt, Germany (2008)
10. Damiano, R., Lombardo, V.: A Unified Approach fo Reconciling Characters and Story in the Realm of Agency. In: Proc. of ICAART 2009, Porto, Portugal (2009)
11. Riedl, M., Young, M.: From linear story generation to branching story graphs. IEEE Journal of Computer Graphics and Applications, 23–31 (2006)
12. Medler, B., Magerko, B.: Scribe: A tool for authoring event driven interactive drama. In: Göbel, S., Malkewitz, R., Iurgel, I. (eds.) TIDSE 2006. LNCS, vol. 4326, pp. 139–150. Springer, Heidelberg (2006)
13. Bratman, M.: Intention, plans, and practical reason. Harvard University Press, Cambridge (1987)
14. Norling, E., Sonenberg, L.: Creating Interactive Characters with BDI Agents. In: Proceedings of the Australian Workshop on Interactive Entertainment, IE 2004 (2004)
15. Rank, S., Petta, P.: Appraisal for a character-based story-world. In: Panayiotopoulos, T., Gratch, J., Aylett, R.S., Ballin, D., Olivier, P., Rist, T. (eds.) IVA 2005. LNCS (LNAI), vol. 3661, pp. 495–496. Springer, Heidelberg (2005)
16. Busetta, P., Ronnquist, R., Hodgson, A., Lucas, A.: JACK Intelligent Agents-Components for Intelligent Agents in Java. Agent Link News Letter 2, 2–5 (1999)
17. Pokahr, A., Braubach, L., Lamersdorf, W.: Jadex: a BDI Reasoning Engine. Multiagent Systems, Artificial Societies and Simulated Organizations 15, 149 (2005)
18. Schank, R.C., Abelson, R.P.: Scripts, Plans Goals and Understanding. Lawrence Erlbaum, Hillsdale (1977)
19. Sengers, P.: Designing comprehensible agents. In: Proceedings of the Sixteenth International Joint Conference of Artificial Intelligence, Stokholm, Sweden (1999)
20. van Fraassen, B.: Values and the heart's command. Journal of Philosophy 70(1), 5–19 (1973)
21. Anderson, E.: Value in Ethics and Economics. Harvard University Press, Cambridge (1993)
22. Egri, L.: The Art of Dramatic Writing. Simon and Schuster, New York (1946)
23. Campbell, J.: The Hero with a Thousand Faces. Princeton University Press, Princeton (1949)
24. McKee, R.: Story. Harper Collins, New York (1997)
25. van Riemsdijk, M., Dastani, M., Winikoff, M.: Goals in Agent Systems: A Unifying Framework. In: Proceedings of AAMAS 2008 (2008)
26. Cohen, P.R., Levesque, H.J.: Intention is choice with commitment. Artificial Intelligence 42, 213–261 (1990)
27. Damiano, R., Pizzo, A.: Emotions in drama characters and virtual agents. In: AAAI Spring Symposium on Emotion, Personality, and Social Behavior (2008)
28. Horty, J.F., Pollack, M.E.: Evaluating new options in the context of existing plans. Artificial Intelligence (2001)
29. Murray, J.: Hamlet on the Holodeck. The Future of Narrative in Cyberspace. The MIT Press, Cambridge (1998)

A Knowledge Management System Using Bayesian Networks

Patrizia Ribino, Antonio Oliveri, Giuseppe Lo Re, and Salvatore Gaglio

Dipartimento di Ingegneria Informatica (DINFO),
Universita' degli Studi di Palermo, Palermo, Italy
{ribino,antonio.oliveri,lore,gaglio}@unipa.it

Abstract. In today's world, decision support and knowledge management processes are strategic and interdependent activities in many organizations. The companies' interest on a correct knowledge management is grown, more than interest on the mere knowledge itself. This paper proposes a Knowledge Management System based on Bayesian networks. The system has been tested collecting and using data coming from projects and processes typical of ICT companies, and provides a Document Management System and a Decision Support system to share documents and to plan how to best use firms' knowledge.

Keywords: Knowledge Management, Ontology, Bayesian Network, Decision Support System, Document Management System.

1 Introduction

Up today decision support and knowledge management processes are strategic and interdependent activities in many organizations. Due to the economy complexity, the firms adopt procedures to change strategies, structures, technologies and operational mechanisms. The great amount of documents and informations generated during working activities requires a correct management to avoid loss of important knowledge, hence the integration between human processes and computer programs becomes more crucial. The adoption of new computer-based information systems, enabling the storage of structured data and the automation of the information-processing activities of the organization, is then fundamental. During the last two decades ad-hoc frameworks known as Knowledge Management Systems (KMS), enabling access, coordination and processing of knowledge assets [4][2][3] have been proposed. KMSs are generally known as information systems which manage organizational knowledge increasing the productivity of operators. This paper proposes an ontology-based knowledge management framework using two Bayesian Networks, to support the decision maker in project planning, document managing and content analysing. The system is composed by two different modules, an expert system for decision support and a document management engine and has been tested using data coming from typical projects and processes of a ICT society. The paper is structured as follows: section 2 provides a vision of the state of the art of KMSs and expert

R. Serra and R. Cucchiara (Eds.): AI*IA 2009, LNAI 5883, pp. 446–455, 2009.

systems; in section 3 we introduce essential aspects of KMSs, Bayesian and Decision Networks; in section 4 we present a detailed description of the system.
Finally, in section 5, we draw some conclusions.

2   State of the Art

During last decade researches on Knowledge Management (KM) and Decision
Support (DS) examined several issues about new strategies and tools to organize, store and share data and individuals' expertises; in the same time the ICT
companies have demonstrated an increasing interest in the development of internal knowledge management instruments, using novel data representation models
and modern AI techniques. A web and ontology-based KMS called WAICENT
(World Agriculture Information Centre) is in use on the United Nations Food
and Agriculture Organization, to improve food security through information use
[12] and Liping Sui [11] and Maya Daneva [10] studied the benefits of a decision
support system within the business management. In addition N. Fenton et.al
[16] and Noothong et.al [17] studied the use of Bayesian networks to make decision about software projects. Some architectures for Document Analysis and
Understanding was proposed, many of them using Prolog sets of rules like in
[14] and in [15]. In this article we introduce a prototypical KMS applied to a
real case representing domain concepts through ontologies; two KM mechanisms
have been implemented, one related to document management and the other
concerning decisional problems about project planning.

3   Overview

KM consists of techniques based on IT tools for information management, to
improve the work teams efficiency applying methods for making knowledge explicit and for sharing firm's professional expertises. A generic KMS supports
the creation and storage of knowledge, makes information from different sources
readily available and manages both explicit and tacit knowledges [5]. To realize
these goals a KMS uses different technologies such as: *Document based, Ontol-
ogy/Taxonomy based* and *AI based.*

An ontology [22] tries to formulate an exhaustive conceptual scheme of a particular application domain, generally through a hierarchical structure containing
all the noteworthy entities and the existing relationships between them.

A Bayesian Network (BN) [20], also called Belief Network, is a graphical
model showing probabilistic relationships among variables of a given problem. In
recent years, BNs have been successfully used in many fields such as DataMining
[18] and Decision Support System [19]. A BN is a directed acyclic graph with
following features:

- A set of random variables are network nodes.
- A set of oriented arcs connecting nodes represents cause/effect relationships.

- A probability table for each node, specifying how the probability of each state of the variable depends on the states of its parents.
- Each node without parents have a prior probabilities table of each state.

The relationship between random variables follows Bayes' rule:

$$p(x|y) = (p(y|x) * p(x))/p(y) \qquad (1)$$

3.1 Decision Network

The BNs provide also decision support for a wide range of problems involving uncertainty and probabilistic reasoning. Decision Networks [23] are an extension of BN, adding: utility nodes, representing variables for the optimization; and decision nodes representing decisions, a finite set of different choices to take allowing the achievement of a desired aim. The values in a utility node represent levels of preference associated with possible choices. Let $C=c_1,...,c_n$ be a set of mutually exclusive choices and N be the associated random variables. The goal is reached optimizing the expected utility function that estimates the preferences among different world states [23].

$$EU(c) = \sum_N U(c,N)P(N|c) \qquad (2)$$

$U(c,N)$ is the utility value for each choice configuration and associated random variable. $P(N|c)$ represents the probability of N conditioned by choice c, when the choice c occurs.

4 Knowledge Management System Architecture

In this paper we present a KMS prototype for the automation of business processes typical of a government ICT society. The starting point of the system is the first release of the prototype called *Kromos*, giving the opportunity to measure the increasement in information sharing and reuse [1]. As Kromos, this system is based on an ontological structure of domain concepts, while the two subsystems were remodeled; as far as the Decision Support System (DSS) for projects planning concerns, we have considered the uncertainty characterizing such process, consequently the DSS has been remodeled adopting a BN; for the Document Management System (DMS) we implemented some probabilistic techniques for information extraction. The aim of this prototype is to improve business processes for planning ICT projects and to obtain an efficient document analysis and retrieval. As the fig.1 shows, the core of the system is the Knowledge Management with the two independent modules, the Document Management and the Inference Engine. The knowledge representation is based on two ontological models, the former reproducing the government offices' structure and the latter modeling the projects developed by the ICT company. Differently from KMS reported in section 3, our system provides reasoning system and a Document Management Bayesian Networks based. In the following sections we will describe the main components of the system architecture.

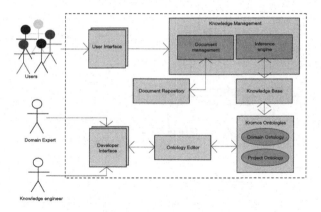

Fig. 1. System architecture

4.1   Knowledge Base and System Ontologies

Ontologies of the proposed system is built using Protégé,, a free and open source platform developed by Stanford University [7]. Since our prototype has been projected as a KMS for a government ICT society dealing with the computerizing of offices activities, it is necessary to reproduce not only the conceptual scheme of the projects but also the structure of the government offices. Our ontology can be considered a collection of two correlated ontologies, a domain and a projects ontology. The *Domain Ontology*, representation of the offices' structure and their activities, is used to characterize the environment in which the system works. The *Project Ontology* is useful to describe ICT company projects; it maps the structure of the project components containing semi-structured explicit knowledge. The Knowledge Base, instead, represents the knowledge container, composed of the set of instances of ontologies previously described.

4.2   Inference Engine: A Decision Network for Project Planning

The project planning process is a complicated trial because it is influenced by many different indeterministic factors (time, costs, resources) that are sources of uncertainty to be estimated in order to optimize the decisional trials. In the last years, the BNs has become a popular representation for encoding uncertain knowledge [21]. In this section a model of DSS based on BNs is proposed. Such DSS will support decision makers during planning and scheduling processes of business projects. The development of the net has been divided in three phases: domain analysis, relationships among the variables discovery and estimation of probability tables. The product is shown in fig.2.

Domain Analysis. Domain Analysis phase concerns the determination of all variables characterizing the domain and the individuation of all possible states of the variables, each representing a node of the BN. Variables set of is the following:

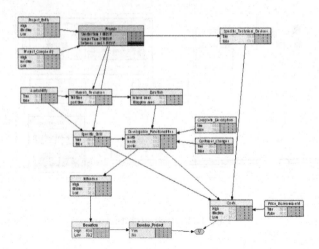

Fig. 2. Projects decision network

- *Project Entity (PE)*: number of modules in a project. Values: High, Medium and Low.
- *Project Complexity(PC)*: how complex the project is. Values: High, Medium and Low.
- *Founds*(F): financing in Millions of Euro. Values: Shorter than 1, Between 1 and 3 and More than 3.
- *Specific Technical Devices (STD)*: availability of specific technical devices. Values: True and False.
- *Availability(A)*: availability of human resources. Values: True and False.
- *Human Resources (HR)*: availability of employees. Values: Full Time and Part Time.
- *Duration (D)*: project duration. Values: High and Low.
- *Specific Skills (SS)*: availability of human resources with specific technical Know-How. Values: True and False.
- *Develop Functionalities (DF)*: project functionalities to be implemented. Values: Many, Mean and Few.
- *Complete Description (CD)*: if the customer has provided an exhaustive product description. Values: True and False.
- *Customer Changes (CC)*: if the customer can change the requisites during development. Values: True and False.
- *Influence (I)*: how important the project is. Values: Many, Mean and Few.
- *Costs (C)*: total cost for project development. Values: High, Medium and Low.
- *Price Increment (PI)*: prices increase (raw material, renewal employment contract etc.). Values: True and False.
- *Benefits (B)*: improvement of human process, of time and costs. Values: High and Low.

Relationships discovery. The Relationships Discovery phase allows the discovery of causal relationships between variables and the influence of other variables. Examining the domain to model, the following set of dependent and independent conditions were discovered:

$$PE, PC, PI, CD, CC, A|\varnothing; \quad F|PE, PC; \quad HR|A, F;$$
$$SS|A, HR; \quad D|HR; \quad DF|SS, CD, CC; \tag{3}$$
$$I|DF; \quad C|SS, DF, STD, PI; \quad B|I;$$

For instance, in this equations the symbol | represents the dependence of the left-side set of variables by the right-side set of variables.

Estimate of the probability tables. The definition of conditional probability distributions for each node is generally an hard task. In our case, the difficulty is that this estimation comes from Project Managers (PM) that don't possess the same knowledge of the projects. Hence, the opinion of a single individual would bring to an unrealistic probability tables evaluation. To avoid this, we combined the opinions by every PM in an unique probability value. We weighted the opinion of each expert, considering the experience degree, product of the qualification of the employee and the number of years he worked in that society (see. eq.5). To intersect all this experience data in a unique value, the schema in fig.3 was adopted.

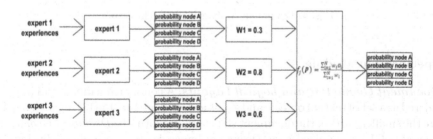

Fig. 3. Combination schema of experts opinion

The opinions about node j, produced by N experts can be combined using a weighed average:

$$f_j(o) = (\sum_{i=1}^{N} w_i o_i)/(\sum_{i=1}^{N} w_i) \tag{4}$$

In this equation O_i indicates the percentage value of the opinion of the expert i, while w_i means the related weight, calculated as it follows:

$$w_i = G_i * A_i/(G_{max} * A_{max}) \quad with \quad G \in [1, G_{max}] \quad and \quad A \in [1, A_{max}] \tag{5}$$

where G indicates the degree of expert experience, while A indicates the number of years he worked in the company.

4.3   Document Management

In companies, the volume of documents produced during working activities grows up rapidly. Documents contain most of the information about projects, people involved and so on. Starting from the analysis of the document structure, it is necessary to develop automatic systems for understanding the content and organizing information. This aim is reached by Document Management component trough two steps, as fig.4 shows. The first step is the *Document Structure Analysis* performed by a *Specification Module*, that captures information about physical and logical document layout through a set of rules. The grouped informations are stored into instances of Project Ontology and Domain Ontology. The second step is the *Document Content Analysis* performed by an *Extraction Module*, that allows text extraction from document structures previously recognized.

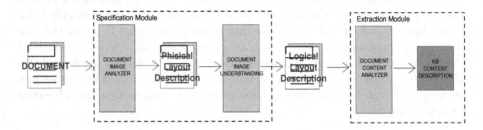

Fig. 4. Document Management Architecture

Specification Module

Document Geometric and Logical Layouts. A common document can be considered as a set of textual objects and graphical elements; as [14][13] suggest, blocks are the smaller parts of the document structure and form the *Document Geometric Layout* (DGL), recognized using *Document Image Analysis* techniques obtaining a digital description of document physical segmentation. The most important phase of the process is the recognition of the *Document Logical Layout* (DLL), performed using *Document Image Understanding* (DIU) techniques. While Geometric Layout recognition extracts geometric structures, DIU maps this in a logical structure, considering the logical relationships between basic blocks using information like position, dimension, type of blocks and so on. The obtained data about geometric and logical layout are stored for a subsequent use.

Logical description rules and the Bayesian recognition. In ICT companies the greatest part of documents is structured or semi-structured. It is possible to define the logical structure using a set of rules to recognize logical units, defined by means of some features. In this paper the DLL recognition is obtained trough a set of rules using a Bayesian Network. Each feature is a node of the net that can assume two or more discrete values, connected to nodes representing

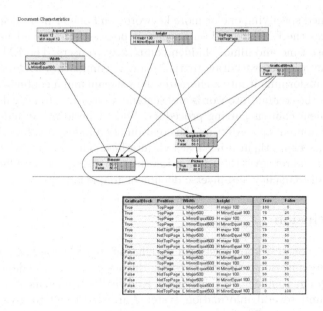

GraficalBlock	Position	Width	height	True	False
True	TopPage	L Major500	H major 100	100	0
True	TopPage	L Major500	H MinorEqual 100	75	25
True	TopPage	L MinorEqual500	H major 100	75	25
True	TopPage	L MinorEqual500	H MinorEqual 100	50	50
True	NotTopPage	L Major500	H major 100	75	25
True	NotTopPage	L Major500	H major 100	50	50
True	NotTopPage	L MinorEqual500	H major 100	50	50
True	NotTopPage	L MinorEqual500	H MinorEqual 100	25	75
False	TopPage	L Major500	H major 100	75	25
False	TopPage	L Major500	H MinorEqual 100	50	50
False	TopPage	L MinorEqual500	H major 100	50	50
False	TopPage	L MinorEqual500	H MinorEqual 100	25	75
False	NotTopPage	L Major500	H major 100	50	50
False	NotTopPage	L Major500	H major 100	25	75
False	NotTopPage	L MinorEqual500	H major 100	25	75
False	NotTopPage	L MinorEqual500	H MinorEqual 100	0	100

Fig. 5. Particular of the Bayesian Network used for Logical Layout recognition

the possible logical units of a document. Oriented arcs connect feature nodes and logical-unit nodes, the latter have probability tables which describe, given a set of feature values, the probability that the blocks can be recognized as a part of a defined logical unit. Probability values of a node can be determined as it follows:

$$Pr = 100 * (N'/N) \qquad (6)$$

considering N, the number of features from which the category-node depends and N' the number of requirements satisfied by the block. Fig.5 shows the part of the net regarding the recognition of a document graphical components.

In figure a graphical element is a logical unit that represent a graphical component, while hight, width and position are features of the element to recognize. All the other components are described in the same way. The entire set of conditional probabilities guarantees that, having information about the characteristics of all geometric blocks, the net can identify the logical units. These information is stored in the system.

Extraction Module

Using the information about logical structure, each block is elaborated by the Extraction Module (EM) for the content analysis. In the actual release the extraction module uses Apache Lucene, an Information Retrieval (IR) engine adapted for the analysis of logical units contents. Lucene is an open-source, high performance text search engine and information retrieval library written in Java, suitable for any application requiring a full text search [6]. Searches in Lucene are performed specifying one or more keywords and one or more fields to search within. Since the EM is based on Lucene, it possesses similar advantages: very fast response time and almost hidden complexity to users. The EM provides: a content elicitation to withdraw textual content eliminating irrelevant information and transforming it into a character data stream and a content tokenization that breaks the content into words and sentences according to lexical analysis for the subsequent content parsing procedure. Then it performs an indexing of the logical unit contents [9]. After an the creation of an index, users need to retrieve documents, seeking them using a search interface. The Lucene system searches in the index and retrieves the relevant information, using a set of different factors such as term frequency and inverse document frequency.

5   Conclusion

In the new economy, knowledge and its efficient exploitation has become a key factor for organizational success, pressing the organizations to adapt themselves to changing environment. In this regard, Knowledge Management strategies are promoted and several information systems are developed to give support to knowledge processes. The proposed system has the aim to improve the growth of organizational knowledge for projects' management, developing a KMS prototype which makes use of a Decision Support System for Decision Support and a Document Management System for Document Analysis. The opportunity to adapt this KMS architecture modeling different domains, gives the chance to use it for different application contexts, increasing information sharing and reuse. The metric to measure the benefits of our system has been essentially the users satisfaction; interviews with Project and Program Managers revealed that the document management system is easy-to-use, with a minimization of latency time on searching information, and the graphic representation of the conceptual structure of their work domain is clear; software reuse was increased by the use of DSS, even if it is still a prototype.

References

1. Oliveri, A., Ribino, P., Gaglio, S., Lo Re, G., Portuesi, T., La Corte, A., Trapani, F.: KROMOS: ontology based information management for ICT societies. In: ICSOFT (2009)
2. Staab, S., Studer, R., Schnurr, H.P., Sure, Y.: Knowledge Processes and Ontologies. IEEE Intelligent Systems, 26–34 (2001)
3. O'Leary, D.E.: Enterprise Knowledge Management, Computer, pp. 54–61. IEEE Computer Society Press, Los Alamitos (1998)
4. Alavi, M., Leidner, D.E.: Review: Knowledge Management and Knowledge Management Systems: Conceptual Foundations and Research Issues,Knowledge Management, Routledge (2005)
5. Takeuchi, H., Nonaka, I.: The knowledge-creating company: How Japanese companies create the dynamics of innovation. Oxford University Press, NY (1995)
6. Gospodnetic, O. and Hatcher, E.: Lucene in Action, Manning (2005)

7. Chaudhri, V.K., Farquhar, A., Fikes, R., Karp, P.D., Rice, J.P.: OKBC: a programmatic foundation for knowledge base interoperability. In: Proc. of the 15 national conf. on AI/Innovative applications of AI (1998)
8. Pirro, G., Talia, D.: An approach to Ontology Mapping based on the Lucene search engine library. In: Proceedings of the 18th Int. Conf. on Database and Expert Systems Applications, USA (2007)
9. Bennett, M.: Contrasting relational and full-text engines. NIE Enterprise Search Newsletter 2 (2004)
10. Daneva, M., Peneva, J., Rashev, R., Terzieva, R.: Knowledge-Based Decision Support System for Competitive Software Audit. In: IEEE Int. Conf. on Systems Man and Cybernetics (1995)
11. Sui, L.: Decision support systems based on knowledge management. In: Proc. of 2005 Int. Conf. on Services Systems and Services Management, vol. 2 (2005)
12. O'Leary, D.E.: A multilingual knowledge management system: A case study of FAO and WAICENT. In: Decision Support Systems. Elsevier, Amsterdam (2008)
13. Klink, S., Dengel, A., Kieninger, T.: Document structure analysis based on layout and textual features. In: Proc. of Int. Workshop on Document Analysis Systems (2000)
14. Niyogi, D., Srihari, S.N.: The use of document structure analysis to retrieve information from documents in digital libraries. In: Proc. SPIE, Document Recognition IV (1997)
15. Dengel, A., Dubiel, F.: Clustering and classification of document structure-a machinelearning approach. In: Proc. of 3rd Conference on Document Analysis and Recognition (1995)
16. Fenton, N., Marsh, W., Neil, M., Cates, P., Forey, S., Tailor, M.: Making resource decisions for software projects. In: Proc. 26th Int. Conf. on Software Engineering (2004)
17. Noothong, T., Sutivong, D.: Software Project Management Using Decision Networks. In: 16 Int. Conf. on Intelligent Systems Design and Applications (2006)
18. Heckerman, D.: Bayesian networks for data mining. Data mining and knowledge discovery, 79–119 (1997)
19. Zhang, S.Z., Yang, N.H., Wang, X.K.: Construction and application of bayesian networks in flood decision support system. In: Proc. of the First Int. Conf. on Machine and Cybernetics (2002)
20. Heckerman, D., et al.: A tutorial on learning with Bayesian networks. Nato Asi Series D Behavioural And Social Sciences (1998)
21. Heckerman, D., Mamdani, A., Wellman, M.P.: Real-world applications of Bayesian networks. Communications of the ACM (1995)
22. Uschold, M., Gruninger, M.: Ontologies: Principles, methods and applications. Knowledge engineering review (1996)
23. Norvig, P., Russell, S.J.: Artificial intelligence: a modern approach. Prentice-Hall, Englewood Cliffs (2003)

Knowledge Discovery and Digital Cartography for the ALS (Linguistic Atlas of Sicily) Project

Antonio Gentile, Roberto Pirrone, and Giuseppe Russo*

Universita' degli Studi Palermo
Viale delle Scienze Ed. 6 P 3
90128 Palermo (IT)
{pirrone,gentile}@unipa.it,
russo@dinfo.unipa.it

Abstract. In this paper the latest developments of the ALS (Linguistic Atlas of Sicily) project are presented. The ALS project has the purpose to define methodologies and tools to allow researches in the socio-linguistic field. Different types of variables (both quantitative and qualitative) are involved.

The whole framework is based on the definition of ontology-based applications for the creation, retrieval, manipulation and browsing of related data. To this aim, some mapping processes have been defined. The framework eventually shows the result in many ways including spatial maps.

The on-going collaboration process is a perfect example a domain hybridizing process, enabling the training on-the-field of a joint group of researchers who, coming from the peculiarly different scientific and cultural domains pertaining to the project, participate to the constitution of the core of a local humanities computing community.

Keywords: Data Abstraction and Mapping, Spatial Databases and GIS, Markup Languages, Ontologies.

1 Introduction

In this work we present the latest developments inside the ALS (Atlante Linguistico Siciliano - Linguistic Atlas of Sicily) project. The ALS project was started in the early nineties from an intuition of Giovanni Ruffino [1]. The ALS is currently a joint effort with research units in Palermo, Catania and Messina leaded by the Dipartimento di Scienze Filologiche e Linguistiche of the University of Palermo. The collaboration with the Dipartimento di Ingegneria Informatica (DINFO) started in 2001, with the development of the first part of the project: the ALSDB [2].

The main goal of the ALS project is the definition of methodologies and models to investigate sociolinguistic variables describing the evolution over space and time of the usage of regional Italian and Sicilian dialect. The core project

---

* Corresponding Author.

R. Serra and R. Cucchiara (Eds.): AI*IA 2009, LNAI 5883, pp. 456–465, 2009.

is related to a new way to look to linguistic that has been defined as linguistic ethnografy (LE) [3]. In a nutshell LE claims that *"language and social life are mutually shaping, and that close analysis of situated language use can provide both fundamental and distinctive insights into the mechanisms and dynamics of social and cultural production in everyday activity"* [4]. The project exploits different type of phenomena related to phonetic, lexical, morpho-syntactic and textual aspects of language. The on-going collaboration process is a perfect example a domain hybridizing process, enabling the training on-the-field of a joint group of researchers who, coming from the peculiarly different scientific and cultural domains pertaining to the project, participate to the constitution of the core of a local humanities computing community. To support the goals of the project a complex framework have been developed. The framework has to deal with complex tasks and has the purpose to facilitate users that are mainly researchers in the linguistic fields to define and prove work hypothesis and to show investigation results. The whole framework has been realized with two main drivers: the easiness for users and the extendibility in terms of functionalities. The first aspect involves mainly the HCI (Human Computer Interaction) with the design of suitable interaction modalities and the Knowledge Management with the definition of proper policies to model and access information. To this aim the knowledge of the system is organized with an ontology-based methodology able to automatically growth within the interaction with users.

The rest of the paper is organized as follows: next section briefly presents the ALS project. The whole architecture is presented in more details in section 3 with particular attention to data flow and knowledge related processes. Sections 4, 5, and 6 show the three main components of the project. Conclusions and future work are discussed in section 7.

2   The ALS (Linguistic Atlas of Sicily) Project

The ALS project development over the years started from the acquisition of data useful to help researchers in the investigations of domain's variables. The acquisition of information was performed through the administration of a questionnaire about population behaviors. For each selected geographic center five families were interviewed. Interviews were transcribed and digitally recorded for a total of 1200. The resulting questionnaire is organized in three sections: a biographic section, a meta-linguistic section, and a linguistic section. The first section is composed of 27 questions focused about statistics, level of instruction (personal and familiar), cultural consumes and other information like quantitative and qualitative relationship between the interviewee and her neighbors. The meta-linguistic section is composed of 35 questions investigating the perception and the self-perception about the two different linguistic codes (Italian language and Sicilian dialect), the evaluation about linguistic differences between different areas, the ideologic attitude that are perceived from and to other people. The third section regards specifically the language and the translation skills between the two codes with tests, reading of written materials in both codes and production of free spoken language. The result of the process is a set of digitally registered materials used for transcriptions and evaluations of sociolinguistic phenomena.

The ALS project is organized according to the next figure (see Fig. 1).

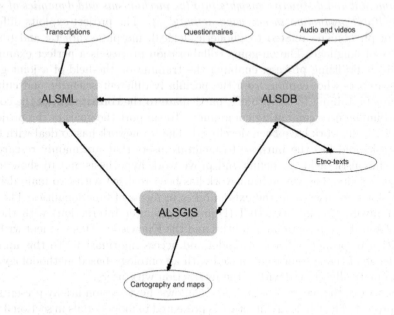

Fig. 1. The three main components of the ALS project and the produced materials for each component

The main components are the ALSDB, the ALSML and the ALSGIS. The ALSDB is the part of the project deputed to the creation, maintenance and fruition of raw data presented as questionnaires, etno-texts or multimedia audio and video. The ALSML is the part of the project deputed to the definition and organization of semi-structured data like transcriptions. The ALSGIS is related to the creation of Geographic Information Systems (GIS). GISs are computer-based systems that allow decision makers to incorporate geographically based data into their analysis.

3   Framework Overview

The whole framework is organized in different systems reflecting the logical organization of the project. The framework has three main subsystems (see Fig. 2).

The system has different repositories according to treated data. The ALSDB was designed and implemented on Microsoft SQL Server^{TM}, although the linguistic section (the third in the questionnaire) is built in the form of a Berkeley database out annotated transcriptions. Transcriptions are obtained from different tools according to data sources. As an example, a tool to obtain transcriptions directly from audio file has been realized. This tool is able to directly load

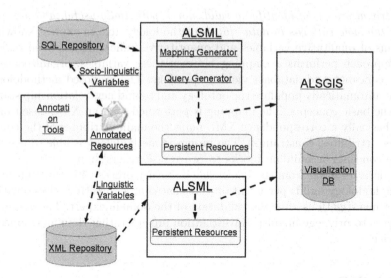

Fig. 2. The ALS framework within the three main subsystems. The whole architecture is realized as a multi-tier framework where low-level data are extracted according to developed functionalities. The framework's repositories are able to collect structured data, semi-structured and multimedia data.

audio and has as output the annotated file in an XML format reflecting the type of annotation required. This software has been realized starting from the Praat software [5] and produces two possible outputs: an XML files with annotations or a text document useful for further investigations. For textual analysis a plug-in compatible with any version of Microsoft Word™ has been realized. It is able to produce, according to a set of incremental rules, annotated files. Three different types of transcription are possible:

- phono-orthographic transcription: it is used to detect mainly linguistic phenomena like the presence/absence of a peculiar feature. It is used mainly at a phrase level.
- phonetic transcription: this transcription is used by phonetists and defines the main phonetic aspects of the interaction at a word level.
- conversational transcription: the conversational transcription investigates the whole interaction and store global aspects like the turn-around between the subjects, the type of interaction and other global features.

The nature of data and the different granularity of researched aspects doesn't allow to define methodologies able to retrieve data in a complete manner: it is impossible for researchers to have the complete knowledge of the data involved in a process. The database ERD for structured data reflects the nature of the project: to extract useful information is necessary to set complex queries. To facilitate users to produce valuable results the entire framework allows to perform knowledge discovery processes. Knowledge Discovery is defined as the *"non-trivial process of identifying valid, novel, potentially useful, and ultimately understandable patterns in data"* [6]. A methodology able to express knowledge in terms of qualitative variables and quantitative indicators has been realized. The approach performs a mapping between these variables, organized as ontology concepts, and database views or portions. The proposed methodology is able to automatically populate the ontology and to find new relationship starting from the basic concepts. The mapping is performed with an XML-based procedure: basically a correspondence XML maps the main attributes of the concept (tables, attributes, constraints). The proposed methodology is able to use these simple concepts as building blocks to prove researchers' hypothesis and to retrieve data resulting from the knowledge discovery process. In addiction to this the methodology builds persistent objects that can be directly accessed without further investigations after the validation of the domain expert. The goal of the project is to privilege linguist's investigation process rather than an engineering approach.

4   ALSDB

In the ALSDB raw data are related to domain symbolic representation. The mapping between ontology concepts and databases has been investigated in literature from different point of views. The mapping between a database and an ontology has a double order of difficulty: first of all, a methodology to define a complete mapping is needed. The methodology has to deal with the uncertainty level. It is so necessary reach a trade-off between researcher autonomy and accuracy of the translation. A correspondence schema has to define correspondences between possibly multiple data schemas. Two possible approaches are pursuable: the Global-as-View (GAV) defining the global schema according to local data and the Local-as-View (GAV) that defines local elements in relation to their position in the global schema [7]. This way, it is possible to define local properties for elements, however it is not possible to define global properties that are the results of the aggregation process. In our methodology the initial correspondence is defined for simple concepts that can be bounded for more complex investigations. The methodology makes use of a Xml formalism to represent both initial mapping of concepts and the results of the investigation that are presented and permanently stored as Xml documents. Some works have been proposed that addresses similar issues [9] [8]. The resulting Xml documents are stored using the iBatis [10] persistence framework. Essentially iBatis is a data mapper. A data mapper moves data between objects that are the quantitative results of the investigation process.

The interface of the system is depicted in the next figure (see Fig. 3).

Users can define new concepts that are the result of a cross investigation of previously defined concepts. Every concepts can be specified with a set of attributes coded as filters in the database. The researchers can specify also the dependency of the new concepts(starting from the concept involved in the generation) and populate the ontology with instances that are obtained from specific

Fig. 3. The Interface of the system: users can retrieve concepts directly from the questionnaire (on the left) or can use concepts resulting of particular investigations (on the right)

values for the tables involved in the mapping process. The methodology is iterative and incremental and the knowledge base of the system growths with the researchers' interaction that can validate and control the new resulting concepts.

5   ALSML

The ALSML is the part of the project deputed to definition of methodologies and tools to deal with semi-structured or textual data that are mostly transcriptions of the third part of the questionnaire. The result is a linguistic annotated transcription where text is lineup with audio. Two different tools have been defined: a tool that is able to lineup directly from audio and a set of tools embedded in the Microsoft Word environment to enrich the transcriptions once are obtained. The result of the transcription process is an XML document. All the documents are stored in an XML native database: the Berkeley DB [12]. The Berkeley XML DB is an XQuery-based [11] DB used to access documents stored in containers and indexed on the bases of their properties. The XML documents are validated with proper schemas reflecting the level of annotation. A common logical structure has been defined for all the produced documents. All the files start with an header section containing general information, like the number of the interview, the linguistic code, the number of question and other information, used to bound the transcription with the ALSDB.

The retrieval process for the documents has the same characteristics of the one described in the ALSDB section. User can choose what are the main features of the investigation she is intend to pursue through a simple interface that incrementally and visually create the corresponding query (see Fig. 4).

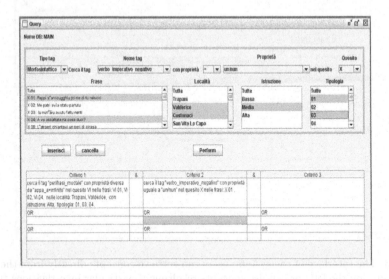

Fig. 4. The interface for transcriptions' querying. Users can choose different combination of tags or phenomena according to the schemas associated to XML documents. They can be combined to produce a more complex pattern starting from the simple ones already defined. The resulting query can be saved and used lately for further investigations.

6   ALSGIS

The ALSGIS is the part of the framework deputed on Geographic Information Systems (GISs) and their intrinsic data organization in homogeneous layers and querying capabilities on a spatial representation. A specific research collaboration was initiated in April 2007 between the Universit degli Studi di Palermo and the Geographic Military Institute (IGM), for the purpose of acquiring IGM specific knowledge in the field. Starting from the IGM DB1000 GIS product, a fully integrated GIS component has been developed leveraging on the initial prototype developed for the socio-variational section of ALS. The final component allows investigations over ALS specific layers, such as ALS interview points, and their clustering in micro-areas, and those resulting from specific querying on the ALSDB, to produce detailed statistical analysis, phonetic data production, and annotated ethno-texts retrieval. The component is based on Map Windows GIS, an open source software that has been modified to incorporate the new layers defined in the investigation process. The interface of the ALSGIS is depicted in the next figure (see Fig. 5):

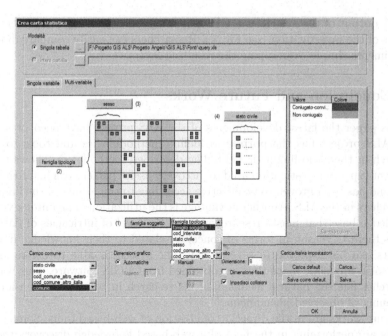

Fig. 5. The Interface for the ALSGIS subsystem. Users are able to retrieve data from the ALSDB and to cluster them according to the performed experiments.

Fig. 6. An example of a particular map. The resulting tables are superimposed as a new layer.

The output of the system is a GIS map where each new layer represents the results of the investigation (see Fig. 6). Researchers can easily compare different experiments simply overlapping the resulting layers.

7   Conclusion and Future Works

In this paper the latest developments of the ALS project have been presented. The ALS project has the purpose to define methodologies and tools to allow research in the socio-linguistic field. The described methodology allows to mix different types of investigation used to discovery useful patterns on heterogeneous data and has been proven to be effective according to the results of the linguistic researchers in the ALS. Another advantage is the utilization of an ontology-based knowledge base that masks researchers from the level of intricacies of data and allows them to reason in terms of concepts related to their knowledge domain. Results can be displayed in several way. The system is able to define and to show socio-linguistic maps that are obtained from a GIS software. This way the researcher has the natural feedback of his research in terms of dimension and stationing of observed phenomena.

Future works are in the direction of creation of web-based tools. Some experiments are performing in the field of a web-based knowledge discovery tool for the ALSDB and a web tool to show geographical information based on Google Maps [13].

Acknowledgments

This work has been partially funded under the Programma di Rilevante Interesse Nazionale (PRIN) 2007, contract no. $2007CY8ETN005$, entitled: "Artificial intelligence methodologies for data processing, management, and analysis for the Linguistic Atlas of Sicily (ALS). Design of intelligent interfaces to the ALS information system." Authors would like to thank the ALS development team lead by prof. Giovanni Ruffino, and are especially indebted with Prof. Maria D'Agostino, Prof. Marina Castiglione, Dr. Luisa Amenta, Dr. Vito Matranga, Dr. Roberto Sottile, Dr. Giuseppe Paternostro and the many Ph.D. students at ALS for the extenuating and thoughtful discussions on the definition of the framework.

References

1. Ruffino, G., D'Agostino, M.: I Rilevamenti Sociovariazionali. Linee Progettuali. Centro studi filologici e linguistici siciliani (2005)
2. Gentile, A., Sorbello, R., Cassara', S., Massara, D.: Informatizzazione dell'Atlante Linguistico della Sicilia. Tech. Report - Dip. Ingegneria Informatica, Universita' di Palermo (2004)
3. Creese, A.: Linguistic Ethnography. In: King, K.A., Hornberger, N.H. (eds.) Encyclopedia of Language and Education, 2nd edn. Research Methods in Language and Education, vol. 10, pp. 229–241 (2008)

4. Rampton, B., Tusting, K., Maybin, J., Barwell, R., Creese, A., Lytra, V.: UK Linguistic Ethnography: A Discussion Paper (2004)
5. Boersma, P., Weenink, D.: Praat: doing phonetics by computer, Version 5.1.05 (2009), http://www.praat.org/
6. Fayyad, U., Fayyad, U., Piatetsky-shapiro, G., Piatetsky-shapiro, G., Smyth, P., Smyth, P.: Knowledge Discovery and Data Mining: Towards a Unifying Framework. In: Proc. 2nd Int. Conf. on Knowledge Discovery and Data Mining, pp. 82–88 (1996)
7. Xu, L., Embley, D.W.: Combining the Best of Global-as-View and Local-as-View for Data Integration. In: Information Systems Technologies and Its Applications - ISTA 123–136 (2004)
8. Sung, S., McLeod, D.: Ontology-driven semantic matches between database schemas. In: ICDEW 2006: Proceedings of the 22nd International Conference on Data Engineering Workshops. IEEE Computer Society, Los Alamitos (2006)
9. Tzvetkov, V., Wang, X.: DBXML - Connecting XML with Relational Databases. In: The Fifth International Conference on Computer and Information Technology, pp. 130–135 (2005)
10. Begin, C., Goodin, B., Meadors, L.: Ibatis in Action. Manning Publications Co., Greenwich (2007)
11. Siméon, J., Chamberlin, D., Florescu, D., Boag, S., Fernández, M.F., Robie, J.: XQuery 1.0: An XML Query Language, http://www.w3.org/TR/2007/REC-xquery-20070123/.
12. Oracle Berkeley DB XML, http://www.oracle.com/database/berkeley-db/xml/index.html
13. Google Maps on-line Documentation, http://code.google.com/intl/it-IT/apis/maps/documentation/

Parameter Tuning of a Stochastic Biological Simulator by Metaheuristics

Sara Montagna and Andrea Roli

DEIS–Cesena, *Alma Mater Studiorum* Università di Bologna, Italy
`sara.montagna, andrea.roli@unibo.it`

Abstract. In this paper we address the problem of tuning parameters of a biological model, in particular a simulator of stochastic processes. The task is defined as an optimisation problem over the parameter space in which the objective function to be minimised is the distance between the output of the simulator and a target one. We tackle the problem with a metaheuristic algorithm for continuous variables, Particle swarm optimisation, and show the effectiveness of the method in a prominent case-study, namely the mitogen-activated protein kinase cascade.

1 Introduction

Modelling and simulating biological intracellular networks is one of the central tasks in Systems Biology. Several approaches have been proposed for reproducing the structure of these complex systems and their dynamic behaviour. A rich literature is available on mathematical and computational, deterministic and stochastic, discrete and continuous models [6,18,4,11].

Although there is a huge amount of studies in this context, a lot of open problems still remain to be solved. *Parameter tuning* is one of these: given the model structure and a set of target data, the goal is to find the values for model parameters so as to reproduce the system behaviour. As in general it is not possible to capture the influence of parameters by theoretical models, this is in fact a resource-intensive task that requires many repeated simulation-analysis runs and the introduction of ad-hoc optimisation algorithms. However, considerably good results have already been achieved and some of the most used frameworks for biochemical network analysis, mainly based on mathematical models, are equipped with tools for parameter tuning – such as CellWare, COPASI, GEPASI, JSim [1].

In [13], we have proposed a novel *computational and stochastic modelling* approach whose simulation is based on the Gillespie's SSA [7]. Tuning parameters in such a kind of systems is even a harder task. To this aim, we formulated this task as an optimisation problem and we devised a parameter tuning module that is illustrated in this paper. The optimisation module makes use of *metaheuristics* to find a parameter configuration such that the simulated system has the desired behaviour. Metaheuristic algorithms combine diverse concepts for exploring the search space and they also apply learning strategies in order to find (near-)optimal solutions efficiently [3]. Metaheuristics are successfully applied

R. Serra and R. Cucchiara (Eds.): AI*IA 2009, LNAI 5883, pp. 466–475, 2009.

to optimisation problems since decades and are particularly effective in tackling problems in which the objective function is rather complex or even an approximation of the actual optimisation criterion. Other works have been published which bring support to the important contribute that the use of metaheuristics can give for solving this optimisation problem [17,2], but they mainly refer to mathematical and deterministic models.

In order to nimbly perform our studies on the biological system, we have built a framework that integrates a *simulator* executing the stochastic model, an *evaluator* estimating the quality of the simulation results, and an *optimiser* finding an optimal parameter setting. Such a framework have been tested with a well known case study —the MAPK cascade— under two different conditions and steady-state responses.

In the reminder of this paper we first explain the role of metaheuristics in parameter tuning in Section 2, then we illustrate the general framework in Section 3. In Section 4 we describe the instance of the framework for our system and we show its application on a notable case-study in Section 5. Finally, we conclude and summarise future work in Section 6.

2   Metaheuristics for Parameter Tuning

Metaheuristics are search strategies upon which approximate algorithms for continuous and combinatorial optimisation problems can be designed. Notable examples of metaheuristics are simulated annealing, tabu search, genetic algorithms and ant colony optimisation [3]. The first two are examples of so-called *trajectory methods*, which are search processes that can be defined as (stochastic) trajectories over a search graph [8]. Conversely, the latter two are prominent cases of *population-based search algorithms*, which perform a search process characterised by an iterative sampling of the search space. The probabilistic model defining the sampling distribution is adapted during search so as to concentrate search around regions containing promising solutions.

The problem of parameter tuning of a system can be cast into an optimisation problem once an error measure (or a performance measure) is defined. Thus, the problem is to finding an assignment to the parameters such that the error is minimised. Depending on the parameter domains (i.e., continuous or categorical parameters), the problem can be continuous, discrete or mixed-integer. The objective function, that we suppose is to be minimised, defines an error landscape which is explored by the search algorithm.

Metaheuristic algorithms, both for discrete and continuous variables, are usually effective in tackling this kind of problems because they can exploit the information provided by the objective function and learn local properties of the error landscape.Metaheuristics have been used to tune parameters in search algorithms. Recent results on this subject are presented in [10] and references therein. In [17], a scatter search algorithm is used to calibrating parameters of a mathematical model based on ordinary differential equations and in [2] a survey on optimisation techniques in computational systems biology is provided.

Our approach is based on the assumption that the model is a black-box that can be controlled by providing as the input a set of parameter values and whose output can be observed. The optimisation process evaluates the output and uses this piece of information to guide search in the parameter space. The choice of the metaheuristic algorithm depends on the problem characteristics and mainly on the parameter domains and the objective function evaluation. In the next section we introduce the framework of our optimisation system and we detail its usage in a case study.

3   A General Framework Architecture

In this section we illustrate the framework that defines our approach for tackling parameter tuning in biological models and, in particular, in our simulation system. The framework describes any situation in which parameters of a system are to be tuned and thus generalises our specific problem. In Section 4 we will detail its actual implementation in our experiments. As from Figure 3, the main entities of the framework are the *model/simulator*, the *evaluator*, the *target* and the *optimiser*.

The *model* can be, in general, any model of a system, possibly stochastic. The *simulator* is responsible for the model execution. In our case, we designed a stochastic model of a biological system, thus its performance is evaluated by collecting statistics on sample executions. For brevity, in the following we will simply refer to the *model/simulator* component as *simulator*.

The *evaluator* is the component in charge of evaluating the performance of the system at a given parameter setting. This component is crucial and it is problem dependent because it has the responsibility of measuring the performance of the model when a particular parameter configuration is chosen. This measure depends on the kind of the system that has to be modelled and it provides the primary information that guides the optimisation/search process. The *evaluator* possibly makes use of a *target* behaviour that the simulator has to reproduce. For example, the target may be an attractor whose characteristics the system has to reproduce in its steady-state.

The fourth component, the *optimiser* has the goal of finding an optimal parameter setting. It is important to remark that the goal is to find any parameter setting that produces in the simulator the desired behaviour, therefore the concept of 'proven optimal solution' is meaningless because the objective of the designer is simply to achieve a model calibration satisfying the requirements. The optimiser can be any optimisation algorithm, but in our framework we consider only metaheuristic algorithms. In this contribution we will illustrate a case-study in which a population-based algorithm for continuous parameter is used.

In the following, we illustrate the details of each component of the framework used in our experiments.

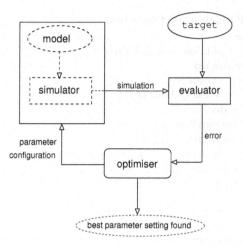

Fig. 1. Framework describing the parameter tuning process of a biological model

4   The Framework in Action

4.1   Model and Simulator

Starting from existing computational models and languages, such as stochastic π-calculus [16] and its implementation in SPiM [14], in [13] we have proposed a stochastic and discrete modelling approach in which a biochemical network can be modelled as a set of chemical reactions among the molecules involved and simulated according to the Gillespie's SSA (stochastic simulation algorithm) [7]. The approach is supplied with a tool that enables the description of a model through a high level language, and its simulation through an engine implementing an optimised version of SSA, so as to exactly foresee the chemical evolution/diffusion of substances in the system.

4.2   Error Evaluation

The error measurement is in general problem dependent. Thus, the designer has to define a proper distance function to evaluate the dynamics of the simulated system in comparison with a target behaviour. For example, the main characteristics of the final attractor can be considered, such as the average values of the output variables, their periods and the oscillation amplitude. Other characteristics can be also extracted in the frequency domain, for example by applying a Fourier transform. In addition, any method for comparing multi-dimensional data series can be used. In this work, we consider two systems of which we want to reproduce the steady-state dynamics, namely either the fixed point or the cyclic attractor. Therefore, once the final attractor target is given, the error evaluation module is designed in such a way that the distance between the attractor of our system and the target attractor is measured. In the first case, the

Algorithm 1. PSO high-level description

> Initialise population
> **while** termination conditions not met **do**
>   **for** $i = 1$ to pop_size **do**
>   **if** $f(\vec{x_i}) < f(\vec{p_i})$ **then** $\vec{p_i} = \vec{x_i}$
> $\vec{p_g} = \text{best}(\{p_j | j \text{ is a neighbour of } i\})$
>   **for** $d = 1$ to n **do**
> $\phi_1 = \text{randomFloat}(0, \phi_{max})$
> $\phi_2 = \text{randomFloat}(0, \phi_{max})$
> $v_{id} = \omega v_{id} + \phi_1(p_{id} - x_{id}) + \phi_2(p_{gd} - x_{id})$
> $v_{id} = \text{sign}(v_{id}) \cdot \text{sign}(|v_{id}|, v_{max})$
> $x_{id} = x_{id} + v_{id}$
>   **end for**
>   **end for**
> **end while**
> **output:** best solution found

Euclidean distance of the vectors representing the average values of the output variables is used, while in the second the vectors represent also features of the cyclic behaviour of the variables, such as peak frequency and amplitude.

4.3 Optimiser: Particle Swarm Optimisation

The metaheuristic algorithm we chose for our experiments is Particle swarm optimisation (PSO), a population-based metaheuristic particularly effective when dealing with problems defined over continuous variables [5]. Population-based algorithms are suitable for this kind of applications because they are very robust against rugged search landscapes, in which there are many local minima. PSO is an optimisation technique inspired by the metaphor of social interaction, for example bird flocking and fish schooling. Besides the metaphor, PSO is defined by formal mathematical models and has been proven to be very effective in solving optimisation problems, mainly continuous ones. The algorithm iteratively samples the search space by a *population* of samples, called *particles*. Particles have their own position and velocity that are updated each iteration by a rule that takes into account the quality of solutions represented by the particles. Positions and velocity are updated trying to gather the swarm toward good solutions, while keeping a form of exploration so as to balance search intensification and diversification. The PSO pseudo-code is reported in Algorithm 1. Particle positions are denoted by n-dimensional vectors $\vec{x_i}$ and velocities by vectors $\vec{v_i}$, for $i = 1, \ldots, \text{pop\_size}$. The best solution found by particle i since the beginning of the search is denoted by $\vec{p_i}$, while the best solution found so far by its neighbouring particles (possibly all the other particles) is denoted by $\vec{p_g}$. Positions and velocity are stochastically adjusted in such a way that each particle searches in the surroundings of the best solution it found and the best solution found by its neighbours. The objective function to be minimised is denoted by $f(\cdot)$. ϕ_{max} and ω are parameters of the algorithm.

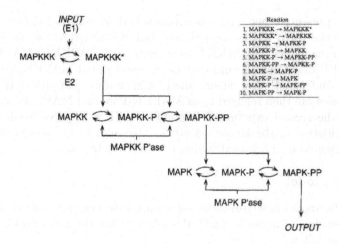

Fig. 2. The MAPK cascade schematic structure, as from [9]

5   The MAPK Cascade as a Case Study

As case study we consider the mitogen-activated protein kinase (MAPK) cascade under two different conditions, without and with a negative feedback loop, that lead to different steady-state responses: fixed point and oscillating attractor, respectively.

5.1   The Biological Background

The MAPK cascades are one of the best known and more often modelled biological pathways [9,12] that involves a series of three proteins kinases constituting portions of signal transduction implicated in diverse biological processes. The cascades always consists of a MAPK kinase kinase (MAPKKK), a MAPK kinase (MAPKK), and a MAPK. MAPKKKs activate MAPKKs that activate MAPKs by phosphorylation as shown in Figure 2. They lead from membrane receptors to cellular targets located in the cell cytoplasm and nucleus, mediating the cells response to external stimuli. The number of reactions that occur in the system is 10, as well as the number of the unknown rate constants, that are the parameters to be set. In a variety of cells MAPK are then involved in feedback loops with MAPK affecting the rate of MAPKKKs phosphorylation. The feedback can be positive or negative depending on whether the terminal kinase stimulates or inhibits the activation of the initial level.

5.2   The Target Data from Literature Models

As target data for the first experimental condition we considered the results of the MAPK model formalised with stochastic π-calculus and implemented in SPiM. The model is inspired from the ODE ones described in [9], validated with the results presented there and it is extensively described in [15]. Both the models assume that MAPKKKs are activated and inactivated by two enzymes —E1 and E2. In the model, MAPKKK* denote activated MAPKKK, MAPKK-P and MAPKK-PP denote singly and doubly phosphorylated MAPKK, respectively. MAPK-P and MAPK-PP denote singly and doubly phosphorylated MAPK. Phosphatases are then referred to as MAPKK-Pase and MAPK-Pase.

As for the second experimental condition with a negative feedback —that brings oscillation in the kinase activities— we refer to the results of an ODE model presented in [12] for extracting the target data set.

5.3   The Model

Our MAPK model is simply deduced from the reactions constituting the biological cascade and it can be easily described within the framework's high-level language as follow:

```
reaction kkkActivation : [kkk,e1] --> [kkkst,e1]   rate r1
reaction kkActivation : [kk,kkkst] --> [kkP,kkkst]   rate r2
reaction kActivation : [k,kkPP] --> [kP,kkPP]   rate r3
reaction kkpActivation : [kkP,kkkst] --> [kkPP,kkkst] rate r4
reaction kpActivation : [kP,kkPP]  --> [kPP,kkPP] rate r5
reaction kkkstDeActivation : [kkkst,e2] --> [kkk,e2] rate r6
reaction kkpDeActivation : [kkP,kkPase]  --> [kk,kkPase] rate r7
reaction kkppDeActivation : [kkPP,kkPase] --> [kkP,kkPase] rate r8
reaction kpDeActivation : [kP,kPase]  --> [k,kPase] rate r9
reaction kppDeActivation : [kPP,kPase]  --> [kP,kPase] rate r10
```

The case with feedback loop is obtained by adding two reactions expressing the effect of MAPK on the activity of the enzyme E1 —namely, MAPK-PP switches-off E1 as follow:

```
reaction e1DeActivation : [kPP,e1] --> [kPP,e1Dis]   rate r11
reaction e1Activation : [e1Dis] --> [e1]   rate r12
```

where $r1, \ldots, r12$ are the model parameters to be tuned. In both the experiments we consider the following initial conditions, where all the missed terms are equal to zero.

```
concentration [e1:1, e2:1, kkPase:1, kPase:1, kkk:10, kk:100, k:100]
```

5.4   Results

In this Section we present and discuss the results of the experiments concerning the MAPK case-study.

With the aim of devising a flexible implementation of the parameter tuning framework, so as it could be used also in other contexts, we defined a clear interface between the components that makes it possible to abstract from the actual implementation. In our prototype system, the *simulator* is the one introduced

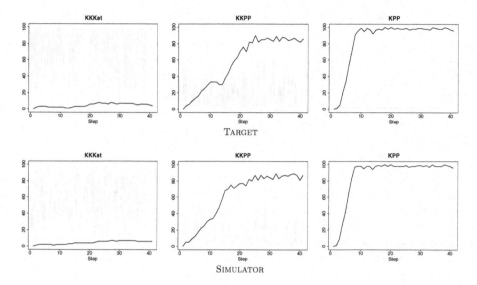

Fig. 3. Concentrations of the three proteins in the MAPK cascade in the target behaviour and in simulation, respectively

in Section 4. The *evaluator* modules extract characteristics of the attractors of a multi-dimensional data series so as to compare the results of our simulator against a target series. The simulator and the evaluation modules have been implemented in C because of efficiency reasons. The comparison of two tentative parameter settings should be done by comparing the sampled distribution of the executions of the stochastic simulator in the two settings.[1] Nevertheless, in a preliminary experimental analysis we observed a negligible variance in the simulator's results, therefore for each parameter configuration we took only one sample of the system execution. The *optimiser* modules have been implemented in Python. Besides a PSO, also a *random search* (RS) has been implemented as a comparison. As for the parameters of the PSO algorithm, we set the number of particles to 100, the number of iterations to 100, $\phi_{max} = 1$ and $\omega = 0.5$. The initial population was generated according to a uniform distribution, but also a beta distribution favouring extremes values was tested.

The first experiment concerns tuning parameters of the simulator reproducing the MAPK model that is characterised by a fixed point attractor, as shown in the uppermost row of Figure 3.[2] In this case, the distance between the output of our system and the target can be evaluated by considering the average value of the variables, once the transient is expired. Therefore, the *evaluator* computes the average values of system variables in our simulation, $\vec{v_s}$, and of the target, $\vec{v_t}$. The distance is computed as the Euclidean distance of the vectors $\vec{v_s}$ and $\vec{v_t}$. The lowermost row of Figure 3 reproduces a typical example of the output

---

[1] For instance, via non-parametric tests.

[2] Detailed data on results are omitted for lack of space and are available upon request.

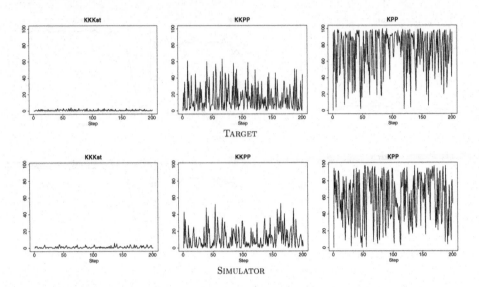

Fig. 4. Concentrations of the three proteins in the MAPK cascade with feedback loop in the target behaviour and in simulation, respectively

corresponding to the best parameter setting returned by the optimiser and we can observe that the steady-state dynamics of our system and the target are substantially the same. RS was not able to find any acceptable parameter setting, proving that metaheuristic search is effective.

The second series of experiments is devoted to the MAPK model with an oscillating attractor state, as shown in Figure 4. We implemented a specific evaluator that measures the similarity of the two attractors as Euclidean distance of attractor features, such as frequency of high and low peaks and their averages. These features capture the essence of oscillating behaviour. However, in this work we only aim at showing the effectiveness of the framework and more sophisticated evaluation modules can be designed. The uppermost row of Figure 4 shows a typical output of our system with a best parameter configuration returned by the *optimiser*. Also in this case the similarity of the attractors is apparent, even though the irregular oscillations are not perfectly captured which is correct according to the stochastic nature of the model. Anyway, by RS it was not possible to find any parameter setting producing a feasible steady-state pattern.

6   Conclusion

In this work, we have presented a framework for parameter tuning, devised mainly for biological models. The problem is defined as an optimisation problem over the parameter space and it is solved through metaheuristic algorithms. We have shown the application of such method to a case-study involving a stochastic simulator and a PSO metaheuristic. Results show that the method is effective in properly tuning parameters. Currently, the framework is in prototypical version and an extension suitable for distribution is planned to be devised. Furthermore, additional optimisation components are under development which implement other metaheuristics, such as simulated annealing and genetic algorithms.

References

1. Alves, R., Antunes, F., Salvador, A.: Tools for kinetic modeling of biochemical networks. Nature Biotechnology 24(6), 667–672 (2006)
2. Banga, J.: Optimization in computational systems biology. BMC Bioinformatics 2(7) (2008)
3. Blum, C., Roli, A.: Metaheuristics in combinatorial optimization: Overview and conceptual comparison. ACM Computing Surveys 35(3), 268–308 (2003)
4. Bower, J.M., Bolouri, H. (eds.): Computational Modeling of Genetic and Biochemical Networks. MIT Press, Cambridge (2004)
5. Clerc, M.: Particle Swarm Optimization. ISTE (2006)
6. De Jong, H.: Modeling and Simulation of Genetic Regulatory Systems: A Literature Review. Journal of Computational Biology 9(1), 67–103 (2002)
7. Gillespie, D.T.: Exact stochastic simulation of coupled chemical reactions. The Journal of Physical Chemistry 81(25), 2340–2361 (1977)
8. Hoos, H., Stützle, T.: Stochastic Local Search Foundations and Applications. Morgan Kaufmann Publishers, San Francisco (2005)
9. Huang, C.-Y.F., Ferrell, J.E.J.: Ultrasensitivity in the mitogen-activated protein kinase cascade. Proceedings of the National Academy of Sciences of the United States of America 93(19), 10078–10083 (1996)
10. Hutter, F., Hoos, H.H., Stützle, T.: Automatic algorithm configuration based on local search. In: Proceedings of the 22nd Conference on Artificial Intelligence (AAAI 2007), pp. 1152–1157 (2007)
11. Kauffman, S.A.: The Origins of Order: Self-Organization and Selection in Evolution. Oxford University Press, Oxford (1993)
12. Kholodenko, B.N.: Negative feedback and ultrasensitivity can bring about oscillations in the mitogen-activated protein kinase cascades. European Journal of Biochemistry 267(6), 1583–1588 (2000)
13. Montagna, S., Viroli, M.: A computational framework for modelling multicellular biochemistry. In: IEEE CEC 2009 Preceedings, Trondheim, Norway, May 18–21 (2009)
14. Phillips, A.: The Stochastic Pi Machine, SPiM (2007), Version 0.05, http://research.microsoft.com/~aphillip/spim/
15. Phillips, A., Cardelli, L., Castagna, G.: A graphical representation for biological processes in the stochastic pi-calculus. In: Transactions on Computational Systems Biology VII. LNCS, pp. 123–152. Springer, Heidelberg (2006)
16. Priami, C.: Stochastic pi-calculus. The Computer Journal 38(7), 578–589 (1995)
17. Rodriguez-Fernandez, M., Egea, J., Banga, J.: Novel metaheuristic for parameter estimation in nonlinear dynamic biological systems. BMC Bioinformatics 7(483) (2006)
18. Szallasi, Z., Stelling, J., Periwal, V. (eds.): System Modeling in Cell Biology - From Concepts to Nuts and Bolts. MIT Press, Cambridge (2006)

Hypotheses about Typical General Human Strategic Behavior in a Concrete Case

Rustam Tagiew

Institute for Computer Science of TU Bergakademie Freiberg, Germany
tagiew@informatik.tu-freiberg.de

Abstract. This work is about typical human behavior in general strategic interactions called also games. This work is not about modelling best human performance in well-known games like chess or poker. It tries to answer the question
'How can we exactly describe what we typically do, if we interact strategically in
games we have not much experience with?' in a concrete case. This concrete case
is a very basic scenario - a repeated zero sum game with imperfect information.
Rational behavior in games is best studied by game theory. But, numerous experiments with untrained subjects showed that human behavior in strategic interactions deviates from predictions of game theory. However, people do not behave in
a fully indescribable way. Subjects deviate from game theoretic predictions in the
concrete scenario presented here. As results, a couple of regularities in the data
is presented first which are also reported in related work. Then, the way of using machine learning for describing human behavior is given. Machine learning
algorithms provide automatically generated hypotheses about human behavior.
Finally, designing a formalism for representing human behavior is discussed.

1 Introduction

'How can we exactly describe what we typically do, if we interact strategically in games
we have not much experience with?' Participants of a strategic interaction (SI) are interested in achieving a certain goal and their success is in part depending on decisions of
other participants. Game theory is the field that studies SI called games. The research
in classical game theory is concentrated on finding a solution for a defined game. A
definition of a game consists of a number of participating agents called players, their
legal actions and a payoff function for every player. The payoff function defines each
players outcome depending on his actions, other players's actions and random events in
the environment. Game theory uses the concept of rationality for motivating solutions
of a game. Rationality means that an agent maximizes his payoff considering what he
knows. The solution of a game is a prediction about the behavior of the players. The
assumption of rationality is the basis for this prediction. However, an agent is not rational, if he can not act irrationally [1]. Free will is the precondition for the rationality.
From the evolutionary point of view, free will is doubtful, because it enables suicide.
But, free will enables also a self-aspired adaptation to hypothetical future situations.
For a game theoretic analysis, the players are rational and they know that other players
are rational too. They also know that other players know that they are rational and so
on. A solution of a game is called an equilibrium. Deviating from an equilibrium does

R. Serra and R. Cucchiara (Eds.): AI*IA 2009, LNAI 5883, pp. 476–485, 2009.
© Springer-Verlag Berlin Heidelberg 2009

not maximize the payoff and is therefore irrational. Not every game has an equilibrium. However, there is at least one mixed strategies equilibrium (MSE) in finite games [2].

The notion game is commonly used for pleasant time spending activities like board games. In game theory, one considers also warfare as game playing. The game structure and the game implementation are two different issues. A board game can have the same game structure as a war. The mentioned pleasant time spending activities are examples, where the game rules are implemented by the players. Why do we invent artificial environments, where we practice self-restriction and try to maximize virtual utilities? We like it to learn and then to perform better in games [3]. Some board games are even developed to train people like Prussian army war game 'Kriegspiel Chess' [4] for officers. Typical human behavior in games is not optimal and deviates from game theoretic predictions [5]. A conceivable reason are the bounded computational resources of humans. Another reason is the (seeming) absence of rationality in human behavior. "British people argue that it is worth spending billions of pounds to improve the safety of the rail system. However, the same people habitually travel by car rather than by train, even though travelling by car is approximately 30 times more dangerous than by train!"[6, p.527–530] Generally spoken if a subject makes mistakes, then he is either irrational or not skilled enough to avoid them (bounded rationality). Both positions are common.

One can say without any doubt that if a human player is trained in a concrete game, he performs close to optimal. But, a chess master does not also play poker perfectly and vice versa. On the other side, a game theorist can find a way to compute an equilibrium for a game, but it does not make a successfull player out of him. There are many games, we can play. For most of them, we are not trained. That is why it is more important to investigate our behavior in general game playing than game playing in concrete game.

This work concentrates on typical human general game playing. It is about the common human deviations from predicted equilibria in games, for which they are not trained. Modeling typical human behavior in general games needs a represention formalism which is not specific to a concrete game. An example-driven development of such a formalism is the challenge addressed in this paper. The example introduced in this work is a repeated two-player zero-sum game with no pure strategy equilibria. Each player has a couple of actions called strategies. Every round of these games is a simultaneous choice of actions. A payoff matrix includes payoffs for both players for every combination of actions. The sum of payoffs is always zero. The solution is to use MSE. A mixed strategy is defined through a distribution over strategies. A player, who uses a mixed strategy, chooses randomly a strategy according to this distribution.

The next section summarizes related work on a formalism for human behavior in games. Then, the conducted experiment is introduced. A preliminary analysis and a black box approach are presented after words. Conclusion with a recommendation for a formalism and future work conclude this paper.

2 Related Work

[5] is a very comprehensive gathering of works in experimental psychology and economics on typical human behavior in games. This work inspired reseach in artificial intelligence [7]. Which led to the creation of network of influence diagrams (NID) as

a representation formalism. NID is a formalism similar to the possible worlds semantics of Kripke models [8] and is a super-set of Bayesian games. The main idea of NID is modelling human reasoning patterns in diverse strategic interaction. Every node of NID is a multi-agent influence diagram (MAID) representing a model of SI of an agent. MAID is an influence diagram (ID), where every decision node is associated with an agent. ID is a Bayesian network (BN), where one has ordinary nodes, decision nodes and utility nodes. In summary, this approach assumes that human decision making can be modelled using BN. This formalism is already applied for modelling reciprocity in the negotiation game Colored Trails [9]. The result of this work is that artificial players adapted to human behavior modelled using BN perform better than standard game theoretical algorithms.

Another independent work is [10]. It investigates repeated zero-sum games no pure strategy equilibria. First, they present a set of existing behavior models - Basic Reinforcement Learning [11], Erev and Roth's Reinforcement Learning [12], Self Tuning Experience Weighted Attraction [13] and Normalized Fictitious Play [14]. Then, they introduce their own model - Regret-Driven-Neuronal Networks. As result, they evaluate these models on existing data of conducted experiments from previous works. On games with more than two strategies, the probability of correct predictions does not exceed 45% for all models.

Both works follow the same approach. First, they construct a model, which is based on theoretical considerations. Second, they adjust the parameters of this model to the experimental data. This makes the human behavior explainable using the concepts from the model.

3   Experiment Setup

The scenario chosen for constructing a formalism for modelling human behavior in games has already been mentioned in [15]. Prevention of both, misconception of game rules and sophistication of subjects, is intended. Roshambo-like games are chosen, because they are fast to explain to test persons and have a simple structure for analysis of basic concepts. On the other side, most people do not train to play them. These games are symmetric, zero sum, two player and the entries of the payoff matrix can be only 1(win), 0(draw) and −1(loss). Unlike the simple Rock-Paper-Scissors, one can also use more than three gestures. The payoff matrices of these games can also be seen as adjacency matrices of directed graphs (through setting −1 to 0). There are 7 distinguishable directed graphs for 3 gestures, 42 for 4, 582 for 5 and so on [16]. All 5 gesture graphs are proved to find variants with only one MSE non-equal distribution. There are 4 such variants (Fig.1). The first variant is an existing American variant of Roshambo with additional gestures 'fire' and 'water'. Three of these graphs with probabilities $\frac{1}{9}$ and $\frac{1}{7}$ and a 4-nodes cyclic graph have been merged to a graph of 9 nodes. Every node has been assigned a plausible name and hand gesture. The resulting game framework is shown on Fig.1. In this framework, the cyclic 4-gesture variant is Paper-Mosquito-Hunter-Monkey.

The feature of gesture recognition has been added to the scenario wit the purpose of making it more situated and similar to real-world physical activities like sport competitions. A cyber glove and a classifier for recognition of gestures are used [17]. The recognition was almost 100% safe for hands of common measure.

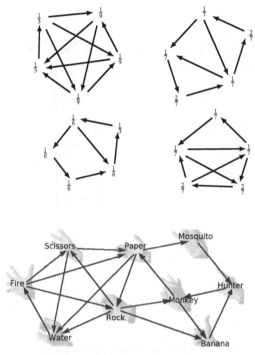

Fig. 1. All nontrivial variants for 5 gestures and the designed game framework [15]

The study was conducted on a thread of 200 one-shot games of 7 kinds using the game framework. A player had a delay for consideration of 6 sec for every shot. If he did not react, the last or default gesture was chosen. A thread lasted $200 * 6$ sec $= 20$ min. None of the games was played for more than 30 rounds. There was no delay between different games. This was aimed to prevent subjects from learning and adjusting strategies. In similar experiments [10], games are played in threads with more than 100 rounds. Every game has got an ID (in brackets). The thread had the following structure - 30 times Rock-Paper-Sci-ssors (31), 30x Hunter-Monkey-Pa-per-Mosquito (41), 30x Paper-Rock-Scissors-Fire-Water (51), 30x Paper-Rock-Monkey-Fire-Water (52), 30x Rock-Banana-Monkey-Paper-Hun-ter (53), 30x Hunter-Monkey-Paper-Mosquito-Banana (54) and finally 20x Scissors-Water-Rock-Monkey-Paper-Fire (61). Every of these games has at least one MSE. An MSE can be considered a high-dimensional vector. The space, in which every vector is an MSE, can be defined as a polytop with a set of vertices. If this polytop has only one vertex, then the game has a single MSE. The games have the following vertices of polytops for MSE - $31 \Rightarrow \{(\frac{1}{3},\frac{1}{3},\frac{1}{3})\}$, $41 \Rightarrow \{(0,\frac{1}{2},0,\frac{1}{2}),(\frac{1}{2},0,\frac{1}{2},0)\}$, $51 \Rightarrow \{(\frac{1}{9},\frac{1}{9},\frac{1}{9},\frac{1}{3},\frac{1}{3})\}$, $52 \Rightarrow \{(\frac{1}{7},\frac{1}{7},\frac{1}{7},\frac{2}{7},\frac{2}{7})\}$, $53 \Rightarrow \{(\frac{1}{7},\frac{1}{7},\frac{1}{7},\frac{2}{7},\frac{2}{7})\}$, $54 \Rightarrow \{(0,\frac{1}{2},0,\frac{1}{2},0),(\frac{1}{4},\frac{1}{4},\frac{1}{4},\frac{1}{4},0)\}$ and $61 \Rightarrow \{(0,\frac{1}{2},0,\frac{1}{2},0),(\frac{1}{4},\frac{1}{4},\frac{1}{4},\frac{1}{4},0)\}$.

Ten computer science undergrates were recruited. They were in average 22,7 years old and 7 of them were male. They had to play the thread twice against another test person. In this way, 2000 one-shot games or 4000 single human decisions are gathered. Every person got € 0.02 for a won one-shot game and € 0.01 for a draw. A won thread was awarded with an additional € 1 and a drawn one with € 0.5. In summary, a person could gain between € 0 and € 10 or in average € 5. The persons, who played against each other, sat in two separate rooms. One of the players used the cyber-glove[1] and the another a mouse as input for gestures. The software architecture used for this experiment is the multi-agent system FRAMASI [15]. The graphical user interface had following features - own last and actual choice, opponents last choice, a graph of the

---

[1] Except of one participant, who had to play both threads with mouse, because of mismatch of the cyber glove size to her hand.

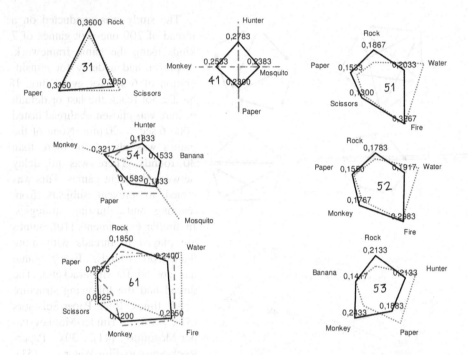

Fig. 2. Radar plots of predicted and observed stategy profiles in different games

current game, a timer and already gained money. According to statements of the persons, they had no problems to understand the game rules and to choose a gesture timely. All winners and 80% of losers attested, that they had fun to play the game.

4   Preliminary Analysis

As it is expected, the distribution over strategies will be nearby a mixture of MSE, equal distribution and best response on opponent's equal distribution [18]. Fig.2 presents radar plots of MSE and observed distribution for every investigated game. Black lines represent the observed distribution. Every edge of a black lined polygon is assigned with a value of average probability. Dashed gray lines represent MSE. There are $10 * 2 * 30 = 600$ recorded human decisions for games from 31 through to 53 and $10 * 2 * 20 = 400$ for game 61. In games 51, 52 and 53, at least the probability for one of the actions deviates with significance at $< 0.1\%$ from MSE towards equal distribution. The high probability of fire in 51 can be explained as best response on equal distribution of the opponent. In game 54, the strategy 'banana' is used although this strategy is dominated. It is irrational to use 'banana' in 54. A possible model to explain this characteristics is quantal response equilibrium (QRE) [19]. But QRE is far from perfect for this need [5].

A human decision in a repeated zero-sum game is assumed to be dependent upon the last rounds. Due to the limitations of human information processing [20] and the number of items in a round, the number of considered previous rounds is set to 3. Tuples are counted, which consists of own actual decision plus own three last decisions plus opponents last three decisions. Persons will produce sequences of turns which are far from being independent and identically distributed [21]. Further, they transform in consequence of this deviation a repeated matrix game to an extensive game, in which one must consider his turns in dependence of own and opponents last turns. In game 31, there are 4 tuples, each of which were observed 4 times. For instance, the tuple {rock, {rock, paper, scissors}, {paper, rock, paper}} was observed in threads of four different persons. The probability for one tuple to be observed in the data of 31 more than three times is about 0.0126%.

The strong significance of the last characteristics gives a motivation to calculate a probabilistic grammar of human behavior. In this grammar, variables $A, B, C, D, \ldots \in$ Gestures have all different values. Patterns are abstract types of sequences and constructed of these variables. For instance, the sequences {paper, rock} and {fire, water} are both of pattern AB. Patterns with length 4 are used. 4 is most reasonable for the size of the data. Theoretical probabilities of these patterns are calculated according to the assumption, that humans use equal distributions. This can be assumed in games 31 and 41. Per game one has $(30 - 3) * 2 * 10 = 540$ sequences and can assign them to patterns. In game 31, the pattern $ABCD$ can not be considered, because there are only three gestures. Each pattern possesses a number of bridges. Bridges are repetitions over 0, 1 or 2 elements in a pattern. Tab.1 shows the results. Many bridges cause that the observed probability is lower than the theoretical. In contrast, the observed probability raises significantly, if there are no or a few bridges.

Table 1. Patterns probabilities for 31 and 41. Boxes are for significance at 5%.

Pattern	Bridges 0 1 2	Game 31 Theoretical	Game 31 Observed avg.	Game 41 Theoretical	Game 41 Observed avg.
ABCD	0 0 0			.0938	.2611
ABCA	0 0 1	.0741	.15	.0938	.1519
ABCB, ABAC	0 1 0	.0741	.1297	.0938	.0926
ABCC, AABC, ABBC	1 0 0	.0741	.0716	.0938	.0741
ABBA	1 0 1	.0741	.0611	.0469	.037
ABAB	0 2 0	.0741	.0593	.0469	.0333
AABA, ABAA	1 1 1	.0741	.0602	.0469	.0213
AABB	2 0 0	.0741	.0463	.0469	.0222
ABBB, AAAB	2 1 0	.0741	.0352	.0469	.0195
AAAA	3 2 1	.037	.0185	.0156	.0056*

*significant at 6%.

5 Black Box Approach

As has been shown in the previous chapter, the behavior observed in the experiment can not be explained using only game theory. Currently, three different approaches can be considered to find an explanation for it: (1) interview the subjects about their behavior, (2) measure their brain activities or (3) find a theory, which explains the data. The third approach has two kinds of procedures. The first procedure is to develop a sophisticated theory and then to prove this theory on the data, as it is done in related work. The second procedure is to find regularities in the data and then to make a hypothesis which explains these regularities.

This work follows the second procedure of the third approach. It can be named as *black box* approach. The black box in this case is the human player. The input are the game rules (game) and previous decisions of players (history). The output is the current decision. Human decision can be denoted by the function HD(Game, History) $\rightarrow$ Decision. Finding a hypothesis which matches the behavior of the black box is a typical problem called supervised learning [22]. There is already a big amount of algorithms for supervised learning. Each algorithm has its own hypothesis space (HS). For a Bayesian learner i.e., the hypothesis space is the set of all possible Bayesian networks. There are many different types of hypothesis spaces - rules, decision trees, Bayesian models, functions and so on. Concrete hypothesis HD^I is a relationship between input and output described by using the formal means of the corresponding hypothesis space.

Which hypothesis space is most appropriate to contain valid hypotheses about human behavior? That is a machine learning version of the question about a formalism for human behavior. The most appropriate hypothesis space contains the most correct hypothesis for every concrete example of human behavior. A correct hypothesis does not only perform well on the given data (training set), but it performs also well on new data (test set). Further, it can be assumed that the algorithms which choose a hypothesis perform alike well for all hypothesis spaces. This assumption is a useful simplification of the problem for a preliminary demonstration. Using it, one can consider the algorithm with the best performance on the given data as the algorithm with the most appropriate hypothesis space. The standard method for measurement of performance of a machine learning algorithm or also a classifier is cross validation.

As it is already mentioned, a machine learning algorithm has to find hypothesis HD^I which matches best the real human behavior function HD. Human decision making depends mostly on a small part of the history due to bounded resources. This means that one needs a simplification function $S(History) \rightarrow$ Pattern. Using function S the function $HD(X,Y)$ is to be approximated through $HD^{II}(X,S(Y))$. The problem for finding the most appropriate hypothesis can be formulated in equation 1. The function match in equation 1 is considered to be implemented through a cross validation run.

$$\arg\max_{HS}(\max_{HD^{II} \in HS}(match(HD(X,Y), HD^{II}(X,S(Y))))) \tag{1}$$

The data of the experiment is transformed to sets of tuples for every game. The tuples have the same structure, as described in the previous chapter. Every tuple has the length $3+3+1 = 7$. The simplification function is a window over three last turns. For a game, there are $|Gestures|^7$ possible tuples. For instance, 2187 different tuples are possible in game 31 and it is the smallest number of possible tuples among the played games. The data for game 31 is the most complete. The decisions in the first three turns of game are not considered. As has already been mentioned, the size of a set is 540 tuples or 340 (only for the last game).

Implementations of classifiers provided by WEKA [23] are used for the cross validation on the sets of tuples. The task is to find a relationship between the last three players's decisions (6 items) and the current decision. There are 45 classifiers available in the WEKA library, which can handle multi-valued nominal classes. Gestures or also strategies in games are nominal, because there is no order between them. These classifiers belong to different groups - rule-based, decision trees, function approximators, baysian learners, instance-based and miscellaneous. A cross validation of all 45 classifiers on all 7 sets of tuples is performed. The number of subsets for crossvalidation is 10.

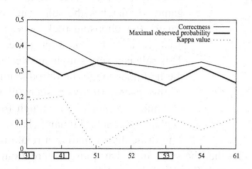

Fig. 3. Average correctness in cross validation. Boxes are for significant gaps between prediction correctness and maximal observed probability.

There is no classifier which performs best on all games. Further even the highest average correctness is very low. Decreasing and increasing the window size in the function S diminishes the performance even more. Fig.3 shows the results. The gap between the highest observed probability of a gesture and the highest average correctness is different depending on the game. It should be mentioned that the highest observed probability on this figure is slightly different from values of fig.2, because the 3 first decisions are not taken into account. The Kappa value is a measure for the deviation of a classifier from random. In game 51, all classifiers completely fail to find a hypothesis in subsets better than 'always fire'. 'Always fire' has a Kappa value of 0. The best classifiers for games with a gap between average correctness in cross validation and maximal probability of a gesture predictions are sequential minimal optimization (SMO) [24] for 31, multinomial logistic regression (L) [25] for 41, single rule classifier for 52, Bayesian networks for 53 and K* (instance based) for 61.

Which classifier is the most robust? One can choose two criteria - highest minimum performance or highest average performance. In game playing conditions, if the correctness of prediction is 5 percentage points higher, one gets a 5% higher payoff. To find the classifier with the most robust usability in game playing conditions, the difference between average correctness and probability of equal distribution ($\frac{1}{|Gestures|}$) is calculated for each classifier and game. SMO has the highest minimum difference and a simple variant of L (SL) has the highest average difference. On the other side, L has the the highest average Kappa value and voting feature intervals classification (VFI) has the highest minimum Kappa value. Fig.4 shows the average correctness of these classifiers on the datasets. Three of these four classifiers have functions as hypothesis space. The problem of functions is that most of them can not be verbalized. Consequently, the first question from the introduction can be answered using natural language. On the

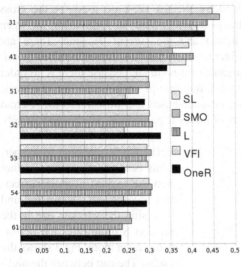

Fig. 4. Chosen results

other side, the success of function based classifiers means that we can not explain our behavior in our natural language.

The single rule classifier (OneR), which is also included in the histogram on fig.4, produces a hypothesis which contains only one single rule. Using this classifier, one can find out that 43.15% of the data in game 31 matches the rule 'choose paper after choosing rock, scissors after rock and rock after paper'. This rule is a very simple answer to the first question in the introduction of this paper. However, such rules of thumb are not exact enough for explaining general human behavior. The difficulty of finding a relationship between input and output is the fact that the same input can cause different outputs. Even using the instance based approach K* which is validated on training data, one achieves only 80.37% correctness in game 31.

6   Conclusion

Is Penrose [26] correct, if he argues that there is something fundamentally non-computational about human thought? His point of criticism is the generalizability. The stra-tegic behavior are only the observable actions, whose origins are tried to be understood as generally as possible. Summarizing the results of this work, it can be said that SMO and L can find the most general hypothesis about regularities of human behavior in the investigated scenario. But the correctness of this hypothesis is still very low. It is only slightly better than in the related work. It is doubtful, wether one can define an algorithm which predicts exactly general human strategic behavior at all. However, one can say that a complex function models human better than a rule of thumb, although this function can not be adequately verbalized. If our behavior can be exactly defined only by a function which can not be put in natural language, then it means that we could have in some cases not enough words to explain own strategic behavior. It is an exciting philosophic issue which is the most important result of this work.

7   Future Work

This work raises a lot of question for future work. Social interaction can be modeled as strategic interaction. In a social interaction it is usefull to have an ability to explain own behavior. But what if you can do it in natural language? Strategography and strategophony are also possible future directions in understanding general human strategic behavior - if we can not verbalize our strategic behavior, can we represent it as images or music?

References

1. Ariely, D. (ed.): Predictably irrational. HarperCollinsPublishers (2008)
2. Nash, J.: Non-cooperative games. Annals of Mathematics (54), 286–295 (1951)
3. Gobet, F., de Voogt, A., Retschitzki, J.: The Psychology of Board Games. Psy. Press (2004)
4. Li, D.H.: Kriegspiel: Chess Under Uncertainty. Premier (1994)
5. Camerer, C.F.: Behavioral Game Theory. Princeton University Press, New Jersey (2003)
6. Eysenck, M.W., Keane, M.T.: Cognitive Psychology. Psychology Press (2005)
7. Gal, Y., Pfeffer, A.: A language for modeling agents' decision making processes in games. In: AAMAS, pp. 265–272. ACM Press, New York (2003)
8. Fagin, R., Halpern, J., Moses, Y., Vardi, M.: Reasoning about Knowledge. MIT Press, Cambridge (1995)
9. Gal, Y., Pfeffer, A.: Modeling reciprocal behavior in human bilateral negotiation. In: AAAI, pp. 815–820. AAAI Press, Menlo Park (2007)
10. Marchiori, D., Warglien, M.: Predicting human interactive learning by regret-driven neural networks. Science 319, 1111–1113 (2008)
11. Erev, I., Roth, A.: Predicting how people play games. American Economic Review 88 (1998)
12. Erev, I., Roth, A., Slonim, L., Barron, G.: Predictive value and the usefulness of game theoretic models. International Journal of Forecasting 18, 359–368 (2002)
13. Ho, T.-H., Camerer, C., Chong, J.-K.: Self-tuning expirience-weighted attraction learning in games. Journal of Economic Theory 133, 177–198 (2007)
14. Ert, E., Erev, I.: Replicated alternatives and the role of confusion, chasing and regret in decisions from experience. Journal of Behavioral Decision Making 20, 305–322 (2007)
15. Tagiew, R.: Towards a framework for management of strategic interaction. In: ICAART, pp. 587–590. INSTICC Press (2009)
16. Harary, F.: The number of oriented graphs. Michigan Mathematical J. 4, 221–224 (1957)
17. Heumer, G., Amor, H.B., Jung, B.: Grasp recognition for uncalibrated data gloves: A machine learning approach. Presence - Teleoperators and Virtual Environ 17, 121–142 (2008)
18. Stahl, D.O., Wilson, P.W.: Experimental evidence on players' models of other players. Economic Behavior & Organization, 309–327 (1994)
19. Richard, M., Palfrey, T.: Quantal response equilibria for normal form games. Games and Economic Behavior 10, 6–38 (1995)
20. Miller, G.A.: The magical number seven plus or minus two: Some limitations on our capacity for processing information. Psychological Review 63, 81–97 (1956)
21. Budescu, D.V., Rapoport, A.: Subjective randomization in one- and two-person games. Journal of Behavioral Decission Making 7, 261–278 (1994)
22. Mitchell, T.M.: Machine Learning. McGraw-Hill, New York (1997)
23. Witten, I.H., Frank, E.: Data Mining. Morgan Kaufmann, San Francisco (2005)
24. Platt, J.C.: Fast training of support vector machines using sequential minimal optimization. In: Advances in Kernel Methods - Support Vector Learning, pp. 185–208. MIT Press, Cambridge (1999)
25. Cessie, S., Houwelingen, J.C.: Ridge estimators in logistic regression. Applied Statistics 41, 191–201 (1992)
26. Penrose, R.: The Emperor's New Mind. Oxford University Press, Oxford (1989)

A Fuzzy Approach to Product Configuration on Standard Databases

Luigi Portinale

Dipartimento di Informatica
Università del Piemonte Orientale, Alessandria, Italy

Abstract. In the present paper, we propose an approach to intelligent retrieval and configuration of component-based products, starting from a set of possibly fuzzy user requirements provided at different levels of detail. A conceptual product model is introduced and its use during the configuration process is discussed. The proposed approach exploits a fuzzy generalization of SQL and a bottom-up (from basic to complex components) configuration process that can be implemented on top of a standard RDBMS. The approach is illustrated through a simple example based on a PC assembly task.

1 Introduction

Several products having a reasonable market on the web are configurable products (e.g. personal computers, travels, music compilations, etc...). The target product is composed by a set of components meeting a given set of (possibly imprecise) requirements. However, a limitation of several web applications to e-commerce is the lack of automatic configuration facilities. In addition, due to the boolean nature of standard SQL retrieval, soft (i.e. imprecise or approximate) requirements are difficult to be dealt with. On the other hand, RDBMS provides efficient data management and query facilities, and real-world applications cannot leave aside the fact that the data of interest are usually stored in a relational database.

Configuration problems are traditionally dealt with in AI either within the constraint satisfaction paradigm [3,10,14] or by structured logic approaches [4,12,7]. Common to every knowledge-based approach to configuration are the following items: ([6]) (1) a set of concepts or domain objects, possibly organized in classes; (2) a set of relations between domain objects, in particular taxonomical and compositional relations; (3) a configuration objective or task, specifying the demand and the constraints a created configuration must fulfill; (4) control knowledge for the configuration process. As recognized in [6], operational processable models are needed. In order to achieve such a goal, different methodologies can be adopted like compilation of models [11], the use of case-based reasoning (CBR) [5,13], the automatic decomposition of the problem during state space search [1]. However, all the above mentioned approaches do not directly address the issue of flexibly exploiting, during the configuration process, the presence of the data of interest in standard relational databases, and

R. Serra and R. Cucchiara (Eds.): AI*IA 2009, LNAI 5883, pp. 486–495, 2009.

the corresponding data management and query facilities. Moreover, most of the proposed frameworks (with the exception of [5]) do not take into consideration the approximate nature of the user requirements. In the present paper, we propose an approach for the definition of an architecture for on-line searching and configuration of products based on the following characteristics: (1) a conceptual model of the configurable products, representing the structural decomposition of the modeled product; (2) a set of approximate, i.e. fuzzy, user requirements on the desired target product; (3) a fuzzy knowledge base providing the semantics for the possible user requirements; (4) a fuzzy extension to the standard SQL query language, through which to implement the retrieval and the composition strategy of the product components, directly on top of a RDBMS. Concerning the last point, we introduce an extended version of SQL, able to deal with fuzzy predicates and conditions, defined over standard attributes of a tables, and based on the **SQLf** language proposed in [2]. We will show how to exploit the conceptual model and the user requirements, in order to automatically derive a set of fuzzy SQL queries able to retrieve the suitable components. A bottom-up strategy on the conceptual model is then introduced with the aim of guiding the application to suitably combine the results of such queries.

The paper is organized as follows: section 2 introduces a short review of fuzzy queries on standard databases, section 3 discusses the conceptual model used for the configuration process while in section 4 the configuration algorithm is presented and discussed through a specific example. Finally, conclusions are drawn in section 5.

2 Fuzzy Queries in Databases

In this work, we are interested in *fuzzy queries* on an *ordinary (non-fuzzy) database.* The implementation of fuzzy extensions to SQL can be actually provided on top of a standard relational DBMS, by means of a translation following the *derivation principle* [2]. In the fuzzy SQL language we consider in this paper (based on **SQLf** [2]), the WHERE condition can be a composite fuzzy formula involving both crisp and fuzzy predicates (i.e. linguistic values defined over the domains of the attributes of interest), as well as crisp and fuzzy operators. We consider, as in [2], the following syntax

SELECT (λ) A FROM R WHERE fc

which meaning is that a set of tuples with attribute set A, from relation set R, satisfying the fuzzy condition fc with degree $\mu \geq \lambda$ is returned. In fuzzy terms, the λ-cut of the fuzzy relation R_f resulting from the query is returned. When the min operator is adopted as *t-norm* for conjunction and the max operator is adopted as *t-conorm* for disjunction, then it is possible to derive from a fuzzy SQL query, an SQL query returning exactly the λ-cut required (see [2] for details).

Example. Consider a generic relation PRODUCT containing the attribute price over which the linguistic term medium is defined. Figure 1 shows a possible fuzzy distribution for medium as well as the distribution of a fuzzy operator

≪ (much less than), defined over the difference $(a - b)$ of the operands, by considering the expression $a \ll b$. Let the retrieval confidence level be $\lambda = 0.8$; it is trivial to verify that if min is the t-norm for conjunction a condition like (price = medium∧price ≪ 100) is satisfied at level at least λ iff $(110 \leq \text{price} \leq 180) \wedge (\text{price} - 100) \leq -18$. The latter condition can be easily translated in a standard WHERE clause of SQL.

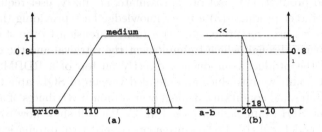

Fig. 1. Fuzzy Distributions

3   Conceptual Modeling

As already mentioned, any knowledge-based approach to product configuration needs to define a suitable conceptual model, where both taxonomic (*is_a*) and partonomic (*composed_by*) relations are of fundamental importance. The model we adopt in our approach is based on the FPC model by Magro and Torasso [7]. The conceptual framework is based on a hierarchy of components with the following basic ontology: *entities* divided into *categories* representing generic components (i.e. generic classes of components) of the final product, and *basic categories* representing classes of components having specific instances associated to them; *composed_by links* connecting a category A to an entity B, meaning that B is a component of A; *is_a links* connecting a category A to an entity B, meaning that B is a subclass of A (i.e. that the set of components in B are a particular typology of the components in A). If B is a component of A or if B is a sub-class of A we say that B is a child of A (i.e. *composed_by* links are from parent to child, while *is_a* link are from child to parent). We also consider cardinality information with respect to a *composed_by* link: in particular every such a kind of link is provided with a *number restriction* representing the minimal and maximum number of sub-components that may occur in the link.

An example of a conceptual model for a (simplified) personal computer system is reported in figure 2[1] (similar examples in the FPC framework are reported in [1,7]). For instance, we read from the model that the PC system is composed by exactly one computational system and exactly one storage system; the computational system can be composed by one or two CPUs, one motherboard (MB) and one set of RAM slots; the storage system of interest is composed by one or two hard-disks and by one (optional) CD-ROM; Different typologies of CPU, MB

---

[1] Dashed squares around some entities will be explained in section 4.

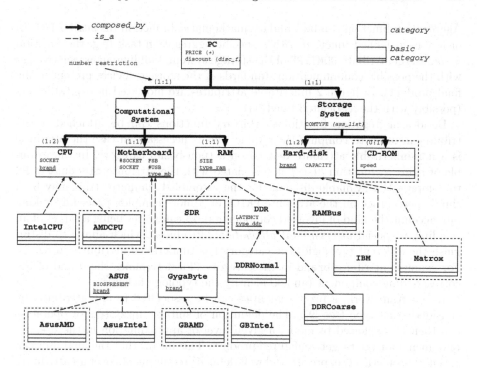

Fig. 2. Example of a conceptual model for a PC

or RAM are possible, as well as different typologies of hard-disks. For example MBs are differentiated into ASUS and GigaByte MBs (GB), which are in turn both distincted into Intel or AMD MBs. The ASUS entity is a category (i.e. there are no specific stored instances of such category), while ASUSAMD is a basic category with specific stored instances associated to such a component. As we will see, instances of non-basic categories are built, during the configuration process, from instances of basic categories. Similar consideration hold for the other entities in fig. 2.

Entities can have attributes. Attributes may be specific to a given entity (i.e. non-inheritable) or inherited from source to target in a *composed_by* link chain and from target to source in a *is_a* link chain. Inheritable attributes are then associated only to categories. In case they are defined on a category having sub-components, an *aggregation function* must be defined; the meaning is that the value of the attribute is determined by combining the values of the inherited attributes of its sub-categories (combination provided by the aggregation function specified). For example the attribute PRICE is defined on the category PC and inherited by every sub-component; however the price of the final product (i.e. the assembled PC) is composed by summing the prices of each component, so the aggregation function sum (+) is defined over PRICE on category PC. The aggregation function is inherited along a *composed_by* link chain (the inheritance of the function is stopped on categories with no sub-components).

Aggregation function can be standard mathematical function (as sum for PRICE) or any user-defined function with a specific aggregation task (e.g. the function *assoc_list* on attribute COMTYPE of the storage system, building an associative list with the possible communication standards of the storage devices present in the final product). In figure 2 inheritable attributes are indicated in capital letters (possibly with the associated function).

Besides aggregation functions, arbitrary functions can be attached to attributes in order to compute their value; this is necessary in case the attribute is a non-inheritable attribute defined on a (non-basic) category. In the example of figure 2, function *disc_f* ia associated with the attribute discount of PC to compute tha available discount for the assembled computer (this may be a simple percentage on the attribute PRICE or a more complex procedure taking into account the kind of involved components). In categories being a target of an *is_a* link, a special *discriminant attribute* is introduced, which aim is to distinguish the different typologies the category has (underlined attributes in fig. 2). For example attribute brand in the category CPU determines the brand of the corresponding component (and so identify the typology below)[2].

In our framework, a fuzzy semantics can be associated to any attribute, by considering its type (see [9]). The definition of suitable fuzzy sets or operators can then be exploited to model a given level of approximation in the user requirements for the target configured product. This means that the final user can specifies both crisp (i.e. precise and well-defined) requirements over an attributes (e.g. "I need an Intel CPU and an hard-disk with a capacity greater than 60Gb "), as well as imprecise (i.e. fuzzy) requirements ("I want a PC with a large memory size and with a price of about $500 ").

The definition of the fuzzy sets associated with the attributes can be made at different levels of details. Indeed, the fuzzy semantics of a linguistic value depends in general from the entity we are considering; for instance, a linguistic value "cheap" for attribute PRICE of category PC has a different semantics (and so a different fuzzy set definition) than the same value defined on the same (inherited) attribute on category CPU, since the price of a component may a have a different order of magnitude than the price of the whole product. This means that, if fuzzy user requirements are allowed on a given attribute of an entity, possible different fuzzy semantics for the linguistic values are allowed for each entity that defines or inherit the attribute, if the entity is part of a *composed_by* chain. Conversely, entities in a *is_a* chain sharing the same attribute (by inheritance), also share the fuzzy semantics for that attribute. For instance, the fuzzy definition of the concept "high capacity" is shared by entities Hard-disk, IBM and Matrox.

---

[2] It is worth noting that the introduction of a typology is a modeling choice, often depending by efficiency reasons: for example the introduction of basic categories IntelCPU and AMDCPU (with *is_a* links to CPU) can be justified when a large number of user queries make a distinction on the brand of the CPU; an alternative could be to make CPU a basic category, with a standard attribute *brand* of type enum('Intel', 'AMD').

To sum up, user requirements represent a set of user constraints on the required configuration that can be finally expressed through a complex possibly fuzzy condition involving both the set of considered entities, as well as the set of attributes. Moreover, as the notion of configuration assumes, the conceptual model is also augmented with a set of *constraints*. We can identify two different kinds of constraints: (1) *model constraints* (MC) representing general constraints imposed on the model and that must be satisfied by any allowed configurations; (2) *user constraints* (UC) which are specified by the user only for the current configuration process (user's requirements). We assume that a MC is an expression according to the following grammar:

```
<mc>    ::= (<mc>) | not <mc> | <mc> and <mc> | <mc> or <mc> | <expr>
<expr>  ::= <expr1> <rel> <expr2>
<expr1> ::= <entity>.<attribute> | #<entity>
<expr2> ::= <entity>.<attribute> | #<entity> | <value>
<rel>   ::= <relational_symbol>
```

where `<entity>` is the name of an entity, `<attribute>` the name of an attribute and $\#E$ represents the cardinality (number of occurrences) of entity E.

User constraints can be either of the same form of MCs or specific condition of selection over component attributes. In the latter case, they can be either crisp/boolean conditions or fuzzy conditions (see [9] and section 2). In the next session we will show some examples.

Finally, in case a user requires $n > 1$ occurences of a given entity (as allowed by the corresponding number restriction), then such an entity E is supposed to be replicated n times ($E(1), \ldots E(n)$).

4   Implementing the Configuration Algorithm

Given the conceptual model described in the previous section, a natural choice is to associate a relation (table) with each basic category; they represents the basic components to be used for the configuration, stored as tuples of the corresponding relation. Data structure corresponding to inner nodes will be created during the configuration process. In the example of figure 2 we will have the following tables (table names represent the path in the hierarchy):

```
PC-CS-CPU-INTELCPU(id1,socket,price)
PC-CS-CPU-AMDCPU(id2,socket,price)
PC-CS-MB-ASUS-ASUSINTEL(id3,biospresent,#sockets,socket,fsb,#usb, price)
PC-CS-MB-ASUS-ASUSAMD(id4,biospresent,#sockets,socket,fsb,#usb, price)
PC-CS-MB-GB-GBAMD(id5,#sockets,socket,fsb,#usb,price)
PC-CS-MB-GB-GBINTEL(id6,#sockets,socket,fsb,#usb,price)
PC-CS-RAM-SDR(id7,size,price)
PC-CS-RAM-RAMBUS(id8,size,price)
PC-CS-RAM-DDR-DDRNORMAL(id9,latency,size,price)
PC-CS-RAM-DDR-DDRCOARSE(id10,latency,size,price)
PC-SS-HD-IBM(id11,capacity,comtype,price)
```

```
PC-SS-HD-MATROX(id12,capacity,comtype,price)
PC-SS-HD-CDROM(id13,speed,comtype,price)
```

where fields id$_i$ work as primary keys. Such fields are then supposed to be available to any entity in the model to identify their components (see below).

Let us suppose to have the following Model Constraints (MC):MC1:"ASUS MBs are incompatible with DDRCOARSE memories"; MC2:"the socket type of CPU and MB must be the same"; MC3:"the number of CPUs must not be greater than the number of sockets in MB"; MC4:"if there are more CPUs they must be the same product"; MC5:"the communication standard of storage devices (HDs and CDROMs) must be the same". We also add a specific user constraint UC1:"CPUs and MB must be of the same brand". Their formalization is reported below:

```
MC1: not(Motherboard.type='Asus' and DDR.type='Coarse')
MC2: CPU.socket = Motherboard.socket
MC3: #CPU ≤ Motherboard.#sockets
MC4: CPU(1).id_cpu = CPU(2).id_cpu (iff #CPU=2)
MC5: Hard-disk.comtype = CDROM.comtype
UC1: CPU.brand = Motherboard.brand
```

Once the model and user general constraints are collected, the configuration can start, by collecting specific user requirements as well. We have implemented a bottom-up configuration algorithm based on the following steps: (1) following initial user requirements, user constraints and model constraints, a part of the general conceptual model is identified and instantiated; (2) starting from basic categories, we ask the user to specify (possibly fuzzy) requirements for the current entity and a fuzzy SQL query is automatically generated; (3) when every child of a given entity has been queried, a view is generated on the parent in the following way: in case of *is_a* links an SQL UNION operation is performed, otherwise an SQL JOIN (checking for constraints) is performed; (4) the procedure is iterated (going to step 2), until a view is generated on the top of the model's hierarchy.

Let us show some details of the above procedure with the model of fig. 2. First of all the user is asked to specify which parts of the final product he/she is interested in; this will provide an instantiation of the whole model with specific cardinalities and whith only some of the whole set of entities involved (i.e. those the user is interested in). Let us suppose that the initial user requirements are the following: "I'm interested in a single CPU system, in IBM disks, in Intel components and I don't want a CDROM"; The first requirement states that #CPU=1, so no replication is needed and constraint MC4 is not activated. Moreover, the system cache the value #CPU=1, since it will use it later on to check MC3 when needed. The remaining requirements prune from the model all the entities surrounded by dashed squares in fig. 2. Now a bottom-up process starts from basic categories, asking the user to specify (possibly fuzzy) requirements over categories, by producing suitable (fuzzy) SQL queries. For example the user

may require an IBM hard-disk with high capacity with a confidence of 0.7: the
following view, representing the possible hard disks the user is looking for is
generated:

```
CREATE VIEW PC-SS-HD AS
SELECT (0.7) id11 as id_hd, brand='IBM', price, comtype
FROM PC-SS-HD-IBM WHERE capacity=high
```

where high is a fuzzy linguistic value suitably defined over the attribute of interest (capacity). Notice that only the *id* field (needed to retrieve the components
with all their inherited attributes) and the (non-inherited) discriminant attribute
are explicitly necessary in such a view. For the sake of convenience, we also store
in the view inherited attributes associated to an aggregation function; this avoids
the need of a further join condition when *composed_by* links are dealt with (see
below). The configuration process switches now to basic categories for RAM
and the user is asked to specify requirements over the RAM (remember that
an inherited attribute keeps its semantics along an *is_a* chain); suppose the user
decides that the price of the RAM has to be not very expensive with a confidence
of 0.7, then the following query (building category DDR) is generated:

```
CREATE VIEW PC-CS-RAM-DDR AS
(SELECT (0.7) id9 as id_ddr, type_ddr='ddrnormal', price
    FROM DDRNormal WHERE price<>very_expensive) UNION
(SELECT (0.7) id10 as id_ddr, type_ddr='ddrcoarse', price
    FROM DDRCoarse WHERE price<>very_expensive)
```

Next step generates the RAM view:

```
CREATE VIEW PC-CS-RAM AS
SELECT id_ddr as id_ram, type_ram='ddr', price FROM PC-CS-RAM-DDR
```

Concerning the CPU, suppose the user wants a CPU with high speed of the
front serial bus (confidence 0.9); the generated query is:

```
CREATE VIEW PC-CS-CPU AS
SELECT (0.9) id1 as id_cpu, brand='Intel', price
    FROM PC-CS-CPU-INTELCPU WHERE fsb=high
```

For the MB, the user asks (confidence 0.9) a high speed front serial bus and a
large number of USB ports; these are the generated views:

```
CREATE VIEW PC-CS-MB-ASUS AS
SELECT (0.9) id3 as id_asus, brand='Intel', price
    FROM PC-CS-MB-ASUS-ASUSINTEL WHERE fsb=high AND #usb=large
CREATE VIEW PC-CS-MB-GB AS
SELECT (0.9) id6 as id_gb, brand='Intel', price
    FROM PC-CS-MB-GB-GBINTEL WHERE fsb=high AND #usb=large
CREATE VIEW PC-CS-MB AS
(SELECT id_asus as id_mb, type_mb='Asus', price FROM PC-CS-MB-ASUS)
UNION
(SELECT id_gb as id_mb, type_mb='gb', price FROM PC-CS-MB-GB)
```

At this stage, all the *is_a* links have been processed and the configuration algorithm starts to deal with compositional links. Concerning the storage system a simple SELECT is sufficient[3]:

```
CREATE VIEW PC-SS AS
SELECT id_hd, price, comtype FROM PC-SS-HD
```

Regarding the computational system a JOIN merging the child components and checking for the available constraints is necessary as follows:

```
CREATE VIEW PC-CS AS
SELECT id_cpu, id_mb, id_ram, (c.price+m.price+r.price) as price
FROM PC-CS-CPU c, PC-CS-MB m, PC-CS-RAM r
WHERE MC1 AND MC2 AND MC3 AND UC1
```

Constraints can be implemented as follows:

```
MC1: NOT(m.type_mb='Asus' AND r.id_ram IN
(SELECT id_ddr FROM PC-CS-RAM-DDR WHERE type_ddr='ddrcoarse'))
MC2: c.socket=m.socket
MC3: m.#sockets >= 1
UC1: c.brand=m.brand
```

Finally, the last view representing the possible PC configurations is generated, by possible imposing further requirements as, for instance, a medium final price with confidence 0.9:

```
CREATE VIEW PC AS
SELECT (0.9) id_cpu, id_mb, id_ram, id_hd (cs.price+ss.price) as price,
    disc_f as discount, (price-discount) as disc_price
FROM PC-CS cs, PC-SS ss
WHERE disc_price = medium
```

The framework and the configuration algorithm we have described have been implemented in a J2EE 3-tier architecture with JDBC interface. More details and some preliminary experiments can be found on [8].

5   Conclusions and Future Works

We have presented an approach to product configuration, based on a hierarchical conceptual model and on fuzzy user requirements on the product features. The approach can be implemented on top of a relational database, exploiting the whole power of an SQL engine to implement both retrieval of product components, component composition and constraint checking. The proposed framework is a first step towards the definition of a flexible configuration architecture, where suitable strategies of system-user interactions can be defined. Indeed, future works will concentrate on such strategies, in such a way to provide the user

---

[3] Notice that if price and comtype would not been stored in the PC-SS-HD view, a join on *id* attributes with PC-SS-HD-IBM would have been necessary to retrieve such aggregated attributes.

with the guarantee of getting a reasonable set of acceptable configurations, by checking when constraints are too strict or too large, by defining suitable policies on the definition of either hard or soft requirements, by tuning the sensitivity of the fuzzy semantics of attributes and by allowing a suitable ranking of the obtained solutions.

References

1. Anselma, L., Magro, D., Torasso, P.: Automatically decomposing configuration problems. In: Cappelli, A., Turini, F. (eds.) AI*IA 2003. LNCS (LNAI), vol. 2829, pp. 39–52. Springer, Heidelberg (2003)
2. Bosc, P., Pivert, O.: SQLf: a relational database language for fuzzy querying. IEEE Transactions on Fuzzy Systems 3(1) (1995)
3. Fleischanderl, G., Friedrich, G., Haselbock, A., Scheiner, H., Stumptner, M.: Configuring large systems using generative constraint satisfaction. IEEE Intelligent Systems 13(4), 59–68 (1998)
4. Friedrich, G., Stumptner, M.: Consistency-based configuration. In: Proc. AAAI Workshop on configuration. AAAI Press, Menlo Park (1999)
5. Geneste, L., Ruet, M.: Fuzzy case-based configuration. In: Proc. ECAI 2002 Workshop on Configuration, Lyon, FR, pp. 1–10 (2002)
6. Krebs, T., Hotz, L., Gunter, A.: Knowledge-based configuration for configuring combined hardware/software systems. In: Proc. PUK 2002, Freiburg, GE (2002)
7. Magro, D., Torasso, P.: Decomposition strategies for configuration problems. AI for Engineering Design, Analysis and Manufacturing 17(1), 51–73 (2003)
8. Portinale, L., Galandrino, M.: On-line product configuration using fuzzy retrieval and J2EE technology. Technical Report TR-INF-2009-05-04-UNIPMN, Computer Science Dept., University of Piemonte Orientale (2009),
   http://www.di.unipmn.it/TechnicalReports/TR-INF-2009-05-04-UNIPMN.pdf
9. Portinale, L., Montani, S.: A fuzzy case retrieval approach based on SQL for implementing electronic catalogs. In: Craw, S., Preece, A.D. (eds.) ECCBR 2002. LNCS (LNAI), vol. 2416, pp. 321–335. Springer, Heidelberg (2002)
10. Sabin, D., Freuder, E.: Configuration as composite constraint satisfaction. In: Proc. Artificial Intelligence and Manufacturing Research Planning Workshop, pp. 153–161. AAAI Press, Menlo Park (1996)
11. Sinz, C.: Knowledge compilation for product configuration. In: Proc. ECAI 2002 Workshop on Configuration, Lyon, FR, pp. 23–26 (2002)
12. Soininen, T., Niemela, I., Tiihonen, J., Sulonen, R.: Representing configuration knowledge with wieight constraint rules. In: Proc. AAAI Spring Symposium on Answer Set Programming. AAAI Press, Menlo Park (2001)
13. Tseng, H.-E., Chang, C.-C., Chang, S.-H.: Apllying case-based reasoning for product configuration in mass customization environments. Expert Systems with Applications 29(4), 913–925 (2005)
14. Veron, M., Aldanondo, M.: Yet another approach to CCSP for configuration problem. In: Proc. ECAI 2000 Workshop on Configuration, Berlin, GE, pp. 59–62 (2000)

Classifying Human Body Acceleration Patterns Using a Hierarchical Temporal Memory

Federico Sassi[1,2], Luca Ascari[2,3], and Stefano Cagnoni[1]

[1] Dip. Ingegneria dell'Informazione, Università di Parma, Italy
federico.sassi@henesis.eu, cagnoni@ce.unipr.it
http://www.dii.unipr.it
[2] Henesis srl, Parma, Italy
ascari@sssup.it
http://www.henesis.eu
[3] Scuola Superiore Sant'Anna, Pisa, Italy
http://www.sssup.it

Abstract. This paper introduces a novel approach to the detection of human body movements during daily life. With the sole use of one wearable wireless triaxial accelerometer attached to one's chest, this approach aims at classifying raw acceleration data robustly, to detect many common human behaviors without requiring any specific a-priori knowledge about movements. The proposed approach consists of feeding sensory data into a specifically trained Hierarchical Temporal Memory (HTM) to extract invariant spatial-temporal patterns that characterize different body movements. The HTM output is then classified using a Support Vector Machine (SVM) into different categories. The performance of this new HTM+SVM combination is compared with a single SVM using real-word data corresponding to movements like "standing", "walking", "jumping" and "falling", acquired from a group of different people. Experimental results show that the HTM+SVM approach can detect behaviors with very high accuracy and is more robust, with respect to noise, than a classifier based solely on SVMs.

1 Introduction

Recent studies demonstrated that being able to monitor and recognize human body movements can greatly improve the quality of rehabilitation therapy [1,2], particularly in elderly people, by detecting commonly repeated dangerous movements that may have severe consequences. The same task can be of interest for insurance companies, as well, who can determine a subject's level of inability based on the type of daily activities he/she is able to perform [3]. In more general contexts, such as autonomous living of elderly people, events which may be potentially harmful, such as a fall, must be detected promptly to be able to contact emergency services in the shortest possible time.

There are many techniques that can be used to assess human movements, like visual motion analysis, foot switches, gait mats, force plates or interviews to the patient; however, most of them have many drawbacks when used for continuous monitoring, being costly, bulky, invasive, or requiring high computation power or complex setups, and are mostly suited for in-laboratory measurements [4].

R. Serra and R. Cucchiara (Eds.): AI*IA 2009, LNAI 5883, pp. 496–505, 2009.

Nonetheless, lately, thanks to the progress of Micro Electro-Mechanical Systems (MEMS), accelerometers have become smaller and cheaper and can be incorporated in small, non-obtrusive continuous monitoring devices that can be attached to the body; often, they can also store the data they have collected [4]. Moreover, progress in low-power wireless data transmission has enabled these devices to stream accelerometer data directly to a remote server for instant evaluation. Recent developments have demonstrated the feasibility of monitoring accelerometer data from many people in the same area using a Wireless Sensor Network approach [5].

We describe a method to detect acceleration patterns related to everyday life, based on a simple and cheap wireless sensor board and on combined use of Hierarchical Temporal Memories (HTM) and Support Vector Machines (SVM). The paper is organized as follows: in section 2 the experimental setup is discussed, with particular regard to the wireless sensor module used to record accelerometer data, to the HTM used to find invariants in data patterns, and to the SVM which perform the final classification. Then, in section 3, we describe the setup of the experiments and the data collection procedure, as well as experimental results of the approach we propose, which are compared to a reference method based on a single SVM.

2   Components

Our approach consists of collecting data by wireless accelerometer sensors and then using such data to train and test our classification method. In this section, after briefly describing its components, we illustrate in detail the global architecture of the system.

2.1   Wireless Triaxial Acceleration Sensor Module

The data collection platform consists of a low-cost wireless module, incorporating a triaxial high resolution accelerometer and a IEEE 802.15.4 RF transmitter (a.k.a. MAC level of a ZigBee network), which continuously sends data to a IEEE 802.15.4-enabled PDA running a custom program. This receives and stores data and permits to label them manually into different classes. The essential features of the module, shown in figure 1, are reported in table 1.

The wireless sensor module was entirely designed and developed by Henesis s.r.l. and is being used for distributed sensing within the Henesis WISnP (Wireless Sensor Network Platform: see http://wisnp.henesis.eu).

Table 1. Wireless Sensor Module: essential features

Accelerometer	ST LIS3LV02DL; BW: up to 640Hz; range: $\pm 2.0g$; max res: $1mg$
Microprocessor	MICROCHIP PIC18F67J11; Max CPU Freq: 40MHz
Transmission range	outdoor: 100m; indoor: 10m (0dBm output power)
Dimensions	h:60 mm X w:39 mm X d:10 mm with $Li - SOCI^2$ ER14250 battery
Memory	on-node flash memory for local storage: 16Mb FLASH, 1Mb EEPROM
On-board sensors	on-node temperature and humidity sensor (SHT11 digital sensor)
Expandability	SPI, I2C, RS232, Analog IOs for additional sensors (eg: heart-rate sensor, ...)

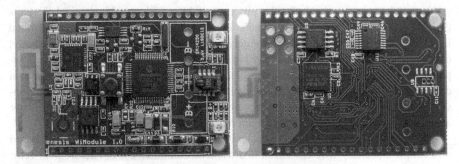

Fig. 1. Front (above left) and rear (above right) sides of the electronic module used for the experiments. The triaxial accelerometer is the integrated circuit in the lower left corner of the rear side, aside the memory banks. Below, the module is applied to the medical-chest-band used to fix it to the thorax of the volunteers.

2.2   Hierarchical Temporal Memory

Hierarchical Temporal Memory is a computational paradigm that takes inspiration from the biological structure and algorithmic properties of the neocortex, which derives from a more general theory, called *Memory-Prediction Framework* [6]. A HTM is not "programmable" in the common sense, but can learn invariants when exposed to spatial-temporal relevant data [7,8,9].

A HTM is structured as a hierarchical network of nodes where the sensory data enter at the bottom level while the outputs of the network are generated by the top nodes, which represent the possible *causes* of the input. Every node in the hierarchy runs the same algorithm, looking for spatial-temporal pattern (*invariants*) in its input and grouping them as *causes*. Every node is trained in an unsupervised manner, in the classical meaning, but *time* is considered to be the supervisor: if two events (inputs) often occur consecutively, they are very likely to belong to the same cause (output).

Nodes at the bottom of the hierarchy receive inputs from the sensors and their output is fed into the nodes that are located in the layer above. Lower level nodes receive input from a limited set of the whole sensory data, while higher level layers have a broader view of the sensory input, as processed by the other nodes. This means that lower level nodes find causes belonging only to a limited time and input scale while higher-level nodes are able to find causes on a larger time scale and broader sensory input. The output of the top level node is similar to a distribution of the "likelihood" of the current input to be generated by the known causes and is still unsupervised. To measure the performance of unsupervised learning, as provided by the top node, it is necessary to add a supervised classifier (like a Support Vector Machine or a KNN) on top of it.

These peculiar characteristics make HTM a very powerful technology to face real-world problems but, at the same time, cause their performance to be strongly dependent on the set of data used to train them. HTMs have been successfully used in vision and speech recognition problems [9].

2.3 Support Vector Machines

Support Vector Machines are a set of methods for supervised learning mainly used for classification purposes. They have the property to minimize the empiric error and to maximize the geometric margin between different classes at the same time. SVMs receive as input a vector of data that is mapped, by a *kernel function*, to a n-dimensional space (hyperspace). Then, all data belonging to the same category are enclosed within a set of hyperplanes defined by the *support vectors*. Finally, a new hyperplane that maximizes the distance between the class-delimiting hyperplanes is built. Therefore, every SVM is defined by its support vectors. In the training phase, the SVM maps data to the hyperspace and searches for the optimal support vectors, then builds the separation hyperplanes. After this stage, classifying a pattern implies mapping it onto the hyperspace and determining the hyperspace region (category) to which it belongs [10].

The validity of using a SVM to classify human motion patterns (or events, when using optical sensors) [11] has been successfully demonstrated. We will use this classifier as the reference against which our proposed approach will be compared to show that SVM can benefit from HTM pre-processing of the input data.

2.4 Hybrid HTM-SVM Classification of Motion Patterns

The automatic recognition of human body movements is a well-known problem which has been tackled by many different approaches, each of which has shown to offer advantages and disadvantages.

In one of them, specific knowledge about the kinematic of the movements that are investigated has been included in the classifier. This approach yields very high accuracy in detecting such events but, at the same time, limits the type of events that can be addressed [3]. Other approaches try to classify the incoming pattern without taking into consideration the biological nature and the consequent variability of those signals [12].

Our approach is deeply rooted in the assumption that humans recognize very easily those motion patterns, even if very noisy, without needing to implement a "classical" classifier inside our brains, but perhaps by finding "invariants" [6] in the signals that our body generates. The final aim of this research is to build a framework within which many different movements from many different people can be recognized and correctly classified, even in noisy situations.

In our method we do not add any a-priori knowledge about the movements we want to classify but the fact that they develop over time. Moreover, we test our solution against very noisy data to prove the robustness of the methodology.

3   Experimental Setup

We aim at recognizing a vast number of different human body movements automatically, using small and non-obtrusive wireless sensors like the one described in 2.1. In doing so, we need to consider the importance of the quality of the data set on the final results. Based on these considerations, our experimental setup aims at:

1. validating the use of a single wireless accelerometer sensor to collect body movements;
2. defining a solid protocol for data acquisition, robust with respect to different people and events;
3. acquiring an extensive dataset but using only a subset during the subsequent phases;
4. exploiting HTM-specific features to handle real-world data.

3.1   Data Acquisition

The location where the accelerometer is placed on the body determines which movements it is able to record best. To study "whole body" movements the ideal placement is as close as possible to the center of mass of the body, e.g. on the sternum or the waist [4]. In our experiment, we firmly attached the wireless accelerator sensor to a medical-chest-band and placed the sensor corresponding to the lower end of the sternum. In this way it would not block any movement and stay stable during data acquisition.

The wireless module is able to sample accelerometer data at rates up to 640Hz. We chose to record data at 160Hz, which is a compromise between the amount of data to be classified and the accuracy of description of fast body movements, like a fall.

During this experiment we collected data from 3 volunteers, all males and in good physical conditions, but with different body height and weight. We asked volunteers to perform some movements while we recorded and labeled them with the correct category using a PDA. Before starting every new movement volunteers were required to stand still, to ease splitting data between events and ensure that the whole event would be recorded. Every movement is saved as an "event", and corresponds to a sequence of data, a few seconds long, from the triaxial acceleration components.

At least 150 events have been acquired for each of the following ten classes of body movements: *Standing, Jumping, Walking forward, Falling, Sitting down, Standing up from sitting, Picking up an object on the ground, Turning left, Turning right, Walking backward*, for a total of 1755 events over 689851 time points at 160Hz, corresponding to more than 71 minutes of recording of continuous movements.

For a preliminary evaluation of the potential of our approach, we decided, at this stage of the research, to use only categories of movements (*Standing, Jumping, Walking Forward* and *Falling*) and to limit the number of events considered for each category. The training set therefore included 10 events per category, in order to keep the complexity and the training time of the classifiers reasonably low, while the test set included 20 events per category.

3.2   Data Preprocessing

Data preprocessing before classification was very light, following the lines stated in section 2.4. To correctly train a HTM, only the input data which correspond to the spatial-temporal patterns to be learned are needed. Therefore one can disregard their label, as training is completely unsupervised. Nevertheless, if one wants to train a supervised classifier on top of a HTM, or as a stand-alone classifier, an accurate pre-classification of the training and the test set is needed to measure performances. To achieve this, we pre-processed data, thresholding their average to detect the onset of an event from a quiet reference condition (like just standing) in which acceleration is virtually absent.

After pre-processing, the input data appear as in figure 2, where one can observe the differences in the accelerometer patterns between different body movements.

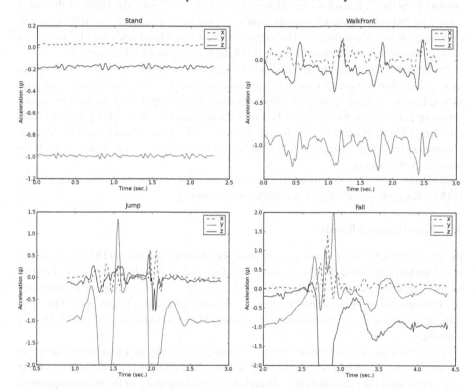

Fig. 2. Plot of triaxial acceleration values from each class considered in this experiment. From top-left clockwise: Stand, Walk forward, Fall and Jump.

As one can observe in figure 2 the three axes are not perfectly aligned, as a consequence of the limited precision and repeatability that can be achieved when the device is applied to the body, and to the physycal properties of the skin; even when standing still, the average value of (x,y,z) is not exaclty (0,-1,0). A possible solution to this problem is to apply a 3D rotation matrix to counteract this drift and to enable the sensor to self-calibrate at any new application. Anyway, at this stage of development and given the peculiarities of the HTM in finding *invariants* in the input space, no such calibrations have been applied, therefore the data we processed were slightly more noisy than they could be.

3.3 HTM Design

We used a 2-level HTM to process triaxial data. In the first level there is one HTM node per axis of the accelerometer data, so every node on this level will be able to learn and recognize patterns occurring in time only on one particular axis. Once these nodes finish training, they output probability-like vectors representing the likelihood with which the input pattern corresponds to the "sequences" stored in memory (as introduced in section 2.2) in the upper level nodes.

The second level is composed of just one single HTM node which takes all three outputs from the first level nodes. This node runs the very same algorithm of the lower level with the sole difference that its inputs are already "sequences", so this node will find "sequences of sequences". The output of the second level node will be fed into a classifier node. Figure 3 represents the HTM network created to this purpose.

In this particular example, the first-level nodes should be able to detect small movements like a slowly increasing acceleration on a single axis, while the second-level node should be able to detect events of longer duration, like a step during walking or the descending part of a jump. As described in section 2.2, one cannot "program" a HTM; thus the HTM's designer should follow this way of thinking when tuning HTM node parameters by which one can set, for example, how far in time a node has to look for sequences of patterns.

Basically, it is very important to stress that the only way to direct the training of a HTM is through a carefully choice of the input data [9].

3.4 Experimental Results

In our experiments we wanted to asses how the introduction of a HTM layer can improve performance of a SVM classifier. The dataset consists of 4 categories of movements: *Standing, Jumping, Walking forward and Falling*. The training set is composed of 10 events per category while the test set includes 20 events per category. The same training set and test set have been used for the HTM+SVM classifier and the SVM-only classifier.

To assess classification performance, we represent results as a *confusion matrix*, in which the row index of each element represents the expected class while the column index represents the predicted class. Therefore, each element $a_{i,j}$ of the matrix represents the frequency with which an element of class i has been classified as belonging to class j.

Table 2 summarizes the results of the classifiers over whole events, which may last several seconds, and produce one classification per sample. The predicted class for the

Table 2. Confusion Matrix for full events:(a) HTM+SVM (b) SVM

	Stand	Jump	WalkFwd	Fall
Stand	96.29%	0%	3.70%	0%
Jump	1.85%	98.14%	0%	0%
WalkFwd	1.85%	0%	98.15%	0%
Fall	0%	0%	0 %	100%

(a)

	Stand	Jump	WalkFwd	Fall
Stand	98.15%	0%	1.85%	0%
Jump	0%	96.29%	3.70%	0%
WalkFwd	0%	0%	100%	0%
Fall	0%	0%	0%	100%

(b)

whole event is the one that has been associated to most samples of the event to be classified. Table 2(a) and table 2(b) report very high classification accuracy on the test set. Notice that this comparison between different classifiers cannot be made in real-time because it requires that data be segmented a-priori into events.

In a more realistic situation the classifier would be exposed to a continuous flow of data, with no prior opportunity of pre-segmenting them. To simulate such a situation, we tested the classifier on a continuous stream of different events. The predicted and expected output of the classifier have been grouped on *sliding* windows of fixed width (3 seconds, corresponding to 480 timepoints, which is about the average length of an event) and with a step of 0.5 seconds (80 timepoints). The classifier output is the class that has been selected most often within the current window; results are shown in table 3.

Table 3. Confusion Matrix using a 3-seconds window: (a) HTM+SVM (b) SVM

	Stand	Jump	WalkFwd	Fall			Stand	Jump	WalkFwd	Fall
Stand	96.81%	1.45%	0.58%	1.16%		Stand	98.84%	0%	1.16%	0%
Jump	3.67%	90.61%	5.71%	0%		Jump	0.41%	86.12%	13.47%	0%
WalkFwd	11.80%	0%	88.19%	0%		WalkFwd	0%	0%	100%	0%
Fall	2.30%	0%	0.66%	97.05%		Fall	0%	0%	2.62%	97.38%
		(a)						(b)		

To justify the slightly lower performances, it can be observed that the classifiers might suffer from the potential co-presence of two different and uncorrelated events within the same sliding window (the length of the window must be correlated with the average duration of the events being studied).

An important feature of classifiers is the capability to handle noisy data correctly. The following matrices describe the results of the classifiers when increasing amounts of Additive White Gaussian Noise (AWGN) are added to the input; in particular, zero-mean noise with standard deviations of 0.01 (table 4), 0.05 (table 5), and 0.1 (table 6) was applied.

Table 4 shows that both classifiers are able to handle small amounts of noise. As noise level increases, the performance of the SVM classifier drops while the performance of the hybrid approach HTM+SVM still assures good accuracy (table 5). This result is mainly obtained thanks to the structure of the temporal algorithms inside every HTM node, that search for invariant patterns over time. When the noise level is further increased (see table 6), HTMs also fail.

Table 4. Confusion Matrix for data+AWGN stdev=0.01:(a) HTM+SVM (b) SVM

	Stand	Jump	WalkFwd	Fall			Stand	Jump	WalkFwd	Fall
Stand	97.39%	0%	2.03%	0.57%		Stand	98.84%	0%	1.16%	0%
Jump	3.26%	90.61%	6.12%	0%		Jump	0.41%	85.71%	13.88%	0%
WalkFwd	11.07%	0%	88.93%	0%		WalkFwd	0%	0%	100%	0%
Fall	2.30%	0%	0.33%	97.38%		Fall	0%	0%	2.62%	97.38%
		(a)						(b)		

Table 5. Confusion Matrix for data+AWGN stdev=0.05:(a) HTM+SVM (b) SVM

	Stand	Jump	WalkFwd	Fall			Stand	Jump	WalkFwd	Fall
Stand	99.42%	0%	0%	0.58%		Stand	0%	0%	99.42%	0.58%
Jump	3.67%	91.43%	4.90%	0%		Jump	0%	84.49%	15.51%	0%
WalkFwd	4.80%	0%	95.20%	0%		WalkFwd	0%	0%	100%	0%
Fall	2.30%	0%	0.33%	97.38%		Fall	0%	0%	2.95%	97.05%
		(a)						(b)		

Table 6. Confusion Matrix for data+AWGN stdev=0.1:(a) HTM+SVM (b) SVM

	Stand	Jump	WalkFwd	Fall			Stand	Jump	WalkFwd	Fall
Stand	15.36%	0.29%	83.768%	0.5797%		Stand	0%	0%	100%	0%
Jump	0%	95.10%	4.90%	0%		Jump	0%	86.12%	13.88%	0%
WalkFwd	0%	1.58%	97.41%	0%		WalkFwd	0%	0%	100%	0%
Fall	0%	0%	2.62%	97.38%		Fall	0%	0%	2.95%	97.05%
		(a)						(b)		

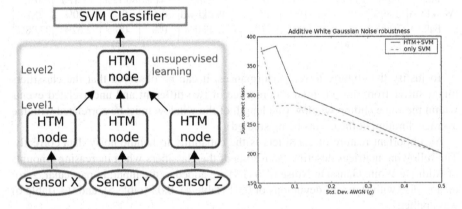

Fig. 3. Architecture of the HTM network used in the experiments

Fig. 4. Comparison of robustness with respect to noise between HTM+SVM and SVM

The plot in figure 4 shows the advantage of using a HTM-based classification approach.

4   Conclusions

Results described in this work suggest that the approach to human body movement classification by means of a triaxial accelerometer and HTM+SVM is promising. In particular, the HTM+SVM architecture proved to be robust to white noise.

Nevertheless, with no or low noise, the proposed architecture seems to be no better than a "simpler" SVM: this seems to indicate that the HTM can be further optimized, both in its topology and in the choice of node parameters.

Our future research will involve a deeper investigation of HTMs, to assess their ability of dealing with "completely unknown" people, and with new events, such as *Sitting down* or *Standing up from sitting* (already recorded, but not studied yet), in which time plays a fundamental role.

Acknowledgment. The authors would like to thank the students Silvia Signorini, Nicola Bonazzi, Fabio Ferramola, Alessandro Gallo and Domenico Mazzitelli for their hard work creating the dataset used in this research.

References

1. Pentland, A.: Healthwear: Medical technology becomes wearable. IEEE Computer, 42–49 (2004)
2. Aggarwal, J., Cai, Q.: Human motion analysis: A review. Computer Vision And Image Understanding 73(3), 428–440 (1999)
3. Zhang, S., Ang Jr., M.H., Xiao, W., Tham, C.K.: Detection of Activities by Wireless Sensors for Daily Life Surveillance: Eating and Drinking. Sensors 9(3), 1499–1517 (2009)
4. Godfrey, A., Conway, R., Meagher, D., OLaighin, G.: Direct measurement of human movement by accelerometry. Medical Engineering & Physics 30(10, Sp. Iss. SI), 1364–1386 (2008)
5. Lee, Y., Kim, J., Son, M., Lee, M.: Implementation of accelerometer sensor module and fall detection monitoring system based on wireless sensor network. In: 29th Int. Conf. IEEE Engineering in Medicine and Biology Soc. EMBS 2007, pp. 2315–2318 (2007)
6. Hawkins, J., Blakelee, S.: On Intelligence. Henry Holt and Company (2004)
7. George, D.: How the brain might work: a hierarchical and temporal model for learning and recognition. PhD thesis, Stanford University (2008)
8. Hawkins, J., George, D.: Hierarchical Temporal Memory - Concepts, Theory, and Terminology. Numenta Inc
9. Numenta: http://www.numenta.com
10. Hsu, C.W., Chang, C.C., Lin, C.J.: A practical guide to support vector classification, http://www.csie.ntu.edu.tw/~cjlin/papers/guide/guide.pdf
11. Grahn, J., Kjellstrom, H.: Using SVM for efficient detection of human motion. In: Proceedings 2nd Joint IEEE International Workshop on VS-PETS, 2005, pp. 231–238 (2005)
12. Mathie, M., Celler, B., Lovell, N., Coster, A.: Classification of basic daily movements using a triaxial accelerometer. Medical & Biological Engineering & Computing 42(5), 679–687 (2004)

Author Index